17.77 (net) Oxford Univ. 9-67 (Swearingen)

D1266748

The Prophet Outcast

TROTSKY: 1929–1940

Trotsky in Mexico, 1940. On one of his last outings

The
Prophet Outcast

TROTSKY: 1929–1940

ISAAC DEUTSCHER

LONDON
OXFORD UNIVERSITY PRESS
NEW YORK   TORONTO
1963

PRINTED IN THE UNITED STATES OF AMERICA

CONTENTS

ILLUSTRATIONS

PREFACE

THIS volume concludes my trilogy about Trotsky and relates the catastrophic *dénouement* of his drama. At the *dénouement*, the protagonist of a tragedy is usually more acted upon than acting. Yet Trotsky remained Stalin's active and fighting antipode to the end, his sole vocal antagonist. Throughout these twelve years, from 1929 to 1940, no voice could be raised against Stalin in the U.S.S.R.; and not even an echo could be heard of the earlier intense struggles, except in the grovelling confessions of guilt to which so many of Stalin's adversaries had been reduced. Consequently, Trotsky appeared to stand quite alone against Stalin's autocracy. It was as if a huge historic conflict had become compressed into a controversy and feud between two men. The biographer has had to show how this had come about and to delve into the complex circumstances and relationships which, while enabling Stalin to 'strut about in the hero's garb', made Trotsky into the symbol and sole mouthpiece of opposition to Stalinism.

Together, therefore, with the facts of Trotsky's life I have had to narrate the tremendous social and political events of the period: the turmoil of industrialization and collectivization in the U.S.S.R. and the Great Purges; the collapse of the German and European labour movements under the onslaught of Nazism; and the outbreak of the Second World War. Each of these events affected Trotsky's fortunes; and over each he took his stand against Stalin. I have had to go over the major controversies of the time; for in Trotsky's life the ideological debate is as important as the battle scene is in Shakespearian tragedy: through it the protagonist's character reveals itself, while he is moving towards catastrophe.

More than ever before I dwell in this volume on my chief character's private life, and especially on the fate of his family. Again and again readers will have to transfer their attention from the political narrative to what common parlance insists on describing as the 'human story' (as though public affairs were not the most human of all our preoccupations; and as if politics were not a human activity *par excellence*). At this stage Trotsky's family life is inseparable from his political fortunes: it gives a new dimension to his struggle; and it adds sombre depth to his drama. The strange and moving tale is told here for the first time on the basis of Trotsky's intimate correspondence with his wife and children, a correspondence to which I have been privileged to obtain unrestricted access. (For this I am indebted to the generosity of the late Natalya Sedova, who two years before her death asked the Librarians of Harvard University to open to me the so-called sealed section of her husband's Archives, the section that by his will was to remain closed till the year 1980.)

.   .   .   .

I would like to comment briefly on the political context in which I have produced this biography. When I started working on it, at the end of 1949, official Moscow was celebrating Stalin's seventieth birthday with a servility unparalleled in modern history, and Trotsky's name seemed covered for ever by heavy calumny and oblivion. I had published *The Prophet Armed* and was trying to complete the first draft of what is now *The Prophet Unarmed* and *The Prophet Outcast* when, in the latter part of 1956, the consequences of the Twentieth Congress of the Soviet Communist Party, the October upheaval in Poland, and the fighting in Hungary compelled me to interrupt this work and turn my whole attention to current affairs. In Budapest raging crowds had pulled down Stalin's statues while in Moscow the desecration of the idol was still being carried out stealthily and was treated by the ruling group as their family secret. 'We cannot let this matter go out of the Party, especially to the Press,' Khrushchev warned his audience at the Twentieth Congress. 'We should not wash our dirty linen before the eyes [of our enemies].' 'The washing of the dirty linen', I then commented, 'can hardly be carried on behind the back of the Soviet people much longer. It will presently have to be done in front of them and in broad daylight. It is, after all, in their sweat and blood that the "dirty linen" was soaked. And the washing, which will take a long time, will perhaps be brought to an end by hands other than those that have begun it—by younger and cleaner hands.'

The Prophet Outcast is appearing after some washing of the 'dirty linen' has already been done in public, and after Stalin's mummy has been evicted from the Red Square Mausoleum. A perceptive Western cartoonist reacted to this last event with a drawing of the Mausoleum in which Trotsky could be seen placed in the crypt just vacated, and next to Lenin. The cartoonist expressed an idea which probably occurred to many people in the U.S.S.R. (although it is to be hoped that the 'rehabilitation' of Trotsky, when it comes, will be carried out in a manner free from cult, ritual, and primitive magic). Meanwhile, Khrushchev and his friends are still exerting themselves to keep in force the Stalinist anathema on Trotsky; and in the controversy between Khrushchev and Mao Tse-tung each side accuses the other of Trotskyism, as if each were bent on providing at least negative evidence of the vitality of the issues raised by Trotsky and of his ideas.

All these events have sustained my conviction of the topicality as well as the historical importance of my theme. But—*pace* some of my critics—they have not significantly affected either my approach or even the design of my work. True, this biography has grown in scale beyond all my original plans: I have produced three volumes instead of one or two. However, in doing so I obeyed solely—and at first reluctantly—the literary logic of the work and the logic of my research, which was unexpectedly growing in scope and depth. The biographical material struggled under my hands, as it were, for the shape and the proportions proper to it, and it imposed its requirements on me. (I know that what I am saying will not exculpate me in the eyes of one critic, a former British Ambassador to Moscow, who says that he has 'always held that the Russian Revolution has never taken place' and who therefore wonders why I should devote so much space to so unreal an event.) As to my political approach to Trotsky, this has remained unchanged throughout. I concluded the first volume of this trilogy, in 1952, with a chapter entitled 'Defeat in Victory', where I portrayed Trotsky at the pinnacle of power. In the Preface to that volume I said that on completing his Life I would consider 'the question whether a strong element of victory was not concealed in his very defeat'. This precisely is the question I discuss in the closing pages of *The Prophet Outcast*, in a Postscript entitled 'Victory in Defeat'.

A Note about Sources and Acknowledgements

The narrative of this volume is based even more strongly than that of the previous volumes on Trotsky's archives, especially on his correspondence with the members of his family. Whenever I refer to *The Archives* in general, I have in mind their Open Section which is accessible to students at the Houghton Library, Harvard University. When I draw on the 'sealed' part of *The Archives* I refer to the 'Closed Section'. A general description of the Open Section was given in the Bibliography of *The Prophet Armed*. The Closed Section is described in the Bibliography attached to the present volume.

Most of the 20,000 documents of the Closed Section consists of Trotsky's political correspondence with adherents and friends; he stipulated that this should be sealed because at the time when he transferred his papers to Harvard University (in the summer of 1940), nearly the whole of Europe was either under Nazi or under Stalinist occupation and the future of many countries outside Europe looked uncertain; and so he felt obliged to protect his correspondents. But there was little or nothing strictly confidential or private in the political content of that correspondence. Indeed, with much of it I had become familiar in the nineteen-thirties—I shall presently explain in what way—so that re-reading it in 1959 I found hardly anything that could startle or surprise me. Trotsky's family correspondence, on the other hand, and even his household papers, also contained in the Closed Section, have revealed to me his most intimate experiences and feelings and have greatly enriched my image of his personality.

Some reviewers of the earlier volumes have complained that my references to *The Archives* are not detailed enough. I can only point out that whenever I cite any document from *The Archives*, I say, *either* in the text *or* in a footnote, by whom the document was written, when it was written, and to whom it was addressed. This is all that any student needs. More detailed annotation might have added impressively to my 'scholarly apparatus', but would be of no use either to the general reader, who has no access to *The Trotsky Archives*, or to the scholar, whom the indications I provide should enable to locate easily any paper I have referred to. Moreover, since I worked on my

earlier volumes *The Archives* have been rearranged so that any
more specific markings I might have given would have become
valueless by now. (e.g. I might have indicated that document
X or Y is in Section B, folder 17, but in the meantime Section
A or B or C—has ceased to exist!) The material is now arranged
in simple chronological order; and as I usually give the date of
any document quoted the student should find the item at a
glance in the excellent two-volume *Index* to *The Archives*,
available at the Houghton Library.

One or two critics have wondered just how reliable are *The
Archives* and whether Trotsky or his followers have not 'doctored
documents'. To my mind the reliability of *The Archives* is overwhelmingly confirmed by the internal evidence, by cross-
reference to other sources, and by the circumstance that *The
Archives* provide Trotsky's critics as well as his apologists with
all the material they may want. Trotsky indeed was above
falsifying or distorting documents. As to his followers, these
have, either from lack of interest or from preoccupation with
other matters, hardly ever looked into the master's Archives.
In 1950 my wife and I were the first students to work on
Trotsky's papers since he had parted with them.

In relating the climate of ideas and describing the parties,
groups, and individuals involved in the inner communist
struggles, of the nineteen-thirties I drew *inter alia* on my own
experience as spokesman of anti-Stalinist communism in
Poland. The group with which I was associated then worked in
close contact with Trotsky. His International Secretariat
supplied us with very abundant documentation, some of it
confidential, with circulars, copies of Trotsky's correspondence,
&c. As writer and debater, I was deeply involved in nearly all
the controversies described in this volume. In the course of the
debates I had to acquaint myself with an enormous political
literature, with Stalinist, Social-Democratic, Trotskyist, Brand-
lerist, and other pamphlets, books, periodicals, and leaflets
published in many countries. Naturally enough, only a small
part of that literature was available to me at the time of writing
—just enough to check the accuracy of my impressions and
memories and to verify data and quotations. My Bibliography
and footnotes do not therefore pretend to exhaust the literature
of the subject.

I have been fortunate in being able to supplement the material drawn from *The Archives* (and from printed sources) by information obtained from Trotsky's widow; from Alfred and Marguerite Rosmer, Trotsky's closest friends in the years of banishment; from Jeanne Martin des Paillères, who transmitted to me papers and correspondence of Leon Sedov, Trotsky's elder son; from Pierre Frank, Trotsky's secretary in the Prinkipo period; from Joseph Hansen, secretary and bodyguard at Coyoacan, and close eye-witness of Trotsky's last days and hours; and from many other people who were Trotsky's adherents at one time or another. (Of those listed here Natalya Sedova, Marguerite Rosmer, and Jeanne Martin died before I completed this volume.)

Outside the circle of Trotsky's family and followers, I am obliged to Konrad Knudsen and his wife, who were Trotsky's hosts in Norway, and to Mr. Helge Krog and Mr. and Mrs. N. K. Dahl for much information and vivid reminiscences about the circumstances of Trotsky's internment and deportation from Norway. I interviewed Mr. Trygve Lie, who was the Minister of Justice responsible for both the admission of Trotsky and his internment; but Mr. Lie, having spoken to me at great length and self-revealingly, then asked me to refrain from quoting him, saying that his memory had misled him and that, in addition, under a contract with an American publisher, he was not allowed to disclose this information otherwise than in his own memoirs. Mr. Lie was good enough, however, to send me the official Report on the Trotsky case which he had submitted to the Norwegian Parliament early in 1937. I have also had the benefit of interviewing Professor H. Koht, Norway's Minister of Foreign Affairs at the time of Trotsky's stay in that country, who was most anxious to establish in detail the truth of the case.

In investigating another important chapter in Trotsky's life, I approached the late John Dewey, who gave me an illuminating account of the Mexican counter-trial and spoke freely about the impression Trotsky made on him; and I am indebted to Dr. S. Ratner, Dewey's friend and secretary, for valuable information about the circumstances in which the old American philosopher decided to preside over the counter-trial. Of many other informants I would like to mention Mr. Joseph Berger, once a member of the Comintern Staff in Moscow who then spent nearly twenty-five years in Stalin's concentration camps —Mr. Berger has related to me his meeting in 1937 with Sergei Sedov, Trotsky's younger son, in the Butyrki prison of Moscow.

My thanks are due to the Russian Research Centre, Harvard University, especially to Professors M. Fainsod and M. D. Shulman for the facilities they offered me, and to Dr. R. A. Brower, Master of Adams House, and his wife, whose pleasant hospitality I enjoyed while working on the Closed Section of *The Trotsky Archives* in 1959. I am greatly obliged to Professor William Jackson and Miss C. E. Jakeman of Houghton Library for their infinitely patient helpfulness and to Mrs. Elena Zarudnaya-Levin for assisting me in reading some of the documents in *The Archives*.

To Mr. John Bell, Mr. Dan M. Davin, and Mr. Donald Tyerman who have read my MS and proofs I am grateful for criticisms and many suggestions for improvements.

My wife's contribution to this volume has been not only that of unfailing assistant and critic—in the course of many years, ever since 1950 when we first pored together over *The Trotsky Archives*, she absorbed the air of this tragic drama; and, through her sensitive sympathy with its *personae*, she has helped me decidedly in portraying their characters and narrating their fortunes.

I. D.

The Prophet Outcast

TROTSKY: 1929–1940

CHAPTER I

On the Princes' Isles

THE circumstances of Trotsky's banishment from Russia contained a foretaste of the years that lay ahead of him. The manner of the deportation was freakish and brutal. For weeks Stalin had delayed it, while Trotsky bombarded the Politbureau with protests denouncing the decision as lawless. It looked as if Stalin had not yet finally made up his mind, or was still consulting the Politbureau. Then, suddenly, the cat and mouse game was at an end: on the night of 10 February 1929, Trotsky, his wife, and elder son were rushed to the harbour of Odessa, and put on board the *Ilyich*, which sailed forthwith. His escort and the harbour authorities were under strict orders which had to be enforced at once, despite the late hour, the gales, and the frozen seas. Stalin would not now brook even the slightest delay. The *Ilyich* (and the ice-breaker that preceded her) had been especially detailed for the task; apart from Trotsky, his family, and two G.P.U. officers, she had not a single passenger on board and carried no cargo. Stalin was at last confronting the Politbureau with a *fait accompli*; he thus cut short all hesitation and prevented the repetition of scenes like those which had occurred when he first asked the Politbureau to authorize the banishment and when Bukharin protested, wrung his hands, and wept in full session, and together with Rykov and Tomsky voted against.[1]

The banishment was effected in the greatest secrecy. The decision was not made public until well after it had been carried out. Stalin was still afraid of commotion. The troops assembled in the harbour were there to prevent any demonstration of protest and any mass farewell such as the Opposition had organized a year earlier, before Trotsky's abduction from Moscow.[2] This time there were to be no witnesses and no

---

[1] See *The Prophet Unarmed* pp. 468–71. Rykov was still Chairman of the Council of People's Commissars, i.e. Soviet Prime Minister in succession to Lenin.

[2] Op. cit., p. 393.

eye-witness accounts. Trotsky was not to travel with a crowd of passengers before whose gaze he might resort to passive resistance. Even the crew were warned to keep to their quarters and avoid all contact with those on board. A nervous mystery surrounded the voyage. Stalin did not yet wish to burden himself with full responsibility. He was waiting to see whether communist opinion abroad would be shocked; and he did not know whether future developments might not compel him to recall his adversary. He took care to stage the deportation so ambiguously that it could be explained away, if need be, or even denied completely—for a few days afterwards communist newspapers abroad were suggesting that Trotsky had gone to Turkey on an official or semi-official mission or that he had gone there of his own accord, with a large suite.[1]

And so suddenly Trotsky found himself on board a bleak and almost deserted ship, heading through gales towards an empty horizon. Even after the year at Alma Ata, this void around him, made even more malignant by the hovering figures of the two G.P.U. officers, was disconcerting. What could it mean? What could it portend? Only Natalya and Lyova were by his side; and in their eyes he could read the same question. To escape the gale and the emptiness they went down to their cabins and stayed there throughout the voyage. The emptiness seemed to creep after them. What did it signify? What was to be the journey's end?

Trotsky was prepared for the worst. He did not think that Stalin would be content to deposit him on the other shore of the Black Sea and let him go. He suspected that Stalin and Kemal Pasha, Turkey's President and dictator, were in a plot against him, and that Kemal's police would seize him from the boat and either intern him or deliver him surreptitiously to the vengeance of White émigrés congregating in Constantinople. The tricks the G.P.U. had played on him confirmed this apprehension: he had repeatedly asked them to release from prison Sermuks and Posnansky, his two devoted secretaries and bodyguards, and to allow them to accompany him abroad; and the G.P.U. had repeatedly promised to do so but had broken the promise. They had evidently decided to put him on shore without a friend to guard him. En route the escorting officers tried to

[1] *Humanité*, February 1929.

reassure him: Sermuks and Posnansky, they said, would join him in Constantinople, and meanwhile the G.P.U. assumed responsibility for his safety. 'You have cheated me once,' he replied, 'and you will cheat me again.'[1]

Baffled and anguished, he recalled with his wife and son the last sea voyage they had made together—in March 1917 when, freed from British internment in Canada, they had set off for Russia on board a Norwegian steamer. 'Our family was the same then,' Trotsky reflects in his autobiography (although Sergei, his younger son, who had been with them in 1917 was not on the *Ilyich*), 'but we were twelve years younger.' More essential than this difference in age was the contrast in the circumstances, on which he makes no comment. In 1917 the revolution called him back to Russia for the great battles to come; now he was driven from Russia by a government ruling in the name of the revolution. In 1917, every day throughout the month spent in British internment, he had addressed crowds of German sailors behind barbed wire, prisoners of war, telling them of the stand taken by Karl Liebknecht in the Reichstag, in jail, and in the trenches against the Kaiser and the imperialist war, and arousing their enthusiasm for socialism. When he was released, the sailors had carried him shoulder high all the way to the camp gate, cheering him and singing the Internationale.[2] Now there was only the void around him and the howling gale. It was ten years now since the defeat of the *Spartakus* and the assassination of Liebknecht; and more than once already Trotsky had wondered whether he too was not fated to suffer 'Liebknecht's end'. A minor incident added a grotesque touch to this contrast. As the *Ilyich* was entering the Bosphorus one of the G.P.U. officers handed him the sum of 1,500 dollars, a grant which the Soviet Government had made their former Commissar of War 'to enable him to settle abroad'. Trotsky could see Stalin's mocking grin; but, being penniless, he swallowed the affront and accepted the money. This was the last wage he received from the state of which he had been a founding father.

[1] Trotsky's messages to the Central Committee, the Executive of the Comintern, and to 'Citizen Fokin, plenipotentiary of G.P.U.', dated 7–12 February 1929. *The Archives; Moya Zhizn*, vol. II, p. 318.

[2] *The Prophet Armed*, p. 247.

.Trotsky would not have been himself if he had brooded over these melancholy incidents. Whatever the future held in store, he was resolved to meet it on his feet and fighting. He would not allow himself to be dispersed in the void. Beyond it there were unexplored horizons of struggle and hope—the past to live up to now and a future in which past and present would live on. He felt nothing in common with those historic personalities of whom Hegel says that once they have accomplished their 'mission in history' they are exhausted and 'fall like empty husks'.[1] He would struggle to break out of the vacuum in which Stalin and events were enclosing him. For the moment he could only record his final protest against expatriation. Before the end of the voyage he delivered to his escort a message addressed to the Central Committee of the Party and the Central Executive Committee of the Soviets. In it he denounced the 'conspiracy' which Stalin and the G.P.U. had entered with Kemal Pasha and Kemal's 'national fascist' police; and he warned his persecutors that a day would come when they would have to answer for this 'treacherous and shameful deed'. Then, after the *Ilyich* had dropped anchor and Turkish frontier guards appeared, he handed them a formal protest addressed to Kemal. Anger and irony broke through his restrained official tone: 'At the gates of Constantinople', he wrote, 'I have the honour to inform you that it is not by my own free will that I have arrived at the frontier of Turkey—I am crossing this frontier only because I must submit to force. Please, Mr. President, accept my appropriate sentiments.'[2]

He hardly expected Kemal to react to this protest, and he was aware that his persecutors in Moscow would not be deterred by the thought that one day they might be called to account for what they were doing. But even if at the moment it seemed vain to invoke history for justice, he could do nothing but invoke it. He was convinced that he spoke not for himself only but for his silent, imprisoned, or deported friends and followers, and that the violence of which he was the victim was inflicted on the Bolshevik Party at large and the revolution itself. He knew that, whatever his personal fortunes, his controversy with Stalin would go on and reverberate through

[1] Hegel, *Philosophie der Weltgeschichte*, p. 78.
[2] *The Archives; Moya Zhizn*, vol. II, p. 317.

the century. If Stalin was bent on suppressing all those who might protest and bear witness, then Trotsky, at the very moment when he was being driven into exile, would come forward to protest and bear witness.

. . . .

The sequel to the disembarkation was almost farcical. From the pier Trotsky and his family were taken straight to the Soviet Consulate in Constantinople. Although he had been branded as a political offender and counter-revolutionary, he was received with the honours due to the leader of October and the creator of the Red Army. A wing of the Consulate was reserved for him. The officials, some of whom had served under him in the civil war, seemed eager to make him feel at home. The G.P.U. men behaved as if they meant to honour the pledge that they would protect his life. They met all his wishes. They went on errands for him. They accompanied Natalya and Lyova on trips to the city, while he stayed at the Consulate. They took care to unload and transport his bulky archives brought from Alma Ata, without even trying to check their contents—the documents and records which he was presently to use as political ammunition against Stalin. Moscow seemed to be still trying to disguise the banishment and soften its impact on communist opinion. Not for nothing did Bukharin once speak of Stalin's genius for gradation and timing: Stalin's peculiar gift for pursuing his aims by slow degrees, inch by inch, showed itself even in details like these.

It showed itself also in the way he had assured himself of Kemal Pasha's co-operation. The Turkish Government informed Trotsky shortly after his arrival that they had never been told that he was to be exiled, that the Soviet Government had simply requested them to grant him an entry permit 'for health reasons', and that, cherishing friendly relations with their northern neighbour, they could not go into the motives for the request and had to grant the visa. Yet Kemal Pasha, uneasy at seeing himself thus turned into Stalin's accomplice, hastened to assure Trotsky that 'it was out of the question that he should be interned or exposed to any violence on Turkish soil', that he was free to leave the country whenever he chose or to stay as long as he pleased; and that if he were to stay, the Turkish Government would extend to him every hospitality and ensure his safety.[1] Despite this respectful sympathy, Trotsky remained convinced that Kemal was hand-in-glove with Stalin. There was, in any case, no knowing how Kemal would behave if Stalin confronted him with further demands—would he risk embroiling himself with his powerful 'northern neighbour' for the sake of a political exile?

The ambiguous situation created by Trotsky's residence in the Soviet Consulate could not last. Stalin was only waiting for a pretext to end it; and it was unbearable to Trotsky as well. 'Protected' by the G.P.U., he remained their virtual prisoner, not knowing whom to fear more: the White émigrés outside the Consulate or his guards inside. He found himself deprived of the sole advantage that exile bestows upon the political fighter: freedom of movement and expression. He was anxious to state his case, to reveal the events that had led to his expulsion, to make contact with followers in various countries, and to plan further action. He could not safely do any of these things from the Consulate. In addition, both he and his wife were ill; and he had to earn his living, which he could do only by writing. He had to settle somewhere, to get in touch with publishers and newspapers; and to start work.

On the day he arrived he sent out messages to friends and well-wishers in western Europe, especially in France. Their response was immediate. 'We need hardly tell you that you can count on us body and soul. We embrace you from the depth of our faithful and affectionate hearts.' Thus Alfred and Marguerite Rosmer wrote to him three days after he had landed.[2] They had been his and Natalya's friends since the First World War, when they were in the Zimmerwald movement. In the early nineteen-twenties Alfred Rosmer had represented the French Communist party on the Executive of the Communist International in Moscow; and for his solidarity with Trotsky he had been expelled from the party. The 'depth of our faithful and affectionate hearts' was no mere turn of phrase with the Rosmers—they were to remain Trotsky's only intimate friends in the years of his exile, despite later disagreements and

---

[1] Quoted from a letter to Trotsky, written on Kemal's order, by the Governor of Constantinople on 18 February 1929. Closed section of the Trotsky *Archives*.

[2] Correspondence between the Rosmers and Trotsky. Ibid.

discords. Boris Souvarine, a former Editor of the theoretical paper of the French Communist party, who alone among all foreign communist delegates in Moscow in May 1924 spoke up in Trotsky's defence, also wrote to offer help and co-operation.[1] Other well-wishers were Maurice and Magdeleine Paz, a lawyer and a journalist, both expelled from the Communist party, and in later years well known as socialist parliamentarians. Addressing him as *'Cher grand Ami'*, they wrote of their anxiety about his precarious position in Turkey, tried to obtain for him entry permits to other countries, and promised to join him shortly in Constantinople.[2]

Through the Rosmers and Pazes Trotsky established contact with western newspapers; and while still at the Consulate, he wrote a series of articles which appeared in the *New York Times*, the *Daily Express*, and other papers in the second half of February. This series was his first public account of the inner party struggle of the last years and months. It was brief, forceful, and aggressive. He spared none of his enemies or adversaries, old or new, least of all Stalin whom he now denounced to the world as he had earlier denounced him to the Politbureau as 'the grave-digger of the revolution'.[3] Even before these articles appeared, he was in trouble with his hosts, who began to urge him to move from the Consulate to a compound inhabited by consular employees, where he would go on living under G.P.U. 'protection'. He refused to move, and the question was shelved until the publication of the articles brought matters to a head. Stalin now had the pretext he needed to bring the banishment into the open. Soviet newspapers spoke of Trotsky having 'sold himself to the world bourgeoisie and conspiring against the Soviet Union'; and their cartoonists depicted *Mister* Trotsky embracing a bag with 25,000 dollars. The G.P.U. declared that they no longer held themselves responsible for his safety and were going to evict him from the Consulate.[4]

For several days Natalya and Lyova, even now solicitously

[1] Souvarine to Trotsky, 15 February 1929, ibid.

[2] Maurice Paz to Trotsky, 18 February 1929, ibid.

[3] The original text bears the date of 25 February 1929. *The Archives; Écrits*, vol. I, pp. 19–52.

[4] Trotsky's correspondence with the G.P.U. representative in Constantinople of 5 and 8 March. *The Archives.*

accompanied by the G.P.U. men, searched breathlessly the suburbs and outskirts of Constantinople for some more or less safe and secluded accommodation. At last they found a house, not in or near the city, but on the Prinkipo Islands, out on the sea of Marmara—it took an hour and a half to reach the islands by steamer from Constantinople. There was a touch of irony in this hurried choice of residence, for Prinkipo, or the Princes' Isles, had once been a place of exile to which Byzantine Emperors confined their rivals and rebels of royal blood. Trotsky arrived there on 7 or 8 March. As he set foot on the shore at Büyük Ada, the main village of Prinkipo, he imagined that he was alighting there as a bird of passage; but this was to be his home for more than four long and eventful years.

.   .   .   .

Trotsky often described this period of his life as his 'third emigration'. The term, not quite precise, reveals something of the mood in which he came to Prinkipo. This was indeed the third time that he had been deported by Russian governments and that he had come to live abroad. But in 1902 and 1907 he had been deported to Siberia or the Polar Region, whence he fled and took refuge in the West; and wherever he came in those days he belonged to that large, active, and dynamic community that was revolutionary Russia in exile. This time he had not chosen to become an émigré; and abroad there was no community of Russian exiles to receive him as one of their own and to offer him the environment and the medium for further political activity. Many new colonies of political émigrés existed; but these formed the counter-revolutionary Russia in exile. Between him and them there was the blood of the civil war. Of those who in that war had fought on his side there was none to join hands with him.

His third exile was therefore different in kind from the previous two. It could not be related to any precedent, for in the long and abundant history of political emigration there hardly ever was a man banished into comparable solitude (except Napoleon who was, however, a prisoner of war). Unconsciously, as it were, Trotsky sought to soften for himself and his family the severity of his present ostracism by relating it to his pre-revolutionary experiences. The memory of those experiences was now comforting. His first period of emigration lasted less than three years—it was interrupted by the *annus mirabilis* of 1905; the second lasted much longer, ten years; but it was followed by the supreme triumph of 1917. Each time history had bounteously rewarded the revolutionary for his restless wait abroad. Was it too much to expect that she would do so again? He was aware that this time the outlook might prove less promising and that he might never return to Russia. But stronger than this awareness was his need for a clear-cut and encouraging prospect, and the optimism of the fighter who even when he courts defeat, or is engaged in a hopeless battle, still looks forward to victory.

This kind of optimism was never to forsake him. But whereas in later years he remained confident of the ultimate triumph of his cause rather than of his chance of living to see it, in the first years of this exile there was still a more personal note to his optimism. He did indeed look forward to his early vindication and return to Russia. He did not consider the political situation there as stable; and, amid the upheavals of collectivization and industrialization, he expected shifts in the nation to produce great shifts in the ruling party as well. He did not believe that Stalinism could achieve consolidation. Was it anything more than a patchwork of incompatible ideas, the shilly-shallying of a bureaucracy not daring to tackle the problems by which it was confronted? He was convinced that the 'interlude' of Stalin's ascendancy must be brought to an end either by a resurgence of the revolutionary spirit and a regeneration of Bolshevism or by counter-revolution and capitalist restoration. This stark alternative governed his thoughts, even if at times he reckoned with other possibilities as well. He saw himself and his co-thinkers as representing the only serious opposition to Stalin, the only opposition that stood on the ground of the October revolution, offered a programme of socialist action, and constituted an alternative Bolshevik government. He did not imagine that Stalin would be able to destroy the Opposition or even to reduce it to silence for long. Here too his hopes fed on pre-revolutionary memories. Tsardom had failed to stifle any opposition, even though it imprisoned, deported, and executed the revolutionaries. Why then should Stalin, who was not yet executing his opponents, succeed where the Tsars had failed?

True, the Opposition had had its ups and downs; but, having deep roots in social realities and being the mouthpiece of the proletarian class interest, it could not be annihilated. As its acknowledged leader he was in duty bound to direct its activity from abroad, as Lenin and indeed he himself had once led their followers from exile. He alone could now speak for the Opposition in relative freedom and make its voice heard far and wide.

In yet another respect, however, his position was unlike what it had been before the revolution. Then he was unknown to the world or known as a Russian revolutionary only to the initiated. This was not his present standing. He had not this time re-emerged from the dimness of an underground movement. The world had seen him as leader of the October insurrection, as founder of the Red Army, as architect of its victory, and as inspirer of the Communist International. He had risen to a height from which it is not given to descend. He had acted his part on a world stage, in the limelight of history, and he could not withdraw. His past dominated his present. He could not lapse back into the protective obscurity of pre-revolutionary émigré life. His deeds had shaken the world; and neither he nor the world could forget them.

Nor could he confine himself to his Russian preoccupations. He was conscious of his 'duty towards the International'. Much of the struggle of recent years had centred on the strategy and tactics of communism in Germany, China, and Britain, and on the manner in which Moscow, for the sake of expediency, emasculated the International. It was unthinkable that he should not carry on this struggle. On the face of it, banishment should have made it easier for him to do so. If, as the champion of internationalism and the critic of Stalinist and Bukharinist 'national narrow-mindedness', he incurred unpopularity in Russia, he had reason to hope for eager response from communists outside Russia, for it was their most vital interest that he sought to advance when to socialism in a single country he opposed the primacy of the international viewpoint. From Moscow and Alma Ata he could not address foreign communists, and Stalin had seen to it that they should either remain ignorant or get only a grossly distorted view of what he stood for. Now at last his enforced stay abroad enabled him to put his case before them.

He still viewed the 'advanced industrial countries of the West', especially those of western Europe, as the main battle-grounds of the international class struggle. In this he was true to himself and the tradition of classical Marxism which he represented in its purity. In fact, no school of thought in the labour movement, not even the Stalinist, yet dared openly to flout that tradition. For the Third as for the Second International western Europe was still the main sphere of activity. The German and the French Communist parties commanded large mass followings, while the Soviet Union was still industrially underdeveloped and extremely weak, and the victory of the Chinese revolution was twenty years off. Just as bourgeois Europe, even in this period of its decline, still ostensibly held the centre of world politics, so the western European working classes still appeared to be the most important forces of proletarian revolution, the most important next to the Soviet Union in the Stalinist conception, and potentially even more important in Trotsky's.

Trotsky, of course, did not believe in the stability of the bourgeois order in Europe. When he arrived at Prinkipo the 'prosperity' which the West enjoyed in the late nineteen-twenties was already nearing its end. But Conservatives, Liberals, and Social Democrats still basked in the sunshine of democracy, pacifism, and class co-operation which were to assure the indefinite continuation of that prosperity. Parliamentary government appeared to be firmly established; and fascism, entrenched only in Italy, seemed a marginal phenomenon of European politics. Yet in his first days in Constantinople Trotsky announced the approaching end of this fools' paradise and spoke of the decay of bourgeois democracy and the ground-swell of fascism: '. . . these post-war trends in Europe's political development are not episodic; they are the bloody prologue to a new epoch. . . . The [first world] war has ushered us into an era of high tension and great struggle; major new wars are casting their shadows ahead. . . . Our epoch cannot be measured by the standards of the nineteenth century, that classical age of expanding [bourgeois] democracy. The twentieth century will in many respects differ from the nineteenth even more than modern times differ from the middle ages.'[1] He had a sense of returning to Europe

[1] *Écrits*, vol. I, p. 47.

on the eve of a decisive turn of history, when socialist revolution
alone could offer the western nations the effective alternative to
fascism. Revolution in the West, he believed, would also free
the Soviet Union from isolation and create a powerful counterbalance to the immense weight of backwardness that had
depressed the Russian Revolution. This hope did not seem vain.
The western labour movement, with its mass organizations
intact and its fighting spirit subdued but not yet deadened, was
still battleworthy. The Communist parties, despite their faults
and vices, still had in their ranks the vanguard of the working
class. Trotsky concluded that what was necessary was to open
the eyes of that vanguard to the dangers and the opportunities,
to make it aware of its responsibilities, to shake its conscience,
and to arouse it to revolutionary action.

This view of the present as well as his own past cast Trotsky
for his peculiar role in exile. He came forward as the legatee of
classical Marxism and also of Leninism, which Stalinism had
degraded to a set of dogmas and to a bureaucratic mythology.
To restore Marxism and to reimbue the mass of communists
with its critical spirit was the essential preliminary to effective
revolutionary action, and the task that he set himself. No
Marxist, except Lenin, had ever spoken with a moral authority
comparable to his, the authority he wielded as both theorist and
victorious commander in a revolution; and none had to act in a
.situation as difficult as his, being on all sides surrounded by
implacable hostility and being caught up in a conflict with the
state which had issued from revolution.

He possessed in abdundance and even superabundance the
courage and energy needed to cope with such a role and to
grapple with such a predicament. All the severe reverses he
had suffered, far from dulling his fighting instincts, had excited
them to the utmost. The passions of his intellect and heart,
always uncommonly large and intense, now swelled into a
tragic energy as mighty and high as that which animates the
prophets and the law-givers of Michelangelo's vision. It was
this moral energy that preserved him at this stage from any
sense of personal tragedy. There was as yet not even a hint of
self-pity in him. When in the first year of exile he concluded his
autobiography with the words: 'I know no personal tragedy',
he spoke the truth. He saw his own destiny as an incident in

the great flux and reflux of revolution and reaction; and it did not greatly matter to him whether he fought in the full panoply of power or whether he did so as an outcast. The difference did not affect his faith in his cause and in himself. When a critic remarked well-meaningly that despite his fall the ex-Commissar of War had preserved the full clarity and power of his thought, Trotsky could only mock the Philistine 'who saw any connexion between a man's power of reasoning and his holding of office'.[1] He felt the fullness of life only when he could stretch all his faculties and use them in the service of his idea. This he was going to do come what might. What sustained his confidence was that his triumphs in the revolution and the civil war still stood out more vividly in his mind than the defeats that followed them. He knew that these were imperishable triumphs. So mighty had been the climax of his life that it over-shadowed the anti-climax and no power on earth could drag him down from it. All the same, tragedy, relentless and pitiless, was closing in on him.

   .   .   .   .

Around 1930 Prinkipo was still as deserted as it probably was when the disgraced brothers and cousins of the Byzantine Emperors lingered away their lives on its shores. Nature itself seemed to have designed the spot to be a regal penitentiary. A 'red-cliffed island set in deep blue', Büyük Ada 'crouches in the sea like a pre-historical animal drinking'.[2] In the blaze of a sunset its purple unfurled gaily and challengingly like a flame over the serene azure; then it burst into a red rage of lonely defiance, gesturing angrily at the remote and invisible world, until at last it sunk resentfully into the dark. The island-ers, a few fishermen and shepherds, dwelling between the red and the blue, lived as their forefathers did a thousand years earlier; and 'the village cemetery seemed more alive than the village itself'.[3] The horn of a motor-car never disturbed the stillness; only the braying of an ass came down from the outlying cliff and field into the main street. For a few weeks in the year noisy vulgarity intruded: in the summer multitudes of holiday makers, families of Constantinople merchants, crowded the

[1] *Moya Zhizn*, vol. II, p. 336.   [2] Max Eastman, *Great Companions*, p. 117.
[3] Quoted from Trotsky's unpublished diary (July 1933). *The Archives*.

beaches and the huts. Then calm returned, and only the
braying of the ass greeted the still and splendid onset of the
autumn.

On the fringe of Büyük Ada, closed in between high hedges
and the sea, fenced off from the village and almost as aloof
from it as the village was from the rest of the world, was
Trotsky's new abode, a spacious, dilapidated villa rented from
a bankrupt pasha. When the new tenants moved in, it was sunk
in cobwebbed squalor. Years later Trotsky recollected the gaiety
and zest for cleanliness, with which Natalya rolled up her
sleeves and made her menfolk do the same to sweep away the
filth, and paint the walls white. Much later they covered the
floors with paint so cheap that many months after their shoes
still stuck to it as they walked. At the centre of the house
was a vast hall with doors opening on a veranda facing the sea.
On the first floor was Trotsky's work room, the walls of which
quickly became lined with books and periodicals arriving from
Europe and America. On the ground floor was the secretariat
with Lyova in charge. An English visitor described 'the dingy
marbles, sad bronze peacock, and humiliated gilt betraying the
social pretensions as well as the failure of the Turkish owner'
—this faded *décor*, designed to give comfort and prestige to a
retired pasha, contrasted comically with the Spartan aura the
place assumed.[1] Max Eastman, who arrived there when the
house was full of secretaries, bodyguards, and guests, compared
it in its 'lack of comfort and beauty' to a bare barrack. 'In these
vast rooms and on the balcony there is not an article of furniture, not even a chair! They are mere gangways and the doors
to the rooms on each side are closed. In each of these rooms
someone has an office table or a bed, or both, and a chair to
go with it. One of them, downstairs, very small and square and
white-walled with barely space for table and chairs, is the
dining-room.' The hedonistically minded American visitor
reflected that 'a man and woman must be almost dead aesthet-
ically' to live in so severe an abode, when 'for a few dollars'
they might have made of it a 'charming home'.[2] No doubt,
the place had none of the cosiness of an American middle-class

---

[1] The *Manchester Guardian*, 17 March 1931. See also Rosmer in the 'Appendice'
to Trotsky's *Ma Vie*, p. 592.
[2] Eastman, loc. cit.

home. Even in normal circumstances it would hardly have occurred to Trotsky or Natalya to set up a 'charming home' with pictures 'for a few dollars'; and their circumstances on Prinkipo were never normal. They sat there all the time as in a waiting-room on a pier, looking out for the ship that would take them away. The garden around the villa was abandoned to weeds, 'to save money' as Natalya explained to the visitor, who half·expected Trotsky to cultivate his little plot of land. Effort and money had to be saved for a desperate struggle in which the Büyük Ada house was a temporary headquarters. Its clean and bare austerity suited its purpose.

.   .   .   .

From the moment of his arrival Trotsky was unreconciled to his isolation and apprehensive of remaining within such easy reach of both the G.P.U. and the White émigrés. Outside his gates two Turkish policemen were posted, but he could hardly entrust his safety to them. Almost at once he began the quest for a visa which he partly described in the last pages of his autobiography.[1]

Even before his deportation from Odessa he had asked the Politbureau to obtain for him a German entry permit. He was told that the German Government—a Social Democratic Government headed by Hermann Mueller—had refused. He was half convinced that Stalin cheated him; and so when soon afterwards Paul Loebe, the Socialist Speaker of the Reichstag, declared that Germany would grant Trotsky asylum, he at once applied for a visa. He was not deterred by the 'malicious satisfaction with which . . . newspapers dwelt on the fact that an advocate of revolutionary dictatorship was obliged to seek asylum in a democratic country'. This lesson, they said, should teach him 'to appreciate the worth of democratic institutions'. The lesson was hardly edifying, however. The German Government first asked him whether he would submit to restrictions on his freedom of movement. He answered that he was prepared to refrain from any public activity, to live in 'complete seclusion', preferably somewhere near Berlin, and devote himself to literary work. Then he was·asked whether

[1] *Moya Zhizn*, vol. II, pp. 318–33. *The Archives.*

it would not be enough for him to come for a short visit, just to undergo medical treatment. When he replied that having no choice he would content himself even with this, he was told that in the government's view he was not so ill as to require any special treatment. 'I asked whether Loebe had offered me the right of asylum or the right of burial in Germany. . . . In the course of a few weeks the democratic right of asylum was thrice curtailed. At first it was reduced to the right of resistance under special restrictions, then to the right of medical treatment, and finally to the right of burial. I could thus appreciate the full advantages of democracy only as a corpse.'

The British House of Commons discussed Trotsky's admission as early as February 1929. The Government made it clear that it would not allow him to enter. The country was just about to have an election and the Labour Party was expected to return to office. Before the end of April two leading lights of Fabianism, Sidney and Beatrice Webb, arrived in Constantinople and respectfully asked Trotsky to receive them.[1] Despite old political animosities he entertained them courteously, eagerly enlightening himself on the economic and political facts of British life. The Webbs expressed their confidence that the Labour Party would win the election, whereupon he remarked that he would then apply for a British visa. Sidney Webb regretted that the Labour Government would depend on Liberal support in the Commons, and the Liberals would object to Trotsky's admission. After a few weeks Ramsay MacDonald did indeed form his second government with Sidney Webb, now Lord Passfield, as one of his Ministers.

Early in June, Trotsky applied to the British Consulate in Constantinople and cabled a formal request for a visa to MacDonald. He also wrote to Beatrice Webb, in terms as elegant as witty, about their talks at Prinkipo and the attraction that Britain, especially the British Museum, exercised on him. He appealed to Philip Snowden, the Chancellor of the Exchequer, saying that political differences should not prevent him from visiting England just as they had not prevented Snowden from going to Russia when Trotsky was in office. 'I hope to be able soon to return you the kind visit you paid me in

---

[1] The Webbs' correspondence with Trotsky is in *The Archives*, Closed Section. The letter in which they ask Trotsky to receive them is dated 29 April 1929.

Kislovodsk', he telegraphed George Lansbury.[1] It was all in vain. However, it was not the Liberals who objected to his admission. On the contrary, they protested against the attitude of the Labour Ministers; and Lloyd George and Herbert Samuel repeatedly intervened, in private, in Trotsky's favour.[2] 'This was a variant', he commented, 'which Mr. Webb did not foresee.' On and off, for nearly two years, the question was raised in Parliament and in the Press. H. G. Wells and Bernard Shaw wrote two statements of protest against the barring of Trotsky; and J. M. Keynes, C. P. Scott, Arnold Bennett, Harold Laski, Ellen Wilkinson, J. L. Garvin, the Bishop of Birmingham, and many others appealed to the Government to reconsider their decision. The protests and appeals fell on deaf ears. 'This "one act" comedy on the theme of democracy and its principles . . .', Trotsky observed, 'might have been written by Bernard Shaw, if the Fabian fluid which runs in his veins had been strengthened by as much as five per cent of Jonathan Swift's blood.'

Shaw, even if his satirical sting was not at its sharpest on this occasion, did what he could. He wrote to Clynes, the Home Secretary, about the 'ironic situation . . . of a Labour and Socialist government refusing the right of asylum to a very distinguished Socialist while granting it . . . to the most reactionary opponents. Now, if the government by excluding Mr. Trotsky could have also silenced him. . . . But Mr. Trotsky cannot be silenced. His trenchant literary power and the hold, which his extraordinary career has given him on the public imagination of the modern world, enable him to use every attempt to persecute him. . . . He becomes the inspirer and the hero of all the militants of the extreme left of every country.' Those who had 'an unreasoning dread of him as a caged lion'

---

[1] The copies of the application, cables, and letters are in *The Archives*, Closed Section. The letter to Beatrice Webb, written in French 'with Rosmer's help' says, *inter alia*: 'Je me souviens avec plaisir de votre visite. Ce fût pour moi une surprise agréable et, bien que nos points de vue se soient révélés irréductibles, ce que nous savions bien du reste, la conversation avec les Webbs m'a montré que celui qui a étudié la désormais classique histoire du trade-unionisme pouvait encore bien tirer profit d'un entretien avec ses auteurs.' Speaking of the attraction Britain had for him, Trotsky mentioned 'ma sympathie déjà ancienne pour le British Museum'.

[2] *The Archives*, Closed Section, British Files. Trotsky's British correspondent who kept him *au courant* with these developments was a cousin of Herbert Samuel. He quoted Samuel himself as the source of the information.

should allow him to enter Britain 'if only to hold the key of his cage'. Shaw contrasted Kemal Pasha's behaviour with Mac-Donald's and found 'hard to swallow an example of liberality set by a Turkish government to a British one'.[1]

Other European governments were no more willing to 'hold the key of his cage'. The French dug up the order of expulsion issued against Trotsky in 1916 and declared it to be still in force. The Czechs at first were ready to welcome him, and Masaryk's Socialist Minister, Dr. Ludwig Chekh, addressing him as 'Most Respected Comrade', informed him, in agreement with Beneš, that the visa had been issued; but the correspondence ended frigidly, with the 'Comrade' addressed as *'Herr'* and with an unexplained refusal.[2] The Dutch, who were giving refuge to Kaiser Wilhelm, would not give it to Trotsky. In a letter to Magdeleine Paz he wrote ironically that, as he did not even know the Dutch language, the government could rest assured that he would not interfere in domestic Dutch affairs; and that he was prepared to live in any rural backwater, incognito.[3] Nor were the Austrians willing to give 'an example of liberality' to others. The Norwegian Government declared that they could not allow him to enter their country, because they could not guarantee his safety. Trotsky's friends sounded out even the rulers of the Duchy of Luxemburg. He found that 'Europe was without a visa'. He did not even think of applying to the United States, for this 'the most powerful nation of the world was also the most frightened'. He concluded that 'Europe and America were without a visa' and, 'as these two continents owned the other three, the planet was without a visa'. 'On many sides it had been explained to me that my disbelief in democracy was my cardinal sin. . . . But when I ask to be given a brief object lesson in democracy there are no volunteers.'[4]

The truth is that even in exile Trotsky inspired fear. Governments and ruling parties made him feel that no one can lead a great revolution, defy all the established powers, and challenge

---

[1] Quoted from the copy of Shaw's letter to Clynes, the Home Secretary, preserved in *The Archives*, ibid. Shaw intervened also with Henderson, the Foreign Secretary, who 'refused to interfere'.

[2] Trotsky's correspondence with Dr. Chekh (Czech), Czechoslovak Minister of Interior. *The Archives*, Closed Section.

[3] Ibid.   [4] *Moya Zhizn*, vol. II, p. 333.

the sacred rights of property with impunity. Bourgeois Europe
gazed with amazement and glee at the spectacle, the like of
which it had not seen indeed since Napoleon's downfall—
never since then had so many governments proscribed one man
or had one man aroused such widespread animosity and
alarm.[1] Conservatives had not forgiven him the part he had
played in defeating the anti-Bolshevik 'crusade of fourteen
nations'. No one expressed their feelings better than Winston
Churchill, the inspirer of that crusade, in a triumphantly
mocking essay on 'The Ogre of Europe'. 'Trotsky, whose frown
meted death to thousands, sits disconsolate, *a bundle of old rags*,
stranded on the shores of the Black Sea.' Presently Churchill
had second thoughts, and when he included the essay in
Great Contemporaries, he replaced the '*bundle of old rags*' by the
words 'Trotsky—*a skin of malice*'. Trotsky's first political
statements made 'on the shores of the Black Sea' showed him to
have remained unshaken as enemy of the established order,
and to be still as defiant and self-confident as he was in the days
when he led the Red Army and addressed the world from the
rostrum of the Communist International. No, no, this was not
'a bundle of old rags'—this was 'a skin of malice'.[2]

Ignorance of the issues that had split Bolshevism magnified
the hatred and the fear. Reputable newspapers could not tell
whether Trotsky's deportation was not a hoax and whether he
had not left his country in secret agreement with Stalin in
order to foster revolution abroad. *The Times* had 'reliable information' that this was indeed the case and saw Trotsky's hand

[1] '. . . Sir Austen Chamberlain [the Foreign Secretary]', Trotsky wrote, 'has,
according to newspaper reports . . . expressed the opinion that regular relations
[between Britain and the Soviet Union] . . . will become perfectly possible on the
day after Trotsky has been put against the wall. This lapidary formula does honour
to the temperament of the Tory Minister . . . but . . . I take the liberty of advising
him . . . not to insist on this condition. Stalin has sufficiently shown how far he is
prepared to go to meet Mr. Chamberlain by banishing me from the Soviet Union.
If he has not gone further, this is not for lack of good will. It would really be too
unreasonable to penalize, because of this, the Soviet economy and British industry.'
Écrits, vol. I, p. 27.

[2] Winston S. Churchill, *Great Contemporaries*, p. 197. My italics. Churchill wrote
the original essay in reply to an article by Trotsky for *John o' London's Weekly*.
Commenting on Churchill's profile of Lenin, Trotsky had pointed out that
Churchill's dates were mostly wrong and that he showed a total lack of insight into
Lenin's character because of the gulf that separated him from the founder of
Bolshevism. 'Lenin thought in terms of epochs and continents, Churchill thinks in
terms of parliamentary fireworks and *feuilletons*.'

behind Communist demonstrations in Germany.[1] The *Morning Post* reported, with circumstantial details, on secret negotiations between Stalin and Trotsky which were to bring the latter back to the command of the armed forces; the paper knew that in connection with this Trotsky's sister had travelled between Moscow, Berlin, and Constantinople.[2] The *Daily Express* spoke of 'this raven perched upon the bough of British socialism'—'Even with the clipped wings and claws, he is not the sort of fowl that we in Britain can ever hope to domesticate.'[3] The *Manchester Guardian* and the *Observer* supported with some warmth Trotsky's claim to political asylum, but theirs were solitary voices. American newspapers saw Trotsky as the 'revolutionary incendiary' and Stalin as 'the moderate statesman' with whom America could do business.[4] The German right wing and nationalist Press was raucous and rabid: 'Germany has enough trouble . . . we consider it superfluous to add to it by extending hospitality to this most powerful propagandist of Bolshevism', said the *Berliner Boersenzeitung*.[5] 'Trotsky, the Soviet-Jewish bloodhound, would like to reside in Berlin', wrote Hitler's *Beobachter*. 'We shall have to keep a watchful eye on this Jewish assassin and criminal.'[6]

The Social Democratic parties, especially those which were in office, felt somewhat disturbed in their democratic conscience, but were no less afraid. When George Lansbury protested at a Cabinet meeting against the treatment of Trotsky, the Prime Minister, the Foreign Secretary, and the Home Secretary replied: 'There he is, in Constantinople, out of the way—it is to nobody's interest that he should be anywhere else. We are all afraid of him.'[7] Beatrice Webb, express-

[1] *The Times*, 10 May 1929.

[2] *Morning Post*, 6–8 July 1929. The report was reproduced in many European papers. See e.g. *Intransigeant* of 8–9 July.

[3] *Daily Express*, 19 June 1929.

[4] See e.g. *The New York American* and *The New York World* of 27 February 1929. 'Stalin, intelligent Russian,' wrote the latter, 'knows that power without money is a shadow, so he leans in the direction of money'; and this should 'interest America's conservative government'.

[5] *Berliner Börsenzeitung*, 1 February 1929.

[6] 9 February 1929. The more 'respectable' *Hamburger Nachrichten* of 25 January 1929 said: 'Stalin is reaping the consequence of his blunder in not having sent Trotsky and the Trotsky crowd into the Great Beyond. . . .'

[7] The source of this information is Lansbury himself. He related it to Trotsky's British correspondent, whom he assured that he remained opposed to the Cabinet decision and that 'anything I can do behind the scenes to advise you, I will'. *The Archives*, Closed Section.

ing admiration for his intellect and 'heroic character', wrote to Trotsky: 'My husband and I were very sorry that you were not admitted into Great Britain. But I am afraid that anyone who preaches the permanence of revolution, that is carries the revolutionary war into the politics of other countries, will always be excluded from entering those other countries.'[1] Historically, this was not quite true: Karl Marx and Friedrich Engels spent most of their lives as refugees in England 'preaching the permanence of the revolution'. But times had changed, and Marx and Engels had not been as fortunate and unfortunate as to turn first from obscure political exiles into leaders of actual revolution and then back into exiles. Trotsky was not greatly surprised by the feeling he evoked. He refused to go about the business of visas more diplomatically, as the Pazes urged him to do; he would not pull strings behind the scenes and refrain from making public appeals.[2] Even while he was seeking a refuge for himself, he was engaged in a battle of ideas. He knew that governments and ruling classes, in their fear of him, were paying him a tribute: they could not view him as a private supplicant; they had to treat him as an institution and as the embodiment of revolution militant.

.   .   .   .

Without waiting for the result of his many requests and the canvassing for visas, Trotsky settled down to work. There was an unusual bustle on Prinkipo in the very first weeks after his arrival. Reporters from all the continents rushed to interview him. Visitors and friends appeared—in a single month, in May, no fewer than seven came from France alone and stayed for weeks, even months. Young Trotskyists arrived to serve as bodyguards and secretaries. German and American publishers called to sign contracts for books and to offer advances on royalties. From everywhere dissident communists wrote to inquire about points of ideology and policy; and presently Trotsky, answering every question systematically and scrupulously filing away mountains of paper, found himself up to his eyes in a correspondence, amazing in volume, which

[1] Beatrice Webb wrote on 30 April 1930 to thank Trotsky for a complimentary copy of *My Life*. She concluded the letter by offering the 'subversive propagandist' help with books, periodicals, and documents.

[2] Magdeleine Paz to Trotsky on 14 June 1929. *The Archives*, Closed Section.

he was to carry on, regardless of circumstances, till the end of his life. He was getting ready the first issue of the *Bulletin Oppozitsii*, the little periodical—it began to appear in July—which was to be his main platform for the discussion of inner party affairs and his most important medium of contact with the Opposition in the Soviet Union. It was not easy to edit it in Büyük Ada and to find Russian printers for it first in Paris and then in Berlin. At the same time he set out to organize his international following.

In addition, during the very first months of his stay on the island, he prepared a number of books for publication. He was anxious to acquaint the world with the 1927 Platform of the Joint Opposition, which was to see the light under the title *The Real Situation in Russia*. He assembled a collection of documents, suppressed in the Soviet Union, which were to make the volume on *The Stalinist School of Falsification*. In *The Third International After Lenin* he presented his 'Critique of the Draft Programme of the Third International' and the message he had addressed to the Sixth Congress from Alma Ata. Shortened and partly garbled versions of these texts had already appeared abroad, which was one more reason why Trotsky was eager to produce the full and authentic statements. *Permanent Revolution* was the small book, also written at Alma Ata, in which he restated and defended his theory in controversy with Radek.

The main literary fruit of the season was, however, *My Life*. Urged by Preobrazhensky and other friends to write his autobiography, he had, at Alma Ata, jotted down the opening parts narrating his childhood and youth; and on Prinkipo he hurriedly went on with the work, sending out chapters, as he completed them, to his German, French, and English translators. His progress was so rapid that one may wonder whether he had not drafted much more at Alma Ata than just the opening parts. Less than three months after he had come to Büyük Ada he was already able to write to the Klyachkos in Vienna, an old Russian revolutionary family with whom he was friendly well before 1914: 'I am still completely immersed in this autobiography, and I do not know how to get out of it. I could have virtually completed it long ago, but an accursed pedantry does not allow me to complete it. I go on looking up references, checking dates, deleting one thing and inserting another. More than once I have felt tempted to throw it all into the fireplace and to take to more serious work. But, alas, this is summer, there is no fire in the fireplaces, and, by the way, there are no fireplaces here either.'[1] In May he had sent to Alexandra Ramm, his German translator, a large part of the work; a few weeks later she already had in hand the chapters on the civil war. But in July his 'accursed pedantry' pestered him again and he went back to rewrite the opening pages of the book. Early in the autumn the whole manuscript had already gone out and fragments were being serialized in newspapers. While he was still fastidiously correcting the German and the French translations, he was getting ready to start the *History of the Russian Revolution*, the first synopsis of which Alexandra Ramm received before the end of November.[2]

Amid this burst of activity he was never free from anxieties about children, grandchildren, and friends he had left 'beyond the frontier'. The sorrow of Nina's agony and death was still fresh with him when Zina's illness—Zina was his elder daughter from his first marriage—disturbed him. He inquired for news from her via Paris, where the Pazes kept in touch with his family in Moscow through a sympathizer on the staff of the Soviet Embassy. Zina suffered from consumption; and the death of her sister, the persecution of her father, the deportation to Siberia of Platon Volkov, her husband, and the difficulty of keeping herself and her two children alive, had strained her mental balance. She tried unsuccessfully to obtain official permission to leave the country and join her father. Trotsky supported her financially; and his well-wishers urged the Soviet Government to grant her an exit permit. Her mother, Alexandra Sokolovskaya, was still in Leningrad, though no one knew how long she would be allowed to stay there; and she took care of Nina's children—their father too, Man-Nevelson, was deported and imprisoned. This was not all: Lyova's wife and child were also left in Moscow, at fate's mercy.

---

[1] The letter was written on 1 June 1929. *The Archives*, Closed Section.

[2] Alexandra Ramm, of Russian origin, was the wife of Franz Pfemfert, editor of a radical weekly *Aktion*. Pfemfert had been expelled from the Communist party as an 'ultra-radical' after the third Congress of the Comintern, when Trotsky's influence was at its height; but he and his wife, disregarding political differences, retained to the end a warm friendship for Trotsky.

Thus, among Trotsky's next of kin no fewer than four families were broken up by the pitiless political conflict. And almost every week brought news about victimization of friends and untold miseries, illnesses in prison, starvation, clashes with jailers, hunger strikes, suicides, and deaths. Trotsky did what he could to arouse protests, especially against the persecution of Rakovsky, until lately the best known and the most respected of Soviet Ambassadors in the West, who was dragged from one place of deportation to another and suffered heart attacks, and from whom there was no news for several months.

Trotsky's vitality got the better of anxiety, worry, and fatigue. He drowned his sorrows in tenacious work and in intercourse with friends and followers; and he sought relief from the strain of work in rowing and fishing in the sky-coloured waters of the Marmara. Even while he rested he was unable to bring his energy to a standstill; he had to expend it in strenuous exertion all the time. As at Alma Ata his fishing was still a matter of elaborate expeditions with heavy boats, stones, and dragnets. He would go out for long trips, accompanied by two Turkish fishermen who gradually became part of the household; and with them he toiled, dragged the nets and stones, and carried back loads of fish. (Eastman, who found Trotsky's 'idea of relaxation' disagreeable, wondered 'if that is the mood in which he will go fishing—intense, speedy, systematic, organized for success, much as he went to Kazan to defeat the White Armies'.[1]) He was unable to use his strength, physical or mental, sparingly; and even chronic ill-health did not seem to impair his sinewy agility. Sometimes he sailed out by himself and, to the alarm of his family and secretaries, disappeared for long periods. A follower who arrived at such a moment wondered whether Trotsky was not afraid that the G.P.U. might lay a trap for him out at sea. Trotsky replied somewhat fatalistically that the G.P.U. were so powerful that once they decided to destroy him he would be helpless anyhow. In the meantime he saw no reason why he should become his own jailer and deny himself the little freedom left to him, and the colour and taste of life.[2]

[1] Eastman, loc. cit.

[2] M. Parijanine describes vividly a fishing escapade with Trotsky far in the waters of Asia Minor: '. . . he was bent on getting his trophy . . . one could sense his secret happiness . . . he is mastering the element.' At nightfall they were caught

The misgivings with which he had arrived in Turkey were somewhat allayed. The Turks behaved correctly, even helpfully. Kemal Pasha was as good as his word, though Trotsky was still incredulous. The police guards, placed at the gates of the villa, attached themselves so much to their ward that they also became part of the household, running errands, and helping in domestic chores. The White émigrés made no attempt to penetrate behind the high fences and hedges. Even the G.P.U. seemed remote and uninterested. This appearance, however, was deceptive: the G.P.U. were anything but aloof. All too often one of their agents, posing as an ardent follower, slipped into Trotsky's entourage as secretary or bodyguard. 'A Latvian Franck stayed at Prinkipo for five months', writes Natalya. 'Later we learned that he was an informer of the Russian Secret Service, just like one Sobolevicius, also a Latvian, who came to us for a short stay only (his brother Roman Well acted as *agent provocateur* in Opposition circles in Paris and central Europe . . .).'[1] The trouble was that not all those who were exposed as *agents provocateurs* necessarily acted that part, whereas the most dangerous spies were never detected. Sobolevicius, for instance, thirty years later imprisoned in the United States as a Soviet agent, confessed that he had indeed spied on Trotsky during the Prinkipo period.[2]

by a great storm. The boat was very nearly overwhelmed; the Turkish gendarme accompanying them was crying with fear; and Trotsky took the oars and struggled vigorously against the tide. Such was his calm, concern for companions, and humour that Parijanine thought of 'Don't fear . . . thou hast Caesar and his fortunes with thee'. They found refuge in an empty hut on a deserted little island. Next morning, left without food, they shot two rabbits. Parijanine, having only wounded his rabbit, killed it off. 'This is not the hunter's way,' Trotsky said, 'one doesn't kill a wounded animal.' In the meantime the Turkish authorities had begun a search; and some peasants came to the rescue. Trotsky received the help with self-irony, recalling Shchedrin's story about two Russian generals lost in an unknown land and unable to procure the barest necessities of life. 'Ah,' sighs one of them, 'if only we could find a *mouzhik* here!' 'And lo, the *mouzhik* appears at once; and in a moment he has done all that was needed'. 'A Léon Trotsky', *Les Humbles*, May-June 1934.

[1] V. Serge, *Vie et Mort de. Trotsky*, pp. 201–2.

[2] See *Hearing before the Subcommittee to Investigate the Administration of the Internal Security Act*, etc. *United States Senate*, 21 November 1957, pp. 4875–6, where Sobolevicius appears under the name of Jack Soble. In his correspondence with Trotsky he used the cover name Senin. His brother Dr. Soblen, also condemned, fled from the U.S.A. to Israel in 1962; but was denied refuge there. Being returned to the United States, via England, he committed two attempts on his life and died in London,

Yet his whole correspondence with Trotsky and the circumstances of their break throw doubt on the veracity of this part
of his confession. Sobolevicius himself broke with Trotsky
after he had openly and repeatedly expressed important
political disagreements, which was not the manner in which an
agent provocateur would behave. Trotsky denounced him in the
end as a Stalinist, but did not believe that he was an *agent
provocateur*. Whatever the truth, both Sobolevicius and his
brother enjoyed Trotsky's almost unqualified confidence
during the first three Prinkipo years. They were no novices
to Trotskyist circles. Sobolevicius had been in Russia as
correspondent of the left Marxist *Saechsische Arbeiterzeitung*,
and there he joined the Trotskyist Opposition in 1927. Both
he and his brother were later not only extremely active in
France and Germany, they also supplied Trotsky with much
useful information and with reference materials for his books;
they helped him to publish the *Bulleten Oppozitsii*; and through
their hands went much of his clandestine correspondence with
the Soviet Union, codes, chemically written letters, cover
addresses, etc.[1]

In an underground organization it is hardly ever possible to
keep out the *agent provocateur* altogether. The organization is
invariably the stool-pigeon's target; and it is just as easy to
err on the side of too much suspicion, which may paralyse the
entire organization, as on the side of too little vigilance. What
made matters worse for Trotsky was that only very few of his
western followers were familiar with the Russian language and
background, and so he was unduly dependent on the few that
were. His work would have been almost impossible without
Lyova's assistance. But this was not enough; and Trotsky
accepted his son's sacrifice with uneasiness, for it was a sacrifice
on the part of a man in his early twenties to condemn himself
to a hermit-like existence on Prinkipo. So Trotsky was all too
often on the look-out for a Russian secretary, and this made it
easier for the stool-pigeon to sneak in. Occasionally friends
forestalled trouble with a timely warning. Thus, early in 1930,
Valentine Olberg, of Russian-Menshevik parentage, posing as a
Trotskyist, tried hard to obtain access to Prinkipo as a secre-

[1] The correspondence between Trotsky, Sobolevicius, and his brother R. Well
(Dr. Soblen) fills two files in the Closed Section of *The Archives*.

tary. But from Berlin Franz Pfemfert and Alexandra Ramm, suspicious of the applicant, informed Trotsky of their fears and Olberg was turned away—in 1936 he was to appear as defendant and witness against Trotsky, Zinoviev, and Kamenev in the first of the great Moscow trials.[1] Such timely warnings were all too rare, however; and in years to come the shadowy figure of the *agent provocateur* was to follow Trotsky like a curse.

. . . .

Trotsky's financial circumstances during the Prinkipo period were much easier than he had expected. His literary earnings were large, life on the island was cheap, and his and the family's needs were extremely modest. As the household increased, with secretaries and long-staying guests always around, and as the correspondence became almost as voluminous as that of a minor government department, the expenses rose to 12,000 and even 15,000 American dollars per year.[2] A wide international readership assured Trotsky of correspondingly high fees and royalties. For his first articles written in Constantinople he received 10,000 dollars, of which he put aside 6,000 as a publication fund for the *Bulletin Oppozitsii* and French and even American Trotskyist papers. Later in the year he received considerable advances on the various editions of *My Life*, 7,000 dollars on the American edition alone. In 1932 the *Saturday Evening Post* paid 45,000 dollars for the serialization of the *History of the Russian Revolution*.[3] When he left the Soviet Consulate in Constantinople, Trotsky borrowed 20,000 French francs from Maurice Paz. A year later he repaid the debt and had no need to borrow any more. When in May 1929 Paz inquired whether he was not in any difficulties, Trotsky

---

[1] Pfemfert's correspondence with Trotsky, April 1930, ibid. Olberg was a member of the *Reichsleitung* of the German Opposition. He aroused suspicion by his insistent inquiries about Trotsky's contacts with followers in the Soviet Union. (See also the correspondence between Olberg and Lev Sedov.) Whether he was an *agent-provocateur* in 1930 or became one later is, as in the case of Sobolevicius, not definitely established. After the rise of Nazism, in 1933–4, Olberg is said to have lived in dire poverty as a political émigré in Czechoslovakia. He may, of course, have acted as a Stalinist stool-pigeon for 'ideological' reasons, without receiving any reward. He was a defendant and one of the Prosecution's chief witnesses in Zinoviev's trial in 1936; and was sentenced to death.

[2] Eastman, op. cit.

[3] These data are drawn from Trotsky's accounts and correspondence with his publishers and literary agents. *The Archives*, Closed Section.

answered that far from this being the case he could now afford
to assist financially his political friends in the West. This, as his
correspondence and preserved accounts show, he did with an
unstinting hand, on which some of the recipients presently
came rather unbecomingly to rely.

.   .   .   .

Long before their defeat Trotsky, Zinoviev, and even
Shlyapnikov had made attempts at organizing their followers
in foreign Communist parties. These efforts were not altogether unsuccessful at first, despite excommunications and
expulsions.[1] The tactical manœuvres and retreats, however, of
the Russian Opposition disorientated communists abroad as
strongly as Stalinist reprisals intimidated them. The final
capitulation of Zinoviev's faction demoralized its foreign
associates. Trotsky's reverses and deportation had not had quite
the same effect. In the eyes of communists not yet fully prepared
to submit to Stalinist dictates, his moral authority stood as high
as ever; and the legend which surrounded his name, the legend
of indomitable militancy and victory, was enriched with its
new note of martyrdom. Yet the Comintern had already
stigmatized Trotskyism with so much brutality and was so
ferociously stamping it out from foreign sections that no
communist could hope to gain any advantage by embracing the
heresy; and few were those prepared to follow the martyr on his
path.

From Prinkipo, Trotsky set out to rally anew his supporters,
past and present. That he had no power to share with them did
not in his eyes render the undertaking hopeless—this made
it in a way even more attractive. Knowing that self-seekers
and bureaucrats would not respond, he appealed only to the
thoughtful and disinterested. Had not the strength of a revolutionary organization always consisted in the depth of the
conviction held by its members and in their devotion rather
than in their numbers? At the turn of the decade Stalin's mastery
of the Comintern was still superficial. Almost anyone who spent

[1] In a letter written to Sobolevicius and Well on 4 November 1929, Trotsky
maintained that the German *Leninbund* carried on its activities for money which its
leaders had received from Pyatakov before the latter's capitulation. The scale of
these activities was so modest that quite a small amount of money would have
enabled them to carry on.

those years in the Communist party can relate from experience
the bewilderment and the reluctance with which cadres and
rankers alike began to conform to the new orthodoxy consecrated in Moscow. Underneath the conformity, still only skin
deep, there was malaise, incredulity, and restiveness; and there
were old Marxist habits of thought and uneasy consciences, to
which Trotsky's fate was a constant challenge. The good party
man considered it his supreme duty to practise solidarity with
the Russian revolution; and so he could not take it upon himself
to contradict the men who now ruled Moscow, who spoke with
the voice of the revolution, and who insisted that the foreign
communist should, at committees and cells, vote for resolutions
condemning Trotskyism. The party man voted as he was
required, but the whole 'campaign' remained to him a sad
puzzle. The venom with which it was pursued vaguely offended
him. He was unable to discern its motive. And sometimes he
wondered why he should be required to add his own modest
endorsement to the awe-inspiring anathemas pronounced from
so far above. Working-class members, except for the very young
and uninformed, recalled the days of Trotsky's glory, his
resounding assaults on world capitalism, and his fiery manifestoes that had stirred so many of them and even brought some
of them into the ranks. The change in the party's attitude
towards the man whom they remembered as Lenin's closest
companion seemed incomprehensible. Yet there was little or
nothing they could do about it. Here and there a few men
disgusted by this or that manipulation of the 'party line'
renounced membership; but most reflected that they should
not perhaps be unduly concerned over what looked like a
feud among the big chiefs, that Russia was anyhow far away
and difficult to understand, but that their own class enemies
were near at home, and against them the Communist party
fought reliably and bravely. They continued to give their
allegiance to the party, but they did so despite and not because
of Stalinism; and for some time yet they shrugged with embarrassment when they heard party officialdom rail against
Trotsky, the 'traitor and the counter-revolutionary'.

Trotsky's hold on the imagination of the left and radical
intelligentsia was still immense. When Bernard Shaw wrote of
him as becoming anew the 'inspirer and hero of all the militants

of the extreme left of every country' he was not as far from the
truth as may have seemed later.[1] We have seen the impressive
list of the celebrities of radical England who spoke up in
Trotsky's defence against their own government. (True,
the British Communist party was less 'infected with Trotskyism' than any other; yet in Trotsky's Prinkipo correspondence
one finds a thick file of extremely friendly and revealing letters
he exchanged with an English communist writer, later notorious for Stalinist orthodoxy.) Among European and American
poets, novelists, and artists, famous or about to gain fame,
André Breton and others of the Surrealist school, Henrietta
Roland Holst, the Dutch poetess, Panait Istrati, whose meteoric
and sad literary career was then at its zenith, Diego Rivera,
Edmund Wilson, the young André Malraux, and many
others, were under his spell. 'Trotsky continued to haunt the
communist intellectuals', says a historian of American communism; and by way of illustration he quotes Michael Gold the
well-known communist writer and editor who even after the
first anathemas on Trotsky 'could not resist extolling Trotsky
[in the *New Masses*] as "almost as universal as Leonardo da
Vinci" '! As late as 1930 Gold wrote, among some tritely
derogatory remarks, that ' "Trotsky is now an immortal part of
the great Russian Revolution . . . one of the permanent legends
of humanity, like Savonarola or Danton".'[2] 'The unbounded
admiration for Trotsky was not confined to Michael Gold',
testifies another American communist man of letters, 'it marked
all the extreme radicals of this country who followed Russian
events. . . .'

In most European countries groups of expelled Trotskyists

[1] Shaw had many times expressed his admiration for Trotsky with unusual
ardour. In one of his letters to Molly Tompkins, for instance, he wrote:

'Yesterday . . . I had with me a bundle of reports of the speeches of our great
party leaders, and a half-crown book by Trotsky. . . . For sheer coarse savage
bloodymindedness it would be hard to beat the orations of Birkenhead, Lloyd
George, and Churchill. For good sense, unaffected frankness, and educated mental
capacity give me Trotsky all the time. To turn from the presidential campaign in
your country and the general election here to his surveys of the position is to move
to another planet.' G. B. Shaw, *To a Young Actress*, p. 78. It was Shaw who first
compared Trotsky, the writer, with Lessing (in terms which he borrowed from
Heines *Zur Geschichte der Philosophie und Religion in Deutschland*). See my Preface to
The Prophet Armed. See also further, p. 369.

[2] Th. Draper, *American Communism and Soviet Russia*, p. 358; and *Roots of American
Communism*, p. 129. See also J. Freeman, *An American Testament*, pp. 383–4.

/

and Zinovievists, led by a few of the founders of the Communist International, were active. It was only five years or so since the Central Committee of the French party had unanimously protested to Moscow against the anti-Trotskyist campaign. Between 1924 and 1929 Alfred Rosmer, Boris Souvarine, and others went on contending against Stalinism.[1] Trotskyist sympathies were alive in the revolutionary-syndicalist circle of Pierre Monatte which had formed one of the constituent elements of the French Communist party but had since become estranged from it. The Zinovievists kept their own *côterie*. In Germany there were the Leninbund and also the Wedding Opposition (so called after Berlin's largest working-class district); but there Zinovievism, as represented by Arkadii Maslov and Ruth Fisher, rather than Trotskyism set the tone of the dissidence. Two important Italian communist leaders, Antonio Gramsci and Amadeo Bordiga, both Mussolini's prisoners, had declared themselves against Stalin: Gramsci, from his prison cell, had sent his declaration to Moscow, where Togliatti, the party's representative with the Comintern Executive, suppressed it.[2] Andrés Nin, the most able

---

[1] *The Prophet Unarmed*, pp. 140–1. In 1926 Pyatakov, then on the staff of the Soviet Embassy in Paris, sought to unite the various anti-Stalinist elements expelled from the French Communist party. In Moscow, Trotsky and Zinoviev were forming the Joint Opposition, and Pyatakov's task was to create a French counterpart to it. He held meetings with Rosmer, A. Dunois, Loriot, Souvarine, Monatte, Paz, and others, and initiated the publication of *Contre le Courant*. But Rosmer and Monatte, hostile towards any idea of a 'bloc' between Trotskyists and Zinovievists, refused to co-operate; and so *Contre le Courant* began to appear as the French organ of the Joint Opposition, under the editorship of the Pazes and Loriot. Rosmer and Monatte continued their anti-Stalinist activities independently.

[2] *Bulletin Oppozitsii*, nos. 17–18, 1930, see also Rosmer's letter to Trotsky of 10 April 1930 in *The Archives*, Closed Section. About this time three members of the Italian Politbureau, Ravazzoli, Leonetti, and Tresso, went over to the Trotskyist Opposition. They were friends and followers of Gramsci; and one of them informed Rosmer about Gramsci's letter to Togliatti and its suppression. In 1961 I asked Togliatti publicly, in the Italian Press, to explain the matter. He answered through a friend of his that Gramsci had indeed urged him in 1926 not to involve Italian communism in the Russian inner-party struggle. (Togliatti had backed Bukharin and Stalin against Trotsky.) Togliatti maintains that Gramsci's letter arrived in Moscow during an inner-party truce; and so, after consulting Bukharin, he decided that it had no relevance to the current situation. When the struggle between Stalin and Trotsky was resumed, the Comintern and the Italian party were nevertheless kept in ignorance about Gramsci's attitude. This attitude accounted for the oblivion to which Gramsci's memory was consigned during the Stalin era. Only after Stalin's death were Gramsci's merits 'rediscovered', and Togliatti initiated something like a posthumous Gramsci cult in the Italian party.

exponent of Marxism in Spain, had thrown in his lot with the Russian Opposition and had for years kept in touch with Trotsky.[1] In Holland Maring-Sneevliet, the first inspirer of Indonesian communism, led a fairly strong group of Dutch left trade unionists opposed to Stalinism. In Belgium Van Overstraeten and Lesoil, ex-chiefs of the Communist party, and their followers strongly entrenched in the large mining district of Charleroi, had also embraced Trotskyism.

The inner party controversy had some repercussions even in Asia. The germs of Trotskyism had been brought to Shanghai, Peking, Quantung, and Wuhan by former students of the Sun Yat-sen University in Moscow, witnesses of Trotsky's struggle over the Chinese issue in 1927. In 1928 they held the first national conference of the Chinese Opposition; and some of them looked forward to an alliance with Mao Tse-tung, on whom the Comintern frowned at this time, because his attitude in 1925–7 had often coincided with Trotsky's and because he was now, at the ebb of the revolution, embarking upon partisan warfare against the Kuomintang. In 1929 Chen Tu-hsiu, the party's leader up to 1927, came out with the Open Letter in which he revealed the sordid inner story of the relations between Moscow, the Kuomintang, and Chinese communism, and acknowledged that Trotsky's criticisms of Stalin's and Bukharin's policy had been only too well founded.[2] The Trotskyist

[1] Nin was in correspondence with Trotsky during the Alma Ata period. *The Archives*.

[2] Trotsky's interest in China was as sustained as his contacts with his Chinese followers were, in the circumstances, close. In the summer or autumn of 1929 Lin Tse (?), an Oppositionist *en route* from Moscow to China, visited him in Prinkipo, and thereafter, until 1940, Trotsky was in almost regular correspondence with several groups in China representing different shades of Opposition. As early as 1929–31 his Chinese followers reported to him the rivalries between Li Li-san, then official party leader, Chu Teh, and Mao Tse-tung, dismissing the former two as 'opportunists' and placing great hopes on Mao. Some of Trotsky's followers were not at all elated over Chen Tu-hsiu's 'conversion to Trotskyism'; they considered him a 'liquidator' and held that he had played out his role. Trotsky, to whom Mao's name could not yet mean much, attached great importance to Chen Tu-hsiu, the 'grand old man' of Chinese Marxism, and tried to reconcile the Chinese Trotskyists with him. Chen Tu-hsiu himself, in a letter to Trotsky of 1 December 1930, explained that he had first acquainted himself with the latter's views on the Chinese Revolution in the summer of 1929, and that no sooner had he done so than he became convinced of their correctness. (*The Archives*, Closed Section. Further reference to this correspondence is made later on pp. 423–24. Chen Tu-hsiu's part in the revolution of 1925–7 is described in *The Prophet Unarmed*, pp. 317–38.)

influence made itself felt in Indo-China, Indonesia, and Ceylon. About the same time Trotsky gained new adherents in America: James P. Cannon and Max Shachtman, members of the Central Committee in the United States, and Maurice Spector, chairman of the Communist party of Canada. Even in remote Mexico a group of communists, encouraged by Diego Rivera, rallied to the cause of the heretics defeated in Moscow.

Trotsky established liaison with all these groups, and tried to weld them into a single organization. Since his deportation from Moscow they had lived on crumbs of his thought and had published, in small papers and bulletins, fragments of his writings, surreptitiously brought out of the Soviet Union. His appearance in Constantinople gave them a fillip; his moral authority was their greatest asset; and they expected him to give life to a world-wide communist opposition to Stalinism. True, his authority was also a liability, for they were becoming accustomed to the constricting roles of disciples and devotees. Trotskyism was already, as Heinrich Brandler put it, a tiny boat overweighted by a huge sail. Even in the Russian Opposition Trotsky's personality had been pre-eminent; but there at least he had been surrounded by associates distinguished in the revolution, men of independent mind, strong character, and rich experience. There were, with one or two exceptions, no men of such weight among his associates outside Russia. He hoped that this weakness of the Opposition would soon be remedied and that new leaders would rise from the ranks. He did not imagine that he would remain the only expatriate leader of the Russian Opposition. He expected that Stalin would banish others beside him, especially Rakovsky and Radek, and that once these had emerged from Russia the international opposition would obtain a 'strong directing centre'.[1] These expectations were not to be fulfilled: Stalin had no intention of strengthening Trotsky's hand by further banishments.

· · · · · · · · · · ·

What, apart from the magic of a personality, did Trotskyism represent at this stage?

[1] *B.O.*, nos. 1–2, July 1929. From now on the initials *B.O.* are used for *Bulletin Oppozitsii*.

At its heart were the principles of revolutionary internationalism and proletarian democracy. Revolutionary internationalism belonged to the heritage of classical Marxism; the
Third International had once rescued it from the failing hands
of the Second; and now Trotsky defended it against both the
Third and the Second Internationals. This principle was no
mere abstraction to him: it permeated his thought and his
political instincts. He never viewed any issue of policy otherwise
than in the international perspective; and the supranational
interest of communism was his supreme criterion. Hence he
saw the doctrine of 'socialism in a single country' as a 'national
socialist' distortion of Marxism and as the epitome of the
national self-sufficiency and arrogance of the Soviet bureaucracy. That doctrine now ruled not only in the Soviet Union,
where at least it met a psychological need; it was also the
official canon of international communism, where it met no
such need. In bowing to the sacred egoism of Stalinist Russia the
Comintern had shattered its own *raison d'être*: an International
hitched to socialism in a single country was a contradiction in
terms. Trotsky pointed out that, theoretically, the conception
of an isolated and self-contained socialist state was alien to
Marxist thinking—it originated in the national-reformist
theory of the German revisionists of the nineteenth century—
and that practically it expressed renunciation of international
revolution and the subordination of Comintern policy to
Stalinist expediency.[1] Upholding the primacy of the inter-

---

[1] Trotsky traced the ancestry of Socialism in a Single Country to G. Vollmar, the
well-known German reformist, who twenty years before Bernstein's 'revisionist'
campaign expounded the idea of the 'isolated socialist state'. (This, we may add,
was a socialist variation on the basic theme of List's economics.) Vollmar's conception, Trotsky pointed out, was more subtle than Stalin's or Bukharin's, because
his isolated socialist state was to be a state like Germany, enjoying technological
ascendancy, not an underdeveloped peasant nation. Vollmar saw in the technological superiority of the isolated socialist state over its capitalist neighbours the
guarantee of its security and success, whereas Bukharin and Stalin (up to 1928)
were satisfied that such a state could flourish even in industrial backwardness. (See
Trotsky, *The Third International After Lenin*, pp. 43–4.) Vollmar also imagined that
a socialist Germany, using the advantages of superior technology and planned
economy, would vanquish its capitalist neighbours through peaceful economic
competition and would thus render revolution in other countries more or less
superfluous. With this idea, Vollmar anticipated not only and not so much the
Stalinist-Bukharinist conception of the 1920s as the Khrushchevite theses of
'economic competition' and 'peaceful transition to socialism' adopted by the XX
Congress of the Soviet Communist party in February 1956.

national interest *vis-à-vis* the national, Trotsky was, however, far from treating the national needs of the Soviet Union with any degree of nihilistic neglect, or from overlooking its specific diplomatic or military interests; and he insisted that the defence of the workers' first state was the duty of every communist. But he was convinced that Stalinist self-sufficiency weakened the Soviet Union, whose ultimate interest lay in overcoming its isolation and in the spread of revolution. He held therefore that at decisive stages of the international class struggle the workers' state should, on a long term view, be prepared to sacrifice immediate advantages rather than obstruct that struggle, as Stalin and Bukharin obstructed the Chinese Revolution in 1925–7. In the coming decade this controversy was to shift to issues of communist strategy and tactics *vis-à-vis* Nazism and the Popular Fronts; but underlying it still was the same conflict between (to use an analogy with contemporary American politics) Trotskyist internationalism and the isolationism which coloured Stalin's policies in the nineteen-twenties and thirties.

On the face of it Trotsky's attitude was, or should have been, much more congenial to communists outside the Soviet Union than was Stalin's, and he had reason to expect that it would meet with the stronger response, for he dwelt on their importance as independent actors in the international class struggle, whereas Stalinism assigned to them the parts of the mere clients of the 'workers' fatherland'.

Trotsky's advocacy of 'proletarian democracy' aimed at freeing the Communist parties from the rigidities of their ultra-bureaucratic organization and at the restitution in their midst of 'democractic centralism'. This principle, too, had been embedded in their Marxist tradition and was still inscribed in their statutes. Democratic centralism had sought to safeguard for the Socialist and later the Communist parties freedom in discipline and discipline in freedom. It obliged them to maintain the strictest concord and unity in action, and allowed them to entertain the widest diversity of views compatible with their programme. It committed minorities to carrying out majority decisions; and it bound the majority to respect the right of any minority to criticize and oppose. It invested the Central Committee of any party (and the leadership of the Internationa

with the power to command effectively the rank and file
during its tenure of office; but it made that Central Committee dependent on the will and the unhampered vote of the
rank and file. The principle had therefore been of great
educative and practical political value for the movement; and
its abandonment and replacement by bureaucratic centralism crippled the International. If in the Soviet party the
monolithic discipline and the over-centralization were part and
parcel of the organic evolution of the Bolshevik monopoly of
power, the extension of this régime to the foreign sections
of the Comintern was wholly artificial and bore no relation
to their national environments and conditions of existence.

Most western Communist parties had been accustomed to act
within the multi-party system where, as a rule, they enjoyed
the formal freedom of criticism and debate. Their leaders now
found themselves in the paradoxical situation that within their
own organization they denied their own followers the rights
which the latter enjoyed outside the organization. By 1930 no
German, French, or other communist could voice dissent from
the party line; they had to accept as gospel all official pronouncements coming from Moscow. Thus every Communist
party became in its own country something like a bizarre
enclave, sharply separated from the rest of the nation not so
much by its revolutionary purpose as by a code of behaviour
which had little to do with that purpose. This was the code of a
quasi-ecclesiastical order which subjected its members to a
mental drill as severe as any that had been practised in any
monastic body since the counter-reformation. It is true that by
means of this drill the Stalinized Comintern achieved extraordinary feats of discipline. But discipline of this type was
destructive of the efficacy of a revolutionary party. Such a
party must be in and of the people among whom it works; it
must not be set apart by the observances of an esoteric cult.
Stalinism, with its devotions, burnt offerings, and incense,
undoubtedly fascinated some intellectuals in search of a creed,
those intellectuals who were later to curse it as the 'God that
failed'. But the cult that captivated them rarely appealed to the
mass of workers, to those 'sturdy proletarians' whom it was
supposed to suit. Moreover, the strange discipline and ritual
tied the party agitators hand and foot when what they needed

was a free and easy approach to those whom they desired to win for their cause. When the European communist went out to argue his case before a working-class audience, he usually met there a Social Democratic opponent whose arguments he had to refute and whose slogans he had to counter. Most frequently he was unable to do this, because he lacked the habits of political debate, which were not cultivated within the party, and because his schooling deprived him of the ability to preach to the unconverted. He could not probe adequately into his opponent's case when he had to think all the time about his own orthodoxy and to check perpetually whether in what he himself was saying he was not unwittingly deviating from the party line. He could expound with mechanical fanaticism a prescribed set of arguments and slogans; but unforeseen opposition or heckling at once put him out of countenance. When he was called upon, as he often was, to answer criticisms of the Soviet Union he could rarely do so convincingly; his thanksgiving prayers to the workers' fatherland and his hosannahs for Stalin covered him with ridicule in the eyes of any sober-minded audience. This ineffectiveness of the Stalinist agitation was one of the main reasons why over many years, even in the most favourable circumstances, that agitation made little or no headway against Social Democratic reformism.

Trotsky set out to shake the Communist parties from their petrifaction and to reawaken in them the *élan*, the self-reliance, and the fighting ardour which were once theirs—and which they could not recover without freedom in their own ranks. Again and again he expounded the meaning of 'democratic centralism' for the benefit of communists who had never grasped it or who had forgotten it. He appealed to them in their own interest, in the name of their own dignity and future, hoping that they would not remain unresponsive. And indeed, if reason, Marxist principle, or communist self-interest had had any say in the matter, his arguments and pleas would not have fallen on deaf ears.

Apart from its fundamental principles, Trotskyism represented also a set of tactical conceptions varying with circumstances. An inordinately large proportion of Trotsky's writings in exile consists of comments on these topics, which are rarely exciting to outsiders, especially after the lapse of time. However, the range of Trotsky's tactical ideas was so wide and his views are in part still so relevant to working-class politics, that what he had to say is of more than historical interest.

It will be remembered that between 1923 and 1928, when the Comintern pursued a 'moderate' line, Trotsky and his adherents criticized it from the left.[1] After 1928 this changed to some extent. Since Stalin had initiated the 'left course' in the Soviet Union, the policy of the Comintern too had, by an automatic transmission to it of every movement and reflex from the Russian party, changed direction. Already at its Sixth Congress, in the summer of 1928, the International began to transpose its watchwords and tactical prescriptions from the rightist to an ultra-left pattern.[2] In the following months the new line was further evolved until it was in every respect diametrically opposed to the old.[3] While in previous years the Comintern spoke of the 'relative stabilization of capitalism', it now diagnosed the end of the stabilization and predicted the imminent and final collapse of capitalism. This was the crux of the so-called Third Period Theory, of which Molotov, who replaced Bukharin as head of the Comintern, became the chief exponent. According to that 'theory', the political history of the post-war era fell into three distinct chapters: the first, one of revolutionary strains and stresses, had lasted till 1923; the second, capitalist stabilization, had come to an end by 1928; while the third, now opening, was to bring the death agony of capitalism and imperialism. If hitherto international communism had been on the defensive, it was time now to pass to the offensive and to turn from the struggle for 'partial demands' and reforms to the direct contest for power.

The Comintern alleged that all the contradictions of capitalism were about to explode because the bourgeoisie would be unable to master the next economic crisis; and that the makings of a revolutionary situation were already evident all over the world, especially in a new radicalism of the working classes, who were shaking off reformist illusions and virtually waiting for the communists to place themselves at their head and lead them into battle. Almost any incident of class conflict now had

[1] See *The Prophet Unarmed*, Chapters II and V.
[2] *Kommunisticheskii Internatsional v Dokumentakh*, (ed. B. Kun), pp. 769–84.
[3] Op. cit., pp. 876–88, 915–25, 957–66.

incalculable revolutionary momentum and could lead to the 'struggle for the street', or, more explicitly, to armed insurrection. 'In the whole capitalist world', *Bolshevik* wrote in June 1929, 'the strike wave is mounting . . . elements of a stubborn revolutionary struggle and of civil war are intertwined with the strikes. The masses of unorganized workers are drawn into the fight. . . . The growth of dissatisfaction and the leftward swing embrace also millions of agricultural labourers and the oppressed peasantry.' 'One must be a dull opportunist or a sorry liberal . . .', Molotov told the Executive of the International, 'not to see that we have stepped with both feet into a zone of the most tremendous revolutionary events of international significance.' These words were not meant as long-term predictions but as topical forecasts and directions for action. Several European Communist parties tried indeed to turn the May Day parades of 1929 and anti-war demonstrations called for 4 August into direct 'struggles for the street', which resulted in fruitless and bloody clashes between demonstrators and police in Berlin, Paris, and other cities.

In accordance with this 'general line', the Comintern also changed its attitude towards the Social Democratic parties. In a truly revolutionary situation, it was said, those parties could only side with counter-revolution; and so no ground was left for communists to seek co-operation or partial agreements with them. As the bourgeoisie was striving to save its rule with the help of fascism, as the era of parliamentary government and democratic liberties was coming to a close, and as parliamentary democracy itself was being transformed 'from the inside' into fascism, the Social Democratic parties too were becoming 'social-fascist'—'socialist in words and fascist in deeds'. Because they concealed their 'true nature' under the paraphernalia of democracy and socialism, the Social Democrats were an even greater menace than plain fascism. It was therefore on 'social fascism' as 'the main enemy' that communists ought to concentrate their fire. Similarly, the left Social Democrats, often speaking a language almost indistinguishable from that of communism, were even more dangerous than the right wing 'social-fascists', and should be combated even more vigorously. If, hitherto, communists were required to form united fronts with the Social Democrats from 'above and below', with leaders and rank and file alike, the Comintern now declared a rigorous ban on any such tactics. 'Only from below' could the united front still be practised—communists were permitted to co-operate only with those of the Social Democratic rank and file who were 'ready to break with their own leaders'. To favour any contact 'from above' was to aid and abet 'social-fascism'.[1]

These notions and prescriptions were to govern the policies of all Communist parties for the next five or six years, almost up to the time of the Popular Front, throughout the fateful years of the Great Slump, the rise of Nazism, the collapse of the monarchy in Spain, and other events in which the conduct of the Communist parties was of crucial importance.

In the previous period, when Trotsky maintained that by its timid policies the Comintern was wasting revolutionary opportunities, he never proposed a reversal of its line as sweeping and extreme as the one now carried out. He therefore criticized the reversal as a 'turn by 180 degrees' and a 'swing from opportunism to ultra-radicalism': the new slogans and tactical prescriptions merely turned the old ones inside out and served to cover up their fiasco. In a devastating comment on Molotov's disquisitions on the Three Periods, Trotsky pointed out that if it was wrong to consider the 'second period', during which the Chinese Revolution and the British General Strike had occurred, as one of stabilization, it was even less realistic to envisage the imminent collapse of capitalism in the 'third period', and to deduce the need for an exclusively offensive policy. The Comintern, he said, had accomplished this 're-orientation' quite mechanically, without any attempt to elucidate what had gone wrong with its old tactics, and without any genuine debate and reappraisal of the issues. Prevented from discussing the rights and wrongs of their own policy, the Communist parties were condemned to veer from extreme to extreme and to exchange, on orders, one set of blunders for another. Their inner régime was no mere matter of organization—it affected the entire policy of the International, making it rigid and unstable at the same time. Nor did the feverish ultra-radicalism of the 'third period' testify to any reawakened revolutionary internationalism in official Moscow. That ultra-radicalism obstructed the growth of communism in the world

[1] Op. cit., pp. 946, 957–66, and *passim*.

not less effectively than did the earlier opportunism, and under-lying it was the same cynical bureaucratic indifference to the international interests of the working class.[1]

Now as before Trotsky expounded the view that the whole epoch opened by the First World War and the Russian Revolu-tion was one of the decline of capitalism, the very foundations of which were shattered. This, however, did not mean that the edifice was about to come down with a crash. The decay of a social system is never a single process of economic collapse or an uninterrupted succession of revolutionary situations. No slump was therefore *a priori* the 'last and final'. Even in its decay capitalism must have its ups and downs (although the ups tended to become ever shorter and shakier and the downs ever steeper and more ruinous). The trade cycle, however it had changed since Marx's time, still ran its usual course, not only from boom to slump but also from slump to boom. It was therefore preposterous to announce that the bourgeoisie had 'objectively' reached its ultimate impasse: there existed no such impasse from which a possessing class would not fight its way out; and whether it would succeed or not depended not so much on purely economic factors as on the balance of political forces, which could be tilted one way or the other by the quality of the communist leadership. To forecast an 'uninterruptedly mounting tide of revolution', to discover 'elements of civil war' in almost any turbulent strike, and to proclaim that the moment had come to pass from defensive to offensive action and armed insurrection was to offer no leadership at all and to court defeat. In class struggle as in war defensive and offensive forms of action could not be separated from and opposed to one another. The most effective offensive usually grows out of successful defence; and an element of defence persists even in armed insurrection, that climax of all revolutionary struggle. During slump and depression the workers had to defend them-selves against attacks on their living standards and against the rise of fascism. To tell them that the time for such defence had passed and that they must be ready for the all-out attack on capitalism was to preach nothing but inaction or surrender, and to preach it at the very top of one's ultra-radical voice.

[1] Trotsky devoted to the criticism of the Third Period Policy a whole issue of the *B.O.*, no. 8 (January 1930), and returned to it in many subsequent issues.

Similarly, to ban all co-operation between Communist and Socialist parties was to invite disaster for the labour movement at large and communism in particular. The notion of the Third Period, Trotsky concluded, was a product of bureaucratic recklessness—'all that had been inaugurated', under the auspices of 'Maestro Molotov', was 'the third period of the Comintern's blunders'.

These early criticisms contained in a nutshell Trotsky's far larger controversy with the Comintern (over the latter's policy during Hitler's rise to power) which was to fill the early nineteen-thirties. Clearly, on these tactical issues Trotskyism now appeared to oppose the Comintern from the right and not, as hitherto, from the left. The change lay not in Trotsky's attitude, which remained consistent with the one which Lenin and he had adopted at the third and fourth Comintern congresses in 1921–2, but in the gyrations of Stalin's 'bureaucratic centralism' and in the 'alternation of its rightist and ultra-left zigzags'. Even so, the position of being Stalin's critic 'from the right' had its inconveniences for Trotsky. Communists accustomed to think of him as Stalin's critic from the left were apt to suspect inconsistency or lack of principle. In fact, the division between Trotskyism and the various rightist quasi-Bukharinist oppositions in the communist camp was blurred, at least in the tactical issues which loomed so large in these controversies. The right oppositions in Europe, of which the Brandlerites were by far the most important—Brandler and Thalheimer had just been expelled from their party—also severely criticized the new ultra-radicalism.[1] Yet what set Trotskyism apart from all other brands of opposition was the intellectual power, the aggressiveness, and the comprehensiveness of its criticism. Brandler and Thalheimer confined themselves to exposing only the latest, the ultra-left, 'zigzag' of the Comintern; Trotsky attacked its entire post-Leninist record. The Brandlerites, concerned mainly with the policies of their national parties, studiously refrained from offending the Soviet leadership: in internal Soviet conflicts they willy-nilly sided with Stalin, endorsing socialism in one country, excusing the bureaucratic régime as fitting Russia's

---

[1] Groups akin to the Brandlerites were those of Warski and Kostrzewa in Poland (who were demoted in 1929 but not yet expelled from the party), of Humbert Droz in Switzerland, and of Lovestone in the United States.

peculiar conditions, and even echoing Moscow's denunciations of Trotskyism.[1] They were convinced that no communist opposition which defied Moscow on principle could evoke response in communist ranks; and they hoped that the Comintern would sooner or later find the Third Period policy impracticable, discard it, and reconcile itself with those of its critics who had shunned an irreparable breach. Against this, Trotskyism insisted that the policies of the various national parties could not be corrected, or their faults remedied, within those parties alone, because the main source of their 'degeneration' lay in Moscow; and that it was therefore the duty of all communists to take the closest interest in domestic Soviet affairs and to oppose on that ground, too, the Stalinist bureaucracy. This call for the intervention of foreign communist opinion in Soviet affairs was peculiar to Trotskyism. It was a challenge, which struck horror in most communist hearts.

Despite the comprehensiveness of its criticism of the Comintern, Trotskyism did not aspire to set up a new communist movement. Now and for several years to come Trotsky was absolutely opposed to the idea of a Fourth International, already canvassed by the Workers' Opposition in the Soviet Union and by some survivors of the Zinovievist opposition in Europe. He declared that he and his adherents owed their loyalty to the Communist International even though they had been expelled from it. They formed a school of thought struggling to regain its place within the general communist movement—only persecution had forced them to constitute themselves into a faction; and a faction, not a rival party, they remained. Their sole purpose was to influence communist opinion, to make it realize that usurpers had seized the reins of the Soviet Government and of the Comintern, and to induce it to strive for the restoration of pristine Marxism and Leninism. They therefore stood for a reform of the International, not for a permanent break with it. Trotsky believed that with all their flaws and vices the Communist parties still represented the militant vanguard of the working classes. The Opposition's

---

[1] The Brandlerite *Arbeiterpolitik* maintained a consistently hostile attitude towards Trotskyism, and Trotsky repaid it in the same coin: 'Just as I do not discuss various trends in materialism with anyone who crosses himself when passing by a church, so I shall not argue with Brandler and Thalheimer', he wrote on one occasion.

place was with that vanguard. If he and his followers were to turn their backs on it, they would voluntarily go out into the wilderness into which Stalin was driving them. True enough, Stalinism did not allow any current of opposition to assert itself within the International; but this state of affairs could not last: critical events inside or outside the Soviet Union would presently stir the dormant *élan* of communism into action again and give the Opposition its chance. Trotsky warned those who stood for a Fourth International that it was not enough for a group of dissidents to raise a new banner in order to become a real factor in politics. Revolutionary movements were not conjured up with banners and slogans, but rose and grew organically with the social class for which they spoke. Each of the Internationals represented a definite stage in the historic experience of the working class and in the struggle for socialism; and no one could ignore with impunity the ties the Second and the Third Internationals had with the masses or the weight of their political traditions. Moreover, the Third International was the child of the Russian Revolution; and the politically conscious workers extended to it the solidarity they felt with the Revolution. They were right in doing so, Trotsky maintained, though they should not allow Stalinism to abuse their loyalty. And so, as long as the Soviet Union remained a workers' state, the workers should not be expected or urged to renounce the Third International.

` On this point, that the Soviet Union, however 'bureaucratically deformed', remained a workers' state, Trotsky was adamant. What, in his view, determined the social character of the Soviet state was the national ownership of the means of production. As long as this, 'the most important conquest of October', was unimpaired, the Soviet Union possessed the foundations on which to base its socialist development. To be sure, its working class had to assert itself against the bureaucracy before it could even begin to make socialism a reality; but, once again, it could not make that into a reality otherwise than on the basis of public ownership. With this preserved, the workers' state was still alive, as a potentiality if not an actuality.

This view was often to be challenged, among others by Trotsky's own disciples; but he was never to compromise over it or to yield an inch from it, even when he revised and modified his other ideas. Thus, during the first half of this term of exile he preached reform, not revolution, in the Soviet Union; whereas in the second half he was to maintain that political revolution was the only answer to bureaucratic absolutism. He was also to revise his conception of the Opposition's role and to proclaim a new Communist Party and a new International. But even then he was never to waver in his insistence that the Soviet Union was a workers' state; he declared the 'unconditional defence of the Soviet Union' against its bourgeois enemies to be the elementary obligation of every member of the Opposition; and he was repeatedly to disown friends and adherents who were reluctant to accept this obligation.[1]

.   .   .   .

The outcome of Trotsky's first attempts to organize his followers in the West was disappointing. He concentrated his attention on France where he had had a more influential following than elsewhere; and in the hope of setting up there a strong base for the Opposition he endeavoured to bring together various Trotskyist and quasi-Trotskyist groups and coteries and to unite these with the Zinovievists and with the syndicalist circle of *Revolution Proletarienne*. At the outset Rosmer warned him about the political depression and demoralization which beset most of these groups. Five years had elapsed since the hey-day of Trotskyism in the French party; in this time the Comintern had managed to restore its influence there and to expel all dissenters and isolate them from the rank and file. The sense of their isolation and the defeats of the Opposition in Russia had disheartened many anti-Stalinists, among whom Rosmer noted a mood of *sauve qui peut* which led them to give up the fight and to wish 'they had never had anything to do with the Opposition'. Even those who withstood this mood were confused and at loggerheads with one another. 'The great misfortune of all these groups', Rosmer went on, 'is that they find themselves outside all action; and this fatally accentuates their sectarian character.'[2]

The truth of Rosmer's observations became evident when Trotsky, disregarding his advice, tried to 'regain' Souvarine

[1] *B.O.*, nos. 3–4, 5, and *passim*; *Écrits*, vol. I, pp. 213–74; *Militant*, December 1929.
[2] Rosmer to Trotsky, 16 April 1929.

and others for the Opposition. Souvarine had once distinguished himself by raising, in Moscow, a lonely voice in
Trotsky's defence; and Trotsky, valuing his journalistic talent,
expected him to be the Opposition's most articulate French
mouthpiece. To his surprise Souvarine displayed intolerable
airs and pretensions. He asked Trotsky to make no public
statements without 'previous agreements with the French
Opposition', that is with himself. Trotsky, anxious to avoid
dissension, answered that he would make no pronouncement
on French issues, but that so far he had spoken in public on
Soviet (and Chinese) affairs only, on which surely he was entitled
to have his say without asking for a French *placet*. Souvarine
replied with an immense epistle, running to over 130 pages,
packed with paradoxes, *bons mots*, odds and ends of shrewd
observation and analysis, but also with incredibly muddled
arguments, all advanced in a tone of venomous hostility
which made a breach inevitable. He asserted that Bolshevism
had 'once for all failed outside Russia', because 'it misunderstood the character of the epoch', underrated the power of
the bourgeoisie, and overrated the militancy of the workers;
it also committed the 'fatal error' of trying to fashion foreign
Communist parties in its own image. This was not a view,
whatever its merits, that Trotsky expected to be advanced by
someone reputed to be his adherent, or that he himself could
accept. He did not agree that Bolshevism was guilty of the 'fatal
errors' Souvarine attributed to it, and he blamed Stalinism,
not Leninism, for the failure of the Comintern. Far more
startling, however, was Souvarine's other reproach which,
despite his talk about Soviet 'state capitalism', had a pro-
Stalinist flavour—namely, the reproach that Trotsky and the
Opposition needlessly 'cultivated a revolutionary intransig-
ence' which prevented them from attending properly to the
'tangible necessities of the Soviet state'. 'There is nothing more
important', these were Souvarine's words, 'for the entire
international workers' movement than the economic success of
the Soviet Union whose state capitalism marks . . . an undeniable advance upon imperialist capitalism. . . .' He went on
to deride the 'useless heroism' which prevented Trotsky and his
associates from serving the Soviet state even if there was no
room for them in the party: 'One can make oneself useful to the

revolution without being a member of the Politbureau or of the Central Committee or even of the party.' Had it not been for their sheer incongruity, these remarks would have sounded like a belated counsel to Trotsky to surrender to Stalin, for nothing short of surrender, if even that, might have enabled him to go on 'serving the revolution' without being a member of the party. Yet in the same breath Souvarine turned with savage sarcasm on Trotsky's loyalty to Bolshevism and Leninism, urging him to emancipate himself from these and 'return to Marx'.[1]

'I do not see anything left of the ties that united us a few years ago', Trotsky wrote back. In what Souvarine said he could not find 'a single reasoning based on Marxist doctrine and . . . the relevant facts'. 'What guides you and suggests your paradoxes to you is the pen of a disgruntled and frustrated journalist.' 'You are treating the party and the International as corpses. You see the great fault of the Russian Opposition in its insistent endeavour to influence the party and to re-enter its ranks. On the other hand you describe the Soviet economy as state capitalist . . . and you demand that the Opposition should lower itself to the role of a servant of that state capitalism. . . . You are crossing to the other side of the barricade.'[2] This brought the correspondence to an end, and Souvarine was forever to remain among Trotsky's adversaries. And although in 1929 he sought to instruct Trotsky 'how to be useful to the revolution' by serving a progressive state capitalism, in later years he was to castigate him from the opposite sin, for seeing any progress at all in the Soviet Union and for thinking that enough was left there of the heritage of the Revolution to be worth defending.

An attempt to come to terms with the syndicalists of the *Revolution Prolétarienne*, of whom Monatte and Louzon were the best known, also came to nothing. Trotsky had once, during the First World War, exercised a strong influence on them, overcoming their characteristic bias against all politics, including those of revolutionary Marxism; later they joined the Communist party only to be expelled from it at the time of the anti-Trotskyist campaign. Their personal attachment to Trotsky was still strong; but their experience with the Comintern confirmed them in their old distaste for politics, and in the

---

[1] Trotsky—Souvarine correspondence. *The Archives*, Closed Section. [2] Ibid.

belief that militant trade union activity, culminating in the general strike, was *the* highway to socialist revolution. Hard as Trotsky tried, he did not manage to bring them back to the Leninist view of the paramount importance of the revolutionary party and induce them to join him in the struggle for a reform of the Comintern.

He fared no better in the mediation which he undertook between his own followers and the Zinovievists. The latter were a tiny sect, but they had a leader of renown in Albert Treint, who had been official chief of the French Communist party in 1924–5. It was Treint who, at the time when Zinoviev was directing the 'Bolshevization', had expelled the Trotskyists from the party, sparing them no denunciation or abuse. For this they bore him a grudge even after he too had been expelled; and they would not hear of making peace with him. Trotsky nevertheless invited him to Prinkipo, in May 1929, and throughout a whole month tried to bring about a reconciliation. But the old resentments were too strong, and Treint, trying to justify his behaviour in 1924, did nothing to assuage them. Trotsky, pressed by his own followers, had to part from Treint; but their parting was more friendly than that with Souvarine, and they remained in amicable though remote relations.

No sooner had Trotsky failed with Souvarine, the syndicalists, and Treint, than he had to deal with discords among the Trotskyists themselves. The story would hardly be worth relating had it not played its part in Trotsky's life and in the eventual failure of Trotskyism as a movement. There were several rival groups and coteries in Paris: the circle of Maurice and Magdeleine Paz who brought out a little periodical, *Contre le Courant*; Rosmer; and the young Trotskyists (with their own papers *Lutte des Classes* and *Vérité*), among whom Pierre Naville and Raymond Molinier formed two antagonistic sets. Of all these men Rosmer alone was a public figure of considerable standing: a member of the small élite of revolutionary internationalists, who had proved themselves in the First World War. Naville was a young writer who had participated in the literary rebellion of the Surrealists, had then joined the Communist party, gained some repute as a Marxist critic of Surrealism, witnessed sympathetically Trotsky's struggle in

Moscow in 1927, and had himself been expelled from the party. He possessed a theoretical education in Marxism, but had little political experience and hardly any ties with the working-class movement. Molinier, on the contrary, was an 'activist', full of energy and enterprise, very much at home in the movement, but not too fastidious in the choice of ways and means and rather crude intellectually. The antithetical types of intellectual and activist often formed a good working partnership when they were carried along by the impetus of practical day-to-day activity in a broad organization; but their antagonism usually wrecked small groups cut off from the mainstream of the movement and remaining 'outside all action'.

When early in the spring of 1929 Maurice and Magdeleine Paz came to Prinkipo, Trotsky urged them to unite their circle with the other groups, to transform *Contre le Courant* into a 'great and aggressive' weekly speaking with the voice of the Opposition, and to launch an ambitious recruiting campaign. He worked out with them the plan of the campaign and promised his own close co-operation. They accepted his suggestions, though not without reservations. On their return to Paris, however, they had second thoughts and refused to launch the great weekly. They saw, they said, no chance for the Opposition to succeed in any drive undertaken on the scale envisaged by Trotsky. Above all, they protested against his 'attempt to impose Rosmer's leadership'; and they spoke disparagingly of the young Trotskyists spoiling for a fight as a bunch of simpletons and ignoramuses. Nothing could be more calculated to convince Trotsky that the Pazes had in them little or nothing of the professional revolutionaries whom he was seeking to gather. They were in truth 'drawing-room Bolsheviks' successful in their bourgeois professions—Maurice, at any rate, was a prosperous lawyer—and indulging in Trotskyism as a hobby. While Trotsky was at Alma Ata they were glad to act as his representatives in Paris and to walk in his reflected glory; but when he emerged from Russia and confronted them in person with his exacting demands, they had no desire to commit themselves seriously. An embarrassing correspondence followed. Trotsky made them feel that he thought of them as philistines: 'Revolutionaries', he wrote to them, 'may be either educated or ignorant people, either intelligent or dull; but there can be no

E

revolutionaries without the will that breaks obstacles, without devotion, without the spirit of sacrifice.'[1]

The Pazes replied in a manner which was not less wounding to Trotsky than his strictures were to them. They dwelt on the strength and attraction of official communism and on the weakness of the Opposition, using the contrast, which was only too real, as an excuse for their lukewarmness. They explained that they would not launch *Contre le Courant* as a weekly because 'the Opposition's journal, if it is not to end in failure, must avail itself of other things besides the scintillating prose and the *nom de bataille* of Comrade Trotsky'—it must have a material and moral base and must be able to 'live with its readers and active sympathizers'. The paper would lack such a base, because the old communists, to whom Trotsky's name had meant so much, had lapsed into apathy; and the young were ignorant and inaccessible to argument. 'Don't give yourself too many illusions about the weight of your name. For five years the official communist Press had slandered you to such an extent that among the great masses there is left only a faint and vague memory of you as the leader of the Red Army. . . .' It was a far cry from the reverence with which the Pazes had a few months earlier addressed Trotsky as 'Cher grand Ami' to the insinuation that he was actuated by egotism and vanity. That his followers were isolated and that Stalinist propagandists made his name odious to the communist rank and file, or sought to bury it in oblivion, Trotsky was not unaware. But this was for him one more reason why his followers should undertake a large-scale counter-attack by which alone they might break through the apathy of the communist rank and file. He concluded that he could do nothing with the Pazes, although the breach with them, following closely upon the rupture with Souvarine, was all the more disagreeable because of the services and the attentions they had given him from the moment of banishment.

What now followed was more than a little pitiable, for Trotsky had at once to deal with the animosities that divided his remaining adherents, Rosmer, and the sets of Naville and Molinier. Molinier had come to Prinkipo with boisterous optimism and with a headful of plans for making Trotskyism into a great political force. He was convinced that the Opposition had golden opportunities in France, because the official party was riddled with discontent and could not remain insensitive to the Opposition's appeal—all the Opposition needed was to act with self-confidence and bold initiative. He had schemes for infiltrating the party with Trotskyists, for mass meetings, newspapers with a large circulation, &c. The implementation of the schemes required much more money than the Opposition could collect from its members; but he had his financial plans too, somewhat vague but not implausible. He was ready to plunge into all sorts of commercial ventures, and he budgeted ahead with the expected profits.[1]

Rosmer and Naville took a more cautious view of the chances, discounted the possibilities of 'mass action' which Molinier held out, and were inclined to content themselves for the beginning with a more modest but steady clarification of the Opposition's ideas and with propaganda among the mature elements of the left. They were afraid that Molinier's ventures might bring discredit on the Opposition; and they distrusted him. *'Ce n'est pas un militant communiste, c'est un homme d'affaires, et c'est un illettré'*, Rosmer said. Unpleasant tales about Molinier were being told in Paris: one was that he had deserted from the army and then before a court martial conducted his defence in a manner unworthy of a communist, describing himself as a conscientious objector of the religious type. Allegations and hints were thrown out about the shady character of his commercial activities, but it was difficult to pin down the allegations to anything specific.

Trotsky, admitting some of Molinier's limitations, nevertheless trusted him implicitly. He was captivated by the man's verve, inventiveness, and courage, qualities he usually valued in followers. There was a streak of the adventurer in Molinier; but there was also genuine revolutionary fervour and unconventionality. It was his unconventionality, Trotsky pleaded, that brought the philistines' displeasure and obloquy on Molinier's head; and he, Trotsky, knew very well that no revolutionary movement could do without such men, in whom some crudeness of thought is compensated for by energy and the will to venture and take risks—how often had he himself had recourse to such men in the years of revolution and civil war!

[1] The Molinier family ran a small bank in Paris, at the Avenue de la République.

Molinier endeared himself to Trotsky by the eagerness with which he did many small yet important chores for him, helping to organize the Prinkipo household and set up the secretariat, keeping an eye on publishing interests in Paris, &c.—he had indeed made himself an indispensable *factotum*. His family, too, his wife Jeanne, and his brother Henri, a modest engineer without political pretensions, all had rendered themselves helpful in the same manner, with the *'énergie Molinièresque'* which greatly pleased Trotsky. They travelled between Paris and Prinkipo and spent much time at Büyük Ada; their relations with Trotsky's family became close and warm. And so Trotsky was anxious to dispel gently Rosmer's doubts and suspicions; all the more so because, much though he valued Rosmer's integrity and judgement, he considered him to be ill-suited for the minutiae of organization and to be too easily disheartened by the petty irritations of factional work, which Molinier took in his stride. With Naville's objections to Molinier, Trotsky had less patience; he chided Naville with 'intellectual haughtiness', 'schematic thinking', political lukewarmness, and reluctance to face 'work among the masses'. Somehow, however, he managed to compose the rivalry for the time being. Rosmer, Molinier, and Naville accepted a 'settlement' and, agreeing to put aside personal dislikes and to work together, returned to Paris with the intention of building up not merely a national but an international organization of the Opposition.[1]

Trotsky was hopeful. True, the 'base' to be set up in France would be narrower than he had expected, but sufficient to become the nucleus of a wider organization. True, also, at this point a dilemma had already presented itself: should the Opposition aim at 'mass action' and come forward with its own agitation and slogans, or should it confine itself to the kind of work that had in the past been carried out, slowly but fruitfully, by small Marxist propagandist circles, expounding patiently their theories and dealing with ideas rather than slogans? But this dilemma did not pose itself clearly or acutely; and so it could be left in the air. The circumstance that the Opposition did not aspire to found a new political party but

---

[1] This account is based on the correspondence between Trotsky, R. Molinier, Naville, V. Serge, L. Sedov, and many others, a correspondence covering the whole of the nineteen-thirties. *The Archives*, Closed Section.

was a faction bent on reforming the old party suggested that it should concentrate on the theoretical propaganda of its ideas. To this form of activity Trotsky the thinker was certainly inclined. But the man of action in him, the great Commissar, and the leader of the Opposition, fretted at its limitations and yearned for the scope and impetus of a mass movement.

In the summer of 1929 Rosmer went on a tour of Germany and Belgium to inspect and rally groups of the Opposition there; and he established contact with Italian, Dutch, American, and other Trotskyists. In detailed reports he kept Trotsky informed about his findings. These were not encouraging, on the whole. Inaction, sectarian squabbles, and personal rivalries, which had so greatly weakened the Opposition in France, had done it great harm elsewhere too. From Trotsky's viewpoint no country was more important than Germany, the main arena of class struggle in Europe, where the Communist party, with a following of several million voters, was stronger than anywhere in the West. Rosmer reported that in Berlin he found several groups, all invoking Trotsky's authority, but frittering away their strength in internecine animosities. The so-called Wedding group comprised the Trotskyists proper, but far more influential was the Leninbund which published the *Fahne des Kommunismus* and was led by Hugo Urbahns. There were also other tiny, 'ultra-left' sects such as the Korschists, so-called after Karl Korsch, a theorist who had in 1923 been Minister of the Communist-Socialist Government of Thuringia. The Zinovievists, Maslov and Fischer, were by far the strongest group; but, paradoxically, after their inspirer had surrendered to Stalin, they themselves took up an extreme anti-Stalinist attitude, similar to that of the survivors of the Workers' Opposition in the Soviet Union; and in their attacks on official communism they went 'much further' than Trotsky was prepared to go. They argued that the Russian Revolution had run its full course, and that the Soviet Union had ushered in an epoch of counter-revolution; that nothing was left there of the proletarian dictatorship; that the ruling bureaucracy was a new exploiting and oppressing class basing itself on the state capitalism of a nationalized economy; that, in a word, the Russian Thermidor was triumphant. They added that even the foreign policy of Stalinism was becoming indistinguishable from that of the Tsarist imperialism. Consequently, no reform could resuscitate the rule of the working class—only another proletarian revolution could achieve that. They also considered it hopeless to aim at a reform of the Third International which was 'a tool of the Russian Thermidorians' and exploited the heroic October legend in order to prevent the workers from facing realities and to harness their revolutionary energy to the engine of a counter-revolution. It went without saying that those who held this view did not feel themselves bound by any solidarity with the Soviet Union, still less by the duty to defend it; and they pointed to the very fact of Trotsky's banishment as conclusive evidence in favour of their attitude. 'The expulsion of Trotsky', they wrote, 'marks the line at which the Russian Revolution has definitely come to a halt.'

Trotsky defended himself against *trop de zèle* on the part of his defenders. In controversies with the Leninbund and the *Révolution Prolétarienne* he elaborated his old argument against those who held that the Soviet Thermidor was an accomplished fact. Once again defining the Thermidor as a bourgeois counter-revolution, he pointed out that this could not occur without civil war. Yet the Soviet Union had not gone through another civil war; and the régime established in 1917 had, despite its degeneration, preserved continuity, which manifested itself in its social structure based on public ownership and in the uninterrupted exercise of power by the Bolshevik party. 'The Russian Revolution of the twentieth century', he wrote, 'is incontestably wider in scope and deeper than the French Revolution of the eighteenth century. The social class in which the October revolution has found its support is incomparably more numerous, homogeneous, compact, and resolute than were the urban plebeians of France. The leadership given to the October revolution has, *in all its currents*, been infinitely more experienced and penetrating than the leading groups of the French Revolution were or could be. Finally, the political, economic, social, and cultural changes the Bolshevik dictatorship has brought about are also incontestably far more profound than those initiated by the Jacobins. If it was impossible to wrest power from the hands of the French plebeians . . . without a civil war—and Thermidor was a civil war in which the *sans culottes* were vanquished—how can anyone think or believe that power could pass from the hands of the Russian proletariat into those of the bourgeoisie peacefully, by way of a quiet, imperceptible bureaucratic change? Such a conception of the Thermidor is nothing but reformism *à rebours*.' 'The means of production', he went on, 'which once belonged to the capitalists remain in the hands of the Soviet state till this day. The land is nationalized. Social elements that live on the exploitation of labour continue to be debarred from the Soviets and the Army.' The Thermidorian danger was real enough, but the struggle was not yet resolved. And just as Stalin's left course and attack on the N.E.P.-man and the kulak had not effaced the Thermidorian danger, so his, Trotsky's, banishment had not obliterated the October revolution. A sense of proportion was needed in the evaluation of facts and in theorizing. The concept of Soviet state capitalism was meaningless where no capitalists existed; and if those who spoke of it denounced state ownership of industry, they renounced an essential prerequisite of socialism. Nor was the bureaucracy a new exploiting class in any Marxist sense, but a 'morbid growth on the body of the working class'—a new exploiting class could not form itself in exercising merely managerial functions, without having any property in the means of production.[1]

The implications of this dispute became apparent when a conflict flared up, in the summer of 1929, between the Soviet Union and China over the possession of the Manchurian Railway. China claimed the railway which the Soviet Government held as a concessionaire. The question arose whose side the Opposition ought to take. The French syndicalists, the Leninbund, and some Belgian Trotskyists held that the Soviet Government should give up the railway (which had been built by Russia in the course of the Tsarist expansion to Manchuria); and in Stalin's refusal to do so they saw evidence of the imperialist character of his policy. To their surprise Trotsky declared that Stalin was right in holding on to the railway and that it was the Opposition's duty to side with the Soviet Union against China.[2] This was, in the first year of his exile, Trotsky's first great controversy with his own followers—we shall see him

---

[1] *Écrits*, loc. cit.; *B.O.*, loc. cit.

[2] Trotsky's role in 1926 as Chairman of the Politbureau's Chinese Commission, concerned *inter alia* with securing Soviet influence in Manchuria, is related in *The Prophet Unarmed*, pp. 322–3.

again, in his last year, during the Soviet-Finnish war of 1939–40, engaged in another, his last, dispute with his own followers, a dispute again centring on the Opposition's attitude towards the Soviet Union; and in that dispute he would again adopt essentially the same view as in 1929.

He saw no reason, he argued, why the workers' state should yield a vital economic and strategic position to Chiang Kai-shek's Government (which had recognized the Soviet concession in Manchuria). He criticized severely Stalin's manner of dealing with the Chinese, his disregard of their susceptibilities, and his failure to appeal to the people in Manchuria —a more considerate and thoughtful policy might have averted the conflict. But once the conflict had broken out, he asserted, communists had no choice but to back the Soviet Union. If Stalin gave up the railway to the Kuomintang he would have yielded it not to the Chinese people but to their oppressors. Chiang Kai-shek was not even an independent agent. If he obtained control of the railway, he would not be able to maintain it but would sooner or later lose it to Japan (or else allow American capital to bring the Manchurian economy under its influence). Only the Soviet Union was strong enough to keep this Manchurian position out of Japan's hands. China's national rights, invoked by the critics, were, in Trotsky's view, not relevant to this case, which was an incident in a complex and many-sided contest between the various forces of world imperialism and the workers' state. He concluded that the time for the Soviet Union to do historic justice and return the Manchurian outpost to China would come when a revolutionary government was established in Peking; and this forecast was to come true after the Chinese revolution. In the meantime, the Soviet Government was obliged to act as the trustee of revolutionary China and keep for it the Manchurian assets.[1]

---

[1] In 1935, Stalin, anxious in view of the approaching war to ward off a Japanese attack on the U.S.S.R., sold the Railway to the Japanese puppet government of Manchukuo. In 1945 the Soviet Union regained control of the Railway; and it was not before September 1952 that Stalin, after some hesitation, ceded it to Mao Tse-tung's government. This was one of Stalin's last important acts of policy. Until that time he had pursued a course of economic penetration of China, and the cession foreshadowed the final abandonment of that course by his successors. In this, as in so many other acts, Stalin and his successors were the reluctant and half-hearted executors of a policy which Trotsky outlined nearly a quarter of a century earlier.

One may imagine the consternation which Trotsky caused
among the zealots of the Opposition. They were puzzled by
his 'inconsistency', thinking that he was missing a great
opportunity to strike at Stalin. He was, indeed, not out to
score points; but his behaviour was consistent with what he was
saying about the Soviet Union as the workers' state. For that
state he felt, as an outcast, the same responsibility that he
had felt as a member of the Politbureau and of Lenin's government. He found the displays of self-righteous indignation over
Soviet policy, in which some of his pupils indulged, wrong-
headed and cheap; and he told them bluntly that he had
nothing in common with 'Trotskyists' who refused to give the
workers' state unshakable, if critical, allegiance.

The rigour with which he stuck to his principles, refusing to
dilute them with demagogy, offended many of his past and
would-be admirers. Indeed, the movement he was sponsoring
was hemmed in, on the one hand, by his severe scrupulousness
about ideas and, on the other, by the unscrupulous ruthlessness
of the Stalinist persecution. The persecution kept his followers
at an impassable distance from the only people in whom his
ideas could strike a chord, the large communist audience in
Europe. His fastidiousness in the choice of his argumentative
weapons was estranging him from the scattered yet growing
anti-Stalinist public consisting of former party members, who
felt tempted to meet the Stalinists on their own ground, to
return blow for blow, to counter villainy with faithlessness and
to match venom with virulence. That public was in no mood to
accept Trotsky's self denying ordnances.

And so, after a year or two of argument and recruiting,
those who followed him on his arduous path were still very
few. New groups came over here or there; another member,
say, of the Italian Politbureau or of the Belgian Central
Committee, or a small band of Czech or even British activists
saw the light and hopefully joined the Opposition. But their
accession failed to change anything in the state of the Opposition. Even though some of the newcomers were until quite
recently influential in the party and had many ties with the
working class, cultivated over the years, they lost influence and
ties once the party expelled them, pursued them with every
imaginable calumny, and chased them away like lepers. They

had against them the authority of Moscow, the prestige of their own party, the hallowed discipline of the proletarian vanguard, an array of massive caucuses, and legions of propagandists and agitators, some of whom were no better than gangsters, but most of whom turned out of a passionate but blind devotion to their cause into the moral assassins of their erstwhile comrades. The new converts to Trotskyism started out with a determination to shake the party they loved and to make it see the light which they themselves, studying Trotsky's writings, had excitedly seen; but soon they found themselves shut in within small, hermetic circles, where they were to accustom themselves to live as noble lepers in a political wilderness. Tiny groups which cannot hitch themselves to any mass movement are quickly soured with frustration. No matter how much intelligence and vigour they may possess, if they find no practical application for these, they are bound to use up their strength in scholastic squabbling and intense personal animosities which lead to endless splits and mutual anathemas. A certain amount of such sectarian wrangling has, of course, always marked the progress of any revolutionary movement. But what distinguishes the vital movement from the arid sect is that the former finds in time, and the latter does not, the salutary transition from the squabbling and the splits to genuine political mass action.

The Trotskyist groups did not lack men of brains, integrity, and enthusiasm. But they were unable to break through the ostracism which Stalinism imposed on them; and, in their beyond-the-pale existence they could never rid themselves of their internal dissensions. Thus, soon after the reconciliation Trotsky brought about among his French followers, the latter fell out again. Rosmer and Naville renewed their complaints against Molinier, charging him with irresponsibility and recklessness, while he reproached them with too little faith and obstructing all plans for action. The puny organization, giving itself the airs and the constitution of a much larger body, had its National Executive and its Paris Committee. On the former, Rosmer and Naville were in a majority, and they proposed to exclude Molinier on the ground that his financial deals threatened to bring the Opposition into disrepute. But Molinier had behind him the Paris Committee and—Trotsky's support. Rosmer implored Trotsky to save the National Executive this embarrassment and to cease sheltering Molinier under his wing.[1] By now Trotsky's attachment to Molinier was little short of infatuation; and his relations with Rosmer became strained and their correspondence somewhat acid. The rivalry also affected the two shadowy international bodies the Opposition had given itself, the International Bureau and the International Secretariat which were equally at loggerheads.[2] In the summer of 1930 Trotsky once again asked his French adherents to come to Prinkipo and settle the differences. They came, patched up another 'peace', and Trotsky sent them back to Paris confident that now at last they would launch in unison the long-delayed drive from which he expected so much. But after a few weeks the quarrel broke out again; and in November Rosmer, hurt by Trotsky's partiality for Molinier, resigned. This was a blow to the organization and to Trotsky personally, who knew that of all his followers in Europe none had Rosmer's qualities or prestige. But he was convinced that Molinier's energy would soon jerk the organization out of the impasse and that then Rosmer would return. Even in resigning Rosmer gave Trotsky proof of a rare disinterested devotion, for he refrained from entering into any controversy, and rather than openly clash with Trotsky withdrew from all factional activity. Yet he resented Trotsky's behaviour so strongly that for several years he refused to meet him or even to exchange views.

Similar dissensions, in which it is well-nigh impossible to disentangle the personal from the political, became a chronic

[1] See the Trotsky-Rosmer correspondence for June and July 1930, and also Trotsky's letters to M. Shachtman, of 18 August 1930, to R. Molinier of January-February 1931, and to the Federation of Charleroi of 28 June 1931. *The Archives*, Closed Section.

[2] The International Bureau, formed at a conference of Trotskyists from several countries, in April 1930, consisted of Rosmer (with Naville as deputy), the American Shachtman, the German Landau, the Spaniard Nin, and the Russian Markin. Under the cover name Markin, L. Sedov (Lyova) represented the Russian Opposition. (He did not, however, participate in the conference.) The Bureau could not function, because Shachtman returned to the States, Nin was imprisoned in Spain shortly after, and Markin could not get out of Prinkipo. An International Secretariat was then formed in Paris, of which Naville was the mainstay, with the Italian Suzo and the American Mill as members. Mill was presently exposed as a Stalinist; and the Secretariat was no more effective than the Bureau. Trotsky then sought to overhaul it with the help of Senin-Sobolevicius and Well. (See Trotsky's letter to Well of 15 December 1931.)

distemper of most, if not all, Trotskyist groups; the French example was infectious if only because Paris was now the centre of international Trotskyism. The personalities were, as a rule, of so little weight, the issues so slight, and the quarrels so tedious that even Trotsky's involvement does not give them enough significance to earn them a place in his biography. With the years his involvement assumed piteous and at times quite grotesque forms. As almost every quarrel shook the entire organization, these triflings devoured much of his time and nerves. He took sides; he acted as arbiter. Being in contact with groups in every corner of the world, he had to deal with an incredibly large number of such altercations; and as he encouraged the various sections of the Opposition to interest themselves in each other's activities, he wrote interminable circulars and epistles explaining, say, to the Belgians why the French fell out, to the Greeks why the German comrades were in disagreement, to the Poles what were the points at issue between different sets of the Belgian or of the American Opposition, and so on, and so forth.[1]

He did all this in the belief that he was educating and training a new levy of communists, new cadres of revolution. The extreme paucity of the Opposition's resources and the feebleness of its organization did not deter him. He held that the worth of a movement lay in the power of its ideas which was bound to prevail eventually; that the chief task was 'to maintain the continuity' of the Marxist school of thought; that only an organization could assure that continuity; and that any organization had to be built in the circumstances that were given and with such human material as was available. Sometimes, the bickering of his followers was enough to drive him to despair and to make him wonder whether his efforts were not wasted. Then he consoled himself with the recollection that Lenin, in the years of his 'factional émigré squabbles', often invoked an image of Tolstoy's which described a man squatting in the middle of a road and making incoherent, maniacal gestures which suggested to passers-by that he was a madman;

---

[1] Of over 300 files, containing about 20,000 documents of the Closed Section of *The Archives* approximately nine-tenths consist of Trotsky's correspondence with his followers. A very large proportion of the Open Section of *The Archives* also consists of his writings on the policy, tactics, and organization of various Trotskyist groups.

but on coming nearer one saw that the queer gesticulation was a purposeful activity—the man was sharpening a knife on a grindstone. And so Trotsky, however purposeless his own dealings with his followers might at times appear, told himself that he was in fact sharpening the mind and the will of a new Marxist generation. He suppressed his distaste at mingling great principles with the pettiest of wrangles, and mustered all his patience and persuasiveness to give freely to his followers. Yet he could not help sensing that the human material with which he was working was quite unlike that with which either he or Lenin had worked before the revolution. Then, whatever the miseries of émigré politics, those involved were genuine and serious fighters, wholly dedicated to their cause and sacrificing to it every interest in life and life itself—human flames of revolutionary enthusiasm. His present followers in the West were made of different stuff: they had in them only little of the passion and heroism that could storm the heavens. They were certainly not or 'not yet' 'genuine Bolsheviks', he reflected; and this accounted for an irreducible psychological distance between him and them. In his thoughts he preferred to dwell with his other friends and disciples, those who were scattered over the prisons and punitive colonies of the Urals and Siberia, and there were fighting, starving, freezing, and wrestling with their problems unto death. Even the most mediocre of the people over there now seemed to him worthier as fighters and closer than almost any of his followers in the West. Sometimes he unwittingly vented this feeling as, for instance, in an obituary on Kote Tsintsadze which he wrote early in 1931. Tsintsadze, a Bolshevik since 1903, head of the Caucasian Cheka during the civil war, and then a leading Oppositionist, had been deported, jailed, and tortured. Ill with tuberculosis, suffering from haemorrhages of the lungs, he fought on, went on hunger strikes, and died in prison. In the obituary, published in the *Bulletin*, Trotsky quoted these prescient words from a letter Tsintsadze had written him at Alma Ata: 'Many, very many of our friends and of the people close to us will have to . . . end their lives in prison or somewhere in deportation. Yet in the last resort this will be an enrichment of revolutionary history: a new generation will learn the lesson.'

'The Communist parties in the West', Trotsky remarked,

'have not yet brought up fighters of Tsintsadze's type'; this was their besetting weakness; and it affected the Opposition as well. He confessed that he was amazed to find how much cheap ambition and self-seeking there was even among Oppositionists in the West. It was not that he deprecated all personal ambition—desire for distinction was often a stimulus to effort and achievement. But 'the revolutionary begins where personal ambition is fully and wholly subordinated to the service of a great idea. . . .' Unfortunately, only too few people in the West had learned to take principles seriously: 'Flirtation with ideas' or dilettante dabbling with Marxism-Leninism was all too common.[1]

It was rarely that Trotsky allowed himself such a complaint. He saw no use in wringing hands over the limitations of the human material produced by history—it was only from this material that the 'new Tsintsadzes' could be formed.

.   .   .   .

Meantime, in the Soviet Union the Opposition was breaking up and the fighters 'of Tsintsadze's type' were either perishing physically or shrinking morally. They were caught in the double vice of the Stalinist terror and of their own dilemmas. Even as early as 1928, while Trotsky was still sustaining their spirit of resistance from Alma Ata, they showed signs of being unequal to the strain. A division of opinion, it will be remembered, arose among them as they watched the end of the coalition between the Stalinists and the Bukharinists and the beginnings of Stalin's left course.[2] These events rendered obsolete some of the Opposition's major demands and battle-cries. The Opposition had called for rapid industrialization and for the gradual collectivization of farming and had charged Stalin with obstruction and with favouring the wealthy farmer. When in 1928 Stalin accelerated the tempo of industrialization and turned against private farming, the Oppositionists first congratulated themselves on the change, in which they saw their vindication; but then they felt themselves robbed of their ideas and slogans and deprived of much of their political *raison d'être.*

[1] *B.O.*, no. 19, March 1931.
[2] See the chapter 'A Year at Alma Ata' in *The Prophet Unarmed.*

Under any régime allowing a modicum of political controversy, a party or faction which has the misfortune of seeing its rivals steal its clothes may still be permitted to assist with dignity at the realization of its own programme by others. The deported Trotskyists were not free even to hint that their clothes had been stolen or to point out, in the hearing of the nation, how worthless and hypocritical had been the accusations the Stalinists had heaped on them when they branded them as 'super industrializers' and 'enemies of the peasantry'. Stalin's left course, which implicitly vindicated the Opposition, sealed its defeat; and the Opposition no longer knew clearly whether or on what ground it was to go on opposing him, especially as up to the middle of 1929, before Stalin decided on 'wholesale collectivization' and the 'liquidation of the kulaks', his policy followed the Opposition's demands quite closely. If it is a galling experience for any party or group to see its programme plagiarized by its adversaries, to the Trotskyists, who in advocating their ideas exposed themselves to persecution and slander, this was a shattering shock. Some began to wonder for the sake of what they should go on suffering and let their next of kin endure the most cruel privations. Was it not time, they asked themselves, to give up the fight and even to reconcile themselves with their strange persecutors?

Those who succumbed to this mood eagerly assented to Radek's and Preobrazhensky's argument that there would be nothing reprehensible in such a reconciliation, and that the Opposition, if it was not merely to grind its axe, should indeed rejoice in the triumph of its ideas, even though its persecutors gave effect to them. It was true, they said, that Stalin showed no willingness to restore within the party the proletarian democracy for which the Opposition had also clamoured; but as he was carrying out so much of the Opposition's programme there was reason to hope that he would eventually carry out the rest of it as well. In any case, Oppositionists would be better able to further the cause of inner-party freedom if they returned to the ranks than if they remained in the punitive colonies, from where they could exercise no practical influence. Whatever it was that they were striving for, they must strive for it within the party, which was, as Trotsky once put it, 'the only historically given instrument that the working class possessed' for furthering the progress of socialism; only through it and inside it could the Oppositionists achieve their purposes. Neither Radek nor Preobrazhensky as yet suggested surrender—they merely advised a more conciliatory attitude, which would make it possible for them to negotiate the terms of their reinstatement.

Another section of the Opposition, for which Sosnovsky, Dingelstedt, and sometimes Rakovsky spoke, rejected these promptings and did not believe that Stalin was in earnest about industrialization and the struggle against the kulaks. They treated the left course as a 'temporary manœuvre' to be followed by sweeping concessions to rural capitalism, the neo-N.E.P., and the triumph of the right wing. They denied that the Opposition's programme was surpassed by events and saw no reason to modify any of their attitudes. The more sanguine were as hopeful as ever that time was working for them. If Stalin were to pursue the left course, they said, its logic would compel him to call off his fight against the left Opposition; and if he were to launch the neo-N.E.P., the subsequent 'shift to the right' would so endanger his own position that again, in order to redress the balance, he would have to come to terms with the Trotskyists. The Opposition would therefore be foolish to try to barter principles against reinstatement, especially to waive its demand for freedom of expression and criticism. This, broadly, was the 'orthodox Trotskyist' view.

The conviction that the Opposition's programme was obsolescent was gaining ground not only among the conciliators, however. It was held with even greater fervour, but for reasons diametrically opposed to Radek's and Preobrazhensky's, by those who formed the most extreme and irreconcilable wing of the Opposition. There the view was already becoming axiomatic that the Soviet Union was no longer a workers' state; that the party had betrayed the revolution; and that the hope to reform it being futile, the Opposition should constitute itself into a new party and preach and prepare a new revolution. Some still saw Stalin as the promoter of agrarian capitalism or even the leader of a 'kulak democracy', while to others his rule epitomized the ascendancy of a state capitalism implacably hostile to socialism.

Up to the end of 1928 these cross currents were not yet so

strong as to destroy the Opposition's outward unity. A ceaseless discussion went on in the colonies; and Trotsky presided over it, holding the balance between the opposed viewpoints. After his banishment to Constantinople, however, the force of the disagreements grew and the opposed groups drifted farther and farther apart. The conciliators eager for reinstatement gradually 'curtailed' the conditions on which they were prepared to come to terms with Stalin, until the conciliation for which they were getting ready became indistinguishable from surrender. On the other hand, the irreconcilables worked themselves up into such a frenzy of hostility towards all that Stalin stood for that they were no longer concerned with changes in his policy or even with what was going on in the country at large; they repeated obsessively their old denunciations of Stalinism regardless of whether these still bore any relation to the facts, old and new. The members of these extreme groups viewed one another as renegades and traitors. The irreconcilables branded their conciliatory comrades in advance as 'Stalin's lackeys', while the latter looked upon the zealots as upon people who had lost their bearings, had ceased to be Bolsheviks, and were turning into *anarchisants* and counter-revolutionaries. The two extreme wings were growing and only the shrinking rump of the Opposition remained 'orthodox Trotskyist'.

Scarcely three months after Trotsky's banishment not a trace was left even of the outward unity of the Opposition. While he was cut off from his followers—it took him a few months to re-establish contacts—Stalin found it all the easier to divide them and demoralize them by means of terror and cajolery. The terror was selective: the G.P.U. spared the conciliators but combed the punitive colonies, picking out the most stubborn Oppositionists and transferring them to jails, where they were subjected to the harshest treatment: placed under military guards; crowded in damp and dark cells unheated in the Siberian winter; kept on a meagre diet of rotten food; and denied reading matter, light, and facilities for communication with their families. They were thus deprived of the privileges which political prisoners had obtained in Tsarist Russia and which the Bolsheviks had, since the end of civil war, granted to anti-Bolshevik offenders. (About this time, as if to mock his former comrades even further, Stalin ordered the release of

F

quite a few Mensheviks and Social Revolutionaries.) As early as March 1929 Trotskyists describing their life at the hard labour prison of Tobolsk compared it with Dostoevsky's haunting image of *katorga* in *The House of the Dead*.[1] If this terror aimed at intimidating and softening the conciliators, it also seemed designed to drive the irreconcilables to demonstrations of such unthinking hostility towards all aspects of the existing régime that it should be easy to brand them as counter-revolutionaries and to drive an even deeper wedge between them and the conciliators.

However, Stalin could not break the Opposition by terror alone—his far more potent weapon was the left course. 'Without severe persecution', Rakovsky remarked, 'the left course would have only brought fresh adherents into the ranks of the Opposition, because it marked the bankruptcy [of the earlier Stalinist policy]. But persecution alone, without the left course, would not have had the effect it has had.'[2] In the months that followed Trotsky's arrival in Constantinople Stalin's hesitation over policy was coming to an end. His break with Bukharin was consummated at the February session of the Politbureau, while Trotsky was *en route* to Turkey. In April the conflict was carried from the Politbureau to the Central Committee, and then to the sixteenth party conference. The conference addressed the nation with a rousing call for a radical speeding up of industrialization and collectivization, a call which reproduced, in part literally, Trotsky's earlier appeals.[3] It became increasingly difficult to maintain, as Trotsky and some of the Trotskyists were still doing, that Stalin's change of policy was a 'temporary manœuvre'. It turned out that Preobrazhensky and Radek who had held all along that Stalin was not trifling with the left course (and that circumstances would not allow him to do so even if he wanted to) had in this point a much better grasp of reality.

At a stroke the Opposition's dilemmas were immensely aggravated. It became almost ludicrous for its members to

---

[1] See the report of 20 March 1929 in *B.O.*, no. 1.

[2] Ibid., no. 7, November–December 1929.

[3] *V.K.P.* (*b*). *Profsoyuzakh*, p. 515. In the resolutions of the conference Trotsky's appeal for socialist competition, now ten years old, was literally, but of course anonymously, reproduced. *K.P.S.S. v. Rezolutsyakh*, vol. II, pp. 496–7; see also my *Soviet Trade Unions*, pp. 95–97.

chew over old slogans, to clamour for more industrialization, to protest against the appeasement of rural capitalism, and to speak of the threatening Neo-N.E.P. The Opposition either had to admit that Stalin was doing its job for it or it had to re-equip itself and 'rearm' politically for any further struggle. Trotsky, Rakovsky, and others were indeed working to bring the Opposition's ideas up to date. But events moved faster than even the most quick-minded of theorists.

The state of the nation not less than changes in official policy contributed to the disarray of the Opposition. This was a time of the gravest emergency. Stalin described it in these terms;[1] but so also did all the leaders of the Opposition, how-ever they differed among themselves. Preobrazhensky, not given to dramatic overstatement, compared the tension of the spring of 1929 with that which had led to the Kronstadt rising, the rising the Bolsheviks had regarded as more dangerous to themselves than any critical phase of the civil war.[2] Radek, speaking of the conflict between Stalinists and Bukharinists in the Central Committee, said that 'the Central Committee looked like the Jacobin Convention on the very eve of the 9 Thermidor', the day that brought the ruin of Jacobinism. Rakovsky described the moment as 'the most fateful since the civil war'.[3] Indeed, there was a complete agreement about this among all observers.

For several years now the gulf between town and country had widened and deepened. The 25–6 millions of small and mostly tiny and archaic farmsteads could not feed the rapidly growing urban population. The towns lived under an almost constant threat of famine. Ultimately, the crisis could be resolved only through the replacement of the unproductive smallholding by the modern large-scale farm. In a vast country accustomed to extensive agriculture, this could be achieved either by the energetic fostering of agrarian capitalism or by collectivization—there was no other choice. No Bolshevik government could act as the foster parent of agrarian capitalism —if it had so acted it would have let loose formidable forces

[1] Stalin, *Sochinenya*, vol. XII, pp. 118ff.
[2] Preobrazhensky, 'Ko Vsem Tovarishcham po Oppozitsii' (*The Archives*), to which reference is made in further pages also; and Rakovsky's report in *B.O.*, loc. cit.
[3] Loc. cit.

hostile to itself and it would have compromised the prospects of planned industrialization.[1] There was thus only one road left, that of collectivization, even though the all-important questions of scale, method, and tempo had still to be resolved. Years of official hesitation had led only to this, that the decisions had now to be taken under conditions far worse than those under which they might have been taken earlier. Stalin's attempts to combine the most contradictory policies, to appease the well-to-do farmers and then to requisition their produce, had infuriated the peasantry. His long-lasting reluctance to press on with industrial development had been no less disastrous. While the country was unable and unwilling to feed the town, the town was unable to supply the country with industrial goods. The peasant, not being able to obtain shoes, clothes, and farm tools, had no incentive to raise his output, still less to sell it. And so both the starving town and the country famished of industrial goods were in turmoil.

The decisions about tempo and scale of industrialization and collectivization were taken in conditions of an acute scarcity of all the human and material elements needed for the two-fold drive. While workers went short of bread, industry was short of skilled labour. It was also short of machinery. Yet machines stood idle for lack of fuel and the raw materials whose supply depended on the rural economy. Transport was disrupted and could not cope with increased industrial traffic. The supply of nearly all goods and services was grievously inadequate to the demand. Inflation was rampant. Controlled prices bore no relation to the uncontrolled ones, and neither reflected genuine economic values.

All the ties and links between the various parts of the body politic were cut, except for the bonds of misery and desperation. Not only had economic intercourse between town and country once again broken down, so had all normal relations between citizenry and state and even between party and state. There

---

[1] Large scale capitalist farming formed the rural background to the industrialization of Britain and the United States; the Junkers' estates and the *Grossbauerwirtschaft* were dominant in Germany's agriculture during her industrial rise. In all these countries large-scale farming had been in existence at the outset of industrialization, whereas in the Russia of the nineteen-twenties it was not. The concentration of farming by any normal processes of capitalist competition would have required much time and much *laissez faire*.

was no extreme of deception and violence to which both the rulers and the ruled were not prepared to go in the scramble. The kulaks, and many 'middle' and even poor peasants, were implacable in their hatred of the 'commissars'. Arson and killings of party agents and agitators were daily occurrences in the villages. The mood of the peasantry communicated itself to the working class among whom newcomers from the country were very numerous. In the twelfth year of the revolution the poverty of the nation and the neglects and the abuses of government provoked a revulsion so bitter and widespread that something great and terrible had to happen or had to be done soon in order either to suppress or to release the pent-up emotions. Under the surface forces were boiling up for what might have become a gigantic explosion of the kind of which, on a small scale, Hungary was to give an example in 1956. Almost cornered, Stalin and his followers fought back with mounting fury.

'The revolution is in danger!' was the cry which the Trotsky-ists raised in their places of deportation and prison cells. Both the 'orthodox' Trotskyists and the conciliators were seized with equal alarm; but whereas the former did not see what course of action was open to them in the conditions in which they were placed and thought that they should keep themselves in readiness for the approaching crisis, the conciliators, on the contrary, felt impelled to 'act at once'; and it was with the cry: 'The revolution is in danger!' that they marched to surrender. The best of them did so from the deep conviction that when the fate of Bolshevism and of the revolution was at stake, it was a crime to cling to factions and to cherish sectional interests and ambitions. The worst among them, the weary opportunists, found in the 'revolution's danger' a convenient pretext for wriggling out of commitment to a lost cause. Those who were neither the best nor the worst, the average conciliators, may not have been aware of their own motives, which were probably mixed or ambivalent.

In April 1929 Preobrazhensky drew the conciliators together with an appeal 'To All Comrades-in-Opposition!'[1] This was an extraordinary document: in it the conciliator for the last time, before surrender had sealed his lips, expressed himself

[1] *The Archives.*

frankly as he looked back on the Opposition's road and turned
his gaze on the tortuous and stony path ahead of him. Preobrazhensky described how the Opposition had been driven to an
impasse by the very triumph of its ideas. He found that many of
his comrades would rather deny the triumph than admit the
impasse. They still behaved as if their forecasts about the Neo-
N.E.P. and the 'shift to the right' had come true; as if there
had been no left course. To be sure, Stalin had initiated the
left course in a manner very different from the one they had
championed. The Opposition wanted industrialization and
collectivization to be carried out in the broad daylight of
proletarian democracy, with the consent of the masses and free
initiative 'from below'; whereas Stalin relied on the force of the
decree and coercion from above. All the same, the Opposition
had stood for what he was doing even if the way he was doing it
was repugnant to them. If they refused to acknowledge this,
they would turn into an Opposition for opposition's sake; and
then to justify themselves they would drift away from their own
principles. He, Preobrazhensky, did not repudiate the Opposition's past: 'In fighting against the Central Committee we have
done our duty.' But the Opposition's present duty was to come
closer to the party and then return to it—and here spoke the
theoretical pioneer of 'primitive socialist accumulation'—in
order 'to hold out together against the pressure of that discontent which must be aroused in a peasant country by a policy
of socialist accumulation and a struggle against agrarian
capitalism'.

Preobrazhensky spoke of the resentment Stalin had aroused,
even among conciliators, by banishing Trotsky 'with the help
of the class enemy' (i.e. of the Turkish Government). The
Oppositionists 'cannot forgive this', he said; but he suggested
that this outrage should not be allowed to obscure considerations of a more general character; and he added that Trotsky
too had confounded the Opposition by carrying the struggle
against Stalin into the bourgeois Press of the West. Preobrazhensky had few illusions about the fate that awaited the
conciliators: he was aware of the blows and humiliations that
would fall on them in the 'difficult, critical years ahead',
although even he could hardly have glimpsed all the mud and
blood through which they were to wade and in which they were

to perish. But he was clear-eyed enough to indicate plainly to his comrades that the course to which he was summoning them would be full of anxiety and torment. His hopes for a genuine and dignified reconciliation, the hopes he had entertained in the previous year, had sagged. He now saw reinstatement as a virtual surrender. 'Those of us', he concluded, 'who have fought in the ranks of the party ten, twenty or more years [Preobrazhensky himself had been a Bolshevik since 1904] will return to it with feelings very different from those with which they once joined it for the first time.' They would go back without their early enthusiasm, as broken-hearted men. They could not even be sure that the Central Committee would agree to reinstate them on any terms. 'Such are all the circumstances of this return and such is the inner party situation that, if readmitted, we shall have to bear responsibility for things against which we have warned and to submit to [methods] to which we cannot give our assent. . . . If we are reinstated we shall, each of us, receive back the *partbilet* [membership card] as one accepts a heavy cross.' Yet for those who wish to serve the cause of socialism effectively nothing was left but to take the cross.

In May, Preobrazhensky was allowed to travel to Moscow in order to try and 'make peace with the party'. At first he sought to obtain favourable terms for the Opposition at large, pleading for a cessation of the terror, for a halt to deportations, for a rehabilitation of party members victimized under Article 58 on the charge of counter-revolutionary activity, and—last but not least—for the rescinding of Trotsky's banishment. He negotiated with Ordjonikidze and Yaroslavsky and other members of the Central Committee and Central Control Commission who acted under Stalin's personal supervision.

To Stalin the capitulation of a large section of the Opposition was important enough because of the effect this was bound to have on the party's morale and on Trotsky's fortunes. Anxious to entice the conciliators and wary of blasting all their hopes at once, he at first feigned readiness to consider some of their desiderata. But he could not in truth accept any. Above all, he could not allow the Oppositionists to say on their reinstatement that they had come back because the party leadership had adopted *their* programme—this would have amounted not merely to a vindication of Trotsky and Trotsky-ism and to a refutation of all the charges against them, but also to an exposure of the lawlessness of the reprisals by which Stalin had overwhelmed them. He could not permit anyone even to allude to the fact that he had taken a leaf—and what a leaf!—out of Trotsky's book. If he did he would have destroyed his own claim to infallibility and power. The capitulators must declare that he, and not they and Trotsky, had been right. They must denounce and recant their own past. They could not be tolerated to come back as misunderstood trail-blazers; they could return only as the remorseful saboteurs of the left course and of all the policies that had consistently led up to it. Even then they must not be allowed to arouse in the party the feelings due to rueful prodigal sons—they could count only on the forgiveness granted to broken sinners and criminals; they must make their way back on their knees. To get them to do this Stalin had to wear down, by slow and stubborn bargain-ing, their mental defences, and induce them to give up one demand after another until they were brought to the point of unconditional surrender. Stalin's behaviour was not surprising: the terms on which Zinoviev, Kamenev, Antonov-Ovseenko, Pyatakov, and so many others had capitulated, and the process by which they had been brought to do it were still fresh in everyone's memory. But such was the power of self-deception that many conciliators who from afar anxiously watched Preo-brazhensky's parleys in Moscow—he was allowed to com-municate with the colonies of deportees—still hoped that they would be spared the indignities inflicted on earlier capitulators.

After a month the result of Preobrazhensky's 'negotiations' was already discernible in the behaviour of his closest comrades. In the middle of June, Radek and Smilga also travelled, under G.P.U. convoy, to Moscow to join Preobrazhensky. Their train stopped at a small Siberian station, where by chance they were met by a group of Oppositionists, who described the encounter in a letter preserved among Trotsky's papers. They spoke only to Radek—Smilga was ill and had to stay in his compartment. Radek told them of the purpose of the journey and made the by now familiar argument for surrender: the nation-wide famine, the shortage of bread felt even in Moscow, the workers' discontent, the threat of peasant risings, the discords in the

Central Committee (where 'Bukharinists and Stalinists were plotting to arrest each other'), &c. The situation, he said, was as grave as in 1919 when Denikin stood at the gates of Moscow and Yudenich stormed Petrograd. They must all rally to the party. On what terms? they asked. Would he demand in Moscow that paragraph 58 of the Criminal Code, the stigma of counter-revolution, be lifted from the deportees? No, he replied; those who persisted in opposition deserved the stigma. 'We ourselves', he shouted, 'have driven ourselves into exile and prison.' Would he demand that Trotsky be brought back? It was only a few weeks since Preobrazhensky had declared that the Opposition 'could not forgive' Trotsky's banishment, and only a few months since Radek himself, the author of the celebrated essay 'Trotsky the Organizer of Victory', had protested to the Central Committee against its causing the 'slow death' of that 'fighting heart of the revolution' and concluded his protest with the words: 'Enough of this inhuman playing with Comrade Trotsky's health and life.' But in the last few weeks the logic of the surrender to Stalin had done its work. And so to their amazement Radek's interlocutors heard this reply: 'I have definitely broken with Lev Davidovich— we are political enemies now. With the contributor to Lord Beaverbrook's papers I have nothing in common.' (Radek himself had often contributed to the bourgeois Press and was to do so again, but in Stalin's interest.)[1] In the very violence of his answer Radek betrayed his guilty conscience. He went on to speak bitterly against the new recruits to the Opposition, the angry young men, who, he alleged, had nothing Bolshevik about them, but joined the Trotskyists from sheer anti-Soviet spite. Once more he appealed to his interlocutors: 'The last party conference has adopted our Platform which has brilliantly proved itself. What can you still have against the party?' Radek's escort provided the answer: while he was arguing, his G.P.U. guards interrupted him, shouting that they would

[1] Trotsky often had to defend himself against this reproach, which was at first made even by his French followers, as Rosmer informed him in a letter of 24 February 1929. Rosmer's and Trotsky's answer was that Marx too had to earn his living by writing for the bourgeois Press. In a special note in the first issue of the *Bulletin Oppozitsii* Trotsky explained his position to Soviet readers and emphasized that even in the bourgeois Press he spoke as a Bolshevik and a Leninist, defending the revolution.

not allow him to agitate against Trotsky's banishment; and they pushed him and kicked him back into the train. Radek burst out with hysterical laughter: 'I? Agitating against Trotsky's banishment!' Then he apologized plaintively: 'I am only trying to persuade these comrades to return to the party'; but the guard would not even listen and kept on pushing him back to the compartment. The year before Radek had scorned Zinoviev and Pyatakov for the 'morbid odour of *Dostoevshchyna*' they and their recantations exuded—now he himself, the prince of pamphleteers, appeared to his erstwhile co-thinkers and co-sufferers as a Smerdyakov descended from Dostoevsky's pages on to the little god-forsaken Siberian station.[1]

After another month of haggling, on 13 July, Radek, Preobrazhensky, Smilga, and 400 other deportees finally announced their surrender.[2] The advantages that Stalin derived from this were many. No event since Zinoviev's and Kamenev's capitulation at the Fifteenth Congress, in December 1927, had done so much to bolster Stalin's prestige. As he was just engaged in a heavy attack on Bukharin's faction, the disintegration of the Trotskyist Opposition relieved him of the need to fight on two fronts simultaneously. Trotsky had often said that in the face of an acute 'danger from the right' Trotskyists and Stalinists would join hands. Well, they were now doing so, but on Stalin's own terms—he was winning them over to his side without and even against Trotsky. Many of the capitulators were men of high talent and experience with whom he would fill industrial and administrative posts from which the Bukharinists were being squeezed out. He knew that the capitulators would throw themselves heart and soul into the industrial drive—many of them were to serve under Pyatakov, the arch-capitulator who was the moving spirit of the Commissariat of Heavy Industry. Radek alone was, as a propagandist, worth more to Stalin than hosts of his own scribes.

Trotsky at once attacked the 'capitulators of the third draft'. (Those of the 'first draft' were Zinoviev, Kamenev, and their followers, and those of the second were Antonov-Ovseenko, Pyatakov, and their friends.) 'They state', Trotsky wrote, 'that the differences between Stalin and the Opposition have almost vanished. How then do they explain the furious character of the

---

[1] *The Archives; B.O.*, no. 6, 1929.   [2] *Pravda*, 13 July 1929.

reprisals? If in the absence of the most irreconcilable and profound differences the Stalinists banish and inflict *katorga* on Bolsheviks, then they do it from sheer bureaucratic banditry, without any political idea. This is how the Stalinist policy presents itself if one looks at it from Radek's viewpoint. How then dare he and his friends raise their voices to advocate unity with political bandits . . . ?' This was not the view that he, Trotsky, took of Stalinist policy; he held that for all its lack of scruple Stalinism had deep political motives for its implacable hostility towards the Opposition; the fundamental differences had lost none of their force. Radek and Preobrazhensky overlooked them or pretended to do so because they broke down morally. Revolution was a great devourer of characters; and every period of reaction took its toll of a tired generation of fighters who knuckled under. But sooner or later the old and weary were replaced by the young who entered the struggle with fresh courage and learned their lessons even from the prostration of their elders. 'We have before us the prospect of a long, tenacious struggle and of a long labour of education.'[1]

In truth, Trotsky received the first news of Radek's surrender with some incredulity; and he attributed Radek's behaviour to 'impulsive character, isolation, and lack of moral support' from comrades. He recalled with warmth of feeling that 'Radek had behind him a quarter of a century of revolutionary Marxist work', and doubted whether he would really be able to make his peace with Stalinism: 'He is too much of a Marxist for that and, above all, he is too internationally minded.' But when *Pravda* came out with Radek's letter of recantation, he found that 'Radek has fallen much lower than I had supposed'. Even now the fall was so incredible that Trotsky imagined that his bargain with Stalin was only temporary and that, having frequently wavered between right and left in the party, he would soon join hands with the Bukharinists. Yet what a tangle this was: 'Radek and a few others with him consider this the most propitious moment for capitulation. Why indeed? Because the Stalinists, you see, are chastising Rykov, Tomsky, and Bukharin. Has it then been our task to make one part of the ruling group chastise the other? Has the approach to fundamental political questions changed? . . . Has the anti-Marxist régime

---

[1] *Écrits*, vol. I, pp. 157–63.

of the Communist International not been maintained? Is there any guarantee for the future?' Radek and Preobrazhensky saw in the first Five Year Plan a radically new departure. 'The central issue', Trotsky replied, 'is not the statistics of this bureaucratic Five Year Plan *per se*, but the problem of the party', the spirit in which the party was led, because this determined also its policy. Was the Five Year Plan, in its formulation and execution, subject to any control from below, to criticism and discussion? Yet on this depended also the results of the Plan. 'The inner party régime is for the Marxist an irreplaceable element of control over the political line . . .'— this had always been the Opposition's essential idea. 'But the renegades usually have, or think that others have, a short memory. One can say with reason that a revolutionary party embodies the memory of the working class: its first and foremost task is to learn not to forget the past in order to be able to foresee the future.' Trotsky still viewed Stalin's left course as a by-product of the Opposition's struggle and pressure; he still thought that Stalin might reverse his policy and that his conflict with Bukharin was, despite all its harshness, only 'superficial'.

Trotsky's arguments did not reach the Oppositionists in the Soviet Union until the autumn; and they could hardly suffice to stop the capitulation stampede. The upheaval in the Soviet Union had already gone deeper, and its impact on the Opposition was far more violent, than he realized. As yet there was in his remarks no hint of the gravity and alarm that one finds in the writings of all, even the most irreconcilable, Oppositionists in Russia. He still viewed the scene of 1929 through the prism of 1928 and was half-unaware of the 'eve of civil war' atmosphere that hung over the country. The full force of the cry 'The Revolution is in Danger' somehow escaped him, as did also the momentum the left course was gathering and the depth of the breach between Stalin and Bukharin. These, however, were the matters that weighed on the minds of all Opposition groups.

The sense that the revolution was threatened by a mortal danger, which the Opposition must ward off jointly with the Stalinists, soon prompted many who had hitherto belonged to its irreconcilable wing to follow in Preobrazhensky's and Radek's footsteps. Ivan Smirnov, the victor over Kolchak and one of Trotsky's closest associates, Mrachkovsky, a fighter of legendary heroism, Byeloborodov, the Commissar in whose home Trotsky found refuge when he left the Kremlin in November 1927, Ter-Vaganyan, Boguslavsky, and many others asked to be reinstated. They began to parley with Stalinist headquarters in a less sombre mood than Preobrazhensky had done, hoping that the general situation would induce Stalin to reinstate them on terms less humiliating.[1] This time the bargaining went on for nearly five months, from June to the end of October, in the course of which Smirnov's group prepared four different political declarations. In an early draft, produced in August and preserved among Trotsky's papers, they gave, as the reasons for their step, agreement with the Five Year Plan, and the 'danger from the right'. But they also advanced clear criticisms of Stalin's policy, saying that insufficient thought was given in the Five Year Plan to the need to raise the depressed standards of living of the workers; that the 'selection of party cadres' was such as to make the expression of critical opinions impossible; and that the doctrine of socialism in one country served as a 'screen for opportunism', as did also the continued official bias in favour of the 'middle' peasant. Having in all these points upheld the Opposition's attitudes, the applicants admitted also its errors. They had been mistaken, they stated, in thinking that the Central Committee would, in the search for a way out of the crisis, turn rightwards and pave the way for the Thermidor—only the behaviour of the Bukharinist minority justified that fear. They agreed that in the present grave circumstances the party leadership should allow no freedom to factions, because only the right elements would benefit. The Trotskyist Opposition should therefore disperse its organization, disband its own leading centre 'which under various names had existed for years', and stop any form of clandestine activity. But they also demanded an end to the reprisals against the Opposition and they pleaded fervently for the recall of Trotsky, 'whose fate is tied to the fate of the working class', and with whose services neither the Soviet Union nor international communism could dispense.[2]

Only slowly, defending every one of their points, did Smirnov and his associates allow their demands to be whittled down.

[1] *Vide* Rakovsky's account in *B.O.*, no. 7, 1929. [2] *The Archives*.

As the year advanced and his difficulties mounted, Stalin was indeed more anxious than before to secure fresh capitulations; and he did not extract from this group a recantation quite as abject as the one he got from Radek and Preobrazhensky. Smirnov and his friends, in softening or dropping their criticisms of Stalin and waiving various demands, still insisted that they be allowed, in the very act of surrender, to call for Trotsky's return—it was mainly over this that the bargaining dragged on for five months. When at last they gave way, they still refused to denounce or renounce Trotsky; and their statement of submission, which appeared with hundreds of signatures in *Pravda* on 3 November 1929, was more restrained and dignified than any previous act of this kind.

The mood of surrender now touched the inner core of the Opposition, the most faithful of Trotskyists. However, Rakovsky, who, gravely ill and suffering from heart attacks, was transferred from Astrakhan to Barnaul, still managed to rally them. Under his inspiration a section of the Opposition as large as that which followed Smirnov stopped just on the brink of capitulation. 'We are fighting for the *whole* programme of the Opposition', Rakovsky declared. Those who made their peace with Stalin, because he was carrying out the economic part of that programme and who hoped that he would carry out the political part as well, were behaving like old-type reformists contenting themselves with the piecemeal realization of their demands. The political ideas of the Opposition were inseparable from its economic desiderata: 'As long as the political part of our programme remains unfulfilled, the whole work of socialist construction is in danger of being blown sky high.' Even more important to Rakovsky was integrity of conviction and honesty in one's attitude towards adversaries. A party leadership which extracted from Oppositionists confessions of imaginary errors merely imitated the Catholic Church, which made the atheist recant on his deathbed—such a leadership 'loses every title to respect; and the Oppositionist who changes his conviction overnight deserves only utter scorn'.[1]

It took Rakovsky's group several months to define its attitude; its 'Open Letter to the Central Committee' was not ready before the end of August. To collect about 500 signatures

---

[1] *B.O.*, no. 6, 1929.

from about ninety places of deportation was not easy; but it was even more difficult to accommodate in the document all the shadings of opinion that could be found among the signatories. The tenor of the Letter, which was in form also an application for reinstatement, bore witness to the prevalence of the conciliatory mood. Like Preobrazhensky and Smirnov, Rakovsky and those who followed him—Sosnovsky, Muralov, Mdivani, Kasparova and others—declared that it was the national emergency and the party's decision to sponsor the first Five Year Plan that prompted them to approach the Central Committee. The success of the Plan, they held, would strengthen the working class and socialism; failure would reopen the door to Thermidor and Restoration. Confronted by the 'gravest conflict between the forces of capitalism and those of socialism', they preferred to dwell on the issues on which they were at one with the party rather than on those on which they were not. To them too the 'danger from the right' was close and acute; and what they still criticized in the party's policy was the lingering desire to appease the 'middle' peasants. They were so whole-heartedly in favour of rapid industrialization that from their punitive colonies they pleaded for higher labour discipline in the factories and for determined action against those who tried to exploit the workers' discontent for counter-revolutionary purposes. But they also held it to be vital for the success of the industrial drive that it should be backed by the mass of the people who still resented the neglect of their living conditions, the run-away inflation, the many unkept official promises, and bureaucratic high-handedness. Having for years championed the course of action the party had taken, the applicants felt that they were entitled to reinstatement, all the more so as they also welcomed the 'left turn' in Comintern policy and admitted the harm of all faction. They regretted the exacerbation of feeling between the Opposition and the Central Committee, to which Trotsky's banishment had contributed so much. 'We appeal to the Central Committee, the Central Control Commission, and the entire party', the statement concluded, 'to ease our way back to the party by freeing the Bolshevik-Leninists, lifting the 58 paragraph, and bringing back Lev Davidovich Trotsky.'

When the statement reached Prinkipo, on 22 September,

Trotsky's satisfaction was mingled with apprehension. He was pleased to see at last a declaration from his followers—the first for many months—which did not ooze utter resignation. Yet he was apprehensive of its tenor. Having by now arranged his contacts with the Soviet Union via Berlin, Paris, and Oslo, he undertook to forward the Letter to those colonies of deportees which had not yet received it. But he added a gloss of his own design to give the statement a sharper edge. He said that he endorsed the Letter because, although it was 'moderate', it was 'not equivocal'. Only those could refuse to sign it who were of the opinion that the Soviet Thermidor was already accomplished, that the party was dead, and that nothing less than a new revolution was necessary in the U.S.S.R. 'Although this opinion has been attributed to us dozens of times, we have nothing in common with it. . . . Despite repression and persecution, we declare that our loyalty to Lenin's party and the October revolution remains unshakeable.' He too acknowledged that with the 'left turn' and the break between Stalin and Bukharin a new situation had arisen: 'If previously Stalin fought the Left Opposition with arguments borrowed from the Bukharinist right, he now attacks the right exclusively with arguments borrowed from the left.' In theory this should have led to a *rapprochement* between the centre and the left; in practice it did not. Stalin's adoption of the Opposition's policy was superficial, fortuitous, or merely tactical; basically they remained poles apart. Stalin conceived the Five Year Plan within the framework of socialism in one country, while the Opposition viewed the whole process of constructing socialism in the context of international revolution. This fundamental difference was as sharp as ever; and while Rakovsky and his friends had declared their solidarity with the new Comintern policy, Trotsky briefly but firmly stated his objections to it. Nevertheless, he agreed that Rakovsky was right in expressing readiness 'to subordinate the struggle we are waging for our ideas to the statutory norms and the discipline of a party that would base itself on proletarian democracy'. They had been willing to defend their views *within* the party at the time when the party was ruled by the right-centre coalition; and they must *a fortiori* be prepared to do so when the right was no longer in control. But to renounce their views because of this would be dishonest and 'unworthy of Marxism and of the Leninist school of thought'.

Trotsky trusted implicitly Rakovsky's integrity and courage; but he sensed the press and the pull of the stampede under which Rakovsky acted. In another gloss he excused Rakovsky's conciliatory tone as designed to 'test openly the inner party régime' in changed political circumstances: 'Was that régime or was it not, after all the recent lessons, capable of making good, at least partly, the immense harm it had done to the party and the revolution?' Was a self-reform of the Stalinist 'apparatus' still possible? Rakovsky's 'reticence, his silence on Stalin's mistakes in the international field, and his emphasis on the recent shifts to the left' were all calculated to facilitate the beginning of such a self-reform. Rakovsky had once again demonstrated that what mattered to the Opposition was the essence, not the form, of things, and the interest of the revolution, not the ambitions of persons or groups. 'The Opposition is ready to take the most modest place inside the party, but only if it can remain true to itself. . . .'[1]

Even while he wrote this, Trotsky wondered how many of those who had signed Rakovsky's statement might yet defect, and in a confidential message he warned Rakovsky that in his quest for conciliation he had gone to the limit and must not go 'even one step further!'. In the same *Bulletin* in which Rakovsky's statement appeared Trotsky published also an anonymous letter from a correspondent in Russia, criticizing Rakovsky for pandering to the capitulators. The writer, one of the few 'optimists' still left, was confident that soon 'Stalin will be on his knees before us as Zinoviev was in 1926'.

At the close of the year only a small minority of the Oppositionists still held out. According to one report not more than about a thousand Trotskyists remained in places of exile and prisons, whereas before the capitulations there were several thousands. Not for the first or the last time Trotsky had to say to himself: 'Friends who set forth at our side, Falter, are lost in the storm!' In the last days of November he wrote to a group of his Soviet disciples:[2] 'Let there remain in exile not 350

---

[1] 'Pismo druzyam' ('Not for Publication') of 25 September 1929. *The Archives; B.O.*, loc. cit.

[2] The letter, dated 26 November 1929, was provoked by a communication from an Oppositionist who was obviously inclined to join the capitulators. *The Archives.*

G

people faithful to their banner, but only 35. Let there remain even three—the banner will remain, the strategic line will remain, the future will remain.' He was ready to struggle on even alone. Did he at this moment think of Adolf Yoffe's farewell message? 'I have always thought', so Yoffe in the hour of his suicide wrote to Trotsky, 'that you have not enough in yourself of Lenin's unbending and unyielding character, not enough of that ability which Lenin had to stand alone and remain alone on the road he considered to be the right road. . . .'[1]

.   .   .   .

Paradoxically, Stalin viewed with some uneasiness the rush of the capitulators to Moscow, much though he benefited from it. Many thousands of Trotskyists and Zinovievists were now back in and around the party, forming a distinctive *milieu*. Stalin did not allow a single one of them to occupy any office of political importance. But the administrators, the economists, and the educationists were assigned to posts on all rungs of the government, where they were bound to exercise an influence. Although Stalin could not doubt their zeal for the left course, especially for industrialization, he knew what value to attach to the recantations he had extracted from them. They remained Oppositionists at heart. They considered themselves the wronged pioneers of the left course. They hated him not merely as their persecutor, but as the man who had robbed them of their ideas. True, he had turned them, politically, into his slaves. But the hidden hatred of slaves can be more dangerous than open hostility; it can lie silently in ambush, follow the master with a thousand eyes, and set upon him when he slips or makes a false step.

The capitulators now had a chance to influence, directly or indirectly, even the Stalinists and Bukharinists, some of whom also were bewildered when they saw Stalin appropriate the ideas and slogans which they had sincerely believed pernicious when Trotsky and Zinoviev had proclaimed them. After all his triumphs over all his opponents, Stalin was therefore at logger-heads with some of his own followers, among whom he began to discover crypto-Trotskyists and crypto-Bukharinists. 'If we

[1] See *The Prophet Unarmed*, p. 382.

were right in 1925-7', such people said, 'when we rejected the Opposition's demand for rapid industrialization and for an offensive against the kulak, and when we branded Trotsky and Zinoviev as the wreckers of the alliance between workers and peasants, then surely we are wrong now. And if we are right now, and if nothing but the left course can save the revolution, should we not have adopted it earlier, when the Opposition urged us to do so?' 'And was it not vile on our part', the most conscientious added, 'to abuse and crush the Opposition?' The answers varied, of course: some drew one conclusion, others another.[1] Enough that as early as the summer and autumn of 1929, while the capitulators were re-entering the party, a few good old Stalinists were being expelled from it, and some even sent to the places of deportation which the capitulators had just vacated. The most notorious cases were those of Uglanov, secretary of the Moscow organization, and other members of the Central Committee, branded as Bukharinists, and of Shatskin, Sten, and Lominadze, eminent propagandists and leaders of the 'young Stalinists', who were all three unmasked as semi-Trotskyists.

These cases revealed something of the ferment in the ruling group itself, a ferment which made it no unmixed advantage for Stalin to have so many capitulators around. Stalin knew that they still looked up to Trotsky as their guide and inspirer and indeed as the true leader of the revolution. Every batch of them, as they negotiated terms of surrender, had asked for Trotsky's return and stuck to this demand even while yielding on all other points of policy and discipline. When at last they were brought to renounce Trotsky, most of them did so with despair in their hearts and tears in their eyes. Few, very few, were those who like Radek perversely quelled their qualms and railed against Trotsky; and Radek's outbursts aroused disgust even among old Stalinists. To most capitulators Trotsky represented all that they had stood for in their better and prouder days. Their débacle and self-abasement had isolated him politically, but threw into fresh relief his moral grandeur. The capitulators, the Bukharinists, and the doubting Stalinists took in avidly every word of his that penetrated into the Soviet Union. At critical

[1] Such discussions went on even as late as 1931, during the writer's stay in Moscow.

moments, when important decisions were pending, the whisper: 'What does Lev Davidovich say about this?' was often heard even in Stalin's antechambers.[1] The *Bulletin* circulated in Moscow—party men returning from assignments abroad, especially members of embassies, smuggled it home and passed it on to friends. Although only very few papers got about in this way—the *Bulletin* seems never to have been printed in more than 1,000 copies—Trotsky's comments and forecasts and the choice morsels of his invective spread quickly by word of mouth. Stalin could not rest on his laurels and contemplate the ferment with equanimity.

. . . .

The Blumkin affair gave him an opportunity to strike. Jacob Blumkin, a high official of the G.P.U.'s foreign department, had a strange career behind him, and stranger still was his present role. Just before the revolution he had, as an adolescent, joined the terrorist organization of the Social Revolutionary Party. Something of a poet, he was a romantic idealist, with a precocious, simple-minded, and boundless devotion to his cause. In October 1917 he was among the Left Social Revolutionaries who made common cause with the Bolsheviks; and he represented his party on the Cheka under Dzerzhinsky—thus as a youngster of twenty—Revolution picks her lovers young!—he was one of the original founders of the Cheka. When his party broke with the Bolsheviks over the peace of Brest Litovsk, Blumkin shared his comrades' fierce conviction that in concluding that peace the Bolsheviks had betrayed the revolution. When his comrades decided to stage a rising against Lenin's government and to force the Soviet Republic into war against Germany, they assigned two men to make an attempt on the life of Count Mirbach, the German Ambassador in Moscow. Blumkin was one of the two. He succeeded; and this event was the signal for the insurrection which Trotsky suppressed. The Bolsheviks seized Blumkin and brought him before Trotsky.

It will be remembered that the Bolshevik party had itself been deeply divided over the Brest Litovsk peace; and so

---

[1] It was in the lobbies of the Central Committee that the writer, to his surprise, repeatedly heard that whisper.

although the Party outlawed the Left Social Revolutionaries, many Bolsheviks felt a warm sympathy for Mirbach's assassin, even though they condemned the deed. Trotsky appealed to the insurgents' revolutionary sentiment and sought to impress on them how misguided their action had been and to convert them to Bolshevik views. When Blumkin was brought before him, he engaged the young and impressionable terrorist in a long and serious argument. Succumbing to superior powers of persuasion, Blumkin repented and asked to be allowed to redeem himself. *Pro forma* he was condemned to death, and the German Government was even informed of his execution; but he was pardoned and given the chance to 'prove his devotion to the revolution'. He undertook to carry out the most dangerous missions for the Bolsheviks; and during the civil war he worked for them behind the lines of the White Guards. The Left Social Revolutionaries considered him a traitor and made several attempts on his life. After one attempt, while he was recovering in a hospital, they threw a hand grenade into his ward; he seized it and flung it out of the window at the very moment of the explosion. Rehabilitated by the Bolsheviks, he then served on Trotsky's military staff, studied at the Military Academy, gained some repute as a writer on military affairs, and was active in the Comintern. After the civil war he rejoined the Cheka or G.P.U. and was a senior officer of its Counter Intelligence Department. His faith in Trotsky knew no bounds; he was attached to the Commissar of War with the whole force of his emotional temperament. He was also in close friendship with Radek, whom he 'adored' and who was more accessible and responsive than Trotsky. When Trotsky and Radek went into opposition, Blumkin made no secret of his solidarity with them. Although the nature of his work prevented him from engaging in the Opposition's activities, he considered it his duty to make his attitude clear to Menzhinsky, the chief of the G.P.U. But, as his skill at counter intelligence was greatly valued, and as he did not participate in the Opposition's work and never committed any breach of discipline, he was allowed to hold his views and remain in his post. He stayed in the party and the G.P.U. even after the Opposition had been expelled.

In the summer of 1929, while travelling on duty from India to Russia, Blumkin stopped at Constantinople where, as

Trotsky maintains, he met Lyova by chance in the street. One may doubt whether this was in fact a chance encounter. It is implausible that Blumkin should have arrived in Turkey without intending to make contact with Trotsky. Having met the son, accidentally or not, he asked for an appointment with the father. Trotsky at first refused, considering the risk too great. But when Blumkin imploringly repeated the request, he agreed to receive him.

Blumkin arrived to pour out his heart to the man before whom eleven years earlier he had stood as Mirbach's assassin. He was, as were most Oppositionists, confused; and he was a prey to a conflict of loyalties. He found it hard to reconcile his position in the G.P.U. with his sentiment for the Opposition. He was torn between the Oppositionists who had capitulated and those who resisted, and between his faith in Trotsky and his friendship for Radek. He did not believe that the breach between the two was irreparable; and in his simple-mindedness he hoped to reconcile them. For hours he remained closeted with Trotsky, relating news from Moscow and listening avidly to Trotsky's arguments about the Opposition's responsibilities and duties and the futility of surrender.

He put before Trotsky his own *cas de conscience* and spoke of his wish to resign from the G.P.U. Trotsky firmly dissuaded him. Difficult as his situation was, Trotsky said, he must go on working loyally for the G.P.U. The Opposition was committed to defend the workers' state; and no Oppositionist should withdraw from any official post in which he acted in the broad interest of the state and not in that of the Stalinist faction. Was the Opposition not on the side of the Soviet Union in the conflict over the Manchurian Railway? Blumkin's activity was directed entirely against the external enemy; and it was perfectly consistent with the Opposition's attitude that he should carry on.

Blumkin accepted the advice and asked Trotsky to give him a message or instructions to Oppositionists at home. He also volunteered to help in arranging contacts and in organizing, with the help of Turkish fishermen, the *Bulletin's* clandestine despatch across the frontier.

Trotsky gave him the message, a copy of which is preserved in *The Archives*. The document contains nothing that could by any stretch of the imagination be described as conspiratorial. Its terms were so general and in part so trivial that it was feckless of Trotsky and Blumkin to take any risk at all in transmitting it. Trotsky forecast that in the autumn Stalin would find himself in great difficulties and that the capitulators would then realize how useless their surrender had been. He appealed, of course, to his followers to hold out, and poured scorn on the faint-hearted. He gave them notice of the attack on Radek he was preparing to publish and reproduced the gist of it. For the nth time he denied the charge, which Radek now echoed, that he was trying to form a new party; and he repeated that the Opposition remained part and parcel of the old party. He gave an account of what he was doing to set up the international organization of the Opposition and explained in humdrum detail the quarrels among the German, French, and Austrian Trotskyists and Zinovievists; he begged the Russians not to be disappointed by all this, but to be confident that the international Opposition would eventually emerge as a vital political force. It is a pathetic thought that the deportees placed such great hopes on this, and that Trotsky had to reassure them. In the whole message there was nothing that he had not said or was not about to say in public, especially in the *Bulletin*.[1] It is, of course, possible to suspect that he gave Blumkin more definitely conspiratorial instructions orally. But, strangely, even the G.P.U. never maintained that he did so; and the inner evidence of his attitude, activity, and correspondence indicates that he had in fact nothing to say to his followers in private that he did not or could not tell them in public. With this message in hand Blumkin departed in high spirits, confident that now he would be able to prove to Radek and others that their charges were

[1] The text of the message (undated) is in *The Archives*, Closed Section, Russian files. I have not been able to ascertain the exact date of Blumkin's visit. On internal evidence it appears to have occurred either in July or in August 1929. Trotsky's message contained in addition these organizational 'instructions': he asked his followers not to send him communications through Urbahns, the leader of the German *Leninbund*, with whom he was in political controversy; and he warned them to beware of one Kharin, an official of the Soviet Embassy in Paris, whom he denounced as a Stalinist *agent-provocateur*. (It was partly through Kharin, it seems, that Trotsky, immediately after the banishment, maintained contact with Russia.) These 'instructions' too had nothing conspiratorial or even confidential about them. In any movement of this kind warnings against an *agent provocateur* are normally given the widest publicity so as to put on guard as many people as possible.

groundless, that Trotsky was as loyal and as great a Bolshevik as ever, and that the Opposition should, under his leadership, restore its unity.

Shortly after his return to Moscow, Blumkin was arrested, charged with treason, and executed. It is not easy to determine how the G.P.U. came to learn about his moves. Some said that he had confided his secret to a woman whom he loved and who, being herself a secret service agent, denounced him. Others maintained that on his return Blumkin went straight to Radek who, fearing to draw suspicion on himself or being anxious to convince Stalin of the sincerity of his own recantation, betrayed his friend. This account gained wide credence and made Radek despised and hated. According to yet another version, upheld by Victor Serge, Radek's role was pitiable rather than sinister. Serge relates that back in Moscow Blumkin felt at once that the G.P.U. knew where he had been and that their agents were shadowing him in order to find out with whom of the Oppositionists he was in touch. Radek was worried about Blumkin's plight and advised him to approach Ordjonikidze, chairman of the Central Control Commission, and make a clean breast of everything. This was the only way, he allegedly said, in which Blumkin could save himself: Ordjonikidze, although a strict disciplinarian, was a conscientious and in his way even a generous man, the only one in the hierarchy who could be expected to treat the case sternly indeed but not without humanity. It was not known, however, whether Blumkin was arrested after or before he approached Ordjonikidze.[1] The whole puzzle may perhaps be explained more simply: the vigilant eye of a member of the Soviet Consulate in Constantinople may have caught sight of Blumkin taking the boat to Prinkipo; or an *agent provocateur* in Trotsky's house may have discovered the identity of the mysterious visitor with whom Trotsky had shut himself up for so many hours.

Blumkin 'carried himself with remarkable dignity' during the interrogation, relates a former G.P.U. officer. 'He went courageously to his execution and when the fatal shot was about to be fired he shouted, "Long live Trotsky!" '[2] More and more

[1] Trotsky's letter to Rosmer, 5 January 1930; *B.O.*, nos. 9 and 10, 1930; and Serge, *Mémoires d'un Revolutionnaire*, pp. 277–9.

[2] A. Orlov, *The Secret History of Stalin's Crimes*, p. 202.

frequently in years to come was this cry to resound amid volleys fired by execution squads.

This was the first execution of its kind. True, other Trotskyists had already paid for their convictions with their lives, perishing from hunger and exhaustion—the year before, for instance, Butov, one of Trotsky's secretaries, died in prison after a long hunger strike. Nevertheless, the rule that the Bolsheviks must never repeat the mortal error of the Jacobins and have recourse in their internecine struggles to execution had hitherto been respected, at least in form. Now that rule was broken. Blumkin was the first party member on whom capital punishment was inflicted for an inner party offence, an offence no graver than being in contact with Trotsky.

Stalin had been apprehensive lest the capitulations should blur the line of division between the Opposition and the party; and Blumkin's venture heightened his apprehension. He could not tolerate a senior G.P.U. officer on active service visiting Trotsky in a comradely manner and mediating between Trotsky and the capitulators—to tolerate this would be to make a mockery of all the official accusations of Trotsky and encourage further contacts. Stalin himself may not have believed in the relatively innocuous character of Blumkin's mission and of Trotsky's message to the Opposition. The thought may have occurred to his suspicious mind that it would be unsafe to assume that Mirbach's assassin would never again vent his simple but strong political passions in a terrorist act. In any case Blumkin's execution was to serve as a warning to others: it was to show them that official charges of counter-revolution must not be trifled with, that paragraph 58 was paragraph 58, and that henceforth comradely connexions with the Prinkipo outcast would be punished with the whole severity of a garbled and perverted law. Curiously enough, no capital punishment was as yet inflicted on the avowed Trotskyists, who from their prisons and punitive colonies were in communication with their leader, who sent him collective greetings on October anniversaries and May Days, and whose names appeared under articles and 'theses' in the *Bulletin Oppozitsii*. For the time being, the warning was meant only for party members, holders of official posts, especially in the G.P.U., and reinstated capitulators. The line of division between party and Opposition was redrawn in blood.

Trotsky learned of the execution from an anonymous Oppositionist who, being still in government service, was on an official mission in Paris.[1] But Moscow was silent; and when a rumour percolated through to the German Press, communist papers denied it. For several weeks Trotsky waited for further information, and in his letters to Russian followers made no allusion to Blumkin—until early in January 1930 a message from Oppositionists in Moscow dispelled all doubts. Trotsky at once disclosed the circumstances of his meeting with Blumkin. He declared that it was Stalin personally who had ordered the execution and that Yagoda gave effect to the order without even referring it to Menzhinsky, the nominal head of the G.P.U. The *Bulletin* published the correspondence from Moscow, the writers of which maintained that it was Radek who had betrayed Blumkin. Trotsky himself, on second thoughts, doubted whether this was so and intimated that Radek had probably acted irresponsibly and stupidly but in good faith. 'Blumkin's misfortune', Trotsky wrote, 'was that he trusted Radek and that Radek trusted Stalin.'

Trotsky enjoined his followers in the West to raise a 'storm of protests'. 'The Blumkin affair', he wrote to Rosmer on 5 January 1930, 'should become the Sacco-Vanzetti affair of the Left Opposition.' Some time earlier the execution in Boston of Sacco and Vanzetti, two Italian-American anarchists, had been the object of a memorable world-wide protest raised by communists, socialists, radicals, and liberals. Trotsky's call found no response. Blumkin's fate did not arouse even a fraction of the indignation that the execution of Sacco and Vanzetti had provoked. It was far easier to arouse the conscience of the left against a miscarriage of justice by the judiciary of a bourgeois state than to move it against a *Justizmord* committed in a workers' state. Barely a few weeks later Trotsky was already having to protest, and to ask others to protest, against two further executions of Oppositionists and against harsh reprisals to which Rakovsky and his friends were subjected.

[1] The news was transmitted to Trotsky by R. Molinier in a letter of 10 December 1929, together with a rather gloomy account of the Opposition's disintegration. *The Archives*, Closed Section.

And once again he failed even to dent the stony indifference of those whom he had hoped to move.[1]

· · · · · · · · · ·

The year 1929 ended in the Soviet Union with an upheaval the violence of which surpassed all expectations. Early in the year Stalin's policy had still been hesitant and uncertain. The industrial drive was gaining momentum, but the government had not yet thrown all caution to the wind; in April, the Sixteenth Party Conference called for speedier collectivization, but proclaimed that the private farms would for many years yet predominate in the rural economy—the Five Year Plan provided for the collectivization of only 20 per cent. of all smallholdings by 1933; the kulak was to pay higher taxes and to deliver more grain, but there was no thought yet of his 'liquidation'. By the end of the year it was as if a whirlwind had swept away these plans and the prudence that had inspired them. The industrial drive burst all bounds: again and again the targets were raised; and the call went out that the Plan must be carried out in four, three, or even two and a half years. On the twelfth anniversary of the revolution Stalin, confronted by the 'difficulties' Trotsky foresaw, the peasantry's refusal to deliver grain, pronounced death sentence on private farming: 'Immediate and wholesale collectivization' was the order of the day; and only four months later he announced that 50 per cent., about 13 million, of the farmsteads had already been collectivized. The whole power of state and party drove the kulaks from the land and forced millions of other peasants to pool all their possessions and accept a new mode of production.[2]

Almost every village became a battlefield in a class war, the like of which had never been seen before, a war which the collectivist state waged, under Stalin's supreme command, in order to conquer rural Russia and her stubborn individualism. The forces of collectivism were small but well armed, mobile,

[1] In no. 10 of the *B.O.* Trotsky named the two executed men as Silov and Rabinovich, saying that they had been charged with 'sabotage of railway transport'. According to Orlov (op. cit., loc. cit.) the real 'crime' of Rabinovich, himself a G.P.U. officer, was that he had informed clandestine Trotskyist circles in Moscow of Blumkin's execution.

[2] See *K.P.S.S. v Rezolutsyakh*, vol. II, pp. 449–69, 593ff.; Stalin, *Sochinenya* vol. XII, pp. 118–35; *Pravda*, 6 January 1930; Deutscher, *Stalin*, 317–22.

and directed by a single will; rural individualism, its great strength scattered, was caught by surprise, and was armed only with the wooden club of despair. As in every war so in this, there was no lack of manœuvres, inconclusive skirmishes, and confused retreats and advances; but eventually the victors seized their spoils and took uncounted multitudes of prisoners, whom they drove into the endless and empty plains of Siberia and the icy wastes of the Far North. As in no other war, however, the victors could neither admit nor reveal the full scope of hostilities; they had to pretend that they carried out a salutary transformation of rural Russia with the consent of the overwhelming majority; and so even after several decades the precise numbers of the casualties, which must have gone into millions, remained unknown.

Such were the suddenness, the magnitude, and the force of the upheaval that few who witnessed it were able to absorb and focus mentally its immensity. Until recently the Trotskyist Opposition could maintain that Stalin, by initiating the left course, was only giving effect to its demands; but the Great Change exceeded those demands to an extent that took away the breath of Trotskyists and Stalinists alike, not to speak of the Bukharinists. Among the Trotskyists, the conciliators showed a clearer awareness of the scope and the finality of events; the resisters still clung to premises and reasonings formed in earlier years. Rakovsky, for instance, treated Stalin's orders for the annihilation of the kulaks as 'ultra-left rhetoric' and asserted that 'the specific weight of the wealthy farms in the national economy will grow even further, despite all the talk about fighting agrarian capitalism'.[1] Just before the twelfth anniversary of the revolution, Trotsky himself claimed that 'the slow development of the rural economy . . . and the difficulties which the countryside experiences favour the growth of the power of the kulaks and the progress of their influence . . .'.[2] He did not imagine that at a stroke, or within a very few years, 25 million private smallholdings could be wiped out by force.

At the beginning of 1930, however, Trotsky began to realize what was happening and in a series of essays devoted to a critique of the Five Year Plan he evolved a new line of attack on Stalin's policy. The new criticism was marked by dialectical

[1] *B.O.*, no. 7, 1929.   [2] *Écrits*, vol. I, p. 76.

duality: he made a sharp distinction between the 'socialist-progressive' and the 'bureaucratic-retrograde' trends in the Soviet Union and illumined their perpetual conflict. He began, for instance, an essay on 'Economic Recklessness and its Perils' with these words:[1]

The success of the Soviet Union in industrial development is acquiring global historical significance. Social Democrats who do not even try to evaluate the tempo which the Soviet economy proves itself capable of attaining deserve but contempt. That tempo is neither stable nor secure . . . but it provides practical proof of the immense possibilities inherent in socialist economic methods. . . . On the basis of the Soviet experience it is not difficult to see what economic power a socialist bloc comprising central and eastern Europe and large parts of Asia would have wielded if the Social Democratic Parties had used the power that the 1918 revolution had given them and carried out a socialist upheaval. The whole of mankind would have had a different outlook by now. As it is, mankind will have to pay for the betrayal committed by the Social Democratic Party with additional wars and revolutions.

Having so emphatically restated his appreciation of the socialist trend in Soviet developments, he attacked Stalin's domestic policy in the same terms in which he had characterized the new Comintern line—as an 'ultra-left zigzag that had come to replace the previous rightist zigzag'. This was consistent with Trotsky's view that Stalin, as a 'centrist', acted under alternate pressures from right and left, a view which properly described Stalin's place in the inner party alignments of the nineteen-twenties, but fitted the realities of later years less well. By and large Trotsky still held that intensive industrialization and collectivization were merely a transient phase of Stalin's policy. He was not aware, and he was never to become fully aware, that in 1929–30 Stalin had gone beyond a point of no return, where he could neither halt the industrial drive in its tracks nor, having destroyed the kulaks, try and make peace with them. This basic error in Trotsky's judgement, to which we shall return later, does not, however, invalidate his specific criticisms, in which he anticipated most of the revisions of policy that Stalin's successors were to carry out after 1953. Just as in the nineteen-twenties Trotsky was the pioneer of primitive socialist

[1] The essay was written in February 1930 and published in the *B.O.*, no. 9.

accumulation, so in the early nineteen-thirties he was the precursor of economic and social reforms that were to be undertaken only several decades later.

He attacked at the outset the rate set by the first Five Year Plan, in its final version, for industrial expansion.[1] From the 'snail's pace', he observed, Stalin had switched over to the 'race-track gallop'. In its early versions the Plan had aimed at an 8–9 per cent rate of annual expansion; and the Opposition's proposal to double the tempo had been decried as fanciful, irresponsible, and dangerous. Now the tempo was trebled. Instead of striving for *optimum* results, Trotsky pointed out, the planners and managers were ordered to strain always for the *maximum*, regardless of the fact that this threw the national economy out of balance, and so reduced the effectiveness of the drive. Production targets grossly exceeded available resources; and so an incongruity arose between manufacturing and primary production, between heavy and light industry, and between investment and private consumption. Worse still was the contrast between the advance of industry and the lag in farming. There is no need to dwell here on these and other disproportions which Trotsky often analysed in detail—it has since become a truism that these disproportions did indeed mark and mar the whole process of industrialization in the Stalin era. But, as happens so often, the truisms of one generation were the dreaded heresy of its predecessor; and communists, but not only they, received Trotsky's criticisms with indignation or derision.

Yet when one re-examines after this lapse of time what Trotsky said on these matters, one is struck by his political restraint rather than by his polemical heat. He usually prefaced almost every piece of criticism by emphatic acknowledgement of the progress achieved under the direction of his adversary, although he insisted that the mainspring of progress lay in the national ownership and planning of industry and that Stalin not only used but also abused these advantages of the Soviet economy. He did not believe that the administrative whip did or

[1] See also Trotsky's 'Open Letter' to party members in *B.O.*, no. 10 (April 1930); his comments on the XVI Congress, ibid., nos. 12–13 (June–July 1930); and 'The Successes of Socialism and the Perils of Recklessness', ibid., nos. 17–18 (November–December 1930).

could accelerate the industrial advance—the whip was all too often the very cause of a halt and a breakdown. National ownership led to central planning and required it; but bureaucratic over-centralization led to the concentration and magnification of the errors committed by those in power, to paralysis of social initiative, and to tremendous wastage of human and material resources. An irresponsible and 'infallible' Leader had to boast all mistakes and reverses out of existence and to flaunt all the time spectacular achievements, unheard-of records, and dazzling statistics. Stalinist planning dwelt on the quantitative side of industrialization to the exclusion of everything else; and the higher the quantity of goods that had to be produced at any price, the lower the quality. For rational planning a comprehensive system of economic coefficients and tests was needed, which would measure continuously not merely the growth of production but changes in quality, costs, the purchasing power of money, comparative rates of productivity, &c. Yet all these facets of the economy were wrapt in obscurity: Stalin conducted the industrial drive 'with all lamps extinguished', amid a complete blackout of vital information.

Trotsky's criticism of the collectivization was even more thoroughgoing. He condemned the 'liquidation of the kulaks' as a monstrosity; and he did so long before the horrors that attended it had become known. In the years when he himself was stigmatized as the 'enemy of the peasantry', he had urged the Politbureau to raise the taxation of well-to-do farmers, to organize farm labourers and poor peasants, to encourage them to form collective farms on a voluntary basis, and to throw the state's resources (agricultural machinery, fertilizers, credit and agronomic assistance) behind the collective farms so as to promote them in their competition with private farming. These proposals had expressed the full extent of his anti-kulak policy; and he had never gone beyond them. It had never occurred to him that a social class as numerous as the rural bourgeoisie could or should be destroyed by decree and violence—that millions of people should be dispossessed and condemned to social, and many also to physical, death. That socialism and private farming were ultimately incompatible and that the capitalist farmer would vanish in a society evolving towards socialism, had, of course, been an axiom of Marxism and Leninism. But Trotsky, like all Bolsheviks until quite recently, envisaged this as a gradual process, in the course of which the smallholder would succumb to the more productive collective method of farming in a way similar to, but far less painful than, the way the independent artisan and small farmer had succumbed to modern industry and large-scale agriculture under capitalism.

There was therefore no element of demagogy in the angry denunciation with which Trotsky met the liquidation of the kulaks. Not only was this to him a malignant and sanguinary travesty of all that Marxism and Leninism had stood for— he did not believe that the *kolkhozy* which Stalin was forcing into existence would be viable. He argued that collectivized agriculture required a technological base far superior to that on which individual husbandry had rested; and such a base did not exist in the Soviet Union: the tractor had not yet replaced the horse.[1] In an expressive simile (of which it could, however, be said that *comparaison n'est pas raison*) he asserted that without modern machinery it was just as impossible to turn private smallholdings into a viable collective farm as it was to merge small boats into an ocean-going liner. Stalin intended, of course, to supply the machinery over the years, as he eventually did. What Trotsky maintained was that collectivization should not outrun the technical means needed for it. Otherwise, the collectives would not be economically integrated; their productivity would not be higher than that of private farming; and they would not bring the peasants the material advantages which could compensate them for the loss of private property.[2] Meantime, before the collectives were technologically in-

---

[1] *Pravda* of 15 January 1930 estimated that 1,500,000 tractors were needed for the full collectivization of Soviet farming. This degree of mechanization was not reached until 1956, when the 'Tractor Park' (calculated in 15 h.p. units) passed the 1,500,000 mark—in actual, nearly 30 h.p., units, it consisted of 870,000 tractors.

The annual output of (15 h.p.) tractors was only a little over 3,000 in 1929 and 50,000 in 1932. The amounts of other available agricultural machinery were altogether negligible. At the beginning of the first Five Year Plan, in 1928, there were fewer than 1,000 lorries on the farms; and there were only 14,000 in 1932. *Narodnoe Khozyaistvo S.S.S.R. v 1958 g.*, Soviet Statistical Yearbook, 1959, pp. 243, 487.

[2] 'The collectivization of *sokhas* [wooden ploughs] . . . is a fraud', Trotsky wrote. His argument was controverted by some Trotskyist economists (see, e.g. Ya. Gref's study on collectivization and overpopulation in *B.O.*, no. 11) and, of course, by the Stalinists, who maintained that the collective farm, even when technologically

tegrated, the peasantry's resentment would show in a decline
or stagnation of agricultural output; and it would threaten to
blow up the collectives from the inside. So acute was Trotsky's
insight into the state of mind of the peasantry that from
Prinkipo he warned Moscow about the coming calamitous
mass slaughter of cattle; and he did this in plenty of time, five
years before Stalin admitted the fact.[1] Even much later Trotsky
remained convinced that the collectivist structure of farming
was chronically in a state of near collapse.

In retrospect it may appear that Trotsky took too black a
view: the collective farms did not collapse, after all. Yet
Stalin's rural policy throughout the nineteen-thirties, with its
whimsical combination of massive terror and petty concessions,
was dictated precisely by the fear of a collapse: only with iron
bands could he hold together the collective farm. The decline
and subsequent stagnation in farm output were all too real,
and became the great theme of official policy twenty-five and
thirty years later.

The state of affairs in the country reacted upon all aspects of
national policy. Industrialization proceeded on a dangerously
narrow and shattered agricultural base, amid famines or a
perpetual dearth of foodstuffs. It was therefore accompanied by
a universal and almost zoologically fierce scramble for the
necessities of life, by widespread discontent, and by low
productivity of labour. The government had continuously to
quell the discontent and to force up productivity by intimidation and subornation. The violent shock of 1929–30 drove the
Soviet Union into a vicious circle of scarcities and terrors from
which it was not to break out for a long time to come.

primitive, would be more productive than the old smallholding. Trotsky's critics
argued from an analogy with British manufacture which, even before the Industrial
Revolution (when it was still manufacture in the strict, etymological sense), was
more productive than individual handicraft, because, as Marx pointed out in *Das
Kapital*, it enjoyed the advantages first of 'simple co-operation,' and then of the
division of manual labour. In strict theory, Trotsky's critics were right: collectivization, even without the prior existence of a technological basis proper to it,
should result in higher productivity, as it did in China for a time, during the middle
nineteen-fifties. Practically, however, and as far as the collectivization of 1929–32
was concerned, Trotsky was right: any advantages which the collective farm might
have obtained from the co-operation and division of manual labour were nullified
by the peasants' resentful attitude towards work and by the initial destruction of
agricultural stock.

[1] *B.O.*, no. 9, 1930.

Stalin had now proclaimed the end of N.E.P. and the abolition of the market economy. Surveying Trotsky's views at an
earlier stage we saw that in these there was 'no room for any
sudden abolition of N.E.P., for the prohibition of private trade
by decree . . .' and that socialist planning 'could not one day
supersede N.E.P. at a stroke, but should develop within the
mixed economy until the socialist sector had by its growing
preponderance gradually absorbed, transformed, or eliminated
the private sector and outgrown the framework of N.E.P.'[1]
Trotsky still stuck to this view. He considered the 'abolition of
N.E.P.' a coinage of the bureaucratic brain—only a bureaucracy which, through long neglect of industrialization and a
faulty approach to the peasantry, had failed to cope with the
forces of the market economy and allowed these to grow out of
control could try to decree the market out of existence. But
'thrown out by the door the market would come back by the
window', Trotsky said. As long as farming was not socialized
organically and securely, and amid an all-round scarcity of
goods, it was impossible to eliminate the play of supply and
demand and to substitute for it the planned distribution of
goods. The spontaneous pressures of the market were bound to
break through in farming first, then in those areas where
farming and industry overlapped, and finally even within the
nationalized sector of the economy, where they would often
upset and distort planning. There was ample evidence of this,
especially during the early nineteen-thirties, in the chaos of
official and unofficial prices of consumer goods, in a fantastic
spread of black markets, in the depreciation of the rouble, and
in a steep fall of the purchasing power of wages. The planners
worked 'without yardstick and scale', unable to assess genuine
values and costs and to appraise productivity. 'Regain yardstick and scale', was Trotsky's insistent advice. Instead of
pretending that they had overcome the pressures of the market,
the planners would do better to acknowledge their existence, to
make allowance for them, and to try and bring them under
control. Even in later years, after the runaway inflation of the
early nineteen-thirties was overcome, these criticisms retained
validity; and here, too, much of what Soviet economists said,
in the first decade after Stalin, about the importance of value

[1] *The Prophet Unarmed*, pp. 100–1.

measurement and cost accountancy sounded like an echo of Trotsky's arguments.

The Stalinist blackout of economic information obscured other crucial questions as well. Who was paying for the industrialization, which social classes—and how much? Which classes and groups benefited from it—and to what extent? In the early nineteen-twenties the leaders of the Opposition, especially Preobrazhensky, had maintained that the peasantry would have to contribute heavily to the investment funds of the nationalized industry. Stalin hoped to ensure through collect-ivization that the peasantry should indeed make this con-tribution, by increasing the output and the supply of foodstuffs and raw materials. But the peasantry foiled him. 'Let my soul perish with the commissars!' was the cry of the smallholder as he left his holding; and, although he did not manage to bring down the pillars upon the collectivist state, he refused to yield up to it a large part of the sinews of industrialization that he was expected to provide. This was what the destruction of farmstock and the decline in output amounted to in practice.

All the heavier was the burden the urban working class was called upon to carry. The major part of industry's huge invest-ment fund was in effect a deduction from the national wages bill. In real terms, a greatly increased working class had to subsist on a shrunken mass of consumer goods while it built new power stations, steel mills, and engineering plants.[1] Ten years earlier Trotsky had said that the working class 'can

---

[1] The urban population of the U.S.S.R. rose in the course of the nineteen-thirties from about 30,000,000 to nearly 60,000,000; and the most intensive rise occurred in the first half of the decade. The gross output of agriculture fell from 124 in 1928 (1913—100) to 101 in 1933, and was only 109 in 1936, while that of cattle farming declined from 137 in 1928 to 65 in 1933 and then rose slowly to 96 in 1936. Throughout the nineteen-thirties the grain crops did not exceed the pre-1913 level or were somewhat below it. (*Narodnoe Khozyaistvo S.S.S.R.*, pp. 350–2.) In 1928, however, the marketable surplus of farm produce amounted to only half the pre-revolutionary volume; and only the requisitions of 1929–32 doubled (approxi-mately) the grain stocks available for urban needs. Supplies of sugar, meat, and fat fell very sharply in the years of the first plan (ibid. p. 302). The output of cotton clothing declined or was stationary between 1928 and 1935 (ibid. p. 274). The same is true of footwear, the scarcity of which was aggravated by the dis-appearance of home industry. (Ibid. p. 293.) Throughout the decade, marked by shortage of labour and materials on which heavy industry had first claim, urban overcrowding, which had been bad enough even earlier, was calamitous. New building provided not more than an average of four square yards of space per new town-dweller.

approach socialism only through the greatest sacrifices, by straining all its strength, and giving its blood and nerves. . . .' Stalin now exacted those sacrifices in blood and nerves. 'There may be moments', Trotsky said in 1923, 'when the government pays you no wages, or when it pays you half your wages and when you, the worker, have to lend the other half to the state in order to enable it to rebuild the nationalized industry.'[1] Stalin now seized that 'other half' of the worker's wages. But whereas Trotsky had excused his proposal by the ruin of the economy after war and civil war and sought to obtain the worker's consent to this method of accumulation, Stalin did what he did after many years of reconstruction and told the worker that his real earnings were doubled and that he was entering the promised land of socialism. For a time inflation concealed the realities from the workers, on whose enthusiasm, endurance, or at least willingness to work the success of the Plan depended.[2]

At the outset the Plan was launched in a spirit which if not egalitarian was nevertheless one of common service and common sacrifice, untarnished by any shocking inequality of rewards. This spirit stirred the fervour of the *Komsomoltsy* and *Udarniki* who rushed to build the *Magnitostroys* and the *Tractorstroys*.[3] But as the first elation sagged and as the great weariness of the workers began to show, the government prodded them on with incentive wages, piece rates, Stakhanovism, rewards for production records, &c. On a par with the bureaucracy and the managers, the labour aristocracy attained

---

[1] *The Prophet Unarmed*, p. 102.

[2] In the resolution of the Central Committee of 10 January 1933 (*K.P.S.S. v Rezolutsyakh*, vol. II, p. 723) the 'average' rise of the incomes of workers and peasants under the first Five Year Plan is given as 85 per cent. In the same period the total sum of retail sales by state-owned and co-operative stores rose from nearly 12 billion to over 40 billion roubles. (*Narodnoe Khozyaistvo v S.S.S.R.*, p. 698.) As apart from bread which was rationed at a fixed price, and perhaps of potatoes, the mass of goods sold was either stationary or rose only to a small extent over these years, it follows that the purchasing power of the rouble, even if measured only in controlled prices, fell to between one-fourth and one-third of that of 1928. In uncontrolled prices the fall was far steeper. Thus, even if the 'average' nominal wage was doubled, the average real wage was in 1932 only half the 1928 wage. It was therefore in a literal sense that, by means of inflation, Stalin took half the worker's wage to finance industrialization.

[3] This enthusiasm was fed by the illusion that the Soviet Union would 'catch up with and surpass' western industrial countries *in two or three years* and so build 'an armoured wall around socialism in one country'. *B.O.*, no. 17–18, 1930.

a markedly privileged status. Henceforth, while Stalin hurled imprecation after imprecation upon 'petty bourgeois levellers', the anti-egalitarian trend gained immense force. Against it Trotsky invoked 'the tradition of Bolshevism which has been one of opposition to labour aristocracy and bureaucratic privilege'. He did not preach levelling. 'It is altogether beyond dispute', he pointed out, 'that at a low level of productive forces and consequently of civilization at large, it is impossible to attain equality of rewards.' He even stated that the egalitarian wages policy of the early revolutionary years had gone too far and impeded economic progress. Yet he held that a socialist government was in duty bound to keep inequality within the limits of what was necessary, to reduce it gradually, and to defend the interests of the great unprivileged mass. 'In the conflict between the working woman and the bureaucrat, we, the Left Opposition, side with the working woman against the bureaucrat . . . who seizes her by the throat. . . .' In the fact that Stalin acted as the protector of privilege, he saw a 'threat to all the conquests of the revolution'.[1]

Trotsky now also redefined his view of proletarian democracy. Only when the toilers were free to express their demands and criticize those in power, he argued, could they arrest the growth of privilege; and from the standpoint of socialism the supreme test 'by which the country's economic condition should be judged is the standard of living of the workers and the role they play in the state'. If in the years of N.E.P. he held that only the strength of proletarian democracy could counterbalance the combined forces of the N.E.P.-men, the kulaks, and the conservative bureaucrats, he now regarded that democracy as the only political setting within which a planned economy could attain its full efficiency. It was therefore a vital economic, and not merely a political, interest of the U.S.S.R. that proletarian democracy be revived. Contrary to a myth of vulgar Trotskyism, he did not advocate any 'direct workers' control over industry', that is management by factory committees or works' councils. This form of management had failed in Russia shortly after the revolution; and Trotsky had ever since been a most determined advocate of one-man management and central control, arguing that management by factory

[1] *B.O.*, no. 23 (August 1931) and no. 27 (March 1932).

committees would become possible only if and when the mass of producers became well educated and imbued with a strong sense of social responsibility. He had also been absolutely opposed to the 'anarcho-syndicalist' schemes of the Workers' Opposition for the transfer of industrial management to trade unions or 'producers' associations'. He did not significantly alter these views when he found himself in Opposition and exile. He conceived proletarian democracy as the workers' right and freedom to criticize and oppose the government and thereby to shape its policies, but not necessarily as their 'right' to exercise direct control over production. He saw in central planning and central direction the essential condition of any socialist economy and of any economy evolving towards social- ism. But he pointed out that the process of planning, to be efficacious, must proceed not only from above downwards but also from below upwards. Production targets must not be decreed from the top of the administrative pyramid, without preparatory nation-wide debates, without careful on-the-spot assessment of resources and capacities, without the preliminary testing of the state of mind of the workers, and without the latter's genuine understanding of the plan and willingness to carry it out. When working-class opinion was not allowed to check, correct, and modify schemes presented by a planning authority, the severe disproportions which characterized the Soviet economy under Stalin were inevitable.[1]

Trotsky turned his criticism against the assumption of national self-sufficiency which underlay Stalin's conduct of economic affairs. Socialism in one country remained to him a 'reactionary, national-socialist utopia', unattainable no matter whether it was to be striven for at racing speed or at the snail's pace. With an emphasis which was sometimes exagger- ated or misplaced, he pointed out that the Soviet Union could not with its own resources and by its own exertions surpass or even reach the productivity of the advanced western capitalism, the productivity which was the *sine qua non* of socialism. The spread of revolution remained in any case the essential con- dition for the achievement of socialism in the U.S.S.R. The Stalinist isolationism affected not only the grand strategy of revolution and of socialist construction but even immediate

[1] Loc. cit.

trade policies: Stalin took no account of the advantages of 'international division of labour', and he virtually ignored the importance of foreign trade for Soviet industrialization, especially after the Great Slump when the terms of trade turned sharply against the Soviet Union. Trotsky then urged Moscow to enhance its trading position by political means and appeal to the many millions of the unemployed workers of the West to raise a clamour for trade with Russia (and for export credits) which would assist Russia but would also help to create employment in the capitalist countries. In his own name and on behalf of his tiny organization, Trotsky published several persuasive manifestoes to this effect; but the idea evoked no response from Moscow.[1]

These detailed criticisms culminated in Trotsky's sustained and passionate protest against the moral discredit that Stalin's policy was bringing upon communism. In 1931 Stalin proclaimed that the Soviet Union had already laid the 'foundations of socialism'—even that it had 'entered the era of socialism'; and his propagandists had to back up this claim by contrasting a fantastically bright image of Soviet society with a crudely overdrawn picture of the miseries of life under decaying capitalism.[2] Exposing the double distortion, Trotsky pointed out that to tell the Soviet masses that the hunger and the privations, not to speak of the oppression, which they endured amounted to socialism was to kill their faith in socialism and to turn them into its enemies. In this he saw Stalin's 'greatest crime', for it was committed against the deepest hopes of the working classes and threatened to compromise the future of the revolution and of the communist movement.[3]

.   .   .   .

We have said that Trotsky's criticism was in all its aspects consistent with the tradition of classical Marxism and also that it anticipated the reforms of the post-Stalin era. The question

---

[1] 'Trotsky urges us to make ourselves more dependent on the capitalist world', Kaganovich said, to which Trotsky retorted that 'autarchy is Hitler's ideal, not Marx's and not Lenin's'. 'Sovetskoe Khozyaistvo v Opasnosti' in *B.O.*, no. 31, November 1932. The value of Soviet exports shrank to one-third and that of imports to one-fourth between 1930 and 1935. Part of this fall was due to adverse trade terms.

[2] See, for instance, *K.P.S.S. v Rezolutsyakh*, vol. II, pp. 717-24. *Pravda, Bolshevik*, and the entire Soviet Press of the nineteen-thirties are full of this contrast.

[3] *B.O.*, loc. cit. and *passim*.

may now be asked whether or to what extent it was relevant to the situation of the nineteen-thirties? Were Trotsky's proposals practicable at the time when he made them? Was a deep divorce between Marxist theory and the practice of the Russian Revolution not an inherent characteristic of that era? And had circumstances not made that divorce inevitable? Only very few questions with which the historian has to deal can tax his confidence in his own judgement as severely as these questions do. Trotsky himself, in his less polemical moods, stressed that the immense difficulties which beset the Soviet Union were rooted in its poverty, backwardness, and isolation. His main charge against Stalin's rule was that it aggravated these difficulties rather than created them; and it was not easy for Trotsky, nor is it for the historian, to draw a line between the 'objective' and the 'subjective' factors of the situation, between the miseries to which the Russian Revolution was heir and those which Stalinist arbitrariness and cruelty produced. Moreover, there was a real 'unity of opposites' here, a dialectical interplay of the objective and the subjective: bureaucratic arbitrariness and cruelty were themselves part and parcel of the Russian backwardness and isolation—they were the backward responses of the inheritors of the revolution to native backwardness.

It was now the commonly (though in part only tacitly) held view of both Trotsky and Stalin that the Soviet Union could achieve rapid industrial ascendancy only through primitive socialist accumulation, a view historically justified by the fact that no underdeveloped nation has, in this century, achieved an advance comparable to Russia's on any other basis. Primitive accumulation, however, presupposed that the workers and peasants should bear more than the 'normal' burden of economic development. Some of the basic disproportions of Stalinist planning were inherent in these conditions. Investment had in any case to expand much faster than consumption. Priority had to be given to heavy over light industry. The theorists of the Opposition had argued that with industrialization the national income would grow so rapidly that popular consumption would rise together with investment, even if not at the same rate. Instead, consumption shrank disastrously in the crucial years of the early nineteen-thirties. Trotsky maintained that this would have been avoided and that the industrial drive would have been carried on under less severe strains and stresses if it had been started several years earlier and in a more rational manner. The argument was plausible; but its truth could not be proved. The Stalinist counter argument, held esoterically rather than stated openly, was also plausible: it was that the Great Change would have been just as cataclysmic even if it had been initiated earlier and more mildly. The threat of famine had hung over urban Russia most of the time since the revolution (and it had recurred periodically before the revolution). Industrialization and the rapid growth of the urban population were, in any case, bound to aggravate it, as long as agriculture remained as fragmented and archaic as it was. Having refused to allow capitalist farming to take charge of the provisioning of the feverishly expanding towns, the Bolsheviks had to opt for collectivization. If they had attempted the gradual collectiviza-tion for which Trotsky stood, so the Stalinist argument went on, they would have had the worst of both worlds: the great mass of the smallholders would have been antagonized anyhow; and progress would have been, as under capitalist farming, too slow to secure the provisioning of the towns during rapid industrialization. Trotsky believed, on the contrary, that it was possible to induce the peasantry to a voluntary and economic-ally sound collectivization; and it is a moot point whether he did not underrate the extent to which any form of collectivism offended the stubborn 'irrationality' of the muzhik's attach-ment to private property. Stalin acted on the Machiavellian principle that nothing was as dangerous for a ruler as to offend and at the same time to seek to propitiate his enemies; and to Stalin his subjects became his enemies. He hurled all the resources of his power against the smallholders; and a whole generation was to labour under the consequences of the economic cataclysm. Yet at this price Stalin, from his view-point, scored an immense political gain: he broke the backbone of the archaic rural individualism which threatened to thwart industrialization. Having made this gain, he could not give it up; he had to defend it tooth and nail.

Trotsky did not believe in the solidity of this Machiavellian achievement; he denied to the end that Stalin had vanquished the peasantry's individualism. Convinced that the latter was still able to destroy the collective farms or to bend them to its own interests and needs, he forecast that a new class of kulaks would rise within the kolkhozy and take command.[1] Here again Trotsky grasped a real tendency; but he overemphasized its strength. The peasantry's acquisitiveness did indeed reassert itself in many ways, and Stalin had to struggle against the resurgence of kulaks in the kolkhozy. By a combination of economic measures and terror, however, he succeeded in keeping the recrudescence of private property within narrow and severely restricted bounds; and the peasantry's individualism was never to recover from the mortal blow he inflicted on it, although its death rattle was to sound in Russia's ears for a quarter of a century.

From exile Trotsky repeatedly implored the Stalinist Politbureau to withdraw from their savage enterprise, to call a halt to the barbarous warfare against rural barbarism, and to revert to the more civilized and humane courses of action to which their Marxist-Leninist heritage committed them. He urged the Politbureau to initiate a great act of reconciliation with the peasantry, to declare before the whole nation that in imposing collectivization they had acted wrongly, and that the peasants who wished to leave the collective farms and resume private farming were free to do so. He had no doubt that this would result in the dissolution of many or perhaps most collective farms; but as, in his view, these were not viable anyhow, little would be lost; and the kolkhozy that survived (if they were supplied with machines, credits, and agronomic assistance and so enabled to offer their members material benefits which were beyond the smallholder's reach) could still become the pioneers of a genuine, voluntary collectivist movement, which would in time transform the whole of agriculture and raise its productivity to the level required by a modern and expanding economy. This, Trotsky proclaimed, was what the Opposition would do if it returned to power.[2]

For the Stalinist Politbureau it was too late to seek such a reconciliation with the peasantry. Ever since the autumn of 1929 all the forces of party and state had been fully engaged in

[1] See, e.g., the chapter on 'Social Contradictions in the Collective Village' in *The Revolution Betrayed*, pp. 128–35.

[2] *B.O.*, no. 29–30, 1932.

the struggle, and an attempt to disengage them for a deep retreat could well end in their rout. So many had from the outset been the victims of the campaign, so bitter were the passions aroused, so much violence had been inflicted on the villagers and so fierce was their urge for revenge, so immense and bloody was the upheaval, that it was more than doubtful whether any rational way out could be found as long as the generation that had experienced the shock held the stage. If the government had proclaimed that the peasants were free to leave the collective farms, the whole agricultural structure would have come down with a crash, and hardly any collectives would have survived. It would then have taken time before private farming got back into its grooves and began to work in its accustomed ways. Meanwhile, production and supply of food would have further declined and industrial development would have suffered a severe setback. Nor was it likely that a mass exodus from the collective farms could proceed peacefully. The peasants would have felt entitled to get their own back on government and party. The reconciliation would have required that the expropriated and the deported be amnestied and indemnified; and one may well imagine the mood in which trainloads of deportees returning from concentration camps would have been received in their native villages. De-collectivization might have let loose violence as furious as that which had accompanied collectivization. Perhaps a new government with a clean record, a government formed by the Opposition, could have sought to appease the country without bringing it to the brink of counter-revolution—this was what Trotsky believed. For Stalin's government any such attempt would have been suicide. Any sign of weakness on its part would have set ablaze the hatreds smouldering in millions of huts. There was nothing left for Stalin but to remain locked in the struggle, even though, as he confessed to Churchill years later, this was more frightful than even the ordeals of the Second World War.[1]

---

[1] 'It was now past midnight . . .' writes Churchill. "Tell me," I asked, "have the stresses of this war been as bad to you personally as carrying through the policy of the Collective Farms?" This subject immediately roused the Marshal. "Oh, no," he said, "the Collective Farm policy was a terrible struggle." . . . "Ten millions [of peasants]", he said, holding up his hands. "It was fearful. Four years it lasted. It was absolutely necessary for Russia. . . ." Winston S. Churchill, *The Second World War*, vol. IV, p. 447.

We have seen that the condition of rural Russia prevented any rational change in industrial policy as well; that a new and huge industrial structure, many times larger than that of pre-revolutionary Russia, had to be mounted on an agricultural base narrower than that of the *ancien régime*; and that for many years the subsistence of an ever-growing mass of town dwellers —their numbers, we know, were to rise from 30 to 60 million people in the nineteen-thirties alone—was to depend on a diminished or highly inadequate stock of foodstuffs. It was beyond the power of any government to correct this disproportion: of any government, that is, not prepared to call a halt to the industrial drive or to slow it down radically and accept the prospect of economic stagnation. If Trotsky and his followers had, at any time after 1929–30, returned to office, they too would have had to reckon with the consequences of the catastrophic destruction and deterioration of agricultural stock; and committed as they were to industrialization they too would have had to suit their policies to these severely restrictive circumstances.

Years earlier Preobrazhensky had asserted that primitive socialist accumulation, which he expected to take place under far less astringent conditions, would be 'the most critical era in the life of the socialist state . . . it will be a matter of life and death that we should rush through this transition as quickly as possible . . .'.[1] How much more was this a matter of life and death for Stalin, who had cut all his avenues of retreat. He rushed through this transition at a murderous pace, paying no heed to warnings and counsels of moderation. Preobrazhensky had urged the Bolsheviks to 'take the productionist and not the consumptionist point of view . . .' because 'we do not live yet in a socialist society with its production for the consumer —we live under the iron heel of the law of primitive socialist accumulation'. How much heavier, how crushingly heavy, that iron heel had now become! How much sterner also was the 'productionist' viewpoint that, after all that had happened and with all his commitments, Stalin had to adopt! Preobrazhensky had foreseen that a relative shortage of consumer goods would in any case accompany accumulation and result in economic

[1] The conclusions of Preobrazhensky's *New Economics* are summarised in *The Prophet Unarmed*, pp. 234–8.

inequality between administrators and workers, and between skilled, unskilled, and semi-skilled labourers; and that this inequality would be necessary in order to promote skill and efficiency; but that it would not produce new and fundamental class antagonisms. Actually, inequality grew in proportion to scarcity; and both surpassed all expectation.

Stalin employed every ideological device to increase, conceal, and justify the gulf between the privileges of the few and the destitution of the many. But ideological prevarication was not enough; and terror held its dreadful vigil over the gulf. Its fierceness corresponded to the tenseness of all social relations. Outwardly the violence of the nineteen-thirties looked like the recrudescent terror of the civil war. In fact it far surpassed it and immensely differed from it in scale and blind force. In the civil war it was the hot breath of a genuine revolutionary anger that struck at the forces of the *ancien régime* which plotted, organized, armed, and fought against the new republic. The agents of the Cheka were freshly recruited from insurgent workers, were steeped in the experience of their class, shared its privations and sacrifices, and relied on its support. Their terror was as discriminating as it could be amid the chaos of civil war: it aimed at the real and active enemies of the revolution, who, even if they were not 'a mere handful', were in any case a minority. And in the stern atmosphere of war communism, it also guarded the utopian Spartan equality of those years.

The terror of the nineteen-thirties was the guardian of inequality. By its very nature it was anti-popular; and being potentially or actually directed against the majority, it was indiscriminate. Yet even this does not fully account for its all-pervasiveness and fury. Mass executions, mass purges, and mass deportations were not needed merely to safeguard differential wage scales or even the privileges of the bureaucracy— far greater inequalities and privileges are normally safeguarded by far milder means. The great burst of violence came with collectivization; it was primarily the need to perpetuate the Great Change in the countryside that perpetuated the terror. Only the presence in the villages of punitive brigades and Political Departments could prevent the peasants from reverting to private farming. Brute force kept in being the kolkhoz which lacked intrinsic economic coherence. The need to bring that force to bear on the great majority of the nation—the peasantry still formed 60 to 70 per cent. of the population—and to bring it to bear at every season of the year, during the ploughing, the sowing, the harvesting, and finally when the farmers were due to deliver their produce to the state—all this resulted in a constant injection of such huge doses of fear into so vast a part of the social organism that the whole body was inevitably poisoned. Once the machine of terror, far more massive than anything hitherto seen, was mounted and set in motion, it developed its own incalculable momentum. Urban Russia could not insulate herself from the convulsions in which rural Russia was caught: the despair and the hatred of the peasantry overflowed into the cities and towns, catching large sections of the working class; and so also overflowed the violence let loose to meet the despair and the hatred.

.   .   .   .

For all their irrational course, the changes of 1929–30 added up to social revolution, quite as irreversible as that of October 1917, although utterly unlike it. What manifested itself in this upheaval was the 'permanence' of the revolutionary process that Trotsky had prophesied—only that the manifestation was so different from what he had expected that he could not and did not recognize it as such. He still thought, as all Bolsheviks had done until quite recently, that revolution was necessary only for the overthrow of feudal and bourgeois rule and the expropriation of landed estates and big capital; but that after this had been accomplished, the 'transition from capitalism to socialism' should proceed in an essentially peaceful and evolutionary manner. In his approach to domestic Soviet issues the author of 'Permanent Revolution' was in a sense a reformist. True, earlier than anyone he had realized that the Soviet Republic would be unable to resolve its inner conflicts and problems within the framework of national reform; and so he looked forward to international revolution to solve them ultimately. His revolutionary approach to the international class struggle and his reformist approach to domestic Soviet issues were the two sides of a single coin. By contrast, Stalin had, up to 1929, been confident that national reform alone could cope with the conflicts of Soviet society. Having found that this was not so, he too had to go beyond the framework of national reform; and he staged another national revolution. What he discarded was the reformist not the nationalist element of his policy. His pragmatic indifference to international revolutionary perspectives and the quasi-revolutionary character of his domestic policy were also two sides of a single coin.

In its own ironic way the historic development now confirmed the essential truth of the idea which underlay Trotsky's scheme, but controverted, at least in part, that scheme. 'Left to itself alone, the working class of Russia', Trotsky had written early in the century, 'will inevitably be crushed by the counter-revolution at the moment when the peasantry turns its back upon the proletariat.' That moment seemed very close, first in 1921 and then again in the late nineteen-twenties, when the peasantry did turn its back upon the Bolsheviks. 'The workers will have no choice', Trotsky had further written, 'but to link the fate . . . of the Russian Revolution with that of the socialist revolution in Europe.' Since 1917 he kept on repeating that Russia could not by herself achieve socialism, but that nevertheless the momentum of her revolution was not yet spent: 1917 had been but the prelude to international revolution. It now turned out that the dynamic force of the Russian Revolution had indeed not yet come to a rest, although its impulse had failed to ignite revolution in Europe. But having failed to work outwards and to expand and being compressed within the Soviet Union, that dynamic force turned inwards and began once again to reshape violently the structure of Soviet society. Forcible industrialization and collectivization were now substitutes for the spread of revolution, and the liquidation of the Russian kulaks was the *Ersatz* for the overthrow of bourgeois rule abroad. To Trotsky, his idea was inseparable from his scheme: only a German, French, or at least a Chinese October would provide the real sequel to the Russian October; the consummation of the revolutionary process in Russia could come only with its internationalization. Historically this was still true; but immediately Stalin acted as the unwitting agent of permanent revolution within the Soviet Union. Trotsky refused to acknowledge this and to accept the *Ersatz* for the real thing.

His view had in it the rationality of classical Marxism. Stalin's Great Change was shot through with irrationality. The classical revolution conceived by Marxism was carried on the high tide of social awareness and of the political activity of the masses; it was the supreme manifestation of their will to live and remake their lives. The upheaval of 1929–30 came at the lowest ebb of the nation's social awareness and political energy—it was a revolution from above, based on the suppression of all spontaneous popular activity. Its driving force was not any social class, but the party machine. To Trotsky whose thought had imbibed and embodied all the rich and varied European tradition of classical revolutions, this upheaval was therefore no revolution at all—it was merely the rape of history committed by the Stalinist bureaucracy. Yet, however 'illegitimate' from the classical Marxist viewpoint, Stalin's revolution from above effected a lasting and as to scale unprecedented change in property relations, and ultimately in the nation's way of life.[1]

.   .   .   .

In the course of our narrative we have repeatedly considered the peculiarity of Russian history which consisted in the state's extraordinary power over the nation. The old Tsarist absolutism had drawn its strength from the primitive, undifferentiated, and formless fabric of Russian society. 'Whereas in the West', Miliukov observed, 'the Estates had created the State, in Russia the State had brought into being the Estates.' Even Russian capitalism, Trotsky added, came into being 'as the child of the state'. The immaturity of Russia's social classes had induced the leaders of the intelligentsia and tiny groups of revolutionaries to substitute themselves for the people and to act as its proxies.[2] After a relatively brief but immense upsurge of Russia's popular energies during the first two decades of this century, the exhaustion of these energies in the civil war and the post-revolutionary disintegration of society produced a similar effect. In 1921–2, with the working class unable to uphold its own class interest, Lenin and his Old Guard assumed the roles of its trustees. The logic of this 'substitutism' led them to

---

[1] See Chapter VIII in Deutscher, *Stalin*.
[2] *The Prophet Armed*, pp. 151ff., 189–90, and *passim*.

establish the political monopoly of the Bolshevik Party, which then gave place to the much narrower monopoly of the Stalinist faction. In order to grasp the further course of events and the struggle between Stalin and Trotsky, we should now briefly re-examine the condition of the various classes of Soviet society a decade after the civil war.

The shrinkage and dispersal of the working class characteristic of the early nineteen-twenties were now a matter of the past. Under N.E.P., as industry recovered, a new working class grew up almost as numerous as the old. After only a few years, by 1932, industrial employment had risen further from 10 to 22 million; and in the course of the decade so many new recruits were drafted to the factories and mines that by about 1940 the working class was nearly three times larger than ever before.[1] Yet, despite this immense growth, the weight of the working class did not make itself felt politically. The workers' direct influence on political life was immeasurably less than it had been in the last years of Tsardom, not to speak of 1917; they were quite unable to assert themselves against the bureaucracy. It was not that in a workers' state they had no need to do so—none other than Lenin insisted, in 1920-1, that the workers needed to defend themselves against their own state; and if they needed to do so in 1921 they needed to do so *a fortiori* in 1931. Yet they remained passive and mute.

What accounted for this phenomenon of a prolonged eclipse of social awareness and paralysis of political will? It could not be terror alone, not even totalitarian terror, for this is effective or ineffective in proportion to the resistance which it meets or fails to meet. There must have been something in the working class itself that was responsible for its passivity. What was it?

The millions of new workers came to industry mostly from the primordially primitive countryside, at first 'spontaneously', driven by rural over-population, and then in the course of that planned transfer of manpower from farm to factory, which the government effected using the collective farms as convenient recruiting centres. The recruits brought with them (into the towns and factory settlements) the illiteracy, the listlessness, and the fatalistic spirit of rural Russia. Uprooted and bewildered by

---

[1] *Narodnoe Khozyaistvo S.S.S.R.*, pp. 656-7. The figures include both workers and employees.

I

unfamiliar surroundings, they were at once caught up in the tremendous mechanism which was to process them into beings very different from what they had been, to break them into the rhythm and discipline of industrial life, to train them in mechanical skills, and to drum into them the party's latest commandments, prohibitions, and slogans. Crowded into huge compounds and barracks, clothed in rags, undernourished, bullied in the workshops, and often kept under quasi-military discipline, they were unable to resist the pressures that bore down on them. Basically, their experience was not very different from that of generations of uprooted peasants thrown into the industrial melting-pots of early capitalism. But whereas under *laissez faire* it was the spontaneous action of the labour market, the fear of unemployment and hunger, that slowly transformed and disciplined the peasant into an industrial worker, in Stalinist Russia it was the state that took care of this and compressed the whole process of transformation into a much shorter time.

So violent was the wrench the industrial recruit suffered, so intensive was the drilling to which he was subjected, so forsaken by God and men did he feel, and so overwhelmed by the hugeness of the forces that shaped his life, that he had neither the mind nor the strength to form any opinion or utter any protest. Sporadically, his resentment found outlet in a drunken brawl, in the stealthy wrecking of a machine, or in the attempt to escape from one factory to another. He tried to fend for himself and improve his own lot without reference to the situation of his class. His atavistic individualism as much as the prohibition of strikes prevented him from associating in self-defence with his fellow-workers and acting in solidarity with them. Stalin, who was stamping out that individualism on its native ground, in the village, encouraged it and played upon it in the industrial workshops, where Stakhanovism and 'socialist competition' excited to the utmost the workers' acquisitiveness and prodded them to compete against one another at the bench.

Thus, while the peasantry was being collectivized, the working class was reduced to such a state that little was left of its traditionally collectivist outlook. 'While our peasantry is being "proletarianized", our working class is becoming completely infected with the peasantry's spirit', observed sadly a deported sociologist of the Opposition.[1] This is not to say that class solidarity and Marxist militancy were completely wiped out. These were still alive in the survivors of the 'October genera-tion' and in quite a few younger people brought up in the nineteen-twenties—as anyone was aware who around 1930 watched the self-sacrificing enthusiasm with which the early *udarniki* set out to build, often on their own bones almost, new steel mills and power plants amid the bare rocks of the Urals or farther to the East. Stalinist propaganda, self-contradictory as it was, continued to inculcate much of the Marxist tradition even while distorting or mutilating it. The workers imbued with that tradition resented the intrusion of peasant individualism into the factories and the scramble for wages and bonuses. But such workers were in a minority and were swamped by the millions of proletarianized muzhiks. Moreover, state and party continually drained the intellectual and political resources of the working class by picking out from its midst the most class-conscious, educated, and energetic individuals in order to fill with them newly created managerial and administrative posts or to draft them into the special brigades whose task it was to collectivize the peasants. Deprived of its élite, the working class was all the more strongly torn by centrifugal forces and split. It was, of course, also deeply divided over collectivization. The drive in the country at first aroused high hope among the proletarians with a strong urban background, who had all along distrusted the rural bourgeoisie. But the labourers who had come from the villages were outraged, filled the towns with tales about the horrors perpetrated in the country, and aroused much sympathy. The sociologist whom we have just quoted observes that in the years of the first Five Year Plan towns were full of people whom he describes as *sans culottes à rebours*. Ever since the French Revolution, he explains, the *sans culotte*, the man without property, had been the enemy of property; but in the Soviet Union, at this time, he was the fiercest defender of property. His presence and mood were felt even in the oldest strongholds of Bolshevism, which was not surprising when, for instance, in the Donetz Coal Basin no fewer than

[1] Ya. Gref in the essay on Collectivization and Overpopulation (*B.O.*, no. 11, 1930). This is a most original, though somewhat dogmatic, analysis of Soviet society during the upheaval.

40 per cent. of the miners were, in 1930, expropriated kulaks and other peasants. In the older layers of proletarian communities the moods ranged from a sullen enmity towards authority to the feeling that party and state did, after all, express the aspirations of the working class and that opposition to them was inadmissible. But there could be no doubt that the mass of the *sans culottes à rebours* and the numerous Lumpenproletarians, displaced peasants who could not fit in with any industrial environment and who filled the suburbs and outskirts with drunkenness and crime, formed potentially a large reserve of cannon-fodder for any 'Thermidorian', counter-revolutionary, or even fascist movement.

In its fragmentation, confusion, and lack of political identity, the new working class partly resembled the proletariat of the early capitalist era, whom Marx had described as a 'class *in* itself' but not '*for* itself'. A class in itself performs its economic function in society, but is unconscious of its place in society, unable to conceive its own corporate and 'historic' interest and to subordinate to it the sectional or private strivings of its members. Marxists had tacitly assumed that once the working class achieved the social self-integration and political awareness that made of it a 'class *for* itself' it would maintain itself indefinitely in that position and would not sink back into immaturity. Instead, the working class of Russia, having overthrown the Tsar, the landlords, and the capitalists, relapsed into the inferior condition of a class unconscious of its interest and inarticulate.

The state of the peasantry was, of course, even worse. The blows that fell on it utterly disorganized and deranged it. Yet before 1929 the peasantry appeared to have achieved a degree of inner cohesion which it had hardly ever attained in the past. In its mass, it seemed, and to some extent was, united in the hostility with which it confronted Bolshevik collectivism. Its antagonism to party and state overshadowed its inner divisions, that is the conflicts between well-to-do and poor farmers. The kulak was at the head of the village community; and farm labourers and *byedniaks*, who had for years watched Bolshevik efforts to come to terms with him, refrained from challenging his position and willy-nilly accepted his leadership. And so the collectivizers, when they first appeared on the scene, found it hard to breach the villagers' solidarity. So inflated had been the kulak's self-confidence and so strongly had the poorer peasants been impressed by it that they did not believe the commissars who threatened the kulak with annihilation to be in earnest. Many thought that it was still safer to side with the kulak and defend the old mode of farming than to follow the call of the commissars. But, as it became clear that the government was in no mood to retreat and that the kulak was indeed doomed, the unity of the village crumbled; the long-subdued but now stoked-up hostility of the poor towards the well-to-do came back into its own. The great mass was torn between conflicting interests, calculations, and sentiments. As the government attacked not only rural capitalism but private farming at large, and as even the poorest farmers were asked to give up their small-holdings, the peasants still tended to remain united in clinging to their possessions. The instinct for property was often as strong in the poorest as it was in the wealthiest peasant; and this instinct and the common sense of humanity were shocked and revolted by the arbitrariness and the inhumanity of the collectivization. Yet these sentiments were disturbed and weakened by the cold reflection of the poor peasants that they might, after all, benefit from the dispossession of the well-to-do and the pooling of the farmsteads; and then, when it was no longer in doubt who was winning the day, many rushed to the victors' band waggon.

The idea of collective farming had, of course, not been alien to rural Russia. The belief that the land was the common good of those who tilled it, not intended by the Creator to enrich some and impoverish others, had once been deeply held; and the *Mir* or *Obshchina*, the primordial rural commune within which the land had been periodically redistributed among members, had survived until shortly before the revolution—it was not till 1907 that Stolypin's government enabled the 'strong farmer' to leave the *Mir* and so to withdraw his possessions from the redistribution and escape its levelling effect. True, since 1917 the peasant's attachment to his own, enlarged, plot of land had grown immensely. Nevertheless, the party agitators were still able to present the kolkhoz as the legitimate successor to the *Mir* and to commend it to the villagers, not as a subversive innovation but rather as the revival, in modified form, of a native institution, which though corroded by capitalist greed and rapacity was still hallowed in memory. Thus the impulses and influences that determined the peasantry's behaviour were intricate and contradictory, with the result that fear and faith, horror and hope, despair and reassurance wrestled in the muzhik's thoughts, leaving him unnerved, resentful yet unresisting, and nourishing his grievances in sluggish submission.

While the peasants were being rapidly reduced to this state, they still took a fiercely insane plunge into dissipation. In the first months of collectivization they slaughtered over 15,000,000 cows and oxen, nearly 40,000,000 goats and sheep, 7,000,000 pigs, and 4,000,000 horses; the slaughter went on until the nation's cattle stock was brought down to less than half what it had been. This great shambles of meat was the main dish at the feast with which the smallholder celebrated his own funeral. The kulak began the carnage and incited others to follow suit. Seeing that he had lost all, that he, the nation's provider, was to be robbed of his property, he set out to rob the nation of its food supply; and rather than allow the collectivizers to drive away his cattle to communal assembly stations, he filled his own larders with the carcasses so as to let his enemies starve. The collectivizers were at first taken aback by this form of 'class warfare' and watched with helpless amazement as the 'middle' peasants and even the poor joined in the butchery, until the whole of rural Russia was turned into an abbatoir.

So began the strange carnival over which despair presided and for which fury filled the fleshpots. An epidemic of orgiastic gluttony spread from village to village, from *volost* to *volost*, and from *gubernia* to *gubernia*. Men, women, and children gorged themselves, vomited, and went back to the fleshpots. Never before had so much vodka been brewed in the country— almost every hut became a distillery—and the drinking was, in the old Slav fashion, hard and deep. As they guzzled and gulped, the kulaks illuminated the villages with bonfires they made of their own barns and stables. People suffocated with the stench of rotting meat, with the vapours of vodka, with the smoke of their blazing possessions, and with their own despair. Such was often the scene upon which a brigade of collectivizers descended to interrupt the grim carouse with the rattle of machine-guns; they executed on the spot or dragged away the crapulous enemies of collectivization and announced that henceforth all remaining villagers would, as exemplary members of the kolkhoz, strive only for the triumph of socialism in agriculture. But after the kulaks and the *podkulachniki*, their helpmeets, had been disposed of, the slaughter of cattle and the feasting went on—there was no way of stopping it. Animals were killed because no fodder was left or because they had become diseased from neglect; and even the *bednyaks* who, having joined the kolkhozes, had every interest in preserving their wealth, went on dissipating it and stuffing their own long-starved stomachs. Then followed the long and dreadful fast: the farms were left without horses and without seed for the sowing; the kolkhozniki of the Ukraine and of European Russia rushed to central Asia to buy horses, and, having returned empty-handed, harnessed the few remaining cows and oxen to the ploughs; and in 1931 and 1932 vast tracts of land remained untilled and the furrows were strewn with the bodies of starved muzhiks. The smallholder perished as he had lived, in pathetic helplessness and barbarism; and his final defeat was moral as well as economic and political.

But the collectivizers too were morally defeated; and, as we have said, the new system of agriculture was to labour under this defeat in years to come. Normally, a revolution does not depend for the success of its constructive task on the social class it has overthrown, be it the landlords or the bourgeoisie; it can rely on the classes that have rallied to its side. The paradox of the rural revolution of 1929–30 was that the realization of its positive programme depended precisely on the vanquished: collective farming could not flourish when the smallholder-turned-kolkhoznik was in no mood to make it work.[1]

The lack of moral and political cohesion among the workers and the peasants made for the apparent omnipotence of the state. If after the civil war bureaucratic rule was established against the background of economic disintegration and the dispersal of the working class,[2] that rule now gained virtually unlimited power from the opposite processes, from economic growth and expansion, which were to give new structure and shape to society, but immediately made society even more

[1] Ya. Gref, op. cit.   [2] See *The Prophet Unarmed*, Chapter I.

shapeless and increased its mental atrophy. In years to come all
the energies of the Soviet Union were to be so intensely occupied
with material progress and the prodigious efforts which this
required that little or no resources were left for the assertion
of any moral and political purposes. And, as the power of the
state was all the greater when it was exercised over a nation
politically reduced to pulp, those in power did all they could to
keep the nation in just that condition.

Yet even the bureaucracy was not truly united by any
common interest or outlook. All the divisions which split the
other classes were reflected in its midst. The old estrangement
between the communist and the non-communist civil servants
was still there; it was sharply revealed in the frequent monster
trials of 'specialists' denounced as saboteurs and 'wreckers'.
Throughout the years of N.E.P. most of these 'specialists' and
their friends had hopefully waited for the moment when the
dynamic force of the revolution would come to rest and Russia
would once again become a 'normal' state. They had indeed
prayed for that Neo-N.E.P. and that Thermidor the spectres of
which haunted the Trotskyists and the Zinovievists; they had
first banked on Stalin and Bukharin against Trotsky; and then
they longed to see Bukharin, or any other 'authentic Thermidorian', prevail against Stalin. These hopes were now frustrated;
and those who had held them, often unable or unwilling to
adjust themselves to the new situation, were in disarray.
In the Bolshevik section of the bureaucracy Bukharinists and
Stalinists were at loggerheads. The former, strongly entrenched
during the years of N.E.P., were tracked down and ejected
from the administration. New men from the working class and
from the young intelligentsia filled their places and the many
other vacancies which were opening all the time. The bureaucracy's composition was therefore highly unstable, and its
outlook heterogeneous. Even the one bond that might have been
expected to unite it, the bond of privilege, was extremely
tenuous when not only individuals but entire groups of the
bureaucracy could be, and frequently were, stripped of all
privileges almost overnight, turned into pariahs, and driven
into concentration camps. And even the strictly Stalinist
elements, the men of the party machine and the leaders of the
nationalized industry, who formed the ruling groups proper,

were by no means exempt from the insecurity in which all the hierarchies trembled under Stalin's autocracy.

Thus the feverish economic expansion, the general unsettlement which accompanied it, the eclipse of social awareness in the masses, and the emaciation of their political will formed the background to the development by which the rule of the single faction now became the rule of a single leader. The sheer multiplicity of the conflicts between the classes and within each class, conflicts which society itself was unable to resolve, called for constant arbitrament, which could come only from the very pinnacle of power. The greater the unsettlement, the flux, and the chaos down below the more stable and fixed that pinnacle had to be. The more enfeebled and devoid of will all social forces were, the stronger and more wilful grew the arbitrator; and the more powerful he became the more impotent were they bound to remain. He had to concentrate in himself all the vigour of decision and action which they lacked. He had to focus in himself the whole dispersed *élan* of the nation. To the extent to which the bulk of the people sunk below the level of higher human aspiration he must appear superhuman. His infallible mind had to dominate their absent-mindedness. His sleepless vigilance had to protect them against all the dangers of which they were unaware and against which they were unable to protect themselves. Everyone had to be blind in order that he, the only seer, might lead. He must be proclaimed the sole trustee of the revolution and of socialism; and his colleagues who had hitherto exercised that trusteeship jointly with him had to renounce all claim to it, and yet had to be crushed as well. To put his pre-eminence beyond any challenge the multitudes had to acclaim him ceaselessly; and he himself had to guard his pre-eminence with the utmost care and see to it that the popular adulation should rise in endless crescendo. Like History's Elect in Hegel, he embodied a great phase in the nation's, and indeed in mankind's, life. But for the obsessive megalomania, which his position bred in him, even this was not enough: the Superman's elbows burst the frame of his time: in him must live and merge past, present, and future: the past with the ghosts of the early Empire-building Tsars incongruously jostling the shades of Marx and Lenin; the present with its tremendous eruptive and creative force; and the future glowing with the fulfilment of mankind's most sublime dreams. The secret of this grotesque apotheosis, however, lay less in Stalin than in the society he ruled: as that society forfeited its own political identity and the sense of its own tremendous movement, that identity and the whole movement of history became personalized in the Leader.

The process by which Stalinist government became Stalin's government was far less distinct and consecutive than the evolution that had led to it, the transformation of the rule of the Bolshevik party into the rule of the Stalinist faction. From the outset the faction's political monopoly had to some extent been Stalin's own, because his supporters had always been far more rigidly disciplined than those of his rivals. He had always been in sole command of his followers in a way neither Trotsky nor Bukharin nor Zinoviev had ever been of theirs. Nevertheless, having crushed all his opponents, Stalin still had to complete a full ascendancy over his own followers. It now turned out that the rule of a single faction no less than that of a single party was a contradiction in terms. Just as in the single party, as long as members could express themselves freely, the various groups and schools of thought formed a shadowy multi-party system incompatible with it, so the single faction tended to reproduce within itself patchy reflections of the factions and schools of thought which it had just suppressed. Stalin had to ferret out the crypto-Trotskyists and crypto-Bukharinists among his own followers. He had to deny all these followers the restricted liberties still left to them. It was now their turn to discover that, having deprived all their opponents of freedom, they had robbed themselves of it as well, and that they had placed themselves at the mercy of their own Leader. Having once proclaimed that the party must be monolithic or it would not be Bolshevik, he now insisted that his own faction must be monolithic or it would not be Stalinist. Stalinism ceased to be a current of opinion or the expression of any political group— it became Stalin's personal interest, will, and whim.

The personalization of all political relations affected Trotsky's position as well. As Stalin was becoming the sole official and orthodox embodiment of the revolution, Trotsky was becoming its sole unofficial and unorthodox representative. This had not been quite the case up to 1929. The Trotskyist Opposition was

in no sense his personal domain, even though he was its outstanding leader. Its directing centre consisted of strong-
minded and independent men: Rakovsky, Radek, Preobrazhensky, Smirnov, Pyatakov, and others, none of whom could
be described as Trotsky's creature; and the rank and file
struggling for freedom within the party preserved it within the
narrower confines of their own faction. In the Joint Opposition,
Zinoviev and Kamenev, though conscious of Trotsky's superiority, were extremely jealous of their own authority and treated
with him on a footing of equality. Not only did he not impose
his dictates but often, as we have seen, he was hamstrung in his
action against Stalin by the concessions he made to his adherents or temporary allies. Until 1929 also Bukharin's school
of thought represented an alternative to both Stalinism and
Trotskyism, an alternative which appealed to many in and out
of the party. Thus, despite the growing concentration of power in
Stalin's hands and the increasing conformism, Bolshevik hopes
and expectations did not as yet focus on any single leader and
policy, but attached themselves to various personalities, teams
of leaders, and various attitudes and shades of attitudes.

The events of 1929–30 changed all this. The Bukharinist
school of thought was wrecked even before it managed to come
out openly against Stalin. It could not go on arguing against
the accomplished facts of the Great Change: it could not resist
the industrial drive or bank on the strong farmer any longer.
The alpha and omega of Bukharinism had been its approach to
the peasantry; and this had become pointless. From the
moment the smallholder vanished the Right Opposition had no
ground to stand on. Therein lay the essential difference between
the defeat of Trotsky and Zinoviev and that of Bukharin and
Rykov: to vanquish the former, Stalin had to steal their political
weapons, while the latter had themselves to throw away their
own weapons as antiquated. This was why Bukharin, Rykov,
and Tomsky, when, in November 1929, they were expelled
from the Politbureau, left with a barely audible whimper,
whereas even Zinoviev and Kamenev had in their time left
with a battle cry.

The capitulation of the Zinovievists and the quietus of
Bukharinism left Stalinism and Trotskyism as the sole con-
tenders for Bolshevik allegiance. But now, by a strangely

parallel though antithetical development, these two factions too were disintegrating, each in its own way, the Trotskyists through endless defections and the Stalinists through doubt and confusion in their own midst. And just as Stalinism, in victory, was being reduced to Stalin's autocracy, so Trotskyism, in defeat, was becoming identified with Trotsky alone. To be sure, even after all the surrenders there were still unrepentant Oppositionists in the prisons and places of deportation; and in the early nineteen-thirties, while Rakovsky guided them, their ranks were at times reinforced by new adherents and by the return of capitulators disillusioned with surrender. Yet, despite such accessions, Trotskyism could not regain the coherence and confidence which it still had even in 1928. At best it was only a loose congeries of splinter groups conscious of their isolation, despairing of the prospects, yet persisting in their allegiance to Trotsky, to what he stood for or was supposed to stand for. They still argued among themselves and produced controversial theses and papers; but these circulated only within prison walls. Even before the terror mounted to the climax of the great purges, the Trotskyists were unable to use the prisons and places of exile as bases for political action in the way revolutionaries had used them in Tsarist times: their ideas did not reach the working class and the intelligentsia. With the years their contact with Trotsky became more and more tenuous until in 1932 even their correspondence ceased altogether. They no longer knew exactly what he stood for; and he could no longer ascertain whether or not his views accorded with theirs. He had no choice but to substitute himself for the Opposition at large; and they had no choice but to acknowledge him, expressly or tacitly, as their sole trustee and by definition the sole trustee of the revolution. His voice alone was now the voice of the Opposition; and the immense silence of the whole of anti-Stalinist Russia was his sounding board.

Thus, against Stalin, the sole trustee of Bolshevism in office, Trotsky stood alone as the proxy of Bolshevism in opposition. His name, like Stalin's, became something of a myth; but whereas Stalin's was the myth of power sponsored by power, his was the legend of resistance and martyrdom cherished by the martyred. The young people who in the nineteen-thirties faced executioners with the cry 'Long Live Trotsky!' often had no more than a mere inkling of his ideas. They identified themselves with a symbol rather than a programme, the symbol of their own anger with all the misery and oppression that surrounded them, of their own harking back to the great promise of October and of their own, rather vague, hope for a 'renascence' of the revolution.

Not only Trotsky's avowed supporters and most of the capitulators viewed him thus. The sense that he represented the sole alternative to Stalinism persisted even among party members who silently carried out Stalin's orders, and outside the party, among politically minded workers and the intelligentsia. Whenever people feared or felt that Stalin was driving them to the brink of catastrophe and whenever even their meekness was shocked by some excess of his brutality, their thoughts went out, if only fleetingly, to Trotsky, of whom they knew that he had not laid down arms and that in foreign lands he continued his lonely struggle against the corruption of the revolution.

Stalin was apprehensively aware of this; and he treated Trotsky as in older times an established monarch treated a dangerous Pretender, or as under the Double and Triple Schisms the Pope treated the Anti-Pope. It was for the role of an Anti-Pope that the ironies of history now cast Trotsky, the legatee of classical Marxism, who was utterly ill-suited for such a role and was neither able nor willing to act it. Throughout a decade crowded with the most momentous and explosive events, the transformation of Soviet society, the great slump in the West, the rise of Nazism, and the rumblings of approaching war—throughout the nineteen-thirties the duel between Stalin and Trotsky remained at the centre of Soviet politics, often overshadowing all the other issues. Not for a moment did Stalin himself slacken, or allow his propagandists and policemen to relax, in the anti-Trotskyist campaign which he carried into every sphere of thought and activity, and which he stepped up from year to year and from month to month. The fear of the Pretender robbed him of his sleep. He was constantly on the look-out for the Pretender's agents, who might be crossing the frontiers stealthily, smuggling the Pretender's messages, inciting, intriguing, and rallying for action. The suspicion that haunted Stalin's mind sought to read the hidden thoughts that the most subservient of his own subjects might have about

Trotsky; and he discovered in the most innocuous of their utterances, even in the flatteries of his courtiers, deliberate and sly allusions to the legitimacy of Trotsky's claims. The bigger Stalin himself looked and talked and the more abjectly Trotsky's old adherents rolled before him in the dust, the more delirious was his obsession with Trotsky, and the more restlessly did he work to make the whole of the Soviet Union share his obsession. The frenzy with which he pursued the feud, making it the paramount preoccupation of international communism as well as of the Soviet Union and subordinating to it all political, tactical, intellectual, and other interests, beggars description: there is in the whole of history hardly another case in which such immense resources of power and propaganda were employed against a single individual.

Morbid though the obsession was, it had a basis in reality. Stalin had not conquered power once and for all; he had to reconquer it over and over again. His success should not obscure the fact that at least up to the end of the Great Purges his supremacy remained unconsolidated. The higher he rose the greater was the void around him and the larger was the mass of those who had reason to fear and hate him and whom he feared and hated. He saw that the old divisions among his opponents, the differences between Right and Left Bolsheviks, were becoming blurred and obliterated; and so he was frightened of those 'Right-Left conspiracies' and 'Trotskyist-Bukharinist blocs' which his police had to unearth or to invent again and again, and the makings of which were indeed inherent in the situation. Finally, his ascendancy over his own faction turned even authentic old Stalinists into potential allies of the Trotskyists, the Zinovievists, and the Bukharinists. Elevated above the whole Bolshevik party, he saw, not without reason, the whole party as one potential coalition against himself; and he had to use every ounce of his strength and cunning to prevent the potential from becoming actual. He knew that if that coalition ever came into being, Trotsky would be its unrivalled leader. Having brought the chiefs of all the oppositions to prostrate themselves before him, he himself worked unwittingly to exalt Trotsky's unique moral authority. He then had to do all he could, and far more than he could, to destroy it. He resorted to ever more drastic means and to ever more absurd slanders; but his efforts were self-defeating. The more loudly he denounced his adversary as the chief or sole prompter of every heresy and opposition, the more strongly did he turn all the mute anti-Stalinist feelings, with which Bolshevik Russia was overflowing, towards the outcast's remote yet towering figure.

CHAPTER II

Reason and Unreason

THROUGHOUT the nineteen-thirties Trotsky's mind battled with the tide of irrationality surging up in world politics. Yet some of his Russian followers feared that, although his criticisms of Stalin's policy were justified and even irrefutable, he somehow failed to make allowance for the irrational element in the situation of the Soviet Union.[1] It was he himself who had maintained a few years earlier, in a controversy with Bertrand Russell, that it was impossible 'to map out the revolutionary road beforehand in a rationalistic manner' and that 'revolution is an expression of the impossibility of reconstructing class society by rationalist methods'.[2] It now turned out that it was impossible to reconstruct society by such methods even after the revolution, under a system which had given up the advantages of capitalism but could not yet avail itself of the advantages of socialism. Most, if not all, of the factors that made for the irrationality of class society—basic conflicts of interests, the fetishism of commodity and money, the inadequacy or absence of social control over productive forces—all these were still intensely at work in the Soviet Union. The Bolshevik aspiration to industrialize and educate Russia, to build up a planned economy, and to achieve control over social chaos became itself infected with the irrationality of the environment to which it was confined. This situation, though it could be explained theoretically and even predicted, gave rise to such monstrous absurdities that the analytical and dialectical mind was at times baffled in its attempts to disentangle reason from unreason.

In the West these were the years of the Great Slump; and history's record of folly and crime was suddenly enlarged by the rise and triumph of Nazism. In one way or another the Nazi triumph from now on overshadows the life of our chief character. Without running too far ahead of the narrative, it may be

---

[1] *B.O.*, no. 11, 1930.   [2] *The Prophet Unarmed*, p. 222.

said here that Trotsky's attempt to arouse the working class of Germany to the danger that threatened it was his greatest political deed in exile. Like no one else, and much earlier than anyone, he grasped the destructive delirium with which National Socialism was to burst upon the world. His commentaries on the German situation, written between 1930 and 1933, the years before Hitler's assumption of power, stand out as a cool, clinical analysis and forecast of this stupendous phenomenon of social psychopathology and of its consequences to the international labour movement, to the Soviet Union, and to the world. What underlines even further the political insanity of the times is with what utter unconcern about the future and venomous hostility the men responsible for the fate of German communism and socialism reacted to the alarm which Trotsky sounded, from his Prinkipo retreat, in these decisive three years. An historical narrative can hardly convey the full blast of slander and derision with which he was met. He represented in effect the self-preservation of the labour movement against the movement itself, which was as if bent on self-destruction. He had to watch the capitulation of the Third International before Hitler as a father watches the suicide of a prodigal and absent-minded child, with fear, shame, and anger—he could not forget that he had been a founding father of the International.

And there was a fierce flash of fate's extravagant cruelty in the inroad which the insanity of the time made even in Trotsky's own family circle.

. . . .

Only a few months had passed since the beginning of the world-wide economic crisis, the Wall Street panic of October 1929, and the whole edifice of the Weimar Republic was shattered. The Great Slump had struck Germany with devastating force and thrown six million workers out of employment. In March 1930 Hermann Müller, the Social Democratic Chancellor, was forced to resign: the Socialist-Catholic coalition on which his government rested had collapsed. The coalition partners could not agree whether or by how much the government should cut the dole it paid out to the unemployed. Field-Marshal Hindenburg, the relic and symbol of the Hohenzollern Empire, now the Republic's President, dissolved

K

Parliament and appointed Heinrich Brüning *Reichskanzler*. Brüning ruled by decree, enforced a rigidly 'deflationary' policy, cut expenditure on social insurance, dismissed government employees *en masse*, reduced wages and salaries, and crushed small businessmen with taxes, thus aggravating the distress and the despair of all. In elections held on 14 September 1930, Hitler's party, which had polled only 800,000 votes in 1928, won six and a half million votes; from the smallest party in the Reichstag it became the second largest. The Communist party, too, increased its vote from about three million to over four and a half. The Social Democrats, who had for years ruled the Weimar Republic, lost; and so did the Deutschnazionale and the other parties of the traditional right wing. The election revealed the instability and the acute crisis of parliamentary democracy.

The leaders of the Weimar Republic refused to read the omens. Conservatives viewed the emergence of the Nazi movement with mixed feelings: disconcerted by their own losses and by the violence of Nazism, they were nevertheless reassured by the rise of a great party which declared implacable war on all working-class organizations; and they hoped to find in Nazism an ally against the left and possibly a junior partner in government. The Social Democrats, frightened by Hitler's threats—he strutted the country proclaiming that 'the heads of Marxists and Jews would soon roll in the sand'—decided to 'tolerate' Brüning's government as 'the lesser of the two evils'. The Communist party exulted in its gains and made light of the huge increase in the vote for Hitler. On the day after the election, the *Rote Fahne*, then the most important communist paper in Europe, wrote: 'Yesterday was Herr Hitler's "great day", but the so-called electoral victory of the Nazis is only the beginning of their end.' 'The 14th of September [*Rote Fahne* repeated a few weeks later] was the high watermark of the National Socialist movement in Germany—what follows now can be only ebb and decline.'

Several months later, after the towns and cities of Germany had had their first taste of the terror of Hitler's Stormtroops, Ernest Thaelmann, the leader of the Communist party, told the Executive of the Comintern in Moscow: 'After 14 September, following the sensational success of the National

Socialists, their adherents all over Germany expected great things from them. We, however, did not allow ourselves to be misled by the mood of panic which showed itself . . . in the working class, at any rate, among the followers of the Social Democratic party. We stated soberly and seriously that 14 September was in a sense Hitler's best day after which there would be no better but only worse days.' The Executive of the Comintern endorsed this view, congratulated Thaelmann, and confirmed its Third Period policy which committed the Communist party to reject the idea of any Socialist-Communist coalition against Nazism and obliged it to 'concentrate fire on the Social-Fascists'.[1]

We know that Trotsky had subjected this policy to severe criticism as early as 1929. In March 1930, six months before the crucial elections, he repeated this criticism in an 'Open Letter' to the Soviet Communist party, where he spoke again of the growing force of fascism all over Europe, but especially in Germany, and insisted on the need for joint Socialist-Communist action.[2] No sooner had the results of the September elections become known than he commented on them in a special pamphlet which he took care to publish in several European languages. 'The first quality of a truly revolutionary party is the ability to face realities', he wrote, dismissing the Comintern's self-congratulations and pointing out that the communist gain of over a million votes was almost insignificant compared with the Nazi gain of nearly six million. The 'radicalization of the masses', of which the Comintern boasted, had benefited counter-revolution rather than revolution. What accounted for the 'gigantic' upsurge of Nazism was 'a profound social crisis', which had upset the mental balance of the lower middle classes, and the inability of the Communist party to cope with the problems posed by that crisis. If communism expressed the revolutionary hopes of the worker, Nazism voiced the counter-revolutionary despair of the *petit bourgeois*. When the

---

[1] The session of the Comintern's Executive took place in April 1931. Manuilsky was the *rapporteur* on the international situation. He expounded the Third Period policy with an uninhibited zeal which only served to underline its absurdity. See *Kommunistische Internationale*, nos. 17–18, 1931.

[2] *B.O.*, no. 10, April 1930. See also his devastating attack on "The Third Period of the Comintern's Blunders", published in *Vérité, Permanente Revolution, Militant*, and other Trotskyist papers in January and February 1930.

party of socialist revolution is in the ascendant it carries with it not only the working class but also large sections of the lower middle class. In Germany, however, the opposite was happening: the party of counter-revolutionary despair had captured the lower middle class and important layers of the working class as well. Comintern analysts consoled themselves with the idea that Nazism was merely a remote aftermath of the crisis of 1923 and of subsequent social tensions. Trotsky argued that far from representing a belated reaction to any crisis of the past, Nazism mobilized forces for a crisis that lay ahead; and that 'the fact that fascism has been able to occupy so strong a starting position on the eve of a revolutionary period, and not at its end, is a source of weakness to communism, not to fascism'. He concluded that 'despite the parliamentary success of the Communist party, proletarian revolution . . . has suffered a serious defeat . . . a defeat which may become decisive'.[1]

In this brochure Trotsky had already outlined an analysis of National Socialism, which he was to develop presently in a series of books and articles. Thirty years later some of his ideas may seem truisms; they were all heresies when he put them forward. In the main, his view of Nazism has retained freshness and originality; it still remains the only coherent and realistic analysis of National Socialism (or of fascism at large) that can be found in Marxist literature. It will therefore not be out of place to summarize his view, which he himself developed mostly in controversial form, in the context of a debate over communist tactics.[2]

The crux of Trotsky's conception lies in his description of National Socialism as 'the party of counter-revolutionary despair'. He saw National Socialism as the movement and ideology of the *wildgewordene Kleinbürger*, the small bourgeois run amok. This set it apart from all other reactionary and counter-revolutionary parties. The forces of conventional reaction worked usually from above, from the top of the social pyramid, to defend established authority. Fascism and National Socialism were counter-revolutions from below, plebeian

---

[1] L. Trotsky, *Écrits*, vol. III, pp. 25–46. *The Archives.*

[2] His most important works on this subject are: *Nemetskaya Revolutsia i Stalinskaya Burokratiya* (published under the title *Was Nun?* in German, and *What Next?* in English) and *Edinstvennyi Put'* (*Der einzige Weg*), essays and articles in *B.O.* and other Trotskyist papers. *Écrits*, vol. III.

movements rising from the depths of society. They expressed the urge of the lower middle class to assert itself against the rest of society. Usually subdued, that urge becomes aggressive in a national catastrophe with which established authority and the traditional parties are unable to cope. During the 'prosperity' of the nineteen-twenties Hitler's party had been on the lunatic fringe of German politics. The slump of 1929 brought it to the fore. The great mass of shopkeepers and white collar employees had hitherto followed the traditional bourgeois parties and had seen themselves as upholders of parliamentary democracy. They now deserted those parties and followed Hitler, because sudden economic ruin filled them with insecurity and fear, and aroused their craving for self-assertion.

The *Kleinbürger* normally resented his social position: he looked up with envy and hatred to big business, to which he so often helplessly succumbed in competition; and he looked down upon the workers, jealous of their capacity for political and trade union organization and for collective self-defence. Marx once described what, in June 1848, had driven the French *petite bourgeoisie* to turn furiously against the insurgent workers of Paris: the shopkeepers, he said, saw access to their shops blocked by the workers' barricades in the streets; and they went out and smashed the barricades. The German shopkeepers of the early nineteen-thirties had no such reason for running amok—no barricades blocked access to their shops. But they were ruined economically; they had cause to blame the Weimar Republic at the head of which they had for years seen the Social Democrats; and they were frightened of the threat of communism, which even if, or because, it did not materialize, kept society in permanent ferment and agitation. In the *Kleinbürger's* eyes big business, Jewish finance, parliamentary democracy, social-democratic governments, communism and Marxism at large, all merged into the image of a many-headed monster which strangled him—all were partners in a sinister conspiracy responsible for his ruin. At big business the small man shook his fists as if he were a socialist; against the worker he shrilled his bourgeois respectability, his horror of class struggle, his rabid nationalist pride, and his detestation of Marxist internationalism. This political neurosis of impoverished millions gave National Socialism its force and impetus. Hitler was the small man writ large, the small man with all these neurotic obsessions, prejudices, and fury. 'Not every *Kleinbürger* run amok can become a Hitler', said Trotsky, 'but there is something of Hitler in every *Kleinbürger* run amok.'

Yet the lower middle class was normally 'human dust'. It had none of the workers' capacity for self-organization, for it was inherently amorphous and atomized; and, despite bluster and threats, it was cowardly wherever it met with genuine resistance. The whole record of European class struggles and the Russian Revolution proved this. The small bourgeoisie could no longer play any independent part—ultimately it had to follow either the upper bourgeoisie or the working class. Its rebellion against big business was impotent—the small artisan and shopkeeper could not prevail against monopolistic capitalist oligarchies. National Socialism in office could not therefore keep any of its 'socialistic' promises. It would reveal itself as an essentially conservative force; it would seek to perpetuate capitalism; it would crush the working class, and hasten the ruin of the same lower middle class which had brought it to power. But in the meantime the lower middle class and its *Lumpenproletarian* fringe were in feverish motion and their imagination was inflamed with the dream of the social and political supremacy which Hitler was to bring them.

This 'human dust', Trotsky argued, is attracted by the magnet of power. It follows in any struggle that side which shows the greater determination to win, the greater audacity, and the ability to cope with a catastrophe like the Great Slump. That was why in Russia, Bolshevism, having assumed the leadership of the working class in 1917, carried also, at decisive moments, the great hesitant and dispersed mass of the peasantry and even part of the small urban bourgeoisie. Similarly, the German working class would still attract to itself the multitudes of the lower middle classes if these felt its strength and determination to win; that is, if socialist and communist policies did not lack direction and purpose. The inflated ambitions of the *Kleinbürger* and the strength of Nazism sprang from the weakness of the working class. The Social-Democratic leaders sought to ingratiate themselves with the middle classes, lower and upper, first by acting, under the Weimar Republic as business managers of the bourgeois state, then by submitting meekly to the Brüning régime, and throughout by defending the social and political *status quo*. Yet it was precisely against the Weimar Republic and its Brüning sequel and against the *status quo* that the lower middle classes were in revolt. Social-Democratic policy therefore contributed decisively to the dangerous estrangement between the organized working class and the small bourgeoisie, the estrangement on which Nazism thrived. The Social-Democrats went on preaching moderation and prudence when moderation and prudence were bankrupt; and they continued to defend the *status quo* when this had become so unbearable that the masses preferred almost anything else, even the abyss into which Hitler was plunging them.

In their ostrich-like behaviour the Social Democrats were true to character. All the greater, Trotsky pointed out, was the responsibility of the Communist party. Yet its leaders were unaware of the magnitude and the nature of the peril. With sham ultra-radicalism they refused to make any distinction between fascism and bourgeois democracy. They maintained that as monopolistic capitalism was bent on rendering bourgeois democracy fascist, all parties standing on the ground of capitalism were bound to undergo this process. All cats then were equally brown: Hitler was a fascist; but so were the leaders of the traditional bourgeois parties, right and centre; so in particular was Brüning, who already ruled by decree; and so even were the Social Democrats, who formed the 'left wing of fascism'. This was no mere abuse of polemical invective, for underlying it was a wrong political orientation and a false strategy. Again and again communist propagandists proclaimed that 'Germany was already living under fascist rule', and that 'Hitler could not make matters worse than they were under Brüning, the Starvation Chancellor'.[1] But, Trotsky countered, in proclaiming that fascism had already won the day they were in fact declaring the battle lost before it had even

[1] Throughout the year 1931 (and in the first half of 1932) these profound diagnoses and prognostications figured almost daily in the *Rote Fahne*; and they were authoritatively supported by the *Internationale Presse Korrespondenz* and the *Kommunistische Internationale* (see also *XI Plenum IKKI*, and *Kommunisticheskii Internatsional*, 1932, nos. 27–30). Not only Molotov, Manuilsky, Pyatnitsky, and other Russian leaders, but such spokesmen of European communism as Togliatti (Ercoli), Thorez, Cachin, Lenski, Kuusinen, and others dutifully reassured themselves and their followers that the only road to salvation was the one along which Thaelmann was guiding the German party.

begun; at any rate, in telling the masses that Hitler would not be worse than Brüning they were morally disarming them before Hitler. Yet it was folly for a working-class party to deny or blur the distinction between fascism and bourgeois democracy. True enough, both were 'only' different forms and methods of capitalist rule; but, circumstances being what they were, the difference of form and method was of the utmost importance. In a parliamentary democracy the bourgeoisie maintained its domination by means of a broad social compromise with the working class, a compromise which necessitated constant bargaining and presupposed the existence of autonomous proletarian organizations, political parties, and trade unions. From the standpoint of the revolutionary Marxist, these organizations formed 'islands of proletarian democracy within bourgeois democracy', strongholds and ramparts from which the workers could fight against bourgeois rule at large. Fascism meant an end to the social compromise and the bargaining between the classes; it had no use for the channels through which that bargaining had been done; and it could not tolerate the existence of any autonomous working-class organization. Drawing a lesson from the evolution of Italian fascism and, no doubt, reasoning also from the experience of the Bolshevik single party system, Trotsky in advance forcefully described Hitler's totalitarian monopoly of power, under which there would be no room for labour parties and independent trade unions. For this reason alone Marxists and Leninists were bound in duty to defend bourgeois democracy, or rather the 'islands of proletarian democracy within it', against fascist attack. In saying that the Social Democrats formed 'the left wing of fascism' and that they would sooner or later 'make a deal with the Nazis', Stalinist propaganda overlooked the objective impossibility of such a deal.[1] (It should be added that the Social Democratic leaders also entertained this illusion; in 1933 they did indeed make a suicidal effort to reach an accommodation with Hitler.[2]) Trotsky had no doubt that

[1] Trotsky *What Next?* Preface and Chapters I–II, *Écrits*, vol. III, pp. 109–13.

[2] Otto Wels, the leader of the Social Democrats in the Reichstag, used one of his last opportunities to speak from the parliamentary rostrum in order to proclaim his party's readiness to support Hitler's government in the field of foreign policy. At this price he hoped to save his party from destruction by the Nazis; but Hitler did not accept the offer.

Hitler would destroy every vestige of the labour movement, reformist as well as communist. His prognostication followed from the view that National Socialism could not but aim at the complete atomization of German society.

It was thus wrong to treat the Brüning régime as fascist, even though it marked the virtual end of the broad compromise between capital and labour on which the Weimar Republic had been based. Brüning was unable to crush the labour movement (and unable also to hold his ground against National Socialism). Apart from the shaky support of the Catholic Centre Party and apart from Social Democratic 'toleration', he could rely only on the normal resources of the bureaucratic establishment. With these alone he could not suppress the organized working class; and so the political structure still remained what it had been under the Weimar Republic. Only the dynamic force of National Socialism could pulverize it. The breakdown of the compromise between the classes had set the stage for a civil war in which Nazism and the labour movement as a whole would be the real antagonists. The Brüning régime was 'like a ball on the top of a pyramid'; it rested on a fleeting equilibrium between the two hostile camps. Meanwhile, the Nazis recruited millions, whipped up hysteria, and mounted an immense striking force; while socialists and communists alike only marked time and virtually sabotaged the mobilization of their own strength.

A few quotations will convey something of the urgency, and even exasperation with which Trotsky argued:

The Brüning régime is a transitional short-lived prelude to catastrophe. . . . The wiseacres who claim that they see no difference between Brüning and Hitler are in fact saying: it makes no difference whether our organizations exist or whether they are already destroyed. Beneath this pseudo-radical verbiage hides the most sordid passivity. . . . Every thinking worker . . . must be aware of this and see through the empty and rotten talk about . . . Brüning and Hitler being one and the same thing. You are blundering! we reply. You are blundering disgracefully because you are afraid of the difficulties that lie ahead, because you are terrified by the great problems that confront you. You give in before the fighting has begun, you proclaim that we have already suffered defeat. You are lying! The working class is split . . . weakened . . . but it is not yet annihilated.

Its forces are not yet exhausted. Brüning's is a transitional régime. It marks the transition to what? Either to the victory of fascism or to the victory of the working class . . . the two camps are only preparing for the decisive battle. If you identify Brüning with Hitler, you identify the situation before the battle with conditions after defeat; you acknowledge defeat beforehand; you appeal in effect for surrender without a battle. The overwhelming majority of workers, of communists in particular, do not want this. The Stalinist bureaucracy does not want it either. But one must take into account not their good intentions with which Hitler will pave the road to his hell. . . . We must expose to the end the passive, timidly hesitant, defeatist, and declamatory character of the policy of Stalin, Manuilsky, Thaelmann, and Remmele. We must show the revolutionary workers that the Communist party still holds the key to the situation but that the Stalinist bureaucracy is attempting to lock with this key the gates to revolutionary action.[1]

The Social Democratic leaders promised to launch a 'major offensive' if and when Hitler tried to seize power; in the meantime they demanded calm and restraint from the workers. The Stalinists bragged that if Hitler seized power the workers would sweep him away. A leading communist parliamentarian, Remmele, said in the Reichstag: 'Let Hitler take office—he will soon go bankrupt, and then it will be our day.' To this Trotsky replied:

The major offensive must be launched before Brüning is replaced by Hitler, before the workers' organizations are crushed. . . . It is an infamy to promise that the workers will sweep away Hitler once he has seized power. This prepares the way for Hitler's domination. . . . Should the German working class . . . permit fascism to seize power, should it evince so fatal a blindness and passivity, then there are no reasons whatsoever to suppose that after the fascists have seized power the same working class will at once shake off its lethargy and make a clean sweep. Nothing like this has happened in Italy [after Mussolini's rise]. Remmele reasons altogether in the manner of those French petty bourgeois phrase-mongers who [in 1850-1] were convinced that if Louis Bonaparte were to place himself above the Republic the people would rise. . . . The people, however, who permitted the adventurer to seize power proved, sure enough, incapable of sweeping him away thereafter . . . historic earthquakes and a war had to occur before he was overthrown. [In exactly the same way was to end this kind of 'struggle' against Hitler, compared

[1] Trotsky, *What Next?* pp. 38–39; *Écrits*, vol. III, pp. 129–30.

with whom Mussolini and Napoleon III would look like some 'mild, almost humanitarian small town apothecaries'.] 'We are the victors of to-morrow', Remmele brags in the Reichstag. 'We are not afraid of Hitler assuming power.' This means that the victory of to-morrow will be Hitler's not Remmele's. And then you may as well carve it on your nose: the victory of the communists will not come so soon. 'We are not afraid' of Hitler's assuming power—what is this if not the formula of cowardice turned inside out? 'We' do not consider ourselves capable of preventing Hitler from assuming power; worse yet: We, bureaucrats, have so degenerated that we dare not think seriously of fighting Hitler. Therefore 'we are not afraid'. What is it that you are not afraid of: fighting against Hitler? Oh, no . . . they are not afraid of Hitler's victory. They are not afraid of refusing to fight. They are not afraid of confessing their own cowardice. Shame![1]

Warning while there was still time, Trotsky expected the socialists and communists to rally. Their situation was far from hopeless; but it was deteriorating rapidly; and he called for nothing less than preparation and readiness for civil war. To the Social Democratic preachers of moderation and to the Stalinists who defied Hitler to seize power, his call sounded like irresponsible and malignant provocation or, at best, the raving of a Quixote. Events were to prove all too grimly on which side were the irresponsibility, the malice, or the quixotry. They were to demonstrate that, of all courses of action open to the German left, civil war that might have prevented Hitler's assumption of power was in fact the least risky, indeed the only one that might have spared Germany and the world the terrors of the Third Reich and the cataclysms of world war. Early in his campaign Trotsky was convinced that a united left could still rout the Nazis almost without a fight, as the Bolsheviks and Mensheviks had routed Kornilov in August 1917, an example he frequently evoked. He argued that a demonstration of socialist-communist strength might still dissolve Hitler's following, that 'human dust' which had assumed the power of an avalanche only because it moved in a political vacuum and met with no coherent resistance. What favoured the left to some extent was also the fact that the traditional right wing had not yet made common cause with Hitler, even though some

[1] Trotsky, *What Next?* pp. 60–62; *Écrits*, vol. III, pp. 143–5.

potentates of German industry and banking were already backing him. In careful surveys of all the strategic and tactical circumstances, Trotsky analysed the ambiguous attitudes of the capitalist oligarchies, the Junkers, the army, the Stahlhelm, and the police, who were all torn between their desire to use Nazism and their fear of it, between their hope to crush labour with Hitler's hands and their apprehension that he might plunge Germany into a bloody civil war the outcome of which could not be foreseen. Hindenburg, the industrial magnates, and the officers' corps were still in a quandary—hence the quarrels and rows between them and the Nazis. Vigorous Social-ist-Communist action was needed to make the quandary even more difficult, to heighten in the eyes of all the conservative leaders the risks of their support to Hitler, to deepen their vacillations and divisions, and to neutralize at least some of them. Disorientation and inaction on the left, by reducing the risks, would only drive the big bourgeoisie, the army, and Hindenburg into Nazi arms.

A 'united front' between socialists and communists could thus still transform the whole political scene. The same mortal menace now hung over both parties, even if neither was aware of it. This alone should have been enough for them to join forces. The very thought was, of course, repugnant to the Social Democratic chiefs. Anti-communism had been the mainspring of their policy ever since 1918, and had caused them to cling to the 'lesser evil' of Hindenburg-cum-Brüning rather than ally themselves with communism against Hitler. Again and again Trotsky showed how by clinging to the 'lesser evil' they were merely opening the gates to the greater evil of Nazism. But this was for him one more reason why the communists should have made of the united front the central issue of all working-class policies. They failed to do so because they were entangled in the Comintern's 'Third Period' line. The Communist party could not even try to open the eyes of the millions of Social Democratic workers to the danger that threatened all of them when its own leaders were blind to the danger; and Moscow's ban on agreement with the Social Democratic party did not permit effective communist approach to that party. The daily Stalinist vituperation against 'social fascists' incessantly deep-ened the division in the working class, provided the Social

Democratic chiefs with a plausible excuse for their anti-communism, and made it all the easier for them to pursue their disastrous course. Only a genuine and convincing communist appeal to the social democratic conscience and self-interest alike, an appeal untiringly repeated in the hearing of the entire working class, could have broken the barriers between the two parties.

Their united front would have had to be not a diplomatic or parliamentary game with empty and insincere cordialities, in the style of the Anglo-Soviet Committee of 1924–6 (or, one may add, of the Popular Front of 1936–8), but joint preparation and organization for common combat. The two parties and their trade unions would have to 'march separately but strike unitedly' and agree among themselves 'how to strike, whom to strike, and when to strike'. For this they had no need to give up any of their principles or seek any ideological accommodation. Communists must never forget that the Social Democrats could at best be only their 'temporary and uncertain allies', who would always be afraid of extra-parliamentary action and might contract out of the struggle at its most critical turn. Yet it was the communists' duty to bring the strongest pressure to bear on them in order to arouse them to action. If they yielded to the pressure, all would be well; if not, millions of their followers would at least see where each party stood and would be more inclined to respond to a purely communist call to action. Already now, in 1930–1, hardly a day passed without scattered but bloody encounters between workers and Stormtroops; but in these the workers' militancy was being dissipated to no purpose. Only sporadically did socialists and communists agree to repel a Nazi attack jointly. Commenting on one such case Trotsky remarked: 'Oh, supreme leaders! Oh, sevenfold sages of strategy! Learn from these workers . . . do as they do! Do it on a wider, on a national scale.' In the course of the year 1931 Hitler's Stormtroops had grown from 100,000 to 400,000. Trotsky urged the German left to raise their own anti-Nazi militias and to concert the mutual defence of their party offices, factory councils, trade unions, &c. With the Russian Red Guards in mind, he wrote: 'Every factory must become an anti-fascist bulwark, with its own commanders and its own battalions. It is necessary to work with a map of fascist barracks and strongholds in every city and every district. The fascists are attempting to encircle the proletarian strongholds. The encirclers must be encircled.'[1]

The chiefs of the German labour movement could not bring themselves to think and act in terms of civil war, partly because Hitler, as he advanced on his road to office, disavowed from time to time any thought of a *coup d'état* and any intention of using violence. He declared that he would assume and exercise office in the constitutional manner; and these assurances had their effect. 'He lulls his antagonists', Trotsky warned, 'in order to catch them napping and deal them a mortal blow at the right moment. His curtsey to parliamentary democracy may help him to set up in the immediate future a coalition in which his party will obtain the most important posts in order to use these later for a *coup d'état*.' 'This military cunning, no matter how plain and simple, secretes a tremendous force because it is calculated to meet the psychological needs of the intermediate parties who would like to settle everything peacefully and lawfully, and—this is far more dangerous—because it satisfies the gullibility of the popular masses.'[2]

Pravda and *Rote Fahne* now spoke of Trotsky as the 'panic-monger', 'adventurer', and 'Brüning's stooge', who urged communists to abandon proletarian revolution, to defend bourgeois democracy, and to forget that 'without a prior victory over social fascism we cannot vanquish fascism'.[3] Not without anger yet with infinite patience Trotsky dealt with even the most preposterous arguments in order to make his views clear to those befuddled by polemical tricks. Untiringly he went on exploding the fallacy that there could be 'no victory over fascism without a prior victory over social fascism', pointing out that, on the contrary, only when fascism had been

---

[1] Trotsky, *Germany, the Key to the International Situation*, p. 41; *B.O.*, no. 27.

[2] *What Next?*, pp. 147–8

[3] An anthology of German Stalinist polemics against Trotsky would make instructive though unendurably monotonous reading. Even a man like W. Münzenberg wrote: 'Trotsky proposes . . . a bloc between the Communist and Social Democratic Parties. Nothing could be as detrimental to the German working class and communism and nothing would promote fascism so much as the realization of so criminal a proposal. . . . He who proposes such a bloc . . . only assists the social-fascist leaders. His role is indeed . . . plainly fascist.' (*Rote Aufbau*, 15 February 1932.) Münzenberg ended this polemical campaign by committing suicide in exile.

defeated could the communists contend effectively against the social democrats, and that proletarian revolution in Germany could develop only out of a successful resistance to Nazism.

It was all to no avail. As late as September 1932, a few months before Hitler became Chancellor, Thaelmann, at a session of the Comintern Executive, still repeated, what Münzenberg had said: 'In his pamphlet on how National Socialism is to be defeated, Trotsky gives one answer only, and it is this: the German Communist Party must join hands with the Social Democratic Party. . . . This, according to Trotsky, is the only way in which the German working class can save itself from fascism. Either, says he, the Communist party makes common cause with the Social Democrats, or the German working class is lost for ten or twenty years. This is the theory of an utterly bankrupt Fascist and counter-revolutionary. This is indeed the worst, the most dangerous, and the most criminal theory that Trotsky has construed in these last years of his counter-revolutionary propaganda.'[1]

'One of the decisive moments in history is approaching', Trotsky rejoined, '. . . when the Comintern as a revolutionary factor may be wiped off the political map for an entire historic epoch. Let blind men and cowards refuse to notice this. Let slanderers and hired scribblers accuse us of being in league with the counter-revolution. Has not counter-revolution become anything . . . that interferes with the digestion of communist bureaucrats . . . nothing must be concealed, nothing belittled. We must tell the advanced workers as loudly as we can: After the "third period" of recklessness and boasting the fourth period of panic and capitulation has set in.' In an almost desperate effort to arouse the communists, Trotsky put into words the whole power of his conviction and gave them once again the ring of an alarm bell: 'Workers-communists! There are hundreds of thousands, there are millions of you. . . . If fascism comes to power it will ride like a terrific tank over your skulls and spines. Your salvation lies in merciless struggle.

[1] Compare *Rote Aufbau*, loc. cit. with *XII Plenum IKKI*, part 3; *Kommunistichesky Internatsional*, 1932, nos. 28–29, pp. 102–3, 111 and *passim*. Thaelmann was serenely confident that 'Germany will of course not go fascist—our electoral victories are a guarantee of this . . . the irresistible advance of communism is a guarantee of this'.

Only a fighting unity with social democratic workers can bring victory. Make haste, communist workers, you have very little time to lose.'[1]

.   .   .   .

To have to let, at such a time, the grass grow under his feet at Prinkipo was for Trotsky more and more painful. Letters and newspapers from the continent reached him with much, sometimes with a fortnight's, delay; it took even longer for his brochures and manifestoes to reach Germany. In 1923, when Germany seemed on the verge of revolution, he had asked the Politbureau to relieve him from his official posts and allow him to go to Germany and direct, as the German party had asked, revolutionary operations there. How much more anxious was he to find himself nearer the scene of action now, when the future of communism and the political fortunes of the world were being decided for decades ahead. In 1931 there was talk about his going on a short lecture tour to Germany; but, of course, nothing came of it. There was no chance of his getting out of Turkey. Worse still, his few followers in the Reich were making no headway. They published a tiny paper *Permanente Revolution*, which appeared once a month, filling its columns with Trotsky's writings, and had almost no impact (although his brochures were quite widely read and discussed). He planned to set up an International Secretariat in Berlin where the brothers Sobolevicius were very active, and whither the *Bulletin Oppozitsii* had already been transferred from Paris. To improve his contact with the Secretariat it was decided that Lyova should leave for Berlin and act there as his father's representative, or, as organizational punctilio demanded, as the 'representative of the Russian Section of the Left Opposition'.

Lyova, we know, had shared with his parents all the vicissitudes of their exile and was Trotsky's right-hand man. Yet relations between father and son had not been unruffled. They were in full political concord, and Lyova's adoration for his father amounted to identification with him. Yet it was this identification that was also a cause of strain. Trotsky had an uneasy feeling that his own personality and interests had

[1] Trotsky, *Germany, the Key*, etc., p. 44.

imposed themselves too overwhelmingly on Lyova, and that he had reduced Lyova to the frustrating part of the great man's little son. Yet he craved the filial devotion. The more lonely he was the more he depended on it. Lyova was the only man with whom he could freely thrash out his ideas and plans and share innermost thoughts, his most trusted critic, and, as he liked to think, his 'link' (in later years, his only link) with the young Russian revolutionary generation. Yet at times Lyova's absolute devotion disturbed him: he wanted greater independence in his son and almost wished for some signs of filial dissent. But dissent, when there was a hint of it, upset him and made him fear estrangement. Seclusion and incessant intercourse deepened the mutual dependence and also heightened the stresses which, though not unnatural between father and son, had in them something of the irritable tension between two prisoners who have shared a dungeon for too long. Trotsky was exacting towards his assistants and secretaries, but his demands were never as severe as those he made on himself and his son. With strangers he was self-controlled and polite; but under great nervous strain his self-control was liable to break down when he was alone with his next of kin. Harsh reproaches would then come down upon Lyova's head on account of 'disorder' in the secretariat, 'sloth and sloppiness', and 'letting down' his father, reproaches which could not but hurt the dedicated, industrious and conscientious young man.[1]

Some relief was therefore mixed with sadness when parents and son agreed on separation. There was probably yet another reason for this decision: Raymond Molinier's wife, Jeanne, had left her husband and chosen to stay with Lyova. Molinier, however, was still a frequent and helpful visitor at Prinkipo; and Lyova's and Jeanne's departure may well have spared them all embarrassing encounters. It was doubtful at first whether Lyova would obtain a German entry permit. (The year before he had in vain applied for a French visa: the French police replied that they knew of his revolutionary activities and did not wish to see him in Paris.) But, having inscribed himself as a student at the *Technische Hochschule* in Berlin, he finally

[1] Relations between father and son are characterized here (and in the following pages) on the basis of the family correspondence, which fills forty folders in the Closed Section of the Trotsky *Archives* and consists of 1,244 items.

L

obtained the German visa in February 1931. The academic purpose of his sojourn was no mere pretext, for at the *Hochshule* he did indeed take with much application courses in physics and mathematics; but his chief preoccupations remained, of course, political.[1]

A few weeks before Lyova's departure, in the middle of January, something occurred that was to affect the life of the entire family: Zina and her five-year-old son Seva arrived from Moscow. For several months she had been expected at Büyük Ada; but hope of her coming had been nearly given up, because the Soviet Government had repeatedly refused her permission to make the journey. Her husband, Platon Volkov, was deported; and she herself had been detained twice because of her involvement with the Opposition. Only after the intervention of western European friends who appealed to Soviet Ambassadors on compassionate grounds—her health had broken down after the death of her sister Nina whom she had nursed to the end—did she obtain the exit permit. But there was a catch. She was allowed to take with her only one child and had to leave behind another, a little daughter, a six- or seven-year-old hostage to Stalin. Alexandra Lvovna, Trotsky's first wife, who, herself under a cloud, was bringing up Nina's two children, took care of this infant too, and urged Zina to leave, join her father, and repair her health abroad.

Zina came to Prinkipo a nervous wreck, though this did not show at once in the flush of reunion. Her father received her with the utmost tenderness. 'In the first period of my stay', she wrote later to her mother in Leningrad, 'he was so soft and attentive to me that I cannot even describe it. . . .' Of all his children she, his first-born, resembled him most. She had the same sharp, dark features, the same fiery eyes, the same smile, the same sardonic irony, the same deep-running emotions, and something also of his untameable mind and of his eloquence. She seemed to have inherited his political passions, his militancy, and craving for activity. 'She was', as

[1] Lyova's mathematical exercise books, densely and neatly filled, with entries dated and marked by his academic teachers, served later as evidence of his alibi in the Mexican counter-trial of 1937. The exercise books are preserved in *The Archives*. In an undated letter to Dr. Soblen (Well), Lyova explained the reasons of organization which impelled him to move to Berlin. (It took seven or eight months before he obtained the German visa.)

her mother put it, 'more public-spirited than family-minded.'[1]

In Trotsky's feeling for her there was a touch of remorse. Ever since those days in 1917 when, addressing multitudes at the *Cirque Moderne* in Petrograd, he had felt the loving eyes of his two adolescent daughters staring up at him from the audience, and fixed on him, he had been aware of Zina's intense emotion for him. Yet she was to him almost a stranger. It was nearly thirty years now since he had left his first wife and their two babies in the eastern Siberian settlement of Verkholensk (the place of his first exile)—nearly thirty years since he had arranged in his bed there the dummy of a man in order to deceive the police and delay their pursuit.[2] It was as if that dummy had deceived the offspring of his first marriage also. In fifteen years, up to 1917, he had seen his daughters only twice or thrice, fleetingly; and he could give them only very little time and attention thereafter, in the years of the revolution, the civil war, and the cruel struggles that followed. His heart went out to them when he was exiled to Alma Ata; but then it was too late: Nina presently died; and Zina was too ill to undertake the journey from Moscow, too ill even to come later to the family's sad farewell meeting on the train when he was being deported from Russia. She arrived at Prinkipo heart-broken yet overwhelmed with joy, love, and pride in her father; she had come not merely as a sick and suffering daughter but as a dedicated follower, hoping to be of use to him, offering her services, and yearning to be admitted to his confidence. They wept together over Nina's death; they talked about friends and comrades and deported relatives; and they argued about politics. She listened all ecstasy and read, with a thrill, the manuscripts of the *History of the Russian Revolution* and his other writings, acquainted herself with the controversies in which he was engaged, absorbed their dramatic gravity, and relished his sarcasm and wit. She was convulsed with laughter when she came across Churchill's essay on the 'Ogre of Europe': and it was as the 'ogre' that she liked to address her father.[3]

---

[1] The quotation is from Sokolovskaya's letter to Trotsky, written after Zina's death. See its summary on pp. 197–8, *The Archives*, Closed Section.

[2] *The Prophet Armed*, p. 55.

[3] These details are drawn from Zina's correspondence. *The Archives*, Closed Section. In *The Archives* I found a picture of her which she presented to her father with the inscription: 'To the ogre'.

The other members of the family also gave her affection and compassion and did their best to make her feel at home. Natalya Ivanovna's position was admittedly delicate; but she had been closer to the children of Trotsky's first marriage than he himself, and had not only tried to overcome estrangement with friendship but had behaved towards them like a second mother. Not deceived by the apparent improvement in Zina's condition, she took her to the doctors and gave meticulous attention to her health. Too sensitive to imagine that the hidden strains could ever vanish altogether, she tried to efface herself whenever she felt that father and daughter would best be left by themselves. Curiously, Lyova's relation with his sister was far more tense. Their characters were discordant. Resembling his mother rather than his father, Lyova was reserved, modest, and even-tempered; he was easily disconcerted by his sister's intensity and impassioned expansiveness, while her feelings were tinged by a jealousy of Lyova's closeness to their father. In the warmth of reunion, and while Lyova was preparing for his move to Berlin, these strains were subdued. The whole family went into raptures over Zina's child, whose chatter and pranks brought an unfamiliar note into the austere and industrious existence of the household. This was, it seems, the first time that Trotsky, who already had five grandchildren, could freely indulge in the sentiments of grandfatherhood.

Shortly after Zina's arrival, in the dead of night, a great fire broke out in the house, consuming most of the family's belongings and Trotsky's library. With difficulty he wrested from the flames his archives and the manuscript of the just completed first volume of his *History*. A suspicion of arson crossed everyone's mind: was this perhaps an attempt by the G.P.U. to destroy the archives? An investigation was opened; witnesses were cross-examined, but nothing was found out. 'All of us felt dejected and were very much disturbed . . .', writes one of Trotsky's secretaries, 'all, except Trotsky himself.' The household moved to a near-by hotel; and 'no sooner were we settled than he laid out his manuscripts on the table, called the stenographer, and began to dictate a chapter of his book, as if nothing had happened during the night'.[1] After a few days they

---

[1] Jan Fraenkel in *Militant*, 2 January 1932. See also *Journal d'Orient*, 8 April 1931.

moved to Kodikoy, an Anglo-American residential suburb on the eastern fringe of Constantinople, into a wooden house surrounded by high barbed wire fences, where the household, complete with secretaries, policemen, and fishermen, stayed for about a year, until the Büyük Ada house was habitable again.

A few months after the move to Kodikoy another fire broke out. Once again the archives were hastily removed; and the family had to bivouac in barns and shacks near by; and once again the thought of an incendiary hand occurred to everyone. But it turned out that the fire was caused by Zina's child playing with matches and a pile of wood, rags, and sawdust in the loft. It was a relief after all the scares; and everyone laughed and teased the 'little G.P.U. agent'.

As the weeks passed Zina's illness came back. Her lungs were diseased; she had to undergo several operations; she could not bear the heat of the eastern Mediterranean; and she was tormented by anxiety over her husband and the child she had left behind. Under the stress of illness and worry, her shaky nervous balance gave way. Hidden tensions and conflicts, probably rooted in the misery of her childhood and nurtured by later experiences, came to the surface. Her behaviour became explosive and incoherent. She gave vent to memories, desires, and grievances that had hitherto lurked beyond the threshold of her consciousness. She was obsessed by the sense of being an unwanted daughter, unwanted by the father whom she adored with all her passion as the life-giving genius of revolution. It was her faith in him, she herself wrote, that kept her alive and gave her the strength to grapple with her predicament—without him life would have been empty. Yet she felt an unsurmountable barrier between herself and him. 'I know, I know', these are words she threw at him, 'that children are not wanted, that they come only as punishment for sins committed.'[1] It was as if the shock she may have suffered as an infant, on that day when she found instead of him the mere dummy of a man in his bed, reverberated in the reproach.

In this emotional turmoil she struggled to suppress her inner resentment at her father's second marriage. Outwardly her attitude towards Natalya Ivanovna was one of affection and care; but there was an unnatural exaltation in it. She walked

[1] Zina's letter of 26 February 1932.

around her stepmother on tiptoe, insistently inquiring and worrying about her, and lavishing caresses and apologies. Yet the resentment was close enough to the surface for father and stepmother to feel it; every now and then it broke through and hit them in their faces. Much though they tried to ignore or soothe it, relations became tense. To avoid making them worse, Trotsky withdrew within himself. The more he did so, the more frustrated was Zina's yearning for his confidence and closeness. She had hoped to work at least as one of his assistants. He, worried about her health and mindful of her possible return to her child in Russia, did not encourage this ambition. He wished her to use her stay abroad for a cure and in the meantime to avoid compromising her politically, as if being his daughter had not already compromised her finally and irretrievably. The worsening of her illness, he felt, necessitated even greater reserve on his part and made work in common almost impossible. He could not take her into his confidence over the affairs of the Opposition in Russia; and it was in these precisely that she was breathlessly interested. At this time his correspondence with his Russian followers was still fairly abundant, part of it being despatched openly but part clandestinely, with coded signatures and addresses. The greatest discretion about the codes had to be exercised; and secrecy had to be redoubled *vis-à-vis* an ill and unbalanced person who on return to Russia might be subjected to inquisitorial interrogation. Elementary rules of underground communications required such safeguards; but the unfortunate woman took these as a slur on her, a sign of her father's distrust. 'To Papa', she often repeated, 'I am a good-for-nothing.' More resentment, more reproach and self-reproach, more gloom, and more and graver mental disturbance made everyone feel worse. In the summer she left home, and in a near-by sanatorium underwent the operations on her lungs. She returned, her physical health somewhat restored, but her misery unrelieved.

Distressed and shaken with pity, Trotsky was a prey to guilt and helplessness. How much easier it was to see in what way the great ills of society should be fought against than to relieve the sufferings of an incurable daughter! How much easier to diagnose the turmoil in the collective mind of the German petty bourgeoisie than to penetrate into the pain-laden recesses of

Zina's personality! How much superior was one's Marxian understanding of social psychology to one's grasp of the troubles of the individual psyche! He watched Zina's features and eyes overcast with insanity—they were his features, his eyes. For him, the prodigy of intellectual lucidity and self-discipline, it was unbearable to see her so incoherent, so distraught. It was as if reason itself had discovered in unreason its closest progeny and its double. Tenderness and horror, compassion and revulsion, pride and humiliation were at odds in him. He was wounded; he was helpless; he grew irritable. Sometimes, when Zina's jealousy burst out to hurt Natalya Ivanovna, he raised his voice demanding tact and courtesy. His raised voice reduced her to utter prostration. Remembering some such scene, she wrote to him a year later: 'Don't shout at me, Papa, don't— your shouting is the one thing I cannot endure; in this I am like my mother.' And she added: 'There is nothing I desire so much, if only I have enough strength to do it, as to soften for Natalya Ivanovna that of which I have turned out without guilt to be guilty towards her.'[1]

With tempers frayed and Zina's illness becoming quite alarming—she began to suffer fits of delirium—she could not stay on. For some time he had thought that she should undergo psychoanalytical treatment, and he had written about this to the Pfemferts in Berlin. She resisted. She had no wish, she said, to submerge herself in the 'filth' of her subconsciousness; and she could not bear the thought that, having overcome so many obstacles and borne so many sacrifices to rejoin her father, she should again be separated from him. She would also have to be separated from her son, for it was very difficult for her to take care of his upbringing. But she yielded to persuasion; and in the autumn of 1931, leaving Seva behind, she went to Berlin. The parting was a torment to both father and daughter. This is how she related it to Lyova: ' "You are an astonishing person, [her father told her in their last talk] I have never met anyone like you." ' 'He said that', she added, 'in an expressive and severe voice'.

This was the voice of reason baffled and thwarted by un-reason.

.   .   .   .

[1] The Russian expression is: '*Bez viny vinovata*.' The letter is undated.

Life in the German capital, when Zina arrived there, was a crescendo of chaos and topsy-turviness. She arrived a few weeks after a plebiscite, arranged on the initiative of Hitler and Goebbels, the purpose of which was to overthrow the Social Democratic *Landesregierung* of Prussia. The Nazis had let loose a savage chauvinistic campaign calling for a 'People's Revolution' against the party that 'had accepted the slavery and humiliation of the Versailles Peace'. The Communist party reacted by addressing to the Social Democratic Ministers of Prussia, Braun and Severing, an ultimatum in which it offered to defend their government if they agreed to certain demands, but threatened to vote against it if the demands were rejected. On the face of it, this was a depature from the 'third period tactics', at least in so far as the communists had made a direct approach to Social Democratic leaders. Actually they 'concentrated fire on the social-fascists'; and when the Prussian Government refused their demands, they called upon the workers to cast their votes against it. Thus, instead of making a united front with the Social Democrats, conditionally or unconditionally, they formed an unavowed but all too real, and unconditional, united front with the Nazis; and to save face they called the enterprise *der Rote Volksentscheid*, the Red Plebiscite.

A fatal and deeply demoralizing ambiguity now appeared in communist policy, which was to persist until Hitler's seizure of power and even thereafter. Not infrequently the same slogans appeared on communist and Nazi banners. The Nazis, seeking to win socially discontented and radical elements, promised that their 'People's Revolution' would settle accounts with finance capital. The Communist party, wary of calling for a proletarian socialist revolution, spoke, instead, of the 'People's Revolution' which would achieve Germany's 'social and national liberation' and break the shackles of Versailles. The spirit of nationalism insinuated itself more and more strongly into its propaganda just at a time when nothing was more urgent in Germany than the need to stem the mounting tide of racial and chauvinist fanaticism. Although the plebiscite went in favour of the Social Democrats, its effect was to deepen the breach in the working class and to make confusion worse confounded.

Trotsky attacked Thaelmann's and the Comintern's 'national communism' with the utmost vigour, exposing the absurdities of the 'Red Plebiscite'. The whole venture, he argued, was all the more repugnant because communists and Nazis remained, and could not help remaining, mortal enemies. In self-justification, the Stalinists pointed out that the Social Democrats were paving the way for Nazism. This was all too true, Trotsky remarked, but if the Social Democrats paved the way for a Nazi victory, should the communists shorten it? It happens sometimes that the parties of revolution and counter-revolution attack the same 'moderate' enemy from their opposite poles. But a Marxist party can afford to do this only when the tide runs in its favour, not when it runs, as it did in Germany, in favour of counter-revolution. 'To go out into the street with the slogan "Down with the government of Brüning and Braun" is a reckless adventure when the whole balance of strength is such that the government of Brüning and Braun can be replaced only by a government of Hitler and Hugenberg. The same slogan would acquire quite a different meaning if and when it presaged the direct struggle for power by the working class.' Even now he did not doubt the good intentions of the Communist party; but 'unfortunately, the Stalinist bureaucracy is trying . . . to act against fascism by using the weapons of the latter. It borrows colours from the political palette of Nazism and tries to outdo Nazism at an auction of patriotism. These are not methods of a principled class struggle, but tricks of a petty market competition . . . a betrayal of Marxism . . . a display of concentrated bureaucratic stupidity.' Those who talked about the 'People's Revolution' and about freeing Germany from the chains of Versailles had forgotten Karl Liebknecht's maxim that for the working class 'the main enemy stands in their own country'. The insinuation of nationalism into communist thinking had begun with Stalin's 'socialism in one country' and it now produced Thaelmann's 'national communism'. 'Ideas have not only their own logic but their own explosive force'; and the lack of scruple with which the Comintern tried to outbid Hitler in nationalist demagogy showed up the 'spiritual emptiness of Stalinism'.[1]

What, according to Trotsky, was at stake was not only all the hard-won achievements of the German labour movement but

[1] 'Protiv Natsjonal-Kommunizma (Uroki Krasnovo Referenduma)', *B.O.*, no. 24. The article was published as a pamphlet in Germany.

the future of civilization: with Nazism the shadow of the dark ages was returning to Europe. Hitler, if victorious, would not merely preserve capitalism but reduce it to barbarism. The enraged *Kleinbürger* 'repudiated not only Marxism but even Darwinism', and to the rationalism and materialism of the eighteenth, nineteenth, and twentieth centuries he opposed the myths of the tenth or eleventh century, the mystique of race and blood. This, their supposed racial superiority, was to boost the pride of Germany's lower middle classes, and give them an imaginary escape from the miseries of their life. In its rabid anti-Marxism and rejection of the 'economic view of history', 'National Socialism descends lower down: from economic materialism to zoological materialism'. Nazism collected 'all the refuse of international political thought . . . to make up the intellectual treasure of the new Germanic Messianism'. It stirred and rallied all the forces of barbarism lurking under the thin surface of 'civilized' class society. It tapped inexhaustible reserves of darkness, ignorance, and savagery. In a memorable phrase, alive with a premonition of the *autos da fé* and gas chambers of the Third Reich, Trotsky thus described the essence of Nazism: 'Everything which society, it if had developed normally [i.e. towards socialism], would have rejected . . . as the excrement of culture is now bursting out through its throat: capitalist civilization is disgorging undigested barbarity —such is the physiology of National Socialism.'[1]

That communist (as well as non-communist) opinion of the early nineteen-thirties was insensitive to such a philosophical-historical view of Nazism need not perhaps surprise the historian. What he must find more difficult to comprehend is how the leaders of the Soviet Union and the great mass of communists all over the world could remain deaf to what Trotsky was saying about the threat to the Soviet Union. In November 1931, ten years before the battle of Moscow, he wrote: 'A victory of fascism in Germany would signify the inevitability of war against the U.S.S.R.'[2] At that time Moscow still saw France as the chief western antagonist of the Soviet Union; and it feared an imminent attack from Japan, which had just embarked upon the invasion of Manchuria. The progress of

[1] *Écrits*, vol. III, pp. 391–9. 'Qu'est-ce que c'est le national-socialisme?'
[2] Op. cit., pp. 100–1.

Nazism had as yet aroused little or no apprehension in Stalin and his advisers, even though Hitler was loudly proclaiming that he was out to destroy Bolshevism and conquer the East. Stalin assumed that these were the ravings of Hitler the 'rebel', but that Hitler the Chancellor would not easily forgo the advantages which Germany derived from her relations with Russia, under the Rapallo Treaty. Stalin expected that Hitler's striving to rearm Germany would bring him into conflict with France and compel him to abate his hostility towards the Soviet Union. It was not for nothing that the Comintern encouraged the German communists to lend ambiguous support to Hitler's campaign against Versailles: that campaign was to divert Hitler from his ambition to lead a western crusade against Bolshevism.

Trotsky struggled against this unawareness of the international implications of Nazism. He did not believe that France was still Russia's chief enemy, as in the years of intervention. 'Not a single one of the normal bourgeois parliamentary governments', he maintained, 'can at present risk a war against the U.S.S.R.: such an undertaking would entail incalculable domestic complications. But once Hitler has seized power . . . and pulverized and demoralized the German working class for many years to come, his will be the only government capable of waging war against the U.S.S.R.'[1] Nor did he believe that the Soviet Union was seriously threatened by Japan. He forecast that by invading Manchuria, Japan would involve herself in a long and exhausting war with China, which would divert Japanese strength from the Soviet Union and hasten revolution in China. 'The basic conditions of the East—immense distances, huge populations, and economic backwardness imply that the whole process [of Japanese conquest] will be slow, creeping, and wasteful. In any case, in the Far East no immediate and grave danger threatens the Soviet Union. The crucial events of the coming period will unfold in Europe, in Germany', where 'the political and economic antagonisms have reached an unprecedented sharpness . . . and the dénouement is close at hand.' And again: 'For many years to come, not only the fate of Germany . . . but the destinies of Europe and the destinies of the entire world will be decided in Germany.' 'Socialist

[1] Loc. cit.

construction in the Soviet Union, the march of the Spanish revolution, the growth of a pre-revolutionary situation in England, the future of French imperialism, the fate of the revolutionary movement in China and India, all these issues reduce themselves . . . to this single question: who is going to win in Germany in the course of the coming months? Communism or fascism?'[1]

Trotsky assumed that for an anti-Soviet crusade Hitler could gain the support of world capitalism, and that this would entail 'a frightful isolation of the Soviet Union and the necessity to fight a life-and-death struggle under the hardest and most dangerous conditions'. 'If fascism were to crush the German working class, this would amount to at least half the collapse of the Republic of the Soviets.' Only if the workers succeeded in barring Hitler's road to power would Germany, the U.S.S.R., and the world be saved from catastrophe. Stalin's policy in Germany was therefore directed against the vital interests of the Soviet Union as well as of German communism. Soviet security and the international proletarian interest were inextricably bound up. For years Stalin and the Comintern had screamed about the imminence of an anti-Soviet crusade; but now, when the peril was real, they were silent. Yet it should be 'an axiom' that a Nazi attempt to seize power 'must be followed by a mobilization of the Red Army. For the workers' state this will be a matter of revolutionary self-defence. . . . Germany is not only Germany. It is the heart of Europe. Hitler is not only Hitler. He is the candidate for the role of a super-Wrangel. But the Red Army is not only the Red Army. It is the instrument of proletarian world revolution.'[2]

A few months later, in April 1932, he restated this idea even more strikingly. Routine-ridden politicians and diplomats, he said, were blind to what was coming, just as they had been on the eve of the First World War. 'My relations with the present government of Moscow are not of such a nature as to permit me to speak in its name or refer to its intentions. . . . With all the greater frankness can I state how, in my view, the Soviet government should act in case of a fascist upheaval in Germany. In their place, I would, at the very moment of receiving telegraphic news of this event, sign a mobilization order calling up

[1] Loc. cit., p. 95.   [2] Ibid., p. 101.

several age groups. In the face of a mortal enemy, when the logic of the situation points to inevitable war, it would be irresponsible and unpardonable to give that enemy time to establish himself, to consolidate his positions, to conclude alliances . . . and to work out the plan of attack. . . .' And again: 'War between Hitlerite Germany and the Soviet Union would be inevitable and this in the short term', in view of which even the question who would attack first was of secondary importance. With an eye to those in France and Britain who hoped to save the *status quo* in the West and the Versailles system by diverting German imperialism eastwards, Trotsky wrote that 'whatever illusions are entertained in Paris one can safely predict that the Versailles system would be one of the first to be consumed in the flames of a war between Bolshevism and fascism'.[1]

The Comintern Press at once branded Trotsky as a 'treacherous warmonger' seeking to embroil Russia and Germany; and to many outside the Comintern too the boldness of his statements seemed reckless. His attitude, however, will not appear quite so reckless if it is remembered that, even in the early nineteen-thirties, with Germany, Britain, and the United States disarmed, the Soviet Union was the greatest military power of the world. But Trotsky did not in fact urge the Soviet Government to wage war against Germany, even a Nazi Germany. In 1933, after Hitler had become Chancellor, Trotsky declared that in the existing circumstances mobilization of the Red Army would serve no purpose. He had advocated it, he explained, on the assumption that Hitler would have to shoot his way to office—he had refused to believe that the German labour movement would allow Hitler to become the master of their country without having to fire a shot. It was in this context of an assumed civil war in Germany that he had insisted on the Red Army's duty to intervene.[2] Admittedly this would have been a hazardous course, but less so than was waiting passively for Hitler's ascendancy and Germany's rearmament. Trotsky's attitude, revolutionary in its political aspect, was in its military aspect similar to that which Winston

[1] Ibid., pp. 104–5.
[2] The article appeared originally in the American *Forum*, 15 April 1932. *Écrits*, vol. III, pp. 233–40. See also 'Hitler i Krasnaya Armija', *B.O.*, no. 34, May 1933.

Churchill was to adopt four or five years later, when he called the British and French Governments to counter Hitler's march into the Rhineland by measures of mobilization and preparation for war. This attitude earned Churchill the unrivalled moral authority he needed to become Britain's leader in the Second World War. Vilification was all it earned Trotsky.

Meanwhile, the Nazi avalanche moved on. In the spring of 1932 Germany was to elect a President, and Hitler posed his candidature. A Socialist-Communist candidate was still sure to poll more votes than Hitler or any other contestant—at the repeated parliamentary elections of that year Communists and Social-Democrats invariably obtained more than 13 million votes. But the Social Democrats decided to uphold the candidature of Hindenburg, the nearly nonagenarian retiring President, whom they had opposed at the previous election as the very symbol of the old Imperial reaction, but behind whose senile back they now sought to shelter. The Communist party called the workers to vote for Thaelmann. Hindenburg was re-elected; and at once he delivered the *coup de grâce* to the parliamentary régime and struck at the Social Democrats. He dismissed Brüning, who had just made a half-hearted attempt to ban Hitler's Stormtroops and had also incurred the enmity of the East Prussian Junkers. Hindenburg's new Chancellor, von Papen, lifted the ban on the Stormtroops; and, on 20 July 1932, he deposed by decree the Social-Democratic government of Prussia which the Nazis had in vain tried to overthrow by plebiscite. The event was remarkable for its tragi-comedy: a lieutenant commanding a section of soldiers turned out of their offices the Prussian Prime Minister and Minister of Interior, who nominally had the whole Prussian police under their orders. Too late and perfunctorily the communists advised the Social Democrats to call a general strike and offered support. Once again the Social Democrats refused to make common cause with their 'enemies on the left'; and they deluded themselves that von Papen and Hindenburg's camarilla (of which General Schleicher was the moving spirit), would somehow outmanœuvre Hitler and keep him at bay. This was a widespread illusion in these last months of the Weimar Republic: von Papen, having so easily seized the Social Democratic 'fortress' in Prussia, seemed very powerful; he appeared to have stolen Hitler's thunder; and the Nazi movement was moment-arily losing impetus.[1]

All the more must one marvel at the accuracy and precision of Trotsky's analyses and predictions. 'The less the workers were prepared to fight', he commented, 'the greater was the im-pression of strength which Papen's government gave. . . .' However, this is not yet the fascist upheaval—that is still to come. Papen will not be able to outmanœuvre Hitler and prevent a Nazi dictatorship, for he does not even have the limited strength Brüning possessed: he is backed only by the most archaic elements of the Prussian bureaucracy. He will not be able to control the fury and the rage of the millions that follow Hitler—only the determination and the militancy of millions of workers might do that. But how could the workers have that determination when they see the Prussian Socialist government allowing itself to be overthrown by a 'flick on the nose', and when the communists, after telling them for years that Germany is already fascist, now call them to rise in general strike against Papen's 'fascist' *coup d'état* and in defence of the 'social-fascist' government of Prussia. Yet, confused though the workers are, the alternative is still a victory of Nazism or a victory of the working class— *tertium non datur*. Papen, Trotsky insisted, will have no more than 'a hundred days'; and so will Schleicher who will follow him as Chancellor. Then the Reichswehr and the Junkers will form a coalition with the Nazis in the hope of taming the latter. It will all be in vain: 'All conceivable [governmental] combinations with Hitler must lead to the absorption of the bureaucracy, the courts, the police, and the army by fascism.' Even now, he held, it was not yet too late for a 'united front' of the workers; but—'how much time has been wasted without purpose, senselessly, and shamefully!'[2]

.   .   .   .

About this time Trotsky was also in controversy with the Comintern over the Spanish revolution. Primo de Rivera's dictatorship came to an end in 1930 and the collapse of the monarchy followed in April 1931. While Germany was

[1] W. L. Shirer, *The Rise and Fall of the Third Reich*, pp. 158–60, 170–2, and *passim*.

[2] *Der Einrige Weg*; see also *B.O.*, nos. 29–30, September 1932.

developing from a bourgeois democracy to an authoritarian régime, in Spain the opposite was happening. Yet in both countries the Comintern clung to the third period policy. While the German party declared that the antagonism between fascism and bourgeois democracy was irrelevant, the Spanish party made light of the conflict between monarchy and republic. In Moscow, Manuilsky told the Comintern Executive in February 1930, after the fall of Primo de Rivera: 'Movements of this kind pass across the historic screen as mere incidents and leave no deep traces in the mind of the working masses. . . . A single strike . . . may be of greater importance than a "revolution" like the Spanish. . . .'[1] The revolution that was to occupy the world for nearly a decade was still referred to in quotation marks. The abdication of King Alfonso caught the party by surprise. When subsequently Spain resounded with the demand for a democratically elected Cortes, the official communists, like the Anarcho-Syndicalists, maintained that the workers and peasants would gain nothing from any parliament; and they favoured the boycott of elections. Yet, at the same time, the Comintern declared that the Spanish revolution, in view of the country's backwardness, must keep within 'bourgeois democratic' limits, and that 'proletarian dictatorship was not on the order of the day'. It is easy to recognize there the Stalinist canon developed as antithesis to Trotsky's Permanent Revolution and applied in China in 1925–7. This canon was to underlie Stalinist policy in Spain through all its phases. At a later stage, in 1936–8, it was invoked to justify the communist coalition with bourgeois republican parties in the Popular Front, the 'moderate' policy of the Communist party and its repressive action against the P.O.U.M., the Trotskyists, and the radical Anarcho-Syndicalists. In the early nineteen-thirties, however, the same canon was incongruously combined with ultra-left tactics and with the rejection of the demands for a Constituent Assembly and democratic liberties, the classical desiderata of bourgeois revolution.

Trotsky asserted that the Spanish revolution would have to pass, as the Russian revolution had done, from the bourgeois

---

[1] Yet later in the year the organ of the Comintern Executive blamed the Spanish Communists for having missed the revolutionary significance of the events. See *Kommunisticheskii Internatsional*, 1930, nos. 34–35,

into the socialist phase, if it was not to be defeated. Of all European countries Spain was closest to pre-1917 Russia in social structure, in the alignment of political forces—and in Spain as in Russia Workers' Councils or Juntas were destined to be the organs of revolution. While insisting on the 'permanence' of the revolution, Trotsky urged the communists to adopt more realistic tactics, to raise or support demands for general franchise, for a Constituent Assembly, for the self-determination of the Catalans and Basques, and, above all, to support the peasantry's struggle for land. The peasants were bound to look to the Cortes for a solution of the land problem; and communists were in duty bound to state their agrarian programme from the parliamentary platform, if only to promote the peasantry's extra-parliamentary action. They could not do this under their 'third period' policy and while they were inclined to ignore and boycott parliament. 'Parliamentary cretinism is a detestable disease, but anti-parliamentary cretinism is not much better', he remarked. Had not the Bolsheviks called for a Constituent Assembly in 1917? In Spain parliamentary politics were bound to be even more important than in Russia, because the rhythm of the revolution would be slower; and the Spanish communists should in their action 'take less into account the Russian experience than that of the great French Revolution. The Jacobin dictatorship was preceded by three parliamentary assemblies'; and something similar might happen in Spain.[1]

The Spanish party was not only disoriented, small, and weak; it was also disorganized by the divisions and splits which were inseparable from Stalinist orthodoxy. It had already expelled several Trotskyist and semi-Trotskyist groups and Andres Nin its founder and one-time leader. The splits were to be the cause of much demoralization in republican Spain in later years, and the baiting of Nin was to end in his assassination. Already in April 1931, only a few days after the overthrow of the monarchy, Trotsky protested in a confidential message to the Politbureau in Moscow against the heresy hunt in Spain. He recalled that in 1917 the Bolsheviks had, under Lenin's guidance, joined hands with all groupings close to them, regardless of past differences—he himself had then entered the Bolshevik

---

[1] *Écrits*, vol. III, pp. 451–71 and *passim*; *B.O.*, nos. 21–22, 1931.

M

party—and they found that this, and their ability to base their unity and discipline on freedom of internal debate, decisively strengthened their hands in the struggle for power. 'Are there any other ways or methods', he asked, 'which would permit the proletarian vanguard of Spain to work out its ideas and to become permeated with the unshakeable conviction of the truth and justice of these ideas—a conviction which alone would enable them to lead the popular masses to the decisive assault on the old order?' The heresy hunts confused and demoralized the ranks and facilitated a fascist victory which would have 'grave repercussions for the whole of Europe and the U.S.S.R.' He asked the Politbureau to advise—'precisely to advise, not to order'—the Spanish communists to call a unity congress; and he offered to advise his followers to co-operate in this. 'The march of events in Spain will daily confirm the need for unity in communist ranks. A grave historic responsibility will burden those who promote the splits.'[1] There was no answer from Moscow to this message; but in it were laid bare the seeds of the defeat which the Spanish revolution was to suffer seven to eight years later.

.   .   .   .

At the height of these controversies Stalin deprived Trotsky of Soviet nationality and of the right ever to return to Russia. *Pravda* published a decree to this effect on 20 February 1932, giving as reason Trotsky's 'counter-revolutionary activity', without specifying his offences. This was an unprecedented reprisal. The Menshevik and Social Revolutionary émigrés, who sat on the leading bodies of the Second International and had, with the material and moral support of that International, conducted their agitation against the Bolsheviks, had not so far been deprived of Soviet nationality. To make good this omission and to conceal somewhat the real target, the decree of 20 February also stripped about thirty Menshevik émigrés of citizenship.

There was a studied malice in this 'amalgam'. Unlike Trotsky, the Menshevik leaders had not been deported: most of them were, in 1921–2, 'advised' to leave if they wished to avoid persecution; and they left. It was Lenin who decided to give them that 'advice'; and Trotsky undoubtedly endorsed the

[1] The letter to the Politbureau was published later in *B.O.* See *Écrits*, vol. III, pp. 447–8.

decision. His hostility towards the Mensheviks remained un-
abated even in exile and led him into a grievous error of judgement only a few months before the decree of 20 February.
In 1931, during the ill-famed trial of the Mensheviks, which
took place in Moscow, Trotsky accepted the prosecution's
charges against them at face value. The defendants Sukhanov,
Groman, and others were accused of economic sabotage and
conspiracy with their émigré comrades. The charges were
based on faked evidence and 'confessions'.[1] What accounted in
part for Trotsky's attitude was the element of truth in the
prosecution's assertion that the chief defendant, Groman,
formerly economic adviser to the State Planning Commission,
had sought to obstruct the first Five Year Plan. Groman had in
fact for a long time backed Stalin's and Bukharin's policy and
had strenuously opposed Trotsky's programme of industrialization. During his trial Trotsky commented that it was with
Stalin's connivance that Groman and his group had 'sabotaged'
the Soviet economy; and that only the 'left course' had brought
Stalin's connivance to an end and the Mensheviks to the dock.[2]
While these circumstances account for Trotsky's acceptance of
the prosecution's case, they do not justify it. Later Trotsky
himself publicly regretted his mistake.[3] But the incident
illustrates how intense his enmity towards the Mensheviks
remained; and one may well imagine with what perverse
pleasure Stalin pilloried both Trotsky and the Menshevik
'saboteurs' in the same decree simultaneously depriving them of
citizenship.

This event followed shortly after the somewhat enigmatic

[1] The Prosecutor alleged that the defendants had taken orders from R. Abramovich, the Menshevik émigré leader, and that the latter had come clandestinely
to Russia to inspect the conspiratorial organization. Abramovich was able to prove
that at the time when, according to the Prosecutor, he was supposed to have travelled in Russia, he was present at sessions of the Executive of the Second International in Brussels and spoke together with Leon Blum, Vanderwelde, and other
Social Democratic leaders from public platforms.

[2] Trotsky's first opinion on the trial of the Mensheviks is in *B.O.*, nos. 21–22,
1931. Thirty years later in July–September 1961, the Menshevik *Sotsialisticheskii
Vestnik* published N. Jasny's reminiscences on Groman which confirmed that
Groman's role in the struggle between the Bolshevik factions had indeed been such
as Trotsky described it, though he was, of course, innocent of the crimes imputed
to him.

[3] See *B.O.*, no. 51, July–August 1936. Trotsky was prompted by Lyova to
admit his mistake; and he did so shortly before the great Zinoviev–Kamenev trial.

'Turkul affair'. On 31 October 1931, *Rote Fahne* published an article alleging that General Turkul, an émigré who had commanded White Guards in the civil war, was about to organize an attempt on Trotsky's life, taking advantage of the fact that Trotsky was not guarded well enough on Prinkipo; and that if the attempt succeeded, the perpetrators would shift the blame on the Soviet Government. These allegations sounded plausible enough; but it was puzzling that the *Rote Fahne*, of all papers, should have come out with them. On Trotsky's prompting, his friends made representations at the Soviet Embassies in Berlin and Paris, reminding the Soviet Government that it had promised to protect his life in exile and asking what it was going to do to honour the pledge. Moscow left the query unanswered; and Trotsky concluded that *Rote Fahne* had had only one purpose: to provide an alibi for Stalin in case of an attempt. His followers then addressed to the Soviet Government a statement, showing clearly the marks of Trotsky's style, which affirmed that 'Stalin was concerned not to prevent the men of the White Guards carrying out their design, but only to prevent them shifting the responsibility for the terroristic act on Stalin and his agents'.[1] Stalin replied indirectly, through the Comintern, chiding Trotsky for the black ingratitude with which he repaid the solicitude he, Stalin, had shown him— the reply suggested that Trotsky's life was indeed threatened by the White Guards.[2] Stalin now punished the 'ingratitude' by rendering Trotsky stateless and depriving him even of the modicum of formal protection that any government owes its subjects in foreign parts.

The reprisal was intended to accomplish what the execution of Blumkin had failed to do, to cut off all contacts between Trotsky and his followers in the Soviet Union. Despite censorship and interception, Trotsky still received much mail from the colonies of deportees and from prisons. In Berlin, Lyova was trying to establish connexions with old comrades who arrived there on official business; and he reported to Prinkipo

---

[1] The message was despatched to Moscow confidentially. Trotsky published it only after he had been deprived of Soviet citizenship. *B.O.*, no. 27, March 1932.

[2] The reply took the form of a secret circular sent out by the Executive of the Comintern to the Central Committees of all Communist parties. A copy of the circular came into Trotsky's possession and is in *The Archives*, Closed Section.

on his successes and failures. Thus, in the spring of 1931, he ran by chance into Pyatakov; but that close friend of earlier years, now 'the Judas, the red-haired'—so Lyova wrote—'turned away his head and pretended not to see me'. Later, in July, while wandering in one of the city's big stores, Lyova met unexpectedly Ivan Smirnov, who since his capitulation had held a high managerial post in Soviet industry. They embraced; Smirnov warmly inquired about Trotsky and all members of his family; and pouring out the capitulator's bitter heart, he spoke about the grim situation and the discontent rife in the Soviet Union. Although disillusioned in the hopes with which he had surrendered to Stalin, he was in no mood to resume the fight; he preferred to wait and see. He said, however, that he and his friends would welcome a 'bloc' with Trotsky and his followers, the immediate purpose of which was to be merely exchange of information. At the very least he wished to keep up a contact with Trotsky; and, as he was about to return to Moscow, he promised to send through a trusted friend a document surveying the state of the Soviet economy and the political moods in the country. They agreed on a password which the messenger was to use. Early in the autumn E. S. Golzman, an old Bolshevik, a capitulator, brought a memorandum from Smirnov, which was to appear in the *Bulletin Oppozitsii* a year later and to reveal, for the first time, the full extent of the destruction of agricultural stock during collectivization, the grave disproportions in industry, the effects of inflation on the whole economy, &c. The memorandum ended with this pregnant conclusion: 'In view of the incapacity of the present leadership to get out of the economic and political impasse, the conviction is growing about the need to change the party leadership.' Lyova and Golzman often met and discussed developments in the Soviet Union.[1]

Smirnov and Golzman spoke not only for themselves but for many capitulators who, timidly yet unmistakably, once again turned their eyes to Trotsky. Their anxiety was aroused by the storm gathering over Germany as well as by the domestic

[1] This account is based on Lyova's correspondence with his father, and on his deposition to the French Commission of Inquiry which, in 1937, conducted investigations preparatory to the Mexican counter-trial. *The Archives*, Closed Section.

situation. They were alarmed by the paralysis of German communism and sympathetically followed Trotsky's campaign. Most of them already thought what Radek was to express later, in 1933, when, speaking to a trusted German communist, he pointed to Stalin's office in the Kremlin and said: 'There sit those who bear the guilt for Hitler's victory.'[1] Seeing no way to change the Comintern's policy, exasperated and frustrated, the capitulators moved some way back towards the Trotskyist Opposition. This did not escape the notice of Stalin, who was more than ever bent on insulating the party from Trotsky's influence. He now regretted banishing Trotsky from Russia, for the banishment enabled Trotsky to broadcast his ideas all over the world. Stalin decided to make good this 'error': Trotsky, deprived of Soviet nationality, was branded as an outcast once and for all. Henceforth, any Soviet citizen trying to communicate with Trotsky would be guilty of association not just with a disgraced leader of a domestic opposition, but with a *foreign* conspirator.

Trotsky replied with an 'Open Letter' to the Presidium of the Central Executive Committee, in whose name the decree of 20 February was published.[2] He exposed the lawlessness of the decree (which he described as a 'consummate amalgam in the Thermidorian style', and 'an impotent and even pitiable' act of Stalin's personal vengeance); and he also drew the balance of a decade of the inner-party struggle. 'Do you think that with this false scrap of paper . . . you will stop the growth of Bolshevik criticism? Prevent us from doing our duty? Intimidate our co-thinkers? . . . The Opposition will step over the decree of 20 February as a worker steps over a dirty puddle on the way to his workshop.' He was aware that this reprisal was not Stalin's 'last word'. 'We know the arsenal of his methods . . . and you know Stalin as well as I know him. Many of you have more than once, in conversations with me or people close to me, given your own estimate of Stalin, and given it without illusions.' He was addressing Stalin's entourage, the 'men of the apparatus'. He appealed to their conscience, but also to their

---

[1] E. Wollenberg, a former editor of the *Rote Fahne* and leader of the *Rotfrontbund* writes in *The Red Army*, p. 278: 'Early in 1933 Zinoviev said to me: "Apart from the German Social-Democrats, Stalin bears the main responsibility to history for Hitler's victory" '.

[2] *B.O.*, no. 27, March 1932.

interest. He sought to persuade them that they too had nothing to gain but much to lose under Stalin's autocracy. He described tellingly the humiliation which together with the whole party they were suffering at Stalin's hands.

You started the fight against 'Trotskyism' under the banner of the Old Bolshevik Guard. To Trotsky's imaginary ambitions of personal leadership, ambitions which you yourselves had invented, you opposed the 'collective leadership of the Leninist Central Committee'. What remains of that collective leadership? What is left of the Leninist Central Committee? The apparatus, independent of the working class and of the party, has set the stage for Stalin's dictatorship which is independent of the apparatus. And now for anyone to take the oath of loyalty to the 'Leninist Central Committee' is almost the same as to call openly for insurrection. Only an oath of loyalty to Stalin may be taken—this is the only permitted formula. The public speaker, the propagandist, the journalist, the theorist, the educationist, the sportsman—all are obliged to include in their speeches, articles, or lectures the phrase . . . 'under Stalin's leadership'; all must proclaim the infallibility of Stalin who rides on the back of the Central Committee. Every party man and Soviet official, from the head of the government to the humble clerk in any backwater, has to swear. . . that in case of any differences arising between the Central Committee and Stalin, he, the undersigned, will support Stalin against the Central Committee.

Stalin was suppressing his own faction which had helped and was still helping him to suppress all his opponents. Within his own faction he had set up a narrower faction of his own, working through secret agents, passwords, codes, &c. He was desperately anxious to destroy the opposition to the end—hence the decree of 20 February—in order to be free to settle accounts with his own followers and his own entourage. The men of the 'apparatus' should therefore in their own interest refuse to do Stalin's bidding—only in this way could they save themselves.

Stalin's strength has always lain in the machine, not in himself. . . . Severed from the machine. . . . Stalin . . . represents nothing. . . . It is time to part with the Stalin myth. It is time that you should place your trust in the working class and its genuine, not its counterfeit, party. . . . You wish to proceed along the [Stalinist] road any further? But there is no road further. Stalin has brought you to an impasse. . . . It is time to bring under review the whole Soviet system and cleanse it ruthlessly from all the filth with which it has overgrown. It is time to carry out at last Lenin's final and insistent advice: 'Remove Stalin!'

It was emphatically to the chiefs of the Stalinist bureaucracy rather than to the Bolshevik rank and file that Trotsky was appealing here. Committed as he was to work for the reform of the ruling party, not for its overthrow, he had to appeal to them, because only the Central Committee, composed almost entirely of Stalinists, could start a reform in a constitutional manner. Trotsky was in effect prompting the chiefs of the old Stalinist faction to initiate—in 1932!—the de-Stalinization that some of them were to carry out twenty-odd years later, after Stalin's death. This appeal, though it was not to be heeded, was by no means pointless, for the conflict between Stalin and his old associates and followers was to end fatally for most of the latter. Trotsky, watching their conflict, was by no means inclined to belittle its significance, even though he played it down in some of his more exoteric writings. This, we know, was the most dangerous and gloomy moment in Soviet history, when the nation came to feel the full force of the calamity in agriculture and of famine, and when inflationary chaos threatened to disrupt its toilsome industrial advance. 'Adversities and frustrations piled up upon one another; Stalin's popularity was at its nadir. He watched tensely the waves of discontent rising and beating against the walls of the Kremlin', so we have described this moment elsewhere.[1] The discontent, it should be added, not only beat against the walls of the Kremlin; it breached them.

The discord between Stalin and his entourage had shown itself as early as 1930, when, in the statement 'On Dizziness from Success', he demonstratively disavowed the use of violence in collectivization and, over the head of the Central Committee, presented himself to the country as the peasantry's sole protector. The Central Committee protested; and Stalin had to tell the nation that the whole Committee and not he alone had called for a halt to the violence. The next dissension was occasioned by Yaroslavsky's temporary eclipse in the same year. Yaroslavsky was a pillar of the Stalinist faction, the most ferocious guardian of its orthodoxy, and the author of a textbook on party history, a feat of falsification which had been

[1] Deutscher, *Stalin*, p. 332.

hailed as a reliable guide through the doctrinal maze of the inner party struggle and had been crammed into the party's mind. It was precisely this textbook that now brought about Yaroslavsky's disgrace. Stalin suddenly found it teeming with heresies and ordered it to be banned. Yaroslavsky, having composed the book in the nineteen-twenties, could not carry falsification to the point that suited Stalin in 1931. The forger of history does not work in a vacuum: the scope he can give himself and the insolence he can afford depend on how large and heavy is the oblivion which time, indifference, and previous falsification have already cast on men and events; and in the nineteen-twenties, Yaroslavsky had to reckon with the fact that many of his readers still had relatively fresh memories of the years of revolution and civil war. In 1931 Stalin required forgeries far more massive. As he grew in sheer power, he required the cloth of history to be cut to his measure ever anew. A few years earlier it was enough for any Stalinist text to denounce Trotsky as a 'deviator' from Bolshevism and to hail Stalin as the reliable interpreter of Leninism. Now the writer of any textbook had to brand Trotsky as one who had always been a rabid counter-revolutionary; depict him as a traitor even at the time when he was President of the Petrograd Soviet and Commissar of War; make people forget that the villain had ever held such exalted posts; clothe Stalin with all splendour of which Trotsky had been stripped; and establish unquestionably the apostolic succession of Marx-Engels-Lenin-Stalin. It was not in the interest of the Stalinist faction at large, but only in that of Stalin's autocracy, that falsification should be carried to such extremes. Yaroslavsky's *History* had represented the Stalinists' viewpoint at the time when they still treated Stalin as their *primus inter pares*: it had therefore extolled Stalinism but had not glorified Stalin himself and the superhuman genius that entitled him to set himself above his own faction. Yaroslavsky had therefore to be struck down. But such was the dismay this caused even among Stalin's henchmen that soon his disgrace had to be lifted.[1]

More dramatic was the deposition, also in 1931, of Ryazanov from the post of Director of the Marx-Engels Institute. The

---

[1] The author was in Moscow at that time and heard many agitated expressions of that dismay from most 'orthodox' party members.

celebrated Marxian scholar had long since withdrawn from political activity and had, despite his old friendship with Trotsky, behaved towards Stalin with complete loyalty, devoting all his energy to the Institute's rich archives and library. Yet by his mere presence at the Institute he kept alive a scholarly tradition of classical Marxism just when Stalin was anxious to turn the Institute into a shrine of his personal cult. Ryazanov was therefore expelled and deported from Moscow under the pretext that he had plotted with the Mensheviks to suppress some of Marx's unpublished writings.[1]

Connected with these affairs was Stalin's notorious attack on the editors of *Proletarskaya Revolutsia*, whom he accused of trafficking in 'Trotskyist contraband'. The journal had published an historical essay on the pre-1914 Bolshevik attitude towards Rosa Luxemburg, duly acknowledging her revolutionary and Marxist merits. There was nothing unusual in this, for ever since Luxemburg's assassination in 1919, communists paid regular and solemn tribute to her memory; after 1924 the anniversaries of Lenin's, Luxemburg's, and Liebknecht's deaths were annually observed in a single solemn celebration of the 'Three L's'. Stalin now denounced Luxemburg's ideas as inherently hostile to Bolshevism and akin to Trotskyism. The kinship was undeniable; but hitherto the Stalinists had fought against the living leader of the Opposition, not against a ghost. Stalin came to suspect that in paying homage to the ghost they slyly aimed at rehabilitating Trotsky.

I think [he wrote] that the editors have been actuated by that rotten liberalism which is now fairly widespread among some Bolsheviks. Some think that Trotskyism is a school of thought within communism, a faction which has, to be sure, committed mistakes, done not a few silly things, and even behaved at times in an anti-Soviet manner; but that it is all the same a communist faction. It is hardly necessary to point out that such a view of

---

[1] Trotsky's defence of Ryazanov is in *B.O.*, nos. 21–22, May–June 1931. As director of the Marx-Engels Institute Ryaznov had done more than anyone to assemble at the Institute the papers of Marx and Engels. He obtained among other things a number of Marx's letters to Kautsky, which Kautsky yielded on condition that some of them, containing strictures of him, would not be published in his lifetime. Ryazanov, bound by his word, refrained from publishing these; and no one held this against him until Stalin needed a pretext for squeezing him out of the Institute and discrediting him.

Trotskyism is profoundly mistaken and harmful. Actually, Trotskyism is the spearhead of the counter-revolutionary bourgeoisie, waging
the struggle against communism. . . . Trotskyism is the vanguard of
the counter-revolutionary bourgeoisie. That is why liberalism towards it . . . borders on crime and on betrayal of the working
class.[1]

It was not only with the 'rotten liberalism' of his own
entourage that Stalin was at loggerheads. He had to contend
with more direct challenges. Within the Central Committee
and around it ever new groups of malcontents formed. The
affairs of Riutin, Slepkov, Syrtsov, and Lominadze had dragged
on for over two years now. All four had in turn been demoted,
denounced, half-rehabilitated, and once again branded as
conspirators. Stalin and the Central Committee could not agree
on just how guilty these men were and what was to be the
measure of their punishment. In 1932 several new 'conspiratorial factions' were unmasked, a group led by A. Smirnov, former
Commissar of Agriculture, Eysmont, a Commissar of Supplies,
and Tolmachev, a Transport Commissar; another group, that
of Konor, Kovarsky, and Vulf, was uncovered in the Commissariat of Agriculture; and 'networks of opposition' were
found to exist in the trade unions and various Commissariats.[2]
The leaders of these groups had not engaged in any real
conspiracy. Those of them who were members of the Central
Committee had merely exercised their statutory right in trying
to persuade their colleagues that Stalin's policies were pernicious, that he was guilty of abusing his power, and that the
Central Committee should depose him as its General Secretary.
They circulated memoranda to this effect and sought to obtain
the moral support of previous oppositions. Thus Riutin sought
Zinoviev's and Kamenev's advice; while Eysmont and Tolmachev appealed to Tomsky and Rykov. In the course of the
years 1931 and 1932 Stalin pressed the Politbureau and the
Central Committee to give him a free hand in dealing with these

---

[1] Stalin, *Sochinenya*, vol. XIII, pp. 98–99.
[2] Popov, N., *Outline History of the C.P.S.U. (b)*, vol. II, pp. 391, 399, 418–19,
434; *K.P.S.S. v Rezolutsyakh*, vol. II, p. 742. The cases of all these 'deviationists'
were the subjects of various 'confessions' in the Moscow trials of 1937–8—see the
Verbatim Reports of the trials. See also Serge, *Mémoires d'un Révolutionnaire*, pp. 280–1,
and *B.O.*, no. 31 and *passim*.

critics. He met with resistance in the Committee; and even the G.P.U. was reluctant to act.[1]

Only after many delays could he, in November 1932 and January 1933, expel some of the malcontents and pronounce a new excommunication on Zinoviev and Kamenev, who were once again banished from Moscow, this time to Siberia. During this, his second deportation, Zinoviev allegedly stated that the greatest mistake of his life, greater even than his opposition to Lenin during the days of the October revolution, had been his decision to desert Trotsky and to capitulate to Stalin in 1927. Soon thereafter Preobrazhensky, Ivan Smirnov, Mrachkovsky, Muralov, Ter-Vaganyan, and many other capitulators were once again expelled and imprisoned; they were persecuted even more cruelly than the Oppositionists who had never surrendered. Towards the end of the year it seemed that the Opposition had regained the ground it had lost since 1927. A contemporary report thus describes the effect of the persecution of the capitulators: 'These old revolutionaries, experienced political leaders, have made an attempt to find a common language with the men of the apparatus. The attempt lasted nearly four years and has ended in failure. When they capitulated the party cells were told that "all the old Bolsheviks had broken with the Opposition". This argument undoubtedly made a great impression. . . . Now the arrest of the [capitulators] is making an even stronger impression, but in the opposite direction: "Well", say many, "the Left Opposition has been right after all, if so many of those who deserted it are now returning to it." '[2] They were not in truth returning of their own accord—Stalin drove them out of the party because he feared their presence there during this early phase of his conflict with his own followers and the disarray in his own entourage. Just at the time of Zinoviev's and Kamenev's second deportation, Nadia Aliluyeva, Stalin's wife, committed suicide: she had broken down under the burden of remorse at the way her husband managed the affairs of party and state.

Such then were the circumstances in which Trotsky urged

---

[1] In his 'secret' speech at the Twentieth Congress N. Khrushchev made public a telegram which Stalin and Zhdanov sent to the Politbureau on 25 September 1936, chiding the G.P.U. for being *'four years behind'* in 'unmasking' Trotskyist-Zinovievist conspiracies. N. Khrushchev, *The Dethronement of Stalin*, p. 12.

[2] See Correspondence from Moscow in *B.O.*, no. 33.

Stalin's entourage to carry out at last Lenin's will and 'remove Stalin'. This was not on his part merely an impulsive reaction to the decree which deprived him of citizenship. He reckoned with the possibility that Stalin's autocratic ambition might at last shock the men of the ruling group and arouse them to act in their self-defence. When one considers that five or six years hence Stalin was to order the execution of 98 out of the 139 members and deputy members of the Central Committee (and of 1,108 out of the 1,966 delegates to the Seventeenth Party Congress) and thus to exterminate the majority of the *Stalinist* 'cadres', nearly three quarters of their élite, one may well admit that Trotsky, in addressing these cadres, had enough reason to invoke not only his, the Opposition's, and the party's interests, but also the dictates of their own self-preservation. 'Save yourselves—this is your last chance!' he said in effect to those Stalinists who were presently to become victims of Stalin's terror. He urged men like Khrushchev and Mikoyan to 'cleanse the Soviet state of the filth with which it was overgrown' twenty-four years before they were ready to start with this, and when there was still far less filth to be cleansed than there would be later. He knew, of course, that even if they decided to act against Stalin they would do so half-heartedly and would be held back by a thousand inhibitions. He nevertheless envisaged a 'united front' with them and offered them his critical support, confident that once the movement against Stalin was started, he and his followers would come to the fore.[1]

He did what he could to give heart to the Stalinist mal-contents. Lyova, who from Berlin was in closer touch with the turmoil in Moscow, was especially eager that he should do so. Reports from Moscow continued to dwell on the exasperation among the Stalinists and on the talk about the need to 'remove Stalin'. But the same reports indicated that the Stalinist malcontents were terrified at the mere thought of Trotsky's return. 'If Trotsky came back', they said, 'he would shoot us all.' Or: 'He will revenge himself for all that we have done to him and his followers and he will put thousands of us before the firing squads.' Stalin played on this fear and whipped it up. 'This indicates along what line we ought to move', Trotsky

---

[1] Ibid., no. 27. During the year 1932 Trotsky often returned to this subject in his correspondence with Lyova.

wrote to his son. 'In no case should we frighten people with slogans or formulas which could be interpreted as expressing any intention . . . to take revenge. The closer the showdown . . . the softer and the more conciliatory should be the manner in which we speak, although we should not, of course, make any concessions of principle.'[1] In the *Bulletin* and in a special leaflet designed for circulation in Russia, Trotsky thus sought to reassure those who feared his revenge:

An end must, of course, be put to the Bonapartist régime of a single leader whom every one is forced to worship—an end must be put to this shameful distortion of the idea of a revolutionary party. But what matters is that the system be changed not that individuals be ostracized. The Stalinist clique assiduously spreads the rumour that the Left Opposition will return . . . sword in hand, and that its first job will be to wreak ruthless revenge on its adversaries. . . . This poisoned lie must be repudiated. . . . Revenge is not a political sentiment. Bolsheviks-Leninists have never been guided by it; least of all shall we be guided by it. We know all too well the . . . causes that have driven tens of thousands of party men into the blind alley. . . . We are prepared to work hand in hand with everyone who is willing to reconstitute the party and forestall a catastrophe.[2]

However, this was the year 1932, not 1953 or 1956. Despite the signs that seemed to augur it, the movement against Stalin did not materialize. The 'men of the apparatus' were unable to act against their chief. The fear of Trotsky's return and revenge was not the most important of the inhibitions that held them back. It was the very decomposition of the Stalinist faction that rendered them incapable. Stalin dominated them by dividing them, setting up rival caucuses and forming his pretorian guard, the members of which knew no loyalty to erstwhile comrades and were willing to promote his personal rule. This was the 'secret staff' working through its own agents with 'secret passwords and codes' which Trotsky had mentioned; and these were the 'quintets', 'sextets', and 'septets' which, according to Khrushchev, Stalin set up within the Politbureau and the Central Committee and through which he reduced the latter to impotence. The arts which had gained him power did not fail to maintain it. He was able to spot any hostile stirring within the Central Committee before it had the

---

[1] Trotsky's letters to Lyova of 17, 24 and 30 October 1932.   [2] *B.O.*, no. 33.

time to spread. No group of malcontents, not even one composed of the most influential Stalinists, could voice any criticism and try to influence others in the hierarchy, for no sooner
had they tried than they were 'unmasked' and stigmatized as
traitors.

Yet the secret caucuses, the 'quintets', the 'sextets', and
Stalin's other conspiratorial devices would have counted for
little if the malcontents had not been paralysed by a fear that
had hamstrung all previous oppositions. They were afraid that
any move against Stalin might become the signal for an explosion of popular discontent and set the stage for a counter-
revolution which would engulf together with Stalin all his
Bolshevik adversaries. This fear haunted Trotsky as well. He
still saw no solution to the dilemma that had beset him in the
nineteen-twenties. Shortly after he had made his dramatic
appeal and concluded it with the words 'Remove Stalin', he
had second thoughts. In October 1932 he wrote to his son:

The slogan 'remove Stalin' is correct in a definite, specific sense
[the sense in which Lenin used it when he advised the Central
Committee to elect another General Secretary]. . . . If we were
strong now . . . there would be no danger at all in advancing this
slogan. But at present Miliukov, the Mensheviks, and Thermidorians of all sorts . . . will willingly echo the cry 'remove Stalin'. Yet,
it may still happen within a few months that Stalin may have to
defend himself against Thermidorian pressure, and that we may
have temporarily to support him. We have not yet left this stage
behind us. . . . This being so, the slogan 'down with Stalin' is
ambiguous and should not be raised as a war cry at this moment. . . .[1]

At the same time Trotsky stated in the *Bulletin*: 'If the bureaucratic equilibrium in the U.S.S.R. [i.e. Stalin's rule] were to
be upset at present, this would almost certainly benefit the
forces of counter-revolution.'[2]

To the Stalinist malcontents in Moscow, not to speak of the
capitulators, this euphemism amounted to advising them to
hold their fire. If even Trotsky thought 'Down with Stalin'
was too hazardous, how much more risky must that cry have

[1] *The Archives*, Closed Section.
[2] *B.O.*, ibid. It is interesting to note that it was partly on Lyova's prompting that
Trotsky came out with this disavowal of the slogan 'Down with Stalin'.

sounded to them. What then were they to do? 'You wish to
proceed along the Stalinist road any further? But there is no
road further', Trotsky had told them in March. 'Stalin has
brought you to an impasse.' They now learned that there was
no way back either, and that all they could do was to try to
survive in the impasse and hope that time and the nation's
progress would lead them out of it. They concluded that in the
meantime they had to bow to the inevitable; and they were to
bow to it for over two decades, till Stalin's death.

.   .   .   .

Zinoviev or Kamenev had once told Trotsky that Stalin
would revenge himself on him and his children and grandchildren 'until the third and fourth generation'. Now indeed
the biblical vengance struck Trotsky's family. The decree
which deprived him of Soviet nationality robbed of it also those
of his relatives who shared his exile; and it forbade them to
return to the Soviet Union. This immediately affected Zina.
She found herself cut off from her husband and younger child,
and without the hope of ever being able to rejoin them.

She had now spent over four months in the German capital.
The unfamiliar city and its political drama at first so engrossed
her that to her doctors' satisfaction she appeared to recover her
balance. The improvement was superficial, and the doctors
may have been misled by a patient too proud to reveal to them
her disturbed mind. She stubbornly resisted psychoanalytical
investigation. 'The doctors have only confused me,' she confessed later, 'but I have confused them, poor creatures, much
more. . . .' Her emotional strains were undiminished. Her
adoration for her father was still at odds with her grievance.
In her thoughts and correspondence she returned to their last
parting: she resented its strange coolness and his remoteness
and Olympian superiority. She brooded over his words: 'You
are an astonishing person, I have never met anyone like you';
and she pined over their uncomprehending severity. She
yearned for warmer contact by correspondence; but he wrote
rarely, more rarely, at any rate, than she wished; and in his
letters, though full of concern for her, she still felt him frigid
and distant.

There was also her discord with Lyova. She could not get

along with him even though there was no one in Berlin closer to her, and even though their father begged them to sustain each other in their plight. She reproached Lyova too with lack of compassion; and the mere sight of him aroused all her agonizing jealousy. 'Every time I see him', she wrote very shortly after she had come to Berlin, 'I suffer a nervous breakdown.'[1] She avoided meeting him; and he was, anyhow, too busy with his political work and the *Hochschule*. His very busyness, which came from his close bonds with their father, excited her envy: she contrasted it with her own 'passivity and uselessness' and despised herself as 'Zina the idler'.

The *ukase* which deprived her of the prospect of a return to Russia sharpened her loneliness and insecurity. Her father advised her to protest at the Soviet Embassy, calmly and moderately: perhaps if they realized in Moscow that she was not engaged in political activity but only trying to repair her health, they might exempt her from the decree.[2] We do not know whether she acted on this advice; she did not, in any case, regain her nationality. Meanwhile, her doctors reached the conclusion that to recover she should rejoin her family in Russia and resume as soon as possible a normal existence in her proper environment. This was precisely what she could not do. An outcast, lonely in the huge and alien city, feeling estranged from one half of her family, and reproaching herself with having abandoned the other, her nervous breakdowns and fits of absent-mindedness became more frequent. She had no choice but to return reluctantly to the psychoanalyst's couch, from which she emerged to stare at the vast political lunacy that was overtaking the nation in whose midst fate had thrown her.

In her letters she described the misery and the torment of Germany, interspersing her descriptions with acute political observations and mordant *Galgenhumor*. When she first wrote to her father to tell him how worried she was at being cut off from Russia and her next of kin there, she told him also that she was quite as much depressed by the 'Red plebiscite' and the confusion and demoralization in the German working class.[3]

[1] See e.g., Zina's letters and postcards to her father of 26 February, 30 May, 7 June 1932.
[2] See Trotsky's correspondence with his children of March 1932.
[3] See her letter of 26 February 1932.

She followed eagerly Trotsky's 'German campaign'; but the
gratification this gave her was spoiled by the sense that she
was excluded from his work and political interests: 'There is
no purpose in corresponding with Papa . . . the doubting
Thomas', she said in a letter. 'He is further and further above
the clouds in the regions of high policy . . . and I am mostly
stuck in psychoanalytical swinishness'.[1] Her own vision of the
political turmoil was heightened by the convulsive insight of
the insane eye. There are phrases in her correspondence as
rich and sarcastic as if they had come from the pen of her own
father. Like a refrain there occurs an image of Berlin, hungry
and drunken, full of the tramping of heavy boots, and swelling
up with despair and bloodthirstiness. 'Berlin is singing . . .
all the time, often in a voice hoarse with drunkenness or
hunger. . . . This is a gay city, very gay indeed. . . . And think
only that old Krylov was so rash as to say that no one would
ever sing on an empty stomach.'[2]

The doom-laden city bewitched her; she became attached
to it as if she belonged to it; she lived through all its tremblings
and fevers. Early in June 1932, when Hitler's Stormtroops,
unscathed by Brüning's ban, re-emerged in riotous triumph,
Lyova urged her to leave Berlin, to go to Vienna, and there, in a
calmer atmosphere, to continue the psychoanalytical cure.
Himself inconvenienced by the police, he feared that she too
would be troubled. She resented the advice, dismissed the fears,
and complained to Prinkipo that Lyova bossed and bullied her.
When her father repeated Lyova's advice, she answered in a
strangely reverential tone, saying that she did not even dare to
protest; but then she dwelt on her fondness for Berlin and
refused to budge. Even her father's and brother's concern
humiliated her. Had not her father said so many times that the
fate of Europe, nay, of mankind, was being decided in Berlin
for decades ahead? Was this not why he had wished Lyova
to be on the spot? Had he not refused to accept a German
Trotskyist as a secretary, saying that it would be a shame if at
such a time a single one of his followers absented himself from
the political battlefield? Why then should she be asked to
leave? She felt rejected and degraded.[3]

As loneliness was grinding her down, the doctors asked that

[1] Letter of 30 May.   [2] Letters of 7 June and 17 August 1932.   [3] Ibid.

at least the child she had left on Prinkipo should be brought to her to occupy her mind and give her some responsibility. But the child, too, was affected by the decree of 20 February: at the age of six Seva was a 'stateless political émigré', officially registered as such—a problem for consular dispensers of travelling permits and visas. Applications were turned down on the ground that he could travel only with one of his parents or grandparents. The child had been badly upset by his mother's absence and by her messages imploring him not to forget her and promising that she would return very soon—it was with difficulty that she was persuaded not to send such messages. Now the expectation of a reunion and the suspense put on edge the child's nerves—and the nerves of the whole family.

In her distress Zina was less and less able to look after herself, even to manage reasonably her monthly allowance and expenses.[1] She reproached herself with being a burden to her father; and she moved to a low-grade boarding-house, where she lived among tramps and rowdies, and often had to stand between them, and separate them when they came to blows. Any attempt by her brother or even father to get her out of such circumstances and to manage her money affairs for her aroused her resentment and provoked nervous attacks. After one breakdown she wrote an angry postcard to her father blaming him for the attack and asking to be left in peace.[2]

Zina's sufferings and the strain they put on Trotsky did something to trouble his relations with Lyova, whom he expected to show more patience and affection towards her. Yet his reliance and dependence on Lyova grew ever stronger and more vulnerable. He lavished praise for the way he managed the *Bulletin* and the political work; and he went on confiding his thoughts, consulting him, and inviting criticism. He was touched by Lyova's self-denial and dedication, of which he had a thousand proofs. (Again and again he remonstrated with Lyova for being over-scrupulous with money accounts and spending his living allowance on the *Bulletin*.[3]) Yet again and again he suspected that the concord in their views and ideas

[1] This is how Lyova described her condition in a letter to his father of 26 November 1932.

[2] Zina to Trotsky, 5 and 24 October 1932.

[3] See, for instance, Trotsky's letter of 11 May 1932.

sprang from filial piety only, that filial piety which he found so gratifying and so irritating. The more tense and weary he became, the more exacting, even whimsical, grew the demands he made on his son. His loneliness and isolation, as Natalya put it, showed itself in the impatience with which he awaited letters from Lyova. When for a few days there was no mail from Berlin he exploded with anger, accused Lyova of indifference, and even insulted him; then he grew angry with himself, full of pity for his son, and even more fretful.[1]

Lyova's pack of personal troubles was also heavy enough. From Moscow his wife wrote harrowing letters about their broken lives and their child's unhappiness. He had gone abroad despite her protests and tears, she reminded him, in order to be with his parents and protect his father; now he was neither with his parents nor with his wife and child. It was no use trying to explain to her what his lot would have been in Russia—she was a simple working woman, ill, poverty-stricken, and in despair; and she threatened to commit suicide.[2] He could do nothing to relieve her plight, except to send her money. Nor did his liaison with Jeanne Molinier turn out to be much happier. Only devotion to his father's cause helped him to get away from his private worries and frustrations. Unflinchingly he carried out the thousand-and-one instructions from Prinkipo; kept in touch with all the scattered Trotskyist groups; harassed the Russian printers to bring out the *Bulletin* on time; saw to it that his father's topical brochures were promptly translated into German and published; bargained with literary agents; and for hours roamed, often hungry, the streets of Berlin in the hope of meeting a countryman on assignment abroad or a western tourist *en route* to Russia, through whom a piece of information could be obtained or a message transmitted. On top of this, he followed pedantically his course in mathematics and physics; and in the small hours of the night he conversed with his parents by correspondence. Nothing made him feel more wretched than his father's ill-humour or any intimation that his efforts did not come up to expectations. He

---

[1] The description is based on Natalya's correspondence, especially her letter to Lyova of 27 July 1932 in the Closed Section of *The Archives*.

[2] Her letter to this effect is among the family correspondence in the Closed Section of *The Archives*.

found it hard to dispel paternal displeasure, to explain himself, to ask for an explanation, or to apologize; it was only to his mother that he grieved and complained.

Natalya, frail and suffering, caught in the dangerous tangle of Zina's emotions, and torn sometimes between husband and son, did what she could. She had enough insight to grasp clearly the predicament of each of them, enough love to feel with each, and enough fortitude to try and sustain each. In her letters she explained to Lyova Zina's problem, and again and again she conveyed to both Lyova and Zina the unbearable tension in which their father lived, presenting all the time a heroic front to a hostile world—what was the wonder that now and then within the family circle his endurance snapped? 'The trouble with father, as you know, is never over the great issues, but over the tiny ones'. In the great problems his patience was infinite; over trivialities he was easily annoyed and even petulant. This, she begged the children, must never make them forget or doubt his deep and passionate love for them. 'Your pain is the pain of all three of us', she wrote to Lyova, imploring him to write more often to father, and to write 'inspiring' letters, and also to give Zina more warmth and attention. Yet at times the blows were too heavy even for Natalya's vigilant fortitude. 'What is to be done—nothing can be done', these resigned words occur not rarely in her letters to Lyova; and once she confessed to him: 'I am writing as you are, with my feelings closed and my eyes closed.'[1]

.   .   .   .

This was the late summer of 1932. It was now three and a half years since Trotsky had arrived in Prinkipo. All this time he had worked hard, pursuing his various interests, neglecting none of his correspondents, filling the pages of the *Bulletin*, and writing, apart from a dozen minor books and brochures, *My Life* and the three large volumes of the *History*. (He sent out the last Appendix closing the third volume to Alexandra Ramm on 29 June.) These had been years of prodigious labour, all the more so because, spurning easy writing, he had repeatedly redrafted almost every chapter of every one of his books, slaving patiently over every page and almost every phrase.

[1] Many of Natalya's letters are undated.

The great toil had tired him. His head was full of new literary plans: he intended to write a History of the civil war, a Life of Lenin, a joint Life of Marx and Engels, and other books. But circumstances did not favour his settling down to a major work; and he needed a rest. More than ever he chafed at his confinement to Prinkipo[1]; and political events made him restless. The trickle of news that was coming out of Russia was just enough to exasperate him. In Germany socialists and communists were moving along their beaten tracks at the very brink of disaster. His campaign was making no impact. The strength of the Trotskyist group there was less than negligible. And in the Opposition's international organization trouble was brewing: in its Berlin Secretariat the brothers Sobolevicius, who had only recently supported him in his controversy against the ultra-left Leninbund, now adopted a disquietingly conciliatory attitude towards Stalinism. Oh, if only he could get away from his enchanted and accursed island and find himself closer to the main currents of political life and to—civilization!

Early in the autumn Danish Social Democratic students invited him to come to Copenhagen and lecture on the fifteenth anniversary of the October Revolution. He had received quite a few such invitations before; but there had never been any chance of his being allowed to appear anywhere in Europe.[2] He doubted whether the Danish Social-Democratic Government would give him a visa, but this time he accepted the invitation. When he received the visa, he was at once ready for the journey. At the back of his mind was a vague hope that he might not need to return, although he was prudent enough to secure the Turkish re-entry permit. He and Natalya also hoped to be able to take Seva to Copenhagen, and from there to send him to Zina. But they could not obtain travelling permits for the child; and they had to leave him at Prinkipo under the care of one of the secretaries.

On 14 November, accompanied by Natalya and three secretaries, Trotsky sailed from Constantinople. He registered as Mr. Sedov, a stateless passenger; but his incognito could not

[1] During all his Prinkipo years Trotsky went out to Constantinople only once or twice to visit the Basilica of St. Sophia and to see a dentist.

[2] *Inter alia* a group of Edinburgh students asked him for permission to put forward his candidature in the elections of a Rector of their University—he politely declined the honour. (*The Archives.*)

shield him from public curiosity—it only thickened the aura of mystery and scandal that surrounded him. *Pravda*, paraphrasing Bernard Shaw, jeered at the 'escaped lion'; and the jeer unintentionally conveyed something of the nervousness with which governments, police headquarters, and the Press of many countries watched his progress. Had he traversed Europe as the head of a real and powerful conspiracy, and had multitudes of followers hailed him, his journey could not have aroused more commotion than it did, when he travelled as an outcast, denied the protection of any government, and accompanied only by an elderly ailing woman and a few young devotees; and when his sole set purpose was to deliver a lecture. Wild rumour ran ahead. Newspapers speculated on the real purpose of his trip; they had no doubt that the lecture was a mere pretext: some said that he was to meet secretly an envoy of Stalin somewhere in Europe; others that he was about to mount his final conspiracy against Stalin. At Greek and Italian ports of call reporters besieged him, but he refused to talk to them. He was not allowed to visit Athens. At Naples he left ship and under police escort visited the ruins of Pompeii. The French forbade him to disembark at Marseilles; out at sea their police ordered him to transfer to a small motor-boat which took him to a forsaken little jetty outside Marseilles, where he landed. He was rushed through France by car and train, with only one hour's stop in Paris, so that reporters who pursued him all the way from Marseilles were able to pick up his trail only at Dunkirk, where he boarded a ship for Denmark. Across France he was followed by the curses of right-wing newspapers, whose leader writers were beside themselves at the thought that the 'traitor of Brest Litovsk', the man who had 'robbed of their savings the widows and orphans' of French rentiers, should have been allowed to set foot on French soil. He tried to calm the excitement and assured reporters that he was on 'a strictly private journey, devoid of all political significance'.[1]

On 23 November he arrived in Denmark and was ordered to disembark at Esbjaerg so as to be 'brought to Copenhagen by a backstairs entrance', as *Politiken* put it. A crowd of communists

[1] His statements to the French Press of the 21 and 22 November 1932. *The Archives.*

had come to boo and hiss him; but, according to the same paper, 'the moment Trotsky showed himself there was a deep silence—the sense of a historic personality and perhaps of a historic occasion'[1]. Reporters noted Trotsky's 'perfect calm' and the nervousness of his secretaries and of the organizers of the trip. He had hardly entered Copenhagen when a member of the Royal family, Prince Aage, echoed by a section of the Press, denounced 'the murderer of the Tsar's family': the Danish Court had not forgotten that the mother of the last Tsar had been a Danish princess. At the same time the Soviet Ambassador expressed his government's concern over the visit. The Social Democrats gave Trotsky a warm welcome; but the warmth did not last. As both the Royal family and the Soviet Embassy continued to vent displeasure, the embarrassed Socialist Ministers became impatient for his early departure.

Trotsky did his best to keep out of public sight. He stayed in somewhat eccentric surroundings, in a villa Raymond Molinier had hired from a famous danseuse who was away on a tour—the rooms were crammed with trinkets and the walls covered with alluring pictures of the absent hostess. Then a newspaper disclosed Trotsky's whereabouts by publishing a photograph of the villa; and so he and his companions hurriedly moved away to a pension in a suburb. There were various minor incidents. Molinier's car, which Trotsky used, vanished mysteriously. After a few hours the police returned it without an explanation and took the owner's . . . fingerprints. There were rumours that Trotsky's enemies were preparing to disrupt the meeting at which he was to lecture. And all the time he was guarded by the police as well as by his followers; only once or twice did he go out for short drives through the city.

The lecture passed without obstruction or disturbance. For two hours, speaking in German, he addressed an audience of about 2,000 people. His theme was the Russian Revolution. As the authorities had allowed the lecture on the condition that he would avoid controversy, he spoke in a somewhat professorial manner, giving his listeners the quintessence of the three volumes of his just concluded *History*. His restraint did not conceal the depth and force of his conviction; the address was a

[1] *Politiken*, 24 November 1932; see also *Berlingske Tidende* and *Informacion*, of the same date.

vindication of the October Revolution, all the more effective because free of apologetics and frankly acknowledging partial failures and mistakes. Nearly twenty-five years later members of the audience still recalled the lecture with vivid appreciation as an oratorical feat.[1] This, incidentally, was the last time that Trotsky addressed any large public meeting in person.

Of his other activities in Copenhagen his interviews and a broadcast in English to the United States may be mentioned. 'My English, my poor English', he said in the broadcast, 'is in no proportion to my admiration for Anglo-Saxon culture.' Against those who, dwelling on retrograde developments in the Soviet Union (and on his own fate), denied the *raison d'être* of the October Revolution, he pointed out that 'in criticism as in creative activity perspective is needed'. The fifteen years since October were only 'a minute on the clock of history'. The American Civil War too had outraged contemporaries. Yet 'out of the Civil War came the present United States, with its unbounded practical initiative, its rationalized technology, its economic *élan*. These achievements . . . will [form] part of the basis for the new society.'[2] He told American interviewers that although the 1929 slump had hit their country so severely, the position of the United States *vis-à-vis* the rest of the capitalist world was strengthened. He declared to French reporters that he would never refuse Stalin his collaboration, if the defence of the Soviet Union required it: '*La politique ne connaît ni ressentiment personnel ni l'ésprit de vengeance. La politique ne connaît que l'éfficacité.*'[3]

Four years later, during the Great Purges and at the trial of Zinoviev, Kamenev, and others, the prosecution was to base a crucial part of its case against Trotsky and the defendants on the allegation that it was from Copenhagen, in this last week of November 1932, that he pulled the strings of a gigantic conspiracy and ordered his adherents to assassinate Stalin, Voroshilov, and other members of the Politbureau, to sabotage industry, to poison masses of Russian workers, and to wreck

---

[1] In 1956, I lectured in Copenhagen, and was approached by quite a few members of the audience who spoke to me about the memorable meeting of 1932 at which they had been present.

[2] He made the statement for the Columbia Broadcasting System. The Danish Radio had refused to broadcast his lecture. *The Archives.*

[3] Ibid.

the country's economic and military power in order to restore capitalism. According to Vyshinsky, the Prosecutor-General, it was in Copenhagen that, in the presence of his son, Trotsky received Golzman, Fritz David, and Berman Yurin, three men who sat behind Zinoviev and Kamenev in the dock, and through them transmitted his orders. There is no need to refute here in detail these accusations and the defendants' 'confessions' by which they were supported. Stalin's successors, who upheld these accusations for twenty years, no longer do so; at the 20th and 22nd Congresses of the Soviet Communist party, Khrushchev, still haunted by Trotsky's ghost, described how such charges were concocted and how such 'confessions' were produced. Even much earlier, during the trials, Trotsky knocked the bottom out of the prosecution's case by exposing its absurdities and contradictions. Thus the Hotel Bristol, which Vyshinsky was imprudent enough to name as Trotsky's headquarters in Copenhagen, did not exist in 1932, having been demolished many years earlier. Lyova, whom Vyshinsky depicted as acting in Copenhagen as chief of staff to the leader of the terrorists, was not with his father in the Danish capital. Trotsky was able to reconstruct every incident of his trip to Denmark from his pedantically systematic records, and also to call numerous eye-witnesses to testify in his favour.[1]

His entourage in Copenhagen was larger than usual. Apart from the three secretaries who had come with him, twenty-five of his followers, Germans, Frenchmen, Italians, and others had arrived, among them Molinier, Naville, Sneevliet, and Gerard Rosenthal, Trotsky's French attorney. A group of students from Hamburg had come to meet him and guard him. Another visitor was Oscar Cohn, an eminent German lawyer, Karl Liebknecht's associate, who acted as Trotsky's attorney in Germany. The presence of so many followers gave Trotsky an opportunity to hold an informal 'international conference', at which they discussed the situation in Germany and the affairs of the various Trotskyist groups. Nothing could be less like a meeting of conspirators than this little gathering of thrilled and rather garrulous devotees of an ineffectual sect. 'Everyone talked endlessly', says the only British participant, 'except Trotsky, who worked hard nearly all the time in his room, either

[1] *The Case of Leon Trotsky*, pp. 135–73 and Closed Section of *The Archives*.

writing or dictating something.'[1] Five years later every one of
those present, if he was not in a Nazi prison or concentration
camp, was to testify that none of the men who, according to
Vyshinsky, took orders from Trotsky in Copenhagen, was there
or could have slipped unnoticed through the numerous guards.
The only man with a Russian connexion whom Trotsky
received was Senin-Sobolevicius. He had come to clear himself
of the suspicion of being a Stalinist agent, and he spent an hour
or two with Trotsky, who treated him not as an *agent provocateur*
but as a political opponent: in their correspondence Sobolevicius had frankly and in part correctly criticized Trotsky for
underrating Stalin's industrial achievement and the lasting
effects of collectivization. As far as one can judge from their
subsequent letters, their meeting in Copenhagen ended in a
patching up of differences. In any case, Sobolevicius was not to
appear as witness at any of the Moscow trials. Nor did he,
apparently, make any other contribution to the trials, for if he
had done so, he would have given the prosecution a description
of Trotsky's surroundings in Copenhagen far more realistic than
that which Vyshinsky presented.

Trotsky's stay in Denmark was thus rather uneventful. After
his public lecture he spoke only once to a small group of the
Danish students who had invited him. His host has recorded this
curious incident:

Trotsky and five or six others were in my home when suddenly
I had a telephone call from a friend, who told me that a newspaper
had just come out with a telegram from Moscow that Zinoviev had
died. Trotsky rose, deeply moved . . . 'I have fought against Zinoviev . . .', he said. 'In some matters I was united with him. I know
his mistakes, but at this moment I will not think about them, I will
think only about the fact that throughout he tried to work for the
labour movement. . . .' Trotsky continued to honour in eloquent
phrases the memory of his dead adversary and co-fighter . . . it was
very moving to hear his solemn speech in this little group.[2]

No outsider, not even Trotsky's friends and secretaries, was

---

[1] The British participant, to whom I am obliged for his impressions, was Mr.
Harry Wicks. He was to try to transmit Trotsky's writings to the U.S.S.R. through
Russian sailors calling at British ports; and Trotsky gave him a letter authorizing
him to do so.

[2] *The Case of Leon Trotsky*, p. 147. The rumour about Zinoviev's death was
denied next day.

aware of the frustration and pain he lived through in Copenhagen. It was galling enough for him to have to cross the whole of Europe, with all the precautions this required and amid all the hostile uproar, only in order to deliver a lecture in Denmark and then to have to go back to Prinkipo. He made piteous efforts to postpone return, if not to escape it. To American journalists he remarked wistfully how much he would have liked to be able for a time to 'watch the world panorama from New York', which would be like surveying a horizon 'from the top of a skyscraper'. 'Is it a Utopian dream, I ask you, to think that I should be able to work in one of the great American libraries for two or three months? The good example set by the Danish Government will not, I hope, be wasted on other countries.'[1] That 'example' was far from edifying, however: the Danish Government refused him any short-term asylum. In vain did Oscar Cohn appeal to Stauning, the Socialist Prime Minister and Cohn's personal friend; in vain did Trotsky himself request Stauning for a prolongation of the visa for a fortnight only so that he and his wife could undergo medical treatment in Copenhagen. In vain did he also appeal for a Swedish visa. This was refused him, allegedly because of objections from the Soviet Ambassador, none other than Alexandra Kollontai, former leader of the Workers' Opposition.

More oppressive than the hermetic hostility into which he had run afresh was the worry about Zina, whose health was going from bad to worse. It was probably during his Danish trip that Trotsky received this lurid letter which sounds like an accusatory farewell: 'You act . . .' she wrote to him, 'too impatiently and therefore sometimes impetuously. Do you know the meaning of something as complex and yet as elementary as instinct—something one must not trifle with . . .? Who says that instinct is blind . . .? That is not true. Instinct has terribly keen eyes which see in the dark . . . and overcome time and space—it is not for nothing that instinct is the memory of generations and begins where life itself starts. It may direct itself to all sorts of purposes. What is most frightful is that it hits infallibly and mercilessly those who are in its way.' She dwelt on the 'premonitions', the 'suspicious imaginings', and the 'terribly sharpened sensitivity' that make out instinct; and she

[1] From a statement for American journalists in *The Archives*.

went on: 'It will not frighten you if I tell you that there was a moment when I felt that something like this touched me; but with a terrible frenzy, I threw myself into the struggle. And no one supported me. The doctors have only confused me . . . do you know what sustained me? *Faith in you.* Despite all that was so plain and obvious, despite everything. . . . And is this not instinct?'[1]

Lyova was to have come to Copenhagen in order, among other things, to consult his parents about Zina; but insuperable passport and visa difficulties detained him in Berlin. Meanwhile, he was sending alarming letters about Zina's behaviour: her mind was getting more and more deranged; she would not be able to look after Seva, if they sent him to her; and she was less and less able to look after herself. He was uneasy about her erratic politics: she had apparently entered into contact with the German Communist party; and he was afraid that she would expose herself to police persecution. 'Don't you see, don't you see', she was telling him, in the days after Papen's resignation, 'that Germany is now heading straight towards a [communist] revolution?'[2] He advised his parents to do their utmost to send her away to Austria. Day after day, and sometimes twice daily, either Trotsky or Natalya anxiously talked with Lyova over the telephone, asking for further news, inquiring whether the doctors too considered it unsafe to entrust Zina with the care of her child, and urging Lyova to come to Copenhagen.

Eight days passed in this way; those days, the world was presently to be told, that Trotsky had used to stage his monstrous conspiracy against the Soviet Government. He spent these days 'conspiring' against the tyranny with which ordinary passport and visa regulations confront the stateless and the homeless. He employed every influence and accidental circumstance, every innocent stratagem and trick of publicity to gain a few more weeks or even days in Denmark, or elsewhere in Europe. Meanwhile, Natalya appealed to Edouard Herriot the French Prime Minister, begging him to allow Lyova to meet her in France while she and Trotsky were on their way

[1] The letter is undated, but internal evidence indicates that it was written in November 1932.

[2] Quoted from Lyova's letter to his parents of 26 November 1932.

back to Turkey. As the eight days, for which Trotsky's Danish visa was granted, were up, he declared that he had missed his boat and was not yet ready to leave. Perhaps, he thought Lyova would arrive while he was waiting for the next boat? Perhaps they would make up their minds whether and how to send the child to Zina? Perhaps, perhaps the heart of some government would melt and a visa would be obtained somewhere on this inhospitable continent? But the Danish Ministry insisted that his time was up and that he must go; and they rushed him out of the country by car so that he should embark before his visa expired. And so, on 2 December, Trotsky, Natalya, and the secretaries left Denmark. This time no one booed or hissed from the quay, and no one had come to say farewell, either.

.   .   .   .

As the ship sailed into Antwerp, the harbour was black with police and cordoned off. Frontier guards came on board to interrogate Trotsky; he refused to answer questions, saying that, as he was not disembarking in Belgium, the interrogation was illegal. There was a wrangle; there were threats of arrest; and none of his companions was allowed to go ashore.

At this moment a memory ten years old came back to him. In 1922, when Dora Kaplan was tried in Moscow for her attempt on Lenin's life, Emil Vanderwelde, the famous Belgian Socialist and President of the Second International, asked to be admitted as Counsel for the defence. His request was granted; and Vanderwelde used the opportunity to attack, in a Soviet court, the Soviet system of government. He did the same in an Open Letter to Trotsky. Having left the Letter unanswered in 1922, Trotsky decided to answer it now, while his ship was in Belgian waters. Vanderwelde had in the meantime been his King's Prime Minister, and even in opposition occupied a most exalted place in Belgian politics.

The government of which I was a member [Trotsky wrote to him] allowed you not only to come to the Soviet Union but even to act in court as attorney for those who had attempted to assassinate the leaders of the workers' first state. In your plea of defence, which we published in our press, you repeatedly invoked the principles of democracy. That was your right. On 4 December 1932 I and my companions stopped in transit at Antwerp harbour. I have no intention of preaching proletarian dictatorship here, or of acting as defence council for any imprisoned Belgian communists and strikers, who, as far as I know, have not made any attempt on the lives of Ministers. [Yet] the part of the harbour where our ship stopped has been thoroughly cordoned off. On both sides, right and left, police boats are on the alert. From our deck we have had the opportunity to review a parade of democracy's police agents. . . . This has been an impressive spectacle! There are more cops and *flics* here—excuse my using such vulgar terms for brevity's sake—than sailors and stevedores. Our ship looks like a temporary prison, and the adjacent part of the harbour like a prison yard.[1]

He knew, of course, that this reception and the vexations which went with it 'were triflings compared with the persecution which militant workers and communists commonly suffered'; he mentioned the facts only to give Vanderwelde the long-overdue answer to his 1922 philippic about Bolshevism and democracy:

I am not mistaken, I trust, in counting Belgium among the democracies. The war [of 1914–18] which you have fought was a war for democracy, was it not? Since the war you have been at the head of Belgium as Minister and Prime Minister. What more has been needed to bring democracy to fruition? . . . Why then does this your democracy reek so much of the old Prussian police state? How can anyone suppose that a democracy, which suffers a nervous shock when a Bolshevik by chance approaches its frontiers, that such a democracy may ever be able to neutralize class struggle and guarantee the peaceful transformation of capitalism into socialism?

Oh yes, he, Trotsky, knew all about the G.P.U. and political persecution in the Soviet Union. But the Soviet Government had at least not boasted of democratic virtues; it openly identified itself with a proletarian dictatorship; and the sole test by which it should be judged was whether it secured the transition from capitalism to socialism.

The dictatorship has its own methods and its own logic, which are rather severe. Not rarely . . . revolutionaries who had established the dictatorship are themselves the victims of its logic. . . . Before class enemies, however, I assume full responsibility not only for the October revolution . . . but even for the Soviet Republic such as it is

[1] *B.O.*, no. 32, December 1932.

to-day, including that government which has banished me and deprived me of Soviet citizenship. [But] you—you are defending capitalism allegedly in the name of democracy. Where then is that democracy? It was in any case not to be found at Antwerp harbour.

For all that, he was leaving the waters of Antwerp 'without the slightest pessimism'. He had before his eyes the picture of 'sturdy, severe Flemish dockers, thickly covered with coal-dust', who, separated from his boat by a police cordon, 'eyed the scene in silence, took the measure of everyone', recognized 'their own', winked ironically at the cops, exchanged friendly smiles with the dangerous passenger on deck, and 'with their gnarled fingers touched their caps' in greeting. 'When the steamer sailed down the Scheldt in the mist, past cranes brought to a standstill by the economic crisis, farewell shouts of unknown yet faithful friends resounded from the quay. Finishing these lines between Antwerp and Flüssingen, I send fraternal greetings to the workers of Belgium.'

.   .   .   .

On 6 December, Trotsky and Natalya alighted in Paris, at the Gare du Nord, where they were again surrounded by a strong police cordon and separated from the crowd of passengers. Waiting for them there was Lyova: Herriot had granted Natalya's request. At the frontier Trotsky had been told that in Marseilles he would have to wait nine days for a boat to Constantinople. He rejoiced at the delay. Molinier rented accommodation near Marseilles; and Trotsky asked friends to come there and spend the few days with him. But no sooner had he arrived at Marseilles than the police told him that he could not stay even a single day and must board at once an Italian cargo vessel which happened to be leaving that night. He embarked under protest; but having found out that the vessel had no passenger accommodation and would be under way for fifteen days, and fearing that he was being led into a trap, he came back on shore. It was midnight. The police tried to force him back but failed. Sparring with gendarmes, the whole party camped in the harbour through the small hours of a wintry and windy night. From the harbour Trotsky addressed telegrams of protest to Herriot, to the Ministry of the Interior, to Blum and Thorez; he also sent a request to Rome

I. Trotsky and his wife, returning from Copenhagen to Prinkipo, 1932

Paul Popper

II. Zina, Trotsky's daughter, shortly before her suicide

for an Italian transit visa. Before dawn the police took him and Natalya to a hotel, warning them to await imminent deportation.

Day came, hours passed, and there was no reply from Herriot or anyone else in Paris. Ironically, Mussolini's Foreign Ministry immediately answered and granted the transit visa. The police then rushed Trotsky and Natalya to the first train departing for Italy. Across the police cordon both embraced Lyova. They had spent only a day with him, a day so full of agitation that they had no chance, as Natalya put it, to have a look at each other, let alone to unburden themselves of the troubles that weighed on their minds—only petty vexations and misunderstandings, arising out of the circumstances, had come between them.

In the train Trotsky and Natalya reflected on the absurdity of it all. They were hurt and weary. It was as if the burdens of their life, the heavy dull-witted spite of governments and gendarmes, Zina's misfortune and uncertainty about her child, had all come down on them at once. Well inside Italy, Natalya wrote to Lyova, 'we long, long sat with Papa in the dark compartment and wept. . . .'[1]

Next morning they awakened in Venice, which they had never seen before; and through tears their eyes opened wide to the lustre and the glory of San Marco.

.   .   .   .

On 12 December they landed at Prinkipo. The 'escaped lion' was back in his 'cage'; but he appeared reconciled to the return. Perhaps his nerves were soothed by the beauty of the island, the courtesy Turkish officials had shown him on the frontier, and the honest faces of the fishermen of Büyük Ada beaming a friendly welcome. The bookshelves and desks, with piles of correspondence and papers, urged him back to his labour. 'It is good to work pen in hand in Prinkipo', he noted later in his diary, 'especially in the autumn and the winter, when the island is empty and woodcocks appear in the park'. Beyond the windows, the sea, with shoals of fish coming right up to the shore, was like an unruffled lake. After all the agitation

[1] Natalya to Lyova, 16 December 1932. Closed Section of *The Archives*. See also Trotsky's statement to the Press made at Brindisi on 8 December. *The Archives*.

o

and uproar of recent weeks, the stillness of the island, never disturbed by a motor horn or a telephone bell, offered a respite and induced reflection.

And so the last weeks of the year passed off quietly and restfully. The only discordant yet minor incident was the final break with Senin-Sobolevicius, who in Berlin had moved a motion dissociating the International Secretariat of the Opposition from one of Trotsky's sharp attacks on Stalin.[1] The incident surprised Trotsky, even though he had months earlier written to Sobolevicius that 'the party is exercising a strong pull on you'. But he had thought that they had come to an agreement in Copenhagen. 'You told me', he wrote to Senin on 18 December, 'that your journey to the Soviet Union had finally convinced you that the Opposition was right.' Even now Trotsky suspected no foul play, but thought that Senin was yielding to 'the party's pull' and that this might lead him to capitulation. 'Capitulation', he warned Senin, 'is political death'; and he advised him to take time off and think matters over. He evidently regretted losing an intelligent and helpful follower; but the break was accomplished, and soon Senin disappeared from Trotsky's horizon.[2]

In these weeks of repose Trotsky found in fishing the old 'diverter of sadness and calmer of unquiet thoughts'. In diary pages, written just before he left Prinkipo, he describes it in a rather Waltonian manner, and draws affectionate character sketches of fellow fishermen, especially of a young, almost illiterate Greek, Kharalambos, with whom he often ventured out.[3] The young Greek 'had angling in his bones'; his forebears, as far as memory reached back, were all fishermen. 'His own world extends approximately to four kilometres around

[1] See Trotsky's correspondence with the brothers Senin—Soblen-Sobolevicius of 15, 16, 18 and 22 December 1932. Trotsky's attack on Stalin ('Obeimi rukami'), to which they objected appeared in *B.O.*, no. 32 in the same month. In it Trotsky had charged Stalin with unprincipled wooing of American capitalism—he based the accusation on an interview Stalin had given to a certain Thomas Campbell, an American engineering expert and author of a book on Russia. Campbell quoted Stalin as saying that the first reason for the break between himself and Trotsky was Trotsky's eagerness to spread revolution to other countries and his, Stalin's, desire 'to limit all his efforts to his own country'. Stalin later denied that he had made this statement, but the denial was rather unconvincing. The brothers Sobolevicius held that Trotsky's attack was unjust and ultra-left.

[2] *The Archives*, Closed Section.

[3] These diary pages, dated 15 July, are in *The Archives*.

Prinkipo. But he knows this world'; and finds in it enough magic to fill his life (as in Walton, 'somewhat like poetry' and somewhat 'like the mathematics that it can never be fully learned'). 'He could read like an artist the beautiful book of Marmara'; and he diverted to it from distant wanderings the mind of the old revolutionary. They talked to each other only in gestures, grimaces, and a few Turkish, Greek, or Russian monosyllables. These were enough for Kharalambos to convey what was going on in the depth of the sea, to tell, by the horizon, the skies, the season and the winds, how the nets should be cast—straight, in spirals, or in semicircles—how weights should be thrown from the boat to bring lobsters into traps, and how the catch should be guarded against dolphins lurking round. The author of *Permanent Revolution* learned eagerly and humbly this 'intricate and primordial art which has not changed for thousands of years'. He noticed 'the annihilating glance' Kharalambos gave him whenever he threw a weight the wrong way. 'From kindness and a sense of social discipline he admits that, on the whole, I do not throw the weights badly. But it is enough that I should compare my work with his and my pride abandons me at once.' It was not so bad, after all, to come back to Kharalambos, to read with him the book of the Marmara, and to write a book of one's own as well.

This idyllic interval ended abruptly and grimly. On 5 January 1933 Lyova informed his parents by cable that Zina had committed suicide. She killed herself a week after her child had at last been brought to her. The child's presence, it seems, far from steadying her nerves, finally shattered them. Among the papers she left was this note written in German: 'I feel the approach of my terrible disease. In this condition I do not trust myself, not even with the handling of my child. *In no circumstances* should he come here. He is very sensitive and nervous. He is also frightened of Frau B. [the landlady]. He is with Frau K. [address follows]. He does not speak a word of German. Telephone my brother.'[1] Her brainstorms had been recurring with ever greater force and frequency; she felt useless even to her child; she had no strength to struggle on; and, on top of all this, the police had just told her that she must leave Germany. These were the last days of General Schleicher's government—

---

[1] The note written in German bears no date.

before the end of the month Hitler was to be acclaimed as Chancellor. Louder than ever Berlin was resounding with the trampling of heavy boots and hoarse and drunken singing; and one song, coarse and cruel, *Die Strassen frei für die braunen Batallionen* drowned all the others. The 'terrific tank' of Nazism was rolling in to crush the German worker. The *Horst Wessel Lied* in her ears, her own country closed to her, and herself torn from her family, driven from Germany, and too sick to look for another refuge, Zina locked and barricaded herself in her room and opened the gas taps. So massive was the barricade she put up that any attempt at saving her was hopeless—her doctor was amazed at the 'rare energy' she had displayed in the very act of dying. And in her last minutes the consciousness of release brought a faint smile to her face, an expression of relief and calm. She was thirty years old.[1]

Lyova's message about the suicide was laconic, but, to quote Trotsky, 'one sensed unbearable moral tension in every line of it' for he 'found himself alone with the corpse of his elder sister. . . .' How was the child to be told what had happened? And how was the news to be broken to Alexandra Sokolovskaya, Zina's mother, in Leningrad? Lyova tried to obtain a telephone connexion with his brother in Moscow. 'Was it because the G.P.U. were disconcerted . . . or because they hoped to overhear some secret—enough that, against all expectations, Lyova obtained the telephone connexion and . . . communicated the tragic news. . . . Such was the last talk of our two sons, the doomed brothers, over their sister's still warm body.'[2]

Six days after Zina's suicide Trotsky wrote an 'Open Letter' to the party leaders in Moscow. He described how the decree of 20 February had broken Zina's spirit: she 'did not choose death of her own will—she was driven to it by Stalin'. 'There was not even a shadow of any political sense in the persecution of my daughter—there was nothing in it but purposeless, naked vengeance.' He ended the letter on a note in which grief stifled anger: 'I am confining myself to this communication

---

[1] A moving description of Zina's death and funeral is in Franz Pfemfert's letter to Trotsky of 20 January 1933 (*The Archives*, Closed Section). Lyova's telegram, ibid.

[2] Quoted from Trotsky's obituary of Lyova, written six years later. *B.O.*, no. 64, March 1938.

without drawing further conclusions. The time for drawing such conclusions will come—a revived party will draw
them.'[1]

From Leningrad, from Zina's mother, came a cry of pain,
reproach, and despair. She had now lost both her children,
both born during their father's first exile and both struck down
during his last exile. 'I shall go mad myself if I do not learn
everything', she wrote to Trotsky on 31 January, asking for an
explanation of all the circumstances. She quoted what Zina had
written her only a few weeks earlier: 'It is sad that I can no
longer return to Papa. You know how I have adored and
worshipped him from my earliest days. And now we are in
utter discord. This has been at the bottom of my illness.' Zina
had complained about his coolness towards her. 'I explained to
her', these are her mother's words, 'that all this comes from your
character, from the fact that you find it so difficult to show your
feelings even when you would like to show them.' (To those
familiar only with the public face of Trotsky, the passionate
rhetorician, his first wife's testimony about his undemonstrative
intimate character may come as a surprise.) Then followed this
poignant reproach: 'Yet have reckoned only with her [Zina's]
physical condition, but she was an adult and a fully developed
being in need of intellectual intercourse.' She had yearned for
political activity and she needed scope, for she had taken after
her father; and—'you, her father, you could have saved her'.
And what, Alexandra asked, had been behind the conflict
between Zina and Lyova, of which Zina had also written?
And why had Trotsky insisted on a psychoanalytic treatment
when 'she was closed in herself—as we both are—and one
should not have pressed her to talk about things she did not
want to talk about!' Yet, as the mother confronted Trotsky with
these reproaches, she softened them with the reflection that if
Zina had remained in Russia, she would have perished anyhow
—she would have died of consumption. 'Our children were
doomed', Alexandra added and described the fear with which
she looked on the grandchildren left with her: 'I do not believe
in life any longer. I do not believe that they will grow up. All
the time I am expecting some new disaster.' And she concluded: 'It has been difficult for me to write and mail this letter.

[1] B.O., no. 33, March 1933.

Excuse my cruelty towards you, but you too should know everything about our kith and kin.'[1]

We do not know whether or how Trotsky answered this
letter—perhaps the wound was too deep for words. Some time
later, apologizing to friends for not having acknowledged condolences, he wrote that he had been struck down by malaria
and 'half deaf'.[2]

.   .   .   .

To the last, Trotsky refused to believe that the German
labour movement was so devoid of any power of self-preservation as to put up almost no resistance to Nazism and to collapse
ignominiously under its first onslaught. For nearly three years
he had argued that it was inconceivable that Hitler should win
without a civil war. The inconceivable had now happened:
on 30 January 1933 Hitler had become Chancellor, before
socialists and communists had even begun to marshal their
immense resources for a fight. A week later Trotsky stated:
'Hitler's accession to power is a terrible blow to the working
class. But this is not yet the final, the irretrievable defeat.
The enemy, whom it was possible to rout while he was still
climbing up, has now occupied a whole series of commanding
posts. He has thus gained a great advantage, but the battle
has not yet been fought.' Even now there was still time, for
Hitler had not yet seized total power; he had to share it with
Hugenberg and the Deutschnazionale. The coalition he headed
was unstable and riddled with contradictions. He still had to
strip his partners of all influence, and to obtain exclusive
control of all the resources of the state. Until then his position
remained vulnerable. Socialists and communists could still
strike back—but it was desperately late: 'what is at stake is the
head of the German working class, the head of the Communist
International and . . . the head of the Soviet Republic!'[3]

We know now from numerous German archives and diaries

---

[1] Alexandra Sokolovskaya's letter, dated 31 January 1933, is in the Closed
Section of *The Archives*.

[2] Trotsky to Franz Pfemfert, 5 February 1933, ibid. According to Pierre Frank,
who was at Büyük Ada then, Trotsky shut himself in his room for several days;
Natalya was with him; and she alone came out now and then. When he emerged
at last, his secretaries noticed how grey his hair had grown during those days.

[3] *B.O.*, no. 33, 1933.

how great indeed was the vulnerability of Hitler's first government, as it came into being.[1] Even a month later, on 5 March,
after the Nazi raid on the Karl Liebknecht House in Berlin and
after the Reichstag fire, in elections held under an unbridled
Nazi terror, the socialists and the communists still polled
12 million votes, not to speak of the nearly 6 million votes
cast for the Catholic opposition to Hitler. We also know of the
quarrels, the rows, and the mutual distrust between Hitler
and his partners, which might well have disrupted their coalition if those millions of socialists and communists had moved
into action. As early as 6 February Trotsky observed that the
working class 'was not conducting any defensive battle but was
retreating, and tomorrow the retreat may well turn into a
panic-stricken rout'. He concluded rather abruptly with this
grave passage:

In order to expose more clearly the historic significance of the
party's decisions . . . in these days and weeks, it is, in my view,
necessary to pose the issue before Communists . . . with the utmost
sharpness and irreconcilability: the party's [continued] refusal to
form a united front and to set up local defence committees, committees which might become Soviets tomorrow, will be nothing less
than a surrender to fascism, an historic crime tantamount to the
liquidation of the party and of the Communist International.
Should such a disaster happen, the working class will have to make its
way towards a Fourth International; and it will have to make it
through mountains of corpses and years of unbearable sufferings and
calamities.[2]

Even before these words appeared in print, the great mass
organizations of German labour, its parties and trade unions, its
many newspapers, cultural institutions, and sports organizations
all lay in ruins.

The great defeat at once affected the fate of Trotsky's
family. The *Bulletin* was banned in Berlin, and Lyova had
to go into hiding and steal across the frontier. On 24 March,
Trotsky wrote to the Pfemferts (whose home the Nazis had
already wrecked): 'We have all the time been very anxious
about L.L. [i.e. Lyova]. German friends think that if he fell into
fascist hands he would not come out alive. I thought the same.
But yesterday we received a telegram from him: "I am moving

[1] A. Bullock, *Hitler*, pp. 229–33ff.   [2] *B.O.* Loc. cit.

to Paris." Let us hope that he will have good luck in completing the move. We have not yet had any further news from him.'[1]

.　.　.　.　.　.　.　.　.　.

In these weeks Trotsky renounced his allegiance to the Third International. In an article under the title 'The Tragedy of the German Proletariat' (and the sub-title: 'The German workers will rise again—Stalinism never!'), he thus summed up the situation: what the labour movement had suffered in Germany was not a temporary reverse or a tactical setback, but a decisive strategic defeat, which would leave the working class prostrated and paralysed for a whole epoch. The Second and Third Internationals alike refused to admit this, spoke of Hitler's 'ephemeral' success, and now, when it was too late, declaimed about a united front. But 'before any decisive struggles become possible in Germany once again, the vanguard of the working class must orientate itself anew, grasp clearly what has happened, fix the responsibility for . . . defeat, clear new roads, and thus regain self-confidence and self-respect'. For years the 'key to the situation' had been in communist hands; it was no longer there. All positions in Germany were lost for years to come; all the more important was it for the labour movement to fortify its strongholds and to fight in the countries surrounding Germany, in Austria, Czechoslovakia, Poland, the Netherlands, and France. 'Austria, most immediately threatened by a fascist upheaval, is now the forward bastion.' It was the height of irresponsibility on the Comintern's part to announce that the German workers were 'on the eve of great battles' because they had cast 5 million votes for the communists. 'Yes, five million communists still managed, each individually, to make their way to the polling booths. But in the factories and the streets their presence is not felt. They are lost, dispersed, demoralized. . . . The bureaucratic terror of Stalinism has paralysed their will even before the gangster terror of fascism has started its work.'[2]

He concluded that Stalinism had had its '4 August', a collapse as ignominious as that which the Second International

---

[1] *The Archives*, Closed Section.
[2] 'The Tragedy of the German Proletariat', dated 14 March, appeared in the May issue of *B.O.* (no. 34).

had suffered at the outbreak of the First World War. Then
Lenin, Trotsky, Rosa Luxemburg, Karl Liebknecht, and their
associates had declared that the Second International was
dead and had proclaimed the idea of the Third International.
The analogy with 4 August suggested that Trotsky would now
proclaim the idea of the Fourth International. He did not yet
do this, however. He called only for the formation of a new
Communist party in Germany. 'The advanced workers of
Germany will henceforth speak about the time when the
Stalinist bureaucracy dominated [German communism] not
otherwise than with burning shame. . . . The official Communist
party of Germany is doomed. From now on it will only dis-
integrate, crumble, and dissolve into nothing.' He still reckoned
with the possibility that the defeat might come as a salutary
shock to the other Communist parties, induce them to delve
into the causes, to find out where the responsibilities lay, and
perhaps to break with Stalinism. Should this happen, then the
Comintern (or a segment of it) might still save its revolutionary
honour and *raison d'être*. But 'in Germany at any rate the sinister
song of Stalinist bureaucracy is at an end. . . . Under the
enemy's terrible blows advanced German workers will have to
build a new party'. It might be argued that it was illogical to
call for a new Communist party, but not for a new International; but the historic development did not proceed altogether
according to the rules of logic; and one should wait and see
whether any Communist parties would draw the lessons from
the German experience.[1]

If Trotsky had any such hopes these were soon dispelled.
The Executive of the Comintern, at its first session after
Hitler's victory, declared that victory devoid of significance.
It asserted that the strategy and tactics of the German party
had been flawless from beginning to end; and it forbade any
Communist party to open any debate over the issue.[2] Not a
single party dared to defy the ban. The spectacle was so shocking that it led Trotsky to state that 'an organization which has
not been wakened up by the thunderbolt of fascism . . . is dead
and cannot be revived'. In July, he declared that it was not
enough to build a new Communist party in Germany; the

[1] Ibid.
[2] *Kommunisticheskij Internatsional*, 1933, no. 36, p. 17; *B.O.*, nos. 36–37, 1933.

time had come to lay the foundations of a new International.[1]

Even now he could not make up his mind whether the new International should extend its activities to the Soviet Union; that is, whether his followers there should cease to consider themselves a faction of the old party and form a new party of their own. For several months he advised them against such a course and insisted that the activities of the Fourth International must stop at the frontiers of the Soviet Union. He still saw in the Bolshevik monopoly of power, abused though it was by Stalin, the *sine qua non* of the revolution's survival. The Opposition, he argued, would be justified in constituting itself an independent party only if it abandoned any hope of reforming the régime and reoriented itself for a revolutionary struggle against Stalinism; this it must not do. A new International could well refrain from working inside the Soviet Union because 'the key to the situation' in the labour movement was no longer in the Soviet Union: the Opposition had hardly any chance of developing its activity there, at any rate in the near future; and so the issue of a new Communist party was academic. Only if and when the new International grew into a vital political force in other countries, could the alignment of forces change in the U.S.S.R. as well. Above all, it would be the advance of revolution in the West, an advance which could not be achieved under Stalin's leadership, that would weaken the stranglehold of Stalinism on the Soviet Union and give fresh strength to the communist opposition.[2]

This was clearly an untenable position; and the logic of his new venture soon got the better of Trotsky once again. It had been inconsistent to advocate a new party in Germany but not a new International; and it was just as inconsistent for the new International to refrain from action within the Soviet Union. And so in October 1933 Trotsky concluded that the Opposition should constitute itself into a new party in the U.S.S.R. as well.[3]

[1] *B.O.* Loc. cit.

[2] Ibid. Pierre Frank relates that during the weeks and months when Trotsky tried to make up his mind on these points his secretaries saw him every day walking in his room for hours, silent, tense and absorbed in his dilemmas. 'His face was profusely covered with sweat; and one sensed the physical exertion of his thought and hesitancy.'

[3] Loc. cit. Trotsky carried out this revision in his essay 'The Class Character of the Soviet State', the writing of which he concluded, according to *The Archives*, on 1 October 1933. In the *B.O.*, nos. 36–37 the date has been misprinted as 1 October 1932.

It had taken him about six months to draw this conclusion. Having done so he had to revise some of the views by which he had stood unflinchingly for ten years. He had ceased to uphold the political monopoly of the ruling party. The new party, if and when it came into existence, was to work not for the reform and the constitutional replacement of the Stalinist government but for its revolutionary overthrow. Did he then still consider the Soviet Union to be a workers' state? Or did he now view its régime as a Thermidorian or Bonapartist variety of counter-revolution? And was the Opposition, or was it not, to remain committed to the unconditional defence of the Soviet Union?

Trotsky argued that after all the experiences of recent years it would be childish to think that it was possible to depose Stalin at a Congress of the Party or of the Soviets. 'No normal constitutional ways are left for the removal of the ruling clique. Only *force* can compel the bureaucracy to hand over power into the hands of the proletarian vanguard.' That vanguard, however, was dispersed and crushed—it would not be able to fight for power in the near future. The question of Reform or Revolution was therefore basically a matter of long term orientation. The Opposition could not claim office unless it had the support of the majority of the working class; and it could not obtain that without previous social shifts at home and radical changes on the international scene, in the first instance, without an advance of revolution outside the Soviet Union. After such shifts and changes 'the Stalinist apparatus would find itself suspended in a vacuum'; and the Opposition, assisted by popular pressure, might be able to win even without revolution or civil war. If Stalin and his adherents, despite their isolation, still continued to cling to power, the Opposition would oust them by means of a 'police operation'. Confronted by an upsurge of political energy in the working class, Stalinism would be utterly weak precisely because it had 'its roots in the working class and nowhere else': only with the acquiescence and submissiveness, if not the active support, of the workers was Stalin strong—without these he could be overthrown by a push.[1]

The Soviet Union, Trotsky reasserted, remained a workers'

[1] *B.O.* Loc. cit.

state. Social ownership of the means of production prevailing, Soviet society was engaged in the transition from capitalism into socialism, even though it paid an exorbitant price for every step forward. The bureaucracy, no matter how privileged, was still only 'a malignant growth on the body of the working class, not a new possessing class'. Privileges and growing social inequality reflected not a new type of exploitation, as the ultra-radicals alleged, but were the consequences of poverty and material scarcities. To some extent, as incentives to efficiency and production, privileges and inequality were 'the bourgeois tools of socialist progress'. Bureaucratic rule, parasitic and tyrannical, might endanger all the conquests of the revolution and provoke counter-revolution; but it might also turn out to be 'the instrument'—a poor and expensive one—'of socialist development'. 'Wasting . . . an enormous portion of the national income, the Soviet bureaucracy is at the same time . . . interested in promoting the economic and cultural growth of the nation: the higher the national income the more abundant is the fund of the bureaucracy's privileges. Yet, the economic and cultural advance of the working masses, achieved on the social foundations of the Soviet state, should undermine the basis of bureaucratic rule.' Thus, twenty years before the end of the Stalin era, Trotsky foresaw that by industrializing the Soviet Union and spreading education among its people, Stalinism might destroy the soil on which it had grown and which nourished it, the soil of primordial poverty, illiteracy, and barbarism.[1]

Having ceased to defend the single party system in the U.S.S.R., Trotsky nevertheless repeated his earlier warning that 'if the bureaucratic equilibrium in the Soviet Union were to be shaken at present, this would almost certainly be to the advantage of counter-revolutionary forces'. He restated his commitment to the unconditional defence of the Soviet Union: '. . . the new International . . . before it can reform the Soviet state, must take upon itself the duty to defend it. Any political grouping which disavows this commitment, under the pretext

---

[1] *B.O.* Loc. cit. 'It is clear', Tròtsky concluded, 'that in this happy historical variant, the bureaucracy would turn out to be only the tool—a poor and expensive tool—of the socialist state.' But he did not take it for granted that this 'happy variant' would materialize.

that the Soviet Union is no longer a workers' state, risks becoming a passive tool of imperialism. . . .' The adherents of the new International, he added, 'must in an hour of mortal danger fight on the last barricade' in defence of the U.S.S.R.[1]

Yet, while he insisted so forcefully that the Soviet Union, judged by its economic structure, remained a workers' state, he now took the view that as a factor of international revolution it was little more than an extinct volcano. 'From the beginning of the First World War, and more explicitly since the October revolution, the Bolshevik party has played a leading role in the global revolutionary struggle. Now this leading position has been lost.' Not only official Bolshevism, that 'parody of the party', but the Bolshevik Opposition as well, was, because of the difficult conditions in which it worked, unable to 'exercise any international leadership'. 'The revolutionary centre of gravity has definitely shifted to the West, where the immediate possibilities for building a new party are much wider.' He proclaimed the idea of the Fourth International in the belief that new impulses for revolution would come from the West, not from the Soviet Union.[2]

We have seen with how much hesitation Trotsky had made up his mind to renounce his allegiance to the Third International. The causes of his hesitancy were not far to seek, for he himself had many times stated his objections to the step he was now taking. It was to the Third International, he had argued, that the revolutionary workers of all countries looked for guidance; it was in it that they saw the legitimate successor to the Second and First Internationals and the very embodiment of the idea of the Russian Revolution; and as long as the Soviet Union remained a workers' state and the Comintern retained its association with it, the class-conscious *élite* of the workers were justified in their loyalty to the Comintern. He was not quite sure that this reasoning had now lost its validity. Nor was it easy for him, in view of the part he had played in the Third International, to announce his final break with it. It is extremely rare for one of the principal architects of a great and vital movement to find in himself the strength to declare that movement worthless. It was far more difficult for Trotsky to turn his back on the Third International than it had been to renounce

---

[1] *B.O.* Loc. cit.   [2] Ibid.

the Second in 1914. Only the Comintern's stunning failure in
Germany brought him to do it. He admitted that there was a
difference between 1914 and 1933. In 1914 the leaders of the
Second International had, by supporting an imperialist war,
betrayed their trust deliberately and with their eyes open;
whereas in 1933 the Comintern had facilitated Hitler's victory
from sheer irresponsibility and blindness. Yet the catastrophe
of 1933 was in other respects even worse than that of 1914.
In the First World War revolutionary Marxism soon recovered
from the blow: Zimmerwald, Kienthal, and the Russian
Revolution registered a powerful protest against the 'social
imperialist' perversion of Marxism. No comparable protest
against the enormity of 1933 had come or was to come from
within the communist movement. Not only had the Comintern's policy contributed to the loss by German labour of all
it had gained in over eighty years of struggle; and not only had
that policy allowed the danger, nay, the certainty of another
world war to come about—in addition, all this had occurred
amid an uncanny indifference and apathy on the part of the
entire movement. What had happened, Trotsky asked, to the
political conscience and understanding of the great mass of
communists?

He concluded that reformism and Stalinism had between
them stultified the minds and destroyed the will of the workers.
That all his own warnings, so clear, so loud, so strikingly confirmed by events, could have gone so unheeded confirmed him
in this conclusion. No one knew better than he himself how
unheeded his warnings had gone, for in a letter to Sobolevicius
he remarked, early in 1932, that the Trotskyist Opposition had
failed to recruit in Germany even 'ten native factory workers'
(and had won over only a few intellectuals and immigrants).[1]
In the First World War at least a few thousand German
workers joined the clandestine *Spartakus* and echoed the denunciation of the '4 August', which Rosa Luxemburg and Karl
Liebknecht voiced from their prison cells. Now, after Hitler's
triumph, all the Communist parties of the world received
the Comintern's self-justifications and self-congratulations in
numbed silence. Was there no spark of intelligence, of
international solidarity, and of responsibility left in all those

[1] Letter to Senin-Sobolevicius, 6 March 1932, *The Archives*, Closed Section.

parties? Trotsky asked again and again. If not, then Stalinism had so irretrievably debased the entire communist movement that to try and reform it was a Sisyphean labour. He had been performing that labour for ten years now; and he refused to go on rolling the heavy rock up the dismal mountain.

It was even more painful for him to renounce finally the Soviet party, the party which Lenin had founded, which had accomplished the revolution, and within which he had achieved greatness. The year before, after the second deportation of Zinoviev, Kamenev, Smirnov, Preobrazhensky, and others, it looked as if the Joint Opposition of 1925–7 was coming back into being. Every message from Moscow indicated that amid the nation-wide turmoil even Stalin's entourage longed to rid itself of him. Since 1932, however, Stalin had once more gained the upper hand. He succeeded in this in part because he once more adopted some of the measures Trotsky had advocated: he gave the economy a 'breathing space' at the end of the first Five Year Plan; he set lower and more realistic targets for the second Plan; he made concessions to the collectivized peasantry. Consequently, the chaos, the turmoil, and the inner party ferment subsided. The German catastrophe instead of weakening Stalin strengthened his hand. Those who realized its implications felt that this was not the time to sap the stability of government in Moscow. The establishment of totalitarian rule in Germany gave a new impetus to the totalitarian trend in the Soviet Union. When the cry *Ein Führer, eine Partei, ein Volk!* thundered over Germany, the Soviet hierarchy and many of the rank and file felt that only under a single leader could the revolution and the Soviet Union survive. In May 1933 Zinoviev and Kamenev once again capitulated and returned from exile. At their first capitulation, in 1927, they had surrendered to Stalinism, but had not gone, and no one expected them to go, on their knees before Stalin's person. When this was required of them in 1932 they could not yet bring themselves to do it. This, however, was what they did in 1933: in their new recantations they glorified Stalin's infallibility and unique genius.

All this occurred while Trotsky was committing himself to the Fourth International, but was not yet ready to call for a new party in the Soviet Union. Stalin's triumphant emergence from the crisis, the new autocratic aura around him, and the spectacle of the latest capitulations impelled Trotsky to sever the last tie which in theory still bound him to the old party. Commenting on Zinoviev's and Kamenev's 'tragic fate', he wrote: 'The future historian, who will wish to show how ruthlessly an epoch of great upheavals devastates characters, will take Zinoviev and Kamenev as his examples . . . the Stalinist apparatus has become a machine for crushing the backbones [of former revolutionaries].' And: 'Like Gogol's hero, Stalin collects dead souls for the lack of living ones.'[1] Trotsky's hope for any regeneration of the Soviet party was now destroyed. It was futile to go on appealing to men with broken backs and to dead souls; and, anyhow, the Marxist-Leninist traditions had gone from a party that could bow to an autocrat. Only in complete independence from it and beyond its confines could Bolshevism have a rebirth.

This, briefly, was Trotsky's case for a new International. Having made it and, after a discussion, obtained for it the endorsement of all his groups, he did not, however, proclaim these groups to be the Fourth International. Aware of their weakness, he contented himself with launching the idea in the hope that it would gain many more adherents presently. He repeated in a way his own experience of the Zimmerwald era, the memory of which is discernible in his writings and behaviour. From the moment when Lenin and he had begun to advocate the Third International in 1915, it took four years of propaganda and preparatory work before they called a foundation congress of the International. Similarly, now there was 'no question of any immediate proclamation of . . . the International, but only of preparatory work. The new orientation means . . . that the talk about "reforming" [the Stalinist organization] and all demands for the reinstatement of expelled Oppositionists should be definitely abandoned. . . . The Left Opposition ceases to think of itself and to act as an [inner-party] Opposition.'[2] It was to take four years exactly before he would be ready to convene a foundation congress.

His hopes for the new International were not as wild in 1933 as they appeared later. Over the German issue the Comintern was in fact utterly discredited, while Trotskyism had scored a

[1] *B.O.*, no. 35, 1933.   [2] *B.O.*, no. 36, 1933.

striking moral victory. If hitherto, so Trotsky thought, all his appeals to European communist opinion had met with all too little response, this had been partly because the main issues of his controversy with Stalin, domestic Soviet affairs and the Chinese revolution, had been too remote from European communists or too obscure. In its latest phase the controversy had centred on Germany, 'the heart of Europe'. Hitler's advent affected immediately every Communist party. It posed problems of life and death. It pointed to war. It threatened communism with extinction. Both he and the Comintern had conducted the argument publicly and with the utmost vigour until the very moment when the differences were tested by events. The outcome of the test was in no doubt. The pros and cons were, or should have been, fresh in everyone's mind: every communist could review and ponder them anew. The conclusion to be drawn was in no doubt either: those who had led the most powerful Communist party of the West to so shameful a débâcle were guilty of incompetence bordering on treason, and had forfeited every title to leadership. By the same token the Opposition had, or should have had, established its claim to leadership.

Some awareness of all this was undoubtedly penetrating into Stalinist ranks. The more spitefully the Comintern had attacked and mocked Trotsky for 'playing the bogey-man', 'exaggerating the Nazi menace', and 'urging a united front with social-fascists', the more did these mockeries rebound on their authors. Embarrassment and shame took hold of many a party cell. Even hardened Stalinists felt a sneaking admiration for Trotsky's clear-sighted and intrepid stand.[1] New Trotskyist and quasi-Trotskyist groups formed themselves among German refugees from Hitler's terror and among Polish, Czech, Dutch, American, and other communists. The groups were small, but their influence could not be ignored. They drew to themselves alert-minded and devoted party members. They assailed the conscience of communism. They forced Stalinism on to the defensive. Only by frantic appeals to party patriotism, threats of expulsion, and actual expulsion could the leaders subdue the

[1] They themselves admitted this 'sneaking admiration' many years later when they were freer to do so; some went out of their way to be able to speak about it to Trotsky's present biographer.

P

malaise in the ranks; and eventually the Comintern could dispel it only by reversing all its attitudes, by throwing overboard the slogans about social-fascism, and by adopting the tactics of the united front (and going beyond them, to the Popular Front). Moreover, the collapse of the Weimar Republic had shaken the Social-Democratic parties too. Their belief in parliamentary democracy had received a rude blow. There was hardly a Socialist party in Europe which did not, under the impact of the German experience, solemnly inscribe some form of 'proletarian dictatorship' in its programme. Inside those parties radical and leftish groups looked up to Trotsky and found his ideas much more rational and alluring than all that official communism could offer. This was indeed the high water mark of his political influence in exile. If he had any chance at all to found an independent Communist party it was now.

Yet the arguments which he himself had so frequently and cogently advanced against the course of action he was now taking had lost none of their strength. It was still true that as long as national ownership in the means of production remained intact in the Soviet Union and as long as the banner of Bolshevism was hoisted over Moscow, the association of international communism with the Soviet Union was indissoluble. To the mass of those who were in sympathy with communism the workers' first state was still the bulwark of international revolution; and the official Communist parties exercised an overwhelming attraction on them. They saw no alternative to the Stalinist leadership which, in their eyes, had come to represent the Russian Revolution and the Bolshevik tradition. The Stalinist bureaucracy had actually succeeded in identifying itself with Leninism and with Marxism at large. Militant French dockers, Polish coalminers, and Chinese guerilla fighters alike saw in those who ruled Moscow the best judges of the Soviet interest and reliable counsellors to world communism. Hence the unreasoning obedience with which they so often accepted the twists and turns and the most preposterous dictates of Stalinist policy. Its opponents appeared to them as the enemies of the Soviet Union and of communism just as to the devout Roman Catholic the enemies of the Holy See were the enemies of Christianity.

All this boded ill for Trotsky's venture. His ideas and

slogans were such that only those who were in sympathy with communism could be sensitive to them—yet these were the people who would be least inclined to rally to a new International. Having for so long remained unimpressed by Trotsky's call for a reform in their parties, they were even less likely to be moved when he urged them to break with these parties.

Nor did or could the aftermath of the German débâcle favour the new International, no matter how discredited the old Internationals were. Each of the old Internationals had arisen on a high tide of the labour movement; and at the moment of formation none of them had had to contend against any established rival.[1] The Fourth International set out to challenge two established and powerful rivals during a deep depression of the movement. In Germany the working class was indeed, as Trotsky had predicted, unable to recover politically for many years to come; but precisely because of this Trotskyism could derive no practical benefit from the moral advantage it had gained over the German issue. Elsewhere in Europe, the working class was to remain in retreat for the rest of the decade, despite the upsurge of its energies in France and Spain in 1936. The long leaden sequence of retreats and defeats produced a moral sickness, amid which even the most persuasive pleas for a new International fell flat. Trotsky argued that the working class needed a new leadership precisely in order to bring the retreat to a halt and to regroup for defence and counter-offensive. But the mass of communists (and of socialists), those of them who had not yet lost heart, felt that they must not swap horses mid-stream. And so the two established Internationals flourished even on their blunders and defeats: their followers, whatever misgivings they felt, refused to look for new leaders and new methods of struggle under the hail of blows Nazism and fascism were inflicting on them. They were prepared to flounder under old and familiar banners, from defeat to defeat, rather than rally to a new standard behind which they could see only the giant but enigmatic or suspect figure of the standard-bearer.

Trotsky was convinced that the Comintern had, as a

[1] For a continuation of this argument, especially for the case of the Polish Trotskyists against the foundation of the Fourth International, see Chapter V, pp. 421–22.

revolutionary organization, played out its role. He was not altogether mistaken. Ten years later Stalin was to disband the organization and declare that it no longer served any purpose; and in those ten years the Comintern was only to add to its German bankruptcy new failures in France and Spain and the ambiguities of its policy under the Stalin-Hitler pact of 1939–41. Yet the movement behind the Comintern was anything but a 'corpse'. All that Stalin did to wreck it morally could not kill it. At the very time he disbanded the Comintern its western European parties were gaining fresh strength from their resistance to the Nazi occupation; and it was still under Stalinist banners, though in implicit conflict with Stalin, that the Yugoslav and the Chinese revolutions were to achieve their victories. No matter how much Stalin had done to degrade all Communist parties to mere pawns, the Yugoslav, the Chinese and some other parties had enough vitality to live their own lives, to wage their own struggles, and to change the fortunes of their countries and of the world. Moreover, they were to take fresh impetus and new revolutionary *élan* from the triumphs of Soviet arms in the Second World War.

The idea that new impulses for revolution would come from the West but not from the Soviet Union was the *leitmotif* of Trotsky's advocacy of the Fourth International. Again and again he asserted that, while in the Soviet Union Stalinism continued to play a dual role, at once progressive and retrograde, it exercised internationally only a counter-revolutionary influence. Here his grasp of reality failed him. Stalinism was to go on acting its dual role internationally as well as nationally: it was to stimulate as well as to obstruct the class struggle outside the Soviet Union. In any case, it was not from the West that the revolutionary impulses were to come in the next three or four decades. Thus the major premiss on which Trotsky set out to create the Fourth International was unreal. Yet, since all his attempts to reform the Comintern had been in vain, he could not, as we have seen, go on with that Sisyphean task. He had to look for another solution. His new task, however, was to prove at least as barren as the old one. Sisyphus had only moved hopefully from one side of his dismal mountain to the other; and there he started to roll his rock again.

.   .   .   .

We have seen how Trotsky, when he turned his back on the Comintern, re-committed his adherents to remain the last-ditch defenders of the Soviet Union. He himself, when he addressed western bourgeois opinion in his articles, sought to arouse it to the fact that the Third Reich spelt world war. As early as the spring of 1933 he urged the Western Powers to enter into an alliance with the Soviet Union. These were the first weeks and months of the Third Reich, when hardly a single western statesman contemplated the idea. Hitler now assumed pacifist postures, and at an International Disarmament Conference accepted, to the relief and delight of official London, Austen Chamberlain's and John Simon's disarmament schemes. On 2 June 1933 Trotsky wrote in an essay on 'Hitler and Disarmament': 'The greatest danger is to under-rate an enemy . . . the leaders of the German labour movement did not wish to take Hitler seriously. . . . The same danger may arise on the plane of world politics.' He noted how ready the British Government was to respond to Hitler's 'moderation' and 'peaceful intentions': 'Diplomatic routine has its advantages as long as things move within familiar grooves. It is at once disconcerted when it has to face new and important facts.' Austen Chamberlain and John Simon 'had expected to meet [in Hitler] a madman brandishing an axe; instead they met a man hiding his revolver in a pocket—what a relief!' This was Hitler's first great diplomatic success. His purpose was to rearm Germany, which had since Versailles recovered its place as Europe's mightiest industrial nation, but was still unarmed. 'This combination of potential power and actual weakness determines both the explosive character of Nazi objectives and the extreme caution of Hitler's first steps leading towards those objectives.' Hitler had endorsed the British disarmament schemes knowing full well that France could not accept them—this gave him the chance to play off Britain against France and to place on the latter the odium for the arms race to follow. 'Hitler's love for peace is not an accidental diplomatic improvisation, but a necessary element in a large manœuvre, designed to turn the balance of power radically in Germany's favour and to prepare the onslaught of German imperialism on Europe and the world.' He forecast that, if Hitler's moves were not countered, they would inevitably lead to world war within five to ten years. 'It is against the Soviet Union that Hitler is eager to march. But should this not prove to be the line of the least resistance, the eruption may well turn in the other direction. . . . Weapons that can be used against the East can just as well be used against the West.'[1] He remarked that he did not consider himself 'called upon to act as guardian of the Treaty of Versailles. Europe needs a new organization. But woe if this job falls into the hands of fascism!'

In statements for the American Press Trotsky urged the United States Government (which in this, the sixteenth, year of the revolution had not yet recognized the Soviet Government) to move closer to the Soviet Union in order to meet threats from Japan and Germany.[2] We do not know whether these promptings had any influence on President Roosevelt's decision, taken shortly thereafter, to establish diplomatic relations with Moscow. But Trotsky's views certainly impressed Stalin's diplomacy, which presently took up the theme of the anti-Nazi alliance. Where the security of his own government was concerned, Stalin was quite willing to benefit from the advice of his adversary, even if he did it often belatedly and always in his own crudely perverse manner.

Meanwhile, the Soviet Government prolonged its Rapallo agreements with Germany; and this tempted ultra-radical anti-Stalinists to denounce yet another of Stalin's 'betrayals'. Trotsky found the issue too serious to make a debating point of it. He did not tire of exposing Stalin's and the Comintern's share of responsibility for Hitler's ascendancy. But he did not deny Stalin the right to act in the diplomatic field from expediency. Two years earlier, we know, he had urged the Soviet Government to mobilize the Red Army if Hitler threatened to seize power; but he had done this imagining that the German left would be up in arms against Nazism, in which case the Red Army would be in duty bound to assist. Hitler's bloodless victory and the total destruction of the German left, Trotsky now pointed out, turned the balance against the Soviet Union,

[1] *B.O.*, no. 35, 1933. The article appeared in the *Manchester Guardian* of 21–22 June 1933 (three weeks after it had been written). It was from this article that Litvinov, Stalin's Foreign Minister, borrowed the much-quoted phrase that 'a gun that can fire to the East can fire westwards as well'.

[2] See, e.g., Trotsky's interview with *The New York World Telegram* given on 4 July 1933.

especially as the Soviet Union was also weakened internally by the Stalinist collectivization. Soviet diplomacy was therefore entitled to bide its time, to parley, and even to seek a temporary accommodation with Hitler. With a somewhat startling disinterestedness Trotsky declared that if the Opposition were to assume office in present circumstances, it would not be able to act differently: 'In its *immediate practical actions* the Opposition would have to start from the existing balance of power. It would be compelled in particular to maintain diplomatic and economic connexions with Hitler's Germany. At the same time it would prepare the *revanche*. This would be a great task, requiring time—a task that could not be accomplished by spectacular gestures, but would demand a radical reshaping of policy in every field.'[1] His judgement remained unclouded by any personal emotion against Stalin, and severely objective.

. . . .

These were Trotsky's last months on Prinkipo. For some time his French friends, especially Maurice Parijanine, his translator, had urged the French Government to cancel the order under which Trotsky had, in 1916, been expelled from France 'for ever', and to grant him asylum. Trotsky was sceptical: he assumed that the Radical government, just formed under Edouard Daladier, would be anxious to improve relations with the Soviet Union and would not tolerate his presence in France. But he did what he could to help. He had just arranged to publish in New York an unflattering character study of Edouard Herriot, written shortly after the nocturnal tussle with the police at Marseilles; and he refrained from publishing lest it provided grist to the opponents of his admission to France. He also wrote to Henri Guernut, the Minister of Education, who as a member of the government pleaded for Trotsky's right to asylum; and he solemnly promised to behave with the utmost discretion in France and to cause the government no trouble.[2]

Weeks passed without a decision, the weeks during which he drafted his ideas on the Fourth International and also wrote a few minor essays on French political and literary topics. Uncertainty about his immediate future caused him to put

---

[1] *B.O.*, no. 35, 1933. [2] *The Archives*, Closed Section.

aside larger literary plans and entailed financial troubles such as he had not known since 1929. The trip to Copenhagen, Zina's illness, Lyova's move to France, and the transfer of the *Bulletin* to Paris had involved him in large expenses just when his income was greatly reduced. In Germany, where his major works had had a wide reading public, the Nazis banned and burned his writings, along with all Marxist and Freudian literature, just after the third volume of the *History of the Russian Revolution* had come off the press. In the United States the *History* did not fare too well either. Already in March he had written to a British admirer: 'The world's financial crisis has become my crisis also, especially as the sales of the *History* are quite pitiable.' He contributed occasionally to the *Manchester Guardian* and other papers, but the fees amounted to little. To speed up the decision about the French visa, he wrote, on 7 July, to Henri Molinier that he would be content with a residence permit that would allow him to stay not in metropolitan France but in Corsica, for even there he would be in closer contact with European politics and somewhat farther from the G.P.U. than on Prinkipo.[1] His French friends, however, demanded asylum for him in France, and their insistence was soon rewarded. Before the middle of July he received the visa. This was by no means an unqualified residence permit: he would be allowed to stay only in one of the southern *départements*; he would not be permitted to come, even for the shortest trip, to Paris; and he would have to keep a strict incognito and submit to stringent police surveillance.

He accepted these terms as an incredible piece of good luck. At last he would be out of his Turkish backwater! And he was going to France, whose way of life and culture were so congenial to him, and which was now the main centre of working-class politics in the West. Preparing, full of hopeful anticipation, for the journey, he yet cast a backward glance on his Prinkipo years. 'Four and a half years ago when we came here', he wrote in his diary, 'the sun of prosperity still shone over the United States. Now those days seem as remote as prehistory, or as a fairy tale. . . . Here on this island of quiet and oblivion echoes from the great world reached us delayed and muffled.' It was not without a tug of emotion that he took leave of the splendour

[1] *The Archives*, Closed Section.

of the Sea of Marmara and the fishing expeditions and that he thought of his faithful fishermen, some of whom, 'their bones saturated through and through with the salt of the sea', had recently found their rest in the village cemetery, while others had, in these years of slump, to struggle harder and harder to sell their catch. 'The house is already empty. The wooden cases are already downstairs; young hands are driving in the nails. The floor of our old and dilapidated villa was painted with such queer paint in the spring that even now, four months later, tables, chairs, and our feet keep sticking to it. . . . Oddly, I feel as if my feet had got somewhat rooted in the soil of Prinkipo.'[1]

Fate had not spared him disappointment and suffering on this island. The shadow of death had darkened for him many a day there, even the hours of the departure. The last thing he wrote on Prinkipo (apart from a farewell message of thanks to the Turkish Government) was an obituary on Skrypnik, the old Bolshevik, a leader of the October insurrection, later a fervent Stalinist, who, having come into conflict with Stalin, had just committed suicide.[2]

Yet, despite all the adversities, the years Trotsky had spent on Prinkipo were the calmest, the most creative, and the least unhappy time of his exile.

[1] 'Pered Otyezdom', 15 July 1933 in *The Archives*.

[2] The obituary bears the date of 15 July; it appeared in *B.O.* in October, nos. 36–37.

CHAPTER III

The Revolutionary
as Historian

IKE Thucydides, Dante, Machiavelli, Heine, Marx, Herzen,
and other thinkers and poets, Trotsky attained his
full eminence as a writer only in exile, during the few
Prinkipo years. Posterity will remember him as the historian
of the October Revolution as well as its leader. No other
Bolshevik has or could have produced so great and splendid
an account of the events of 1917; and none of the many writers
of the anti-Bolshevik parties has presented any worthy counterpart to it. The promise of this achievement could be discerned
in Trotsky very early. His descriptions of the revolution of 1905
provide till this day the most vivid panorama of that 'general
rehearsal' for 1917. He produced his first narrative and
analysis of the upheavals of 1917 only a few weeks after the
October insurrection, during the recesses of the Brest Litovsk
peace conference; and in subsequent years he went on working
at his historical interpretation of the events in which he had
been a protagonist. There was in him a twofold *vis historica*:
the revolutionary's urge to make history and the writer's
impulse to describe it and grasp its meaning.

All banished men brood over the past; but only a few,
very few, conquer the future. Hardly any one among them
however, has had to fight for his life, morally and physically,
as Trotsky fought. Stalin at first inflicted exile on him in
the way the Romans used to inflict it—as a substitute for the
death penalty; and he was not to remain content with the
substitute. Even before Trotsky was assassinated physically, his
moral assassins were at work for years, first effacing his name
from the annals of the revolution and then reinscribing it as
the eponym of counter-revolution. Trotsky the historian was
therefore doubly embattled: he defended the revolution against
its enemies; and he defended his own place in it. No writer

has ever created his major work in similar conditions, designed to inflame all his passions, to rob him of every calm thought, and to distort his vision. In Trotsky all passions were aroused, but his thought remained calm and his vision clear. He often recalled Spinoza's maxim: 'Neither weep nor laugh but understand'; but he himself could not help weeping and laughing; yet he understood.

It would not be quite right to say that as historian he combined extreme partisanship with rigorous objectivity. He had no need to combine them: they were the heat and the light of his work, and as heat and light belonged to each other. He scorned the 'impartiality' and 'conciliatory justice' of the scholar who pretends 'to stand on the wall of a threatened city and behold at the same time the besiegers and the besieged'.[1] His own place was, as it had been in the years 1917–22, within the revolution's threatened city. Yet his involvement in the struggle, far from blurring his sight, sharpens it. His antagonism to Russia's old ruling classes and their willing and unwilling supporters makes him see clearly not only their vices or weaknesses but also such feeble and ineffective virtues as they possessed. Here, as in the best military thinking, extreme partisanship and scrupulously sober observation indeed go hand in hand. To the good soldier nothing is of greater importance than to get a realistic picture of the 'other side of the hill', unclouded by wishful thinking or emotion. Trotsky, the commander of the October insurrection, had acted on this principle; and Trotsky the historian does the same. He achieves in his image of the revolution the unity of the subjective and the objective elements.

His historical writing is dialectical as is hardly any other such work produced by the Marxist school of thought since Marx, from whom he derives his method and style. To Marx's minor historical works, *The Class Struggle in France*, *The 18th Brumaire of Louis Bonaparte*, and *The Civil War in France*, Trotsky's *History* stands as the large mural painting stands to the miniature. Whereas Marx towers above the disciple in the power of his abstract thought and gothic imagination, the disciple is superior as epic artist, especially as master of the graphic

---

[1] Trotsky referred in particular to L. Madelin, 'the reactionary and therefore fashionable' French historian. Preface to *History of the Russian Revolution*, Vol. I.

portrayal of masses and individuals in action. His socio-political analysis and artistic vision are in such concord that there is no trace of any divergence. His thought and his imagination take flight together. He expounds his theory of revolution with the tension and the *élan* of narrative; and his narrative takes depth from his ideas. His scenes, portraits, and dialogues, sensuous in their reality, are inwardly illumined by his conception of the historical process. Many non-Marxist critics have been impressed by this distinctive quality of his writing. Here, for instance, is what a British historian, A. L. Rowse, says:

> The real importance of Trotsky's *History* does not lie in his power of word painting, either of character or of scene, though indeed his gift is so brilliant and incisive that one is continually reminded of Carlyle. There is something of the same technique, the same mannerism even, in the way the rapid lights shift across the scene and particular odd episodes are brought out in singular sharpness of relief and made to bear general significance; something of the same difficulty in following the sequel of events—the lights are so blinding—one may add. But where Carlyle had but his magnificent powers of intuition to rely on, Trotsky has a theory of history at his command, which enables him to grasp what is significant and to relate things together. The same point can be illustrated more appositely by comparison with Winston Churchill's *The World Crisis*, for the two men are not dissimilar in character and gifts of mind. But here again one notices the difference; for Mr. Churchill's history, for all its personality, its vividness, and vitality, points which it has in common with Trotsky—has not a philosophy of history behind it.[1]

The remark about the similarity between Trotsky and Churchill is correct: at their opposite poles the two men represent the same blend of realism and romanticism, the same pugnacity, the same inclination to look, and to run, ahead of their class and milieu, and the same urge to make and to write history. One need not deny Churchill a 'philosophy of history' even if he holds it only instinctively; but it is true that Trotsky's is a fully formed and elaborate theory. What is important is that his theoretical *Weltanschauung* permeates his sensitivity, amplifies his intuition, and heightens

---

[1] A. L. Rowse, *End of an Epoch*, pp. 282–3.

his vision. And, although he has in common with Carlyle the intensity and dazzling brilliance of imagery, he also has the compactness and clarity of expression and the balance of the greatest classical historians. He is indeed the only historian of genius that the Marxist school of thought has so far produced and so far—rejected.[1]

.   .   .   .

Of Trotsky's two major historical works, *My Life* and *The History*, the former is, of course, the less ambitious. He wrote it too early in a sense, though if he had not written it in 1929, or shortly thereafter, he might not have written it at all. It tells in the main one half of his story, that of his revolutionary triumph; it only sketches the beginning of the other half, which was still unfolding. He concluded the book after a few months in exile, only five years or so after the struggle between him and Stalin had begun in earnest. The conflict was still too fresh, and in relating it he was handicapped by tactical considerations and lack of perspective. What he was to live through in the coming eleven years was not only to be of great weight in itself, but was to reflect back upon all his earlier experience: the whole of his life was to take on the glow of tragedy from its grave and gloomy epilogue. He concluded *My Life* with a statement defying those who spoke of his tragedy: 'I enjoy the spectacle every scene of which I understand . . .' he repeated after Proudhon. 'What makes others wither, elevates . . . inspires, and fortifies me; how then . . . can I lament destiny . . .?'[2] Would he have repeated these words a few years later? In a sense, if it were to be held that tragedy necessarily includes the protagonist's penance, there was indeed no tragedy in Trotsky—there was no penance in him to the end. Like Shelley, who could not bear that his Prometheus should end by humbling himself before Jupiter, Trotsky was 'averse from a catastrophe so feeble'. His was the modern tragedy of the precursor in conflict with his contemporaries, the tragedy an example of which he himself saw in Babeuf—only that his was a far larger drama, of far

---

[1] This is true, however, only to the extent to which it may be permissible to characterize the communist movement under Stalin and Khrushchev as Marxist.
[2] *Moya Zhizn*, vol. II, p. 338.

greater catastrophic force. Yet even of this kind of tragedy there is no premonition in his autobiography, which consequently leaves the impression of a certain superficiality in the writer's view of his own fortunes, the superficiality characteristic of the protagonist of a tragedy just before disasters assault him from all sides.

The least convincing part of *My Life* is in the last chapters where he relates his struggle with Stalin. Even there he gives us a wealth of insight, incident, and characterization; but he does not go to the root of the matter and he leaves Stalin's ascendancy only half explained. He portrays Stalin too much as villain *ex machina*; and he views him still as he had viewed him years earlier, as too insignificant to be his antagonist, let alone to dominate the stage of the Soviet state and of international communism for full three decades. 'To the leading group of the party (to wider circles he was not known at all) Stalin always seemed a man destined to play second and third fiddle', he says; and he suggests that although Stalin had come to play first fiddle he would soon, very soon, play out his tune.[1] It may be recalled that Lenin in his will described Stalin as one of the 'two most able men of the Central Committee', the other being Trotsky, and warned the party that the animosity of these two men was the gravest danger to the revolution. Trotsky could not gloss over the wider political reasons for Stalin's ascendancy, and he shows Stalin as the incarnation of the party machine and of the new bureaucracy greedy for power and privileges. Yet he could not explain convincingly why the leading cadre of Bolsheviks first assisted in the usurpation and then connived in it, and why all this led to such extraordinary forms of the inner party struggle. As autobiographer no less than as leader of the Opposition, Trotsky virtually ignores the intrinsic connexion between the suppression by Bolshevism of all parties and its self-suppression, of which Stalin was the supreme agent. He does not see why the party should have turned against itself the weapons it had wielded, far less savagely, against its enemies; and that it did so appears to him to be the result of a mere 'conspiracy'.[2]

Yet *My Life* remains an autobiographical masterpiece. François Mauriac rightly compares its opening chapters with

[1] Op. cit., vol. II, p. 247.   [2] Ibid., pp. 227–34.

Tolstoy's and Gorky's descriptions of childhood.[1] Trotsky has the same 'childlike' freshness of the eye and the same almost inexhaustible visual memory, the same power in the evocation of atmosphere and mood, and the same seeming ease in bringing characters and scenes to life. With one or two small strokes describing a grimace, a gesture, or the glimmer of an eye he conveys the inwardness and moral flavour of a human being. In this manner he portrays entire galleries of relatives, domestic servants, neighbours, schoolmasters, and so on. Here are a few examples, although his prose is too close-textured for any excerpt to be even remotely as vibrant with life as it is in its context. He describes his headmaster at his school in Odessa: 'He never looked at the person with whom he talked; he moved about the corridors and the classrooms noiselessly on rubber heels. He spoke in a small, hoarse, falsetto voice which, without being raised, could be terrifying. . . . A humanity-hater by nature . . . he seemed even-tempered, but inwardly was in a state of chronic irritation.' One of the masters was 'thin, with a prickly moustache on a greenish-yellow face; his eyeballs were muddy, his movements as sluggish as if he had just awakened. He coughed noisily and spat in the classroom . . . he would stare beyond his pupils. . . . Several years later he cut his throat with a razor.' Another master: 'A large and imposing man with gold-rimmed glasses on a small nose, with a manly young beard around his full face. Only when he smiled did it suddenly appear . . . that he was weak-willed, timid, torn within himself. . . .' And yet another: 'A huge German with a large head and a beard which reached to his waist line, he carried his heavy body, which seemed a vessel of kindliness, on almost childlike limbs. He was a most honest person and suffered over the failures of his pupils. . . .'[2]

We are made to see the 'seal of doom' on the families of landowning neighbours, who 'were all progressing with extraordinary rapidity, and all in the same direction, towards downfall'. To one of these families 'the whole countryside had once belonged; but now their scion lives by writing petitions, complaints, and letters for the peasants. When he came to see us he used to hide tobacco and lumps of sugar up his sleeve, and his wife did the same. With dribbling lips she would tell us stories of her youth, with its serfs, its grand pianos, its silks, and its perfumery. Their two sons grew up almost illiterate. The younger, Victor, was an apprentice in our machine shop.' And here is a glimpse of a Jewish landlord: he 'had received an education of the aristocratic kind. He spoke French fluently, played the piano. . . . His left hand was weak, but his right hand was fit, he said, to play in a concert. . . . He would often stop in the midst of playing, get up, and go to the mirror. Then, if no one was present, he would singe his beard on all sides with his burning cigarette—this was his idea of keeping his beard tidy.' And behind these galleries of bankrupt landlords and upstart farmers, emaciated labourers and diverse relatives, there is always the breath of the Ukrainian steppe: 'The name of Falz-Fein [a landlord, the 'king of sheep'] rang like the sound of the feet of ten thousand sheep in motion, like the bleating of countless sheep, like the sound of the whistle of a shepherd in the steppes . . . like the barking of many sheepdogs. The steppe itself breathed this name both in summer heat and winter cold.'[1]

From the environment of his childhood Trotsky takes us to the first revolutionary circles of Nikolayev, the prisons of Odessa and Moscow, the colonies of exiles in Siberia; and then he shows us the galaxy of *Iskra*'s editors, the schism at the second Congress of the party, and the birth of Bolshevism. In the whole literature about that period there is not a single memoir or eye-witness account that fixes so graphic a picture of the schism as that which we get from *My Life*. The fact that Trotsky had been a Menshevik in 1903 but wrote as a Bolshevik has much to do with his rendering of the atmosphere and his portrayal of the personalities. In retrospect he sides with Lenin; but he also has to do justice to himself, to Martov, Axelrod, and Zasulich, and to explain why they all went against Lenin. Unlike nearly all Bolshevik and Menshevik memoirists, he shows each of the opposed groups from the inside; and although he now condemns politically the Mensheviks and himself, he does it with understanding and sympathy. Even before he introduces us to the political controversy, he makes us feel the underlying clash of characters:

[1] Op. cit., vol. I, chapter II.

Working side by side with Lenin, Martov, his closest companion in arms, was already beginning to feel not quite at his ease. They still addressed each other in second singular, but a certain coolness was beginning to creep into their relations. Martov lived much more in the present. . . . Lenin, although firmly entrenched in the present, was always trying to pierce the veil of the future. Martov evolved innumerable, often ingenious, guesses, hypotheses, and propositions which he himself promptly forgot; whereas Lenin waited until the moment when he needed them. The elaborate subtlety of Martov's ideas made Lenin shake his head. . . . One can say that even before the split . . . Lenin was 'hard' and Martov 'soft'. And they both knew it. Lenin would glance at Martov, whom he highly esteemed, with a critical and somewhat suspicious look; and Martov, feeling this glance, would look down and his thin shoulders would twitch nervously. When they met and talked afterwards, at least in my presence, one missed the friendly inflection and the jests. Lenin would look beyond Martov as he talked, while Martov's eyes would grow glassy under his drooping and never quite clean pince-nez. And when Lenin spoke to me of Martov, there was a peculiar intonation in his voice: 'Who said that? Julius?'—and the name Julius was pronounced in a special way, with a slight emphasis, as if to give warning: 'A good man, no question about it, even a remarkable one, but much too soft.'[1]

One has at once the sense of destiny coming at this moment between the two 'closest comrades in arms', and of defeat suspended over Martov's frail and untidy figure. Trotsky does not forget how much as a young man he owed to Martov; and so, even as he passes his final judgement on him, he does it with sorrowful warmth: 'Martov [was] . . . one of the most tragic figures of the revolutionary movement. A gifted writer, an ingenious politician, a penetrating thinker, he stood far above the . . . movement of which he became the leader. But his thought lacked courage; his insight was devoid of will. Sheer doggedness was no substitute. His first reaction to events always tended to be revolutionary. In his second thoughts, however, lacking the support of an active will, he usually slid back.' The lack of active will is depicted here as the basic infirmity crippling a daring mind and noble character. How different is this sketch of Plekhanov drawn with discreet antipathy:

[1] Op. cit., vol. I, pp. 175–6.

Q

. . . he apparently sensed something. . . . At least he told Axelrod referring to Lenin: 'Of such stuff the Robespierres are made.' Plekhanov himself did not play an enviable part at the Congress. Only once did I see and hear him in all his power. That was at a session of the Programme Commission. With a clear, scientifically exact scheme of the Programme in mind, sure of himself, his knowledge and superiority, with a gay ironic sparkle in his eyes, his greying moustache alert and bristling, with slightly theatrical, lively and expressive gestures, Plekhanov as Chairman illumined the entire large gathering with his personality, like a live firework of erudition and wit.[1]

How devastating is this seemingly flattering picture of the man, with his self-satisfaction and vanity breaking through his brilliance, and with the suggestion of the firework about to fizzle out in darkness.

No less suggestive and memorable are the character sketches of the leaders of European socialism in the pre-1914 era: August Bebel, Karl Kautsky, Jean Jaures, Victor Adler, Rudolf Hilferding, Karl Renner, and many others. In a brief, often humorous passage, dealing with an outwardly trivial incident, Trotsky tells us about the time and the men more than do many learned volumes. He relates, for instance, how in 1902, after his first escape from Siberia, he stopped, penniless, hungry, but full of the importance of his mission, in Vienna and called at Social Democratic headquarters to ask the celebrated Victor Adler for help in arranging his further journey to London. It is Sunday: the offices are closed. On the staircase he meets an old gentleman 'looking none too amiable', whom he tells that he must see Adler. 'Do you know what day it is?' the gentleman replies sternly. 'It is Sunday', and tries to by-pass the intruder. 'No matter, I want to see Adler.' At this the accosted man 'replies in the voice of one who is leading a battalion to the attack: "I am telling you Doctor Adler cannot be seen on a Sunday." ' Trotsky tries to impress the old man with the urgency of his business; but he thunders back: 'Even if your business were ten times as important— do you understand?—even if you brought the news—you hear me?—that your Tsar was assassinated, that a revolution had broken out in your country—do you hear me?—even that

[1] Op. cit., vol. I, p. 189.

would not give you the right to disturb the Doctor's Sunday rest.' This was Fritz Austerlitz, famous editor of the *Arbeiterzeitung*, the 'terror of his office', who in 1914 was to become a most chauvinistic war propagandist.[1]

On that staircase the young revolutionary, freshly emerged from the Russian underground, ran straight into the embodiment of the orderly, hierarchical, routine-ridden bureaucracy of European socialism. In a few sentences he relates his meeting with Adler, whom he managed to reach after all: 'A short man, with a pronounced stoop, almost a hunch, and with swollen eyes in a tired face.' Trotsky apologized for disturbing the Sunday rest. ' "Go on, go on," Adler replied with seeming sternness, but in a tone which encouraged instead of intimidating me. One could see intelligence emanating from every wrinkle of his face.' Told about the strange staircase encounter, Adler wondered: 'Who could it have been? A tall man? And did he speak to you like that? He shouted? Oh, that was Austerlitz. You say he shouted? Oh, yes, it was Austerlitz. Do not take it to heart. If you ever bring news of revolution in Russia you may ring my bell even at night.' These few lines at once confront us with another element of European pre-1914 socialism: the sensitive intelligence of the old pioneer leader, who, however, becomes gradually the glorified prisoner of the party's sergeant-major. The book is strewn with hundreds of such laconic and expressive incidents and dialogues.

When he comes to the climax of his life, the October Revolution and the Civil War, Trotsky describes it with the utmost restraint, with sparse, often pointillistic, touches. This, to take a random illustration, is how he shows the current of popular feeling underlying the brief triumph of reaction in the hungry and stormy July days of 1917, when Bolshevism seemed to be down and out, and Lenin, branded as German spy, had gone into hiding. Trotsky takes us into the canteen of the Petrograd Soviet:

I noticed that Grafov (a soldier in charge of the canteen) would slip me a hotter glass of tea, or a sandwich better than the rest, avoiding looking at me. He obviously was in sympathy with the Bolsheviks but had to hide this from his superiors. I began to look

[1] Op. cit., vol. I, p. 165.

about me more attentively. Grafov was not the only one: the whole lower staff of the Smolny—porters, messengers, watchmen—were unmistakably with the Bolsheviks. Then I felt that our cause was half won. But so far only half.[1]

A child's remark, a glimpse of Lenin's 'soiled collar' on the day after the October rising, the view of a long, dark, crowded corridor in the Smolny, alive like an anthill, a grotesque episode occurring in the middle of a decisive battle, and a terse dialogue—it is mostly through such details that he conveys the colour and the air of an historic scene. His artistry is in his indirect approach to events too immense to be depicted frontally (in an autobiography) and too big for big words.

It has been said of *My Life* that it shows up Trotsky's egotism and 'self-dramatization'. Autobiography being 'egotistical' by definition, this criticism amounts to saying that he should not have indulged in it. He himself had his 'Marxist' scruples, which lingered on even while he was putting the title to the book. 'If I had been writing these memoirs in different circumstances,' he apologizes, 'although in other circumstances I should hardly have written them at all—I should have hesitated to include much of what I say in these pages.' But he was compelled to counter the avalanche of Stalinist falsification which covered every part of his life story. 'My friends are in prison or in exile. I am obliged to speak of myself. . . . It is a question not merely of historical truth but also of a political struggle that is still going on.' He was in the position of a man in the dock, charged with every imaginable and unimaginable crime, who tries to vindicate himself by giving the court a full account of his doings and is then shouted down for his preoccupation with himself.

This is not to deny that there was an unmistakable streak of self-centredness in Trotsky. It belonged to his artistic nature; it developed during the pre-revolutionary years, when he walked by himself, neither a Bolshevik nor a Menshevik; and Stalinist vilification, forcing him into an intensely personal self-defensive attitude, brought it to the fore. Yet, of his self-dramatization' one would be entitled to speak only if his autobiography, or any biography of him, could at all make his life appear more dramatic than it actually was. To the

[1] Op. cit., vol. II, pp. 36–7.

extent to which in *My Life* he was not yet conscious of the tragic quality of his fate, it would be more correct to say that he under-dramatized himself. Nor, as we shall see later, can there be any question of his having over-stated his role in the revolution. In both *My Life* and the *History*, his real hero is not himself but Lenin, in whose shadow he deliberately placed himself.

Others have criticized *My Life* for its lack of introspection and the author's failure to reveal his subconscious mind. True enough, Trotsky produces no 'interior monologue'; he does not dwell on his dreams or complexes; and he observes an almost puritanical reticence about sex. This is, after all, a political autobiography, political in a very wide sense. Still, the author's respect for the rational core of psychoanalysis shows itself in the care he takes with the description of his childhood, where he does not omit such possible clues for the psychoanalyst as experiences and 'accidents' of the infantile years, toys, &c. (The narrative begins with the words: 'At times it has seemed to me that I can remember suckling at my mother's breast....') He gives this incidental explanation of his caution about Freudian introspection: 'Memory is . . . not disinterested', he says in the Preface. 'Not rarely it suppresses or relegates to an obscure corner episodes which go against the grain of the individual's controlling vital instinct. . . . This, however, is a question for "psychoanalytical" criticism, which is sometimes ingenious and instructive, but more often whimsical and arbitrary.' He had gone into the subject of psychoanalysis deeply and sympathetically enough to know its pitfalls; and he had neither the time nor the patience for 'whimsical and arbitrary' guesses about his subconscious. Instead, he offered a self-portrait remarkable for its conscious integrity and human warmth.

As a political work *My Life* failed to achieve its immediate purpose: it made no impression on the communist public at whom it was primarily aimed. To average party members the mere reading of it was an impiety; and they did not read it. The few who did felt offended or antagonized. They were either committed to the Stalin cult, and the book only confirmed for them Stalinist imputations about Trotsky's personal ambition; or they were shocked to see that a leader of the revolution should at all engage in self-portraiture. 'Here is

Trotsky, the Narcissus, in the act of self-adulation' was a typical comment. And so communists overlooked the rich historical material Trotsky put before them, his insights into revolution, and his interpretation of Bolshevism from which they might have drawn many lessons for themselves. On the other hand, the book found a wide bourgeois reading public, which admired its literary qualities, but had little or no use for its message. '*Mein Leid ertönt der unbekannten Menge, Ihr Beifall selbst macht meinem Herzen bang . . .*', Trotsky might have said of himself.

. . . .

The *History* is his crowning work, both in scale and power and as the fullest expression of his ideas on revolution. As an account of a revolution, given by one of its chief actors, it stands unique in world literature.

He introduces us to the scene of 1917 with a chapter 'Peculiarities of Russia's Development' which sets the events in deep historical perspective; and one recognizes in this chapter at once an enriched and mature version of his earliest exposition of Permanent Revolution, dating back to 1906.[1] We are shown Russia entering the twentieth century without having shaken off the Middle Ages or passed through a Reformation and bourgeois revolution, yet with elements of a modern bourgeois civilization thrust into her archaic existence. Forced to advance under superior economic and military pressure from the West, she could not go through all the phases of the 'classical' cycle of western European progress. 'Savages throw away their bows and arrows for rifles all at once, without travelling the road which lay between those two weapons in the past.' Modern Russia could not enact a Reformation of her own or a bourgeois revolution under bourgeois leadership. Her very backwardness impelled her to advance politically all at once to the point western Europe had reached and to go beyond it—to socialist revolution. Her feeble bourgeoisie being unable to cast off the burden of a semi-feudal absolutism, her small but compact working class, eventually supported by a rebellious peasantry, came forward as the leading revolutionary force. The working class could not content itself with a revolution resulting in the

[1] See *The Prophet Armed*, chapter VI.

establishment of a bourgeois democracy—it had to fight for the realization of the socialist programme. Thus by a 'law of combined development' the extreme of backwardness tended towards the extreme of progress, and this led to the explosion of 1917.

The 'law of combined development' accounts for the force of the tensions within Russia's social structure. Trotsky, however, treats the social structure as a 'relatively constant' element of the situation which does not account by itself for the events of the revolution. In a controversy with Pokrovsky, he points out that neither in 1917 nor in the preceding decade did any fundamental change occur in Russia's social structure—the war had weakened and exposed that structure but not altered it.[1] The national economy and the basic relations between social classes were in 1917 broadly the same as in 1912–14, and even in 1905–7. What then accounted directly for the eruptions of February and October, and for the violent ebb and flow of revolution in between? The changes in mass psychology, Trotsky replies. If the structure of society was the constant factor, the temper and the moods of the masses were the variable element which determined the flux and reflux of events, their rhythm and direction. 'The most indubitable feature of a revolution is the direct intervention of the masses in historic events. The revolution is there in their nerves before it comes out into the street.' The *History* is therefore to a large extent a study in revolutionary mass psychology. Delving into the interconnexion between the 'constant' and 'variable' factors, he demonstrates that what makes for revolution is not merely the fact that the social and political institutions have long been in decay and crying out to be overthrown, but the circumstances that many millions of people have for the first time heard that 'cry' and become aware of it. In the social structure the revolution had been ripe well before 1917; in the mind of the masses it ripened only in 1917. Thus, paradoxically, the deeper cause of revolution lies not in the mobility of men's minds, but in their inert conservatism; men rise *en masse* only when they suddenly realize their mental lag behind the times and want to make it good all at once. This is

[1] Preface to vol. I and introduction to vols. II and III of the *History*.

the lesson the *History* drives home: no great upheaval in society follows automatically from the decay of an old order; generations may live under a decaying order without being aware of it. But when, under the impact of some catastrophe like war or economic collapse, they become conscious of it, there comes the gigantic outburst of despair, hope, and activity. The historian has therefore to 'enter into the nerves' and the minds of millions of people in order to feel and convey the mighty heave that overturns the established order.

The academic pedant burrowing in mountains of documents in order to reconstruct from them a single historical incident may say that no historian can 'enter into the nerves' of millions. Trotsky is aware of the difficulties: the manifestations of mass consciousness are scrappy and scattered; and this may lead the historian to arbitrary constructions and false intuitions. But he points out that the historian can nevertheless verify the truth or untruth of his image of mass consciousness by certain severely objective tests. He must follow faithfully the internal evidence of the events. He can and must check whether the motion of mass consciousness, as he sees it, is consistent with itself; whether every phase of it follows necessarily from what went before it, and whether it leads clearly to what comes after it. He must further consider whether the flow of mass consciousness is consistent with the movement of events: are the moods of the people reflected in the events and do they in turn reflect these? If it be argued that the answers to such questions must be vague and subjective, Trotsky replies by referring, in the Marxist manner, to practical action as the final criterion. He points out that what he is doing as an historian, he and other Bolshevik leaders did while they were making the revolution: relying upon analysis and observation they made guesses about the state of mind and the moods of the masses. All their crucial political decisions rested on those 'guesses'; and the course of the revolution is there to show that, despite trial and error, these had been broadly correct. If in the heat of battle the revolutionary was able to form an approximately correct image of the political emotions and thoughts of millions, there is no reason why the historian should not be able to form it after the event.

The manner in which Trotsky depicts the mass in action

has much in common with Eisenstein's method in the classical *Potemkin*. He picks out of the crowd a few individuals, exposes them in a moment of excitement or apathy, and lets them express their mood in a phrase or gesture; then he shows us the crowd again, a dense and warm crowd, swayed by a tidal emotion or moving into action; and we recognize at once that this is the emotion or action which the individual phrase or gesture had foreshadowed. He has a peculiar gift for over-hearing the multitudes as they think aloud and for letting us hear them for ourselves. In conception and image he leads perpetually from the general to the particular and back to the general; and the passage is never unnatural or strained. Here one is again reminded of the comparison between Trotsky and Carlyle; but the comparison lights up a contrast rather than a similarity. In the Histories of both much of the ethos depends on the mass scenes. Both make us feel the elemental force of an insurgent people, so that we view it as if we were watching landslides or avalanches on the move. But whereas Carlyle's crowds are driven only by emotion, Trotsky's think and reflect. They are elemental; yet they are human. Carlyle's mass is enveloped in a purple haze of mysticism, which suggests that the revolutionary people of France are God's blind scourge bringing retribution upon a sinful ruling class. His mass fascinates us and repels us. He 'enters into its nerves', but only after he has worked himself up into a frenzy—he himself is all nerves and hallucinatory fever. Trotsky draws his mass scenes with not less imaginative *élan*, but with crys-talline clarity. He lets us feel that here and now men make their own history; and that they do it in accordance with the 'laws of history', but also by acts of their consciousness and will. Of such men, even though they may be illiterate and crude, he is proud; and he wants us to be proud of them. The revolution is for him that brief but pregnant moment when the humble and downtrodden at last have their say. In his eyes this moment redeems ages of oppression. He harks back to it with a nostalgia which gives the re-enactment a vivid and high relief.

He does not, however, overstate the role of the masses. He does not oppose them to the parties and leaders as, for instance, does Kropotkin, the great anarchist historian of the French

Revolution, who seeks to prove that every advance of the revolution is due to spontaneous popular action and every setback to the scheming and the 'statesmanship' of politicians. Trotsky sees the masses as the driving force of the upheaval, yet a force which needs to be concentrated and directed. Only the party can provide direction. 'Without a guiding organization the energy of the mass would dissipate like steam not enclosed in a piston box. But nevertheless what moves things is not the piston or the box but the steam.' The great contrast which he draws between the two revolutions of 1917 is based on this idea. The February revolution was essentially the work of the masses themselves, whose energy was powerful enough to force the Tsar to abdicate and to bring the Soviets into existence, but then dissipated before having solved any of the great issues, allowing Prince Lvov to become the head of the government. The October Revolution was primarily the work of the Bolsheviks who focused and directed the energy of the masses.

The relationship between classes and parties is much more complex in Trotsky's presentation, however, than any mechanistic simile might suggest. He shows the subtle interplay of many objective and subjective factors. What guides a party in its action is basically a definite class interest. But the connexion between class and party is often involved and sometimes ambiguous; in a revolutionary era it is also highly unstable. Even if a party's behaviour is ultimately governed by its nexus with one particular class, it may recruit its following from another, a potentially hostile, class. Or it may represent only one phase in the development of a social milieu, a phase to which some leaders remain mentally fixed, while the milieu has left it far behind. Or else a party may be ahead of its class and expound a programme which the latter is not yet ready to accept, but which events will force it to accept; and so on and so forth. In a revolution the traditional political balance collapses, and new alignments take shape abruptly. Trotsky's *History* is a grand inquiry into the dynamic of these processes.

.   .   .   .

We have said that Trotsky does not disguise his hostility

towards the enemies of the October Revolution. To put it more accurately, he confronts them before the tribunal of history as Counsel for the Prosecution; and there he inflicts upon them for a second time the defeat he had inflicted on them in the streets of Petrograd. As a rule this is not a role that fits the historian. Yet in history as in law it happens that the Counsel for the Prosecution may present the fullest possible truth of a case—namely, when he charges the men in the dock with offences they have actually committed; when he does not exaggerate their guilt; when he enters into their conditions and motives and gives due weight to mitigating circumstances; when he supports every count of the indictment with ample and valid evidence; and, finally, when the defendants, having full freedom to refute the evidence, not only fail to do so, but loudly quarrelling among themselves in the dock only confirm it. Such is the manner in which Trotsky discharges his duty. When his *History* was published, and for many years thereafter, most of the chiefs of the anti-Bolshevik parties, Miliukov, Kerensky, Tseretelli, Chernov, Dan, Abramovich, and others were alive and active as émigrés. Yet none of them has exposed a single significant flaw in the fabric of fact which he presented; and none, with the partial exception of Miliukov, has seriously attempted an alternative account.[1] And so (since no History worthy of the name has so far been produced in the Soviet Union either), Trotsky's work is still, in the fifth decade after October, the only full-scale history of the revolution. This is no accident. All the other major actors, again with the partial exception of Miliukov, were so entangled in their contradictions and failures as to be incapable of presenting in full their own more or less coherent versions. They refused to go back as historians to the fatal battlefield where every landmark and indeed every inch of land reminded them of their disgrace. Trotsky revisits the battlefield, his conscience clear and his head up.

Yet his story has no real villains. He does not, as a rule,

[1] Miliukov, however, himself partly renounced his own work as being from an historical viewpoint inadequate. Miliukov, *Istorya Vtoroi Russkoi Revolutsii*, Preface. The main, or rather the only, point of fact on which Kerensky seeks to refute Trotsky is in reiterating the old accusation that Lenin and the Bolshevik party were spies in German pay. Kerensky, *Crucifixion of Liberty*, pp. 285ff.

depict the enemies of Bolshevism as corrupt and depraved men. He does not strip them of their private virtues and personal honour. If they nevertheless stand condemned, it is because he has shown them as defending indefensible causes, as lagging behind the times, as elevated by events to heights of responsibility to which they had not risen mentally and morally, and as perpetually torn between word and deed. The villainy he exposes lies in the archaic social system rather than in individuals. His determinist view of history allows him to treat adversaries, not indulgently indeed, but fairly, and at times generously. When he depicts an enemy in power he shows him complacent, talking big, throwing his weight about; and he crushes him with irony or indignation. Not rarely, however, he stops to pay a tribute to an adversary's past achievement, integrity, even heroism; and he sighs over the deterioration of a character worthy of a better destiny. When he describes a broken enemy, he dwells on the necessity of what had happened and exults in its historic justice; but sometimes the exultation subsides and he casts a commiserating glance—usually his last glance—at the prostrate victim.

He never paints the enemies of the revolution blacker than they have painted one another. Often he paints them less black, because he dissects their mutual animosities and jealousies and makes allowance for exaggeration in the cruel insults they exchanged. He treats the Tsar and the Tsarina no more mercilessly than Witte, Miliukov, Denikin, and even more orthodox monarchists have treated them. He even 'defends' the Tsar against Liberal critics, who have held that by means of timely concessions the Tsar might have averted the catastrophe. Nicholas II, Trotsky argues, made quite a few concessions, but could not yield more ground than self-preservation permitted. As in Tolstoy's *War and Peace*, so in Trotsky the Tsar is a 'slave of history'. 'Nicholas II inherited from his ancestors not only a giant empire, but also a revolution. And they did not bequeath him one quality which would have made him capable of governing an empire, or even a province, or a county. To that historical flood which was rolling its billows each one closer to the gates of his palace, the last Romanov opposed only a dumb indifference.'[1] He draws

---

[1] Trotsky, op. cit., vol. I, p. 71.

a memorable analogy between three doomed monarchs: Nicholas II, Louis XVI, and Charles I, and also between their Queens. Nicholas's chief characteristic is not just cruelty, of which he was capable, or stupidity, but 'meagreness of inner powers, a weakness of the nervous discharge, poverty of spiritual resources'. 'Both Nicholas and Louis XVI give the impression of people overburdened by their job, but at the same time unwilling to give up even a part of those rights which they are unable to use.' Each went to the abyss 'with the crown pushed down over his head'. But, Trotsky remarks, 'would it be any easier . . . to go to an abyss which you cannot escape anyway with your eyes wide open?' He shows that at the decisive moments, when the three sovereigns are overtaken by their fate, they look so much like each other that their distinctive features seem to vanish, because 'to a tickle people react differently, but to red hot iron alike'. As for the Tsarina and Marie Antoinette, both were 'enterprising but chickenheaded' and both 'see rainbow dreams as they drown'.[1]

And here is how he portrays the Cadets, the Mensheviks, and the Social Revolutionaries. Miliukov: 'Professor of history, author of significant scholarly works, founder of the Cadet Party . . . completely free from that insufferable, half-aristocratic and half-intellectual, political dilettantism which is proper to the majority of Russian Liberal men of politics. Miliukov took his profession very seriously and that alone distinguishes him.' The Russian bourgeoisie did not like him because 'prosaically and soberly, without adornment [he] expressed the political essence of the Russian bourgeoisie. Beholding himself in the Miliukov mirror, the man of the bourgeoisie saw himself grey, self-interested, and cowardly; and, as so often happens, he took offence at the mirror'. Rodzianko, the Tsar's Lord Chamberlain who became one of the leaders of the February régime, cuts a grotesque figure: 'Having received power from the hands of conspirators, rebels, and tyrannicides, [he] wore a haunted expression in those days . . . sneaked on tiptoe round the blaze of the revolution, choking from the smoke and saying: "Let it burn down to the coals, then we will try to cook up something." '[2]

Trotsky's Mensheviks and Social Revolutionaries have, of

[1] Ibid., pp. 108–18.   [2] Ibid., pp. 197–8.

course, little in common with the faceless counter-revolutionary phantoms usually shown in Stalinist and even post-Stalinist literature. Each of them belongs to his species, but has his individual traits of character. Here is a thumbnail sketch of Chkheidze, the Menshevik President of the Petrograd Soviet: 'He tried to consecrate to the duties of his office all the resources of his conscientiousness, concealing his perpetual lack of confidence in himself under an ingenious jocularity. He carried the ineradicable imprint of his province . . . mountainous Georgia . . . the Gironde of the Russian revolution.' The 'most distinguished figure' of that Gironde, Tseretelli, had for many years been a hard labour convict in Siberia, yet

remained a radical of the southern French type. In conditions of ordinary parliamentary routine he would have been a fish in water. But he was born into a revolutionary epoch and had poisoned himself in youth with a dose of Marxism. At any rate, of all the Mensheviks, Tseretelli . . . revealed the widest horizon and the [strongest] desire to pursue a consistent policy. For this reason he, more than any other, helped on with the destruction of the February régime. Chkheidze wholly submitted to Tseretelli, although at moments he was frightened by that doctrinaire straightforwardness which caused the revolutionary hard labour convict of yesterday to unite with the conservative representatives of the bourgeoisie.[1]

Skobelev, once Trotsky's disciple, looks like an undergraduate 'playing the role of a statesman on a home-made stage'. And as for Lieber:

If the first violin in the orchestra . . . was Tseretelli, the piercing clarinet was played by Lieber, with all his lung power and blood in his eyes. This was a Menshevik of the Jewish Workers' Union (the Bund) with a long revolutionary past, very sincere, very temperamental, very eloquent, very limited, and passionately desirous of showing himself an inflexible patriot and iron statesman . . . beside himself with hatred of Bolsheviks.

Chernov, the ex-participant in the Zimmerwald movement, now Kerensky's Minister:

A well-read rather than educated man, with a considerable but unintegrated learning, Chernov always had at his disposition a boundless assortment of appropriate quotations, which for a long

[1] Trotsky, op. cit., vol. I, p. 243.

time caught the imagination of the Russian youth without teaching them much. There was only one single question which this many-worded leader could not answer: Whom was he leading and whither? The eclectic formulas of Chernov, ornamented with moralisms and verses, united for a time a most variegated public who at all critical moments pulled in different directions. No wonder Chernov complacently contrasted his methods of forming a party with Lenin's 'sectarianism'. . . . He decided to evade all issues, abstaining from the vote became for him a form of political life. . . . With all the differences between Chernov and Kerensky, who hated each other, they were both completely rooted in the pre-revolutionary past—in the old, flabby Russian society, in that thin-blooded and pretentious intelligentsia, burning with a desire to teach the masses of the people, to be their guardian and benefactor, but completely incapable of listening to them, understanding them, and learning from them.[1]

What distinguishes Trotsky's Bolsheviks from all other parties is precisely the ability to 'learn from the masses' as well as to teach them. But it is not without reluctance and inner resistance that they learn and rise to their task; and when Trotsky concludes with an apotheosis of the revolution and its party, he leaves us wondering for just how long the Bolsheviks will go on 'learning from the masses'. The party he shows us is very different from the 'iron phalanx' which, in the official legend, marches steadfastly and irresistibly, free from all human frailty, towards its predetermined goal. It is not that Trotsky's Bolsheviks lack 'iron', determination and audacity; but they possess these qualities in doses appropriate to the human character and distributed rather unevenly among leaders and rankers. We see them in their finest moments, when isolated, insulted, and battered, they hope and struggle on. In selfless devotion to a cause none of their adversaries is their equal. Greatness of purpose and character is ever present in their picture. But we see them also in disarray and confusion, the leaders shortsighted and timid, the rankers groping tensely and awkwardly in the dark. Because of this Trotsky has been accused of presenting a caricature of Bolshevism. Nothing is further from the truth. His picture is superbly true to nature precisely because he exposes all the weaknesses, doubtings, and waverings of Bolshevism. At the decisive

[1] Ibid., pp. 244–6.

moment the hesitancy and the divisions are subdued or overcome, and doubt gives place to confidence. That the party had
to struggle with itself as well as with its enemies in order to
rise to its role does not derogate from its accomplishment—
it makes the accomplishment all the greater. Trotsky does not
detract from the political honour even of Zinoviev, Kamenev,
Rykov, Kalinin, and the others who shrank from the great
leap of October; if his narrative brings discredit upon them,
it is only because after the event they posed as the unflagging
leaders of the iron phalanx.

The *History* highlights two great 'inner crises' of Bolshevism
in the year of the revolution. In the first Lenin, just returned
from Switzerland, presents his April Theses and politically
'rearms' his party for warfare against the February régime;
in the second, at the penultimate stage of the revolution, the
advocates and opponents of insurrection confront each other
in the Bolshevik Central Committee. In both crises the limelight rests for a long time on a narrow circle of leaders. Yet
the scenes engrave themselves on our mind as deeply as do the
broader, majestic images of the February rising and of the
October Revolution or as does the sombre interval of the July
days, when the movement is shown at its nadir. In both crises
we are made to feel that it is on the few members of the Central
Committee that the fate of the revolution hangs: their vote
decides whether the energies of the masses are to be dissipated
and defeated or directed towards victory. The problem of
masses and leaders is posed in all its acuteness; and almost at
once the limelight is focused even more narrowly and intensely
on a single leader—Lenin.

Both in April and October, Lenin stands almost alone, misunderstood and disavowed by his disciples. Members of the
Central Committee are on the point of burning the letter in
which he urges them to prepare for insurrection; and he
resolves to 'wage war' against them and if need be to appeal,
disregarding party discipline, to the rank and file. 'Lenin did
not trust the Central Committee—without Lenin . . .', Trotsky
comments; and 'Lenin was not so wrong in his mistrust'.[1] Yet
in each crisis he eventually won the party for his strategy and
threw it into battle. His shrewdness, realism, and concentrated

[1] Op. cit., vol. III, p. 131.

will emerge from the narrative as the decisive elements of the historic process, at least equal in importance to the spontaneous struggle of millions of workers and soldiers. If their energy was the 'steam' and the Bolshevik party the 'piston box' of the revolution, Lenin was the driver.

Here Trotsky is grappling with the classical problem of personality in history; and here he is perhaps least successful. His factual account of Lenin's activity is irreproachable. At no stage is it possible to say that here, at this or that point Lenin did not act and the other Bolsheviks did not behave as Trotsky tells us they did. Nor is he out to present Lenin as a self-sufficient maker of events. 'Lenin did not oppose the party from outside, but was himself its most complete expression', he assures us; and he repeatedly demonstrates that Lenin merely translated into clear formulas and action the thoughts and moods which agitated the rank and file, and that because of this he eventually prevailed. Leader and mass act in unison. There is a deep concord between Lenin and his party, even when he is at cross purposes with the Central Committee. Just as Bolshevism had not made its historic entry by chance, so Lenin's part was not fortuitous: he was 'a product of the whole past . . . embedded in it with the deepest roots. . . .' He was not 'a demiurge of the revolutionary process'; but merely a link, 'a great link', in a chain of objective historic causes.[1]

However, having placed Lenin as a link in this chain, Trotsky then intimates that without the 'link' the 'chain' would have fallen to pieces. He asks what would have happened if Lenin had not managed to return to Russia in April 1917—'Is it possible . . . to say confidently that the party without him would have found its road? We would by no means make bold to say that. . . .' It is quite conceivable, he adds, that 'a disoriented and split party might have let slip the revolutionary opportunity for many years'. If in the *History* Trotsky expresses this view with caution, he dots the i's elsewhere. In a letter he wrote to Preobrazhensky from Alma Ata he says: 'You know better than I do that had Lenin not managed to come to Petrograd in April 1917, the October Revolution would not have taken place.' In his French Diary he makes the

[1] Op. cit., vol. I, pp. 341–2.

R

point categorically: 'Had I not been present in 1917 in Petrograd the October Revolution would still have taken place—*on the condition that Lenin was present and in command.* If neither Lenin nor I had been present in Petrograd, there would have been no October Revolution: the leadership of the Bolshevik Party would have prevented it from occurring—of this I have not the slightest doubt!'[1] If Lenin is not yet a 'demiurge of history' here, this is so only in the sense that he did not make the revolution *ex nihilo*: the decay of the social structure, the 'steam' of mass energy, the 'piston box' of the Bolshevik party (which Lenin had designed and engineered)—all these had to be there in order that he should be able to play his part. But even if all these elements had been there, Trotsky tells us, without Lenin the Bolsheviks would have 'let slip the revolutionary opportunity for many years'. For how many years? Five—six? Or perhaps thirty—forty? We do not know. In any case, without Lenin, Russia might have continued to live under the capitalist order, or even under a restored Tsardom, perhaps for an indefinite period; and in this century at least world history would have been very different from what it has been.

For a Marxist this is a startling conclusion. The argument admittedly has a flavour of scholasticism, and the historian cannot resolve it by reference to empirical evidence: he cannot re-enact the revolution, keep Lenin out of the spectacle, and see what happens. If the issue is nevertheless pursued a little further here, this is done not for the sake of the argument but for the light it throws on our chief character. On this point the views of Trotsky, the historian, are closely affected by the experience and the mood of Trotsky, the leader of the defeated Opposition—it is doubtful whether earlier in his career he would have expressed a view which goes so strongly against the grain of the Marxist intellectual tradition.

Of that tradition Plekhanov's celebrated essay *The Role of the Individual in History* is highly representative—like Plekhanov's other theoretical writings it exercised a formative influence on several generations of Russian Marxists. Plekhanov discusses the issue in terms of the classical antinomy of

---

[1] *Trotsky's Diary in Exile*, pp. 53–54. The letter to Preobrazhensky, written in 1928, is in the Trotsky *Archives*.

necessity and freedom. He does not deny the role of the personality; he accepts Carlyle's dictum that 'the great man is a beginner': 'This is a very apt description. A great man is precisely a beginner because he sees *farther* than others and desires things *more strongly* than others.' Hence the 'colossal significance' in history and the 'terrible power' of the great leader. But Plekhanov insists that the leader is merely the organ of an historic need or necessity, and that necessity creates its organ when it needs it. No great man is therefore 'irreplaceable'. Any historic trend, if it is deep and wide enough, expresses itself through a certain number of men, not only through a single individual. In discussing the French Revolution, Plekhanov asks a question analogous to that which Trotsky poses: what would have been the course of the revolution without Robespierre or Napoleon?

Let us assume that Robespierre was an absolutely indispensable force in his party; but even so he was not the only one. If the accidental fall of a brick had killed him in, say, January 1793, his place would, of course, have been taken by someone else; and although that other person might have been inferior to him in every respect, events would have nevertheless taken the same course as they did with Robespierre. . . . The Gironde would probably not have escaped defeat, but it is possible that Robespierre's party would have lost power somewhat earlier . . . or later, but it would have certainly fallen. . . .[1]

What Trotsky suggests is that if a brick had killed Lenin, say in March 1917, there would have been no Bolshevik revolution in that year and 'for many years after'. The fall of the brick would consequently have diverted a tremendous current of history in some other direction. The discussion about the individual's role turns out to be a debate over accident in history, a debate with a close bearing on the philosophy of Marxism. Plekhanov concludes his argument by saying that such accidental 'changes in the course of events might, to some extent, have influenced the subsequent political . . . life of Europe', but that 'in no circumstances would the final outcome of the revolutionary movement have been the "opposite" of what it was. Owing to the specific qualities of their minds and

[1] G. Plekhanov, *Izbrannye Filosofskie Proizvedenya*, vol. II, p. 325. (In English: *The Role of the Individual in History*, pp. 46–47.)

their characters, influential individuals can change the *individual features of events and some of their particular consequences*, but they cannot change their general *trend*, which is determined by other forces'. Trotsky implies that Lenin's personality changed not merely the 'individual features of events', but the general trend—without Lenin the social forces that made that trend or contributed to it would have been ineffective. This conclusion accords ill with Trotsky's *Weltanschauung* and with much else besides. If it were true that the greatest revolution of all time could not have occurred without one particular leader, then the leader cult at large would by no means be preposterous; and its denunciation by historical materialists, from Marx to Trotsky, and the revulsion of all progressive thought against it would be pointless.

Trotsky evidently succumbs here to the 'optical illusion' of which Plekhanov speaks in his argument against historians who insist that Napoleon's role was decisive because no one else could have taken his place with the same or a similar effect. The 'illusion' consists in the fact that a leader appears irreplaceable because, having assumed his place, he prevents others assuming it.

> Coming forward [as the 'saviour of order'] . . . Napoleon made it impossible for all other generals to play this role; and some of them might have performed it in the same or almost the same way as he did. Once the public need for an energetic military ruler was satisfied, the social organization barred the road to this position . . . for all other gifted soldiers. . . . The power of Napoleon's personality presents itself to us in an extremely magnified form, for we credit him with the power of the social organization which had brought him to the fore and held him there. His power appears to us quite exceptional because other powers similar to his did not pass from the potential to the actual. And when we are asked: 'What would have happened if there had been no Napoleon?' our imagination becomes confused, and it seems to us that without him the social movement upon which his strength and influence were based could not have taken place.[1]

Similarly, it may be argued, Lenin's influence on events appears to us greatly magnified because once Lenin had assumed the post of the leader, he prevented others from

[1] Plekhanov, op. cit., pp. 325–6. (English ed., loc. cit.)

assuming it. It is, of course, impossible to say who might have taken his place had he not been there. It might have been Trotsky himself. Not for nothing did revolutionaries as important as Lunacharsky, Uritsky, and Manuilsky, discussing, in the summer of 1917, Lenin's and Trotsky's relative merits, agree that Trotsky had at that time eclipsed Lenin—and this while Lenin was there, on the spot; and although Lenin's influence on the Bolshevik party was decisive, the October insurrection was in fact carried out according to Trotsky's, not to Lenin's, plan. If neither Lenin nor Trotsky had been there someone else might have come to the fore. The fact that among the Bolsheviks there was apparently no other man of their stature and reputation does not prove that in their absence such a man would not have emerged. History has indeed a limited number of vacancies for the posts of great chiefs and commanders; and once the vacancies are filled, potential candidates have no opportunity to develop and achieve 'self-fulfilment'. Need it be held that they would not have achieved it in any circumstances? And could Lenin's or Trotsky's part not have been played by leaders smaller in stature, with this difference perhaps that the smaller men instead of 'allowing destiny to direct' them would have been 'dragged' by it?

It is a fact that almost every great leader or dictator appears irreplaceable in his lifetime; and that on his demise someone does fill his place, usually someone who to his colleagues appears to be the least likely candidate, a 'mediocrity' 'destined to play second or third fiddle'. Hence the surprise of so many at seeing first Stalin as Lenin's successor and then Khrushchev as Stalin's heir, the surprise which is a by-product of the optical illusion about the irreplaceable colossus. Trotsky maintains that only Lenin's genius could cope with the tasks of the Russian Revolution; and he often intimates that in other countries too the revolution must have a party like the Bolshevik and a leader like Lenin in order to win. There is no gainsaying Lenin's extraordinary capacity and character, or Bolshevism's good fortune in having him at its head. But have not in our time the Chinese and the Yugoslav revolutions triumphed under parties very different from that of the Bolsheviks of 1917, and under leaders of smaller, even much smaller, stature? In each case the revolutionary trend found or created its organ in such human material as was available. And if it seems implausible to assume that the October revolution would have occurred without Lenin, this is surely not as implausible as is the opposite assumption that a brick falling from a roof in Zurich early in 1917 could have altered the fortunes of mankind in this century.

Let us add that this last view accords so ill with Trotsky's basic philosophy and conception of the revolution that he could not uphold it consistently. Thus, in the *Revolution Betrayed*, written a few years later, he asserts:

> The quality of the leadership is, of course, far from being a matter of indifference . . . but it is not the only factor, and in the last analysis is not decisive. . . . The Bolsheviks . . . conquered . . . not through the personal superiority of their leaders, but through a new correlation of social forces. . . . [In the French Revolution too] in the successive supremacy of Mirabeau, Brissot, Robespierre, Barras, and Bonaparte, there is an obedience to objective law incomparably more effective than the special traits of the historic protagonists themselves.[1]

As indicated, Trotsky's 'optical illusion' about Lenin sheds a light on himself and his state of mind in these years rather than on Lenin. He produced the *History* after the orgy of the Stalinist 'personality cult' had begun; and his view of Lenin was a negative reflex of that cult. He appealed against the 'irreplaceable' Stalin to the 'irreplaceable' Lenin. Moreover, in view of the apathy and amorphousness of Soviet society, the leader did indeed loom incomparably larger in those years than in 1917, when the whole mass of the nation was seething with political energy and activity. On the one hand Stalin was emerging as autocrat; on the other, Trotsky was of necessity exercising a sort of ideal, moral autocracy as sole mouthpiece of the Opposition. He too, in his defeat, loomed as an individual exceptionally, even uniquely, large. As historian, he projected the leader's huge apparition back on to the screen

---

[1] Trotsky, *The Revolution Betrayed*, pp. 87–88. Characteristically, Sidney Hook in his reaction against Marxism (and Trotskyism) leaned heavily on the subjectivist note in Trotsky's treatment of Lenin, and concluded that the October Revolution 'was not so much a product of the whole past of Russian history as a product of one of the most event-making figures of all time'. Hook, *The Hero in History*, pp. 150–1.

of 1917, and drew this self-defensive moral: 'From the extra-ordinary significance which Lenin's arrival acquired, it should only be inferred that leaders are not accidentally created, that they are gradually chosen out and trained up in the course of decades, that they cannot be capriciously replaced, that their mechanical exclusion from the struggle gives the party a living wound, and in many cases may paralyse it for a long period.'[1] In his Diary he draws the moral even more explicitly:

> . . . I think that the work on which I am engaged now [the opposition to Stalin and the foundation of the Fourth International], despite its extremely insufficient and fragmentary nature, is the most important work of my life—more important than 1917, more important than the period of the civil war, or any other. . . . I cannot speak of the 'indispensability' of my work, even in the period of 1917 to 1921. But now my work is 'indispensable' in the full sense of the word. There is no arrogance in this claim at all. The collapse of the two Internationals has posed a problem which none of the leaders of these Internationals is at all equipped to solve. The vicissitudes of my personal fate have confronted me with this problem and armed me with important experience in dealing with it. There is now no one except me to carry out the mission of arming a new generation with the revolutionary method. . . . I need at least about five more years of uninterrupted work to ensure the succession.[2]

He needed to feel that the leader, whether Lenin in 1917 or he himself in the nineteen-thirties, was irreplaceable—from this belief he drew the strength for his solitary and heroic exertions. And now, when alone of a whole Bolshevik genera-tion he spoke against Stalin, no one indeed was in a position to take his place. But, precisely because he was alone and irreplaceable did so much of his labour run to waste.

Quite apart from the pros and cons of this argument, Trotsky's feelings towards Lenin need further elucidation. The opinions of two contemporaries may be cited. 'Trotsky is prickly and imperative. Only in his relations with Lenin, after

---

[1] *History of the Russian Revolution*, vol. I, p. 342. There is, however, a *non sequitur* in this moral, for if leaders are 'not accidentally created' they are not accidentally (or 'capriciously') eliminated either.

[2] *Diary in Exile*, p. 54.

their union, did he show always a touching and tender deference. With the modesty characteristic of truly great men, he recognized Lenin's priority',[1] thus wrote Lunacharsky in 1923, at the beginning of the anti-Trotsky campaign. Krupskaya, speaking, in the early nineteen-thirties, to a famous foreigner, a non-communist, and knowing that she was eavesdropped upon and that her words would be reported to Stalin, also remarked on Trotsky's 'domineering and difficult character', but added: 'He loved Vladimir Ilyich very deeply; on learning of his death, he fainted and did not recover for two hours.'[2] This love and recognition of Lenin's priority are evident in all of Trotsky's post-revolutionary utterances on Lenin. As early as September 1918, after Dora Kaplan's attempt on Lenin's life, he paid this tribute to the wounded leader:

All that was best in Russia's revolutionary intellectuals of earlier times, their spirit of self-denial, their audacity, their hatred of oppression—all this is concentrated in the figure of this man. . . . Supported by Russia's young revolutionary proletariat, utilizing the rich experience of a world-wide workers' movement, he has risen to his full stature . . . as the greatest man of our revolutionary epoch. . . . Never yet has the life of any one of us seemed to us so secondary in importance as it does now, at a moment when the life of the greatest man of our epoch is in peril.[3]

There was not a hint of sycophancy in these words. Lenin was not yet surrounded by any cult; and Trotsky was more than once yet to voice strong disagreement with him. In 1920, on the occasion of Lenin's fiftieth birthday, he published an essay, more restrained in tone, on Lenin as a 'national type' embodying the best sides of the Russian character.[4] In exile, after he had left Prinkipo, he began to work on Lenin's full-scale biography, of which he finished only the few opening chapters. His failure to complete this work is partly made good by a wealth of biographical sketches he had written and published in the early nineteen-twenties. These deal with two decisive periods of Lenin's life, the years 1902–3 and 1917–18, and give a portrait throbbing with life and suffused with the tenderness of which Lunacharsky spoke.[5]

[1] Lunarharsky, *Revolutsionnye Siluety.*   [2] *Memoirs of Michael Karolyi*, p. 265.
[3] Trotsky, *Lénine*, pp. 211–18.   [4] Op. cit., pp. 205–10.   [5] Op. cit.

What Trotsky admired in Lenin was his '*tseleustremlennost*', his being completely geared up to his great purpose, his *tension vers le but*—but also the personality, in which high-mindedness is matched by zest for life, gravity of purpose by rich humour, fanatical devotion to principle by supple-ness of thought, ruthlessness and cunning in action by delicate sensitivity, high intellect by simplicity. He shows the 'greatest man of our epoch' as a fallible being; and so he demolishes the Stalinist icon of Lenin. Yet he himself approaches Lenin bare-headed, as it were, and, unabashed, reveres him. But he does not genuflect. He pays a manly tribute not to an idol but to the man as he knew him. Even while he depicts Lenin's heroic character, he does not make a demigod of him. He presents a lifesize and workaday figure, not a solemn statue. He employs the most ephemeral genre, the journalistic sketch, to create an enduring picture; and his sketches of Lenin have far greater artistic effect than those drawn by two eminent contemporary novelists, Gorky and Wells. He watches Lenin avidly from every angle: he catches his mind at work; the way he constructs an argument; his appearance and manner on the platform; his gesticulation and the movements of his body; the tone of his laughter; even his practical jokes. We see Lenin's brow clouded with indignation and anger; we observe him playing gently with a dog in a dramatic moment, while he is making up his mind on a grave issue; we catch a glimpse of him as he races like a schoolboy across the Kremlin Square to the gov-ernment's conference room eager to play an amusing trick on his fellow Commissars. And all the time there is in the painter's searching eye a glimmer of love for the 'prosaic genius of the revolution'.

There is also a flicker of remorse in the painter's eye. Trotsky had spent by Lenin's side, in close partnership, only about six years, his best, his epoch-making years. He had passed the earlier thirteen or fourteen years in factional struggle against Lenin, assailing him with ferocious personal insults, as 'slovenly attorney', as 'hideous caricature of Robes-pierre, malicious and morally repulsive', as 'exploiter of Russian backwardness', 'demoralizer of the Russian working class', &c., insults compared with which Lenin's rejoinders were restrained, almost mild. Though Lenin had never, since 1917, even hinted at all this, the invective had been too wounding not to have left any scar. Even between 1917 and 1923, when they were in the closest political union, their relations lacked a note of personal intimacy—there was a certain reserve in Lenin.[1] Trotsky, in his 'touching deference', made tacit and tactful amends. In his writings he is still, perhaps only half consciously, anxious to compensate Lenin posthumously for all the abuse. He admits that in 1903, when he broke with Lenin, the revolution was still largely a 'theoretical abstraction' to himself, while Lenin had already fully grasped its realities. Again and again he speaks of the inner resistance he had to overcome, while 'moving towards Lenin'. But having overcome it and rejoined Lenin, he placed himself in his shadow; and there he still keeps himself as historian. He relates conscientiously all their differences; but his memory shrinks from the recollection. It shortens instinctively the time of their separation, softens the harshness of the antagonism, and dwells with delight on the years of friendship, seeking to extend them as it were backwards and forwards. Sometimes in reverie he seems to relive his life in continuous undisturbed harmony with Lenin. He thinks of writing a book about the intimate, fruitful, and lifelong friendship of Marx and Engels, his ideal of friendship which it was not given him to achieve in his own life. Eleven years after Lenin's death he notes in the Diary:

> Last night . . . I dreamt I had a conversation with Lenin. Judging by the surroundings, it was on a ship on the third class deck. Lenin was lying in a bunk, I was standing or sitting near him. . . . He was questioning me anxiously about my illness. 'You seem to have accumulated nervous fatigue, you must rest. . . .' I answered that I had always recovered from fatigue quickly, but . . . that this time the trouble seems to lie in some deeper processes. . . . 'Then you should *seriously* (he emphasized the word) consult the doctors

[1] *The Prophet Armed*, pp. 92–93. When I remarked to Natalya Sedova on the lack of the note of personal intimacy between Lenin and Trotsky and suggested that the wounding character of Trotsky's pre-revolutionary polemics had rendered this impossible, she replied that she had never thought about it in this way. On reflection, however, she added: 'Perhaps such was indeed the reason for a certain reserve on Lenin's part. Those old factional struggles were conducted in a savage and beastly manner (*Eto byla zverinnaya borba*)'.

(several names . . .).' I answered that I had already had many consultations . . . but looking at Lenin I recalled that he was dead. I immediately tried to drive away this thought. . . . When I had finished telling him about my therapeutic trip to Berlin in 1926, I wanted to add, 'This was after your death'; but I checked myself and said, 'After you fell ill . . .'.[1]

Dream and reverie shield Trotsky in his vulnerability; and in wish-fulfilment he sees himself protected by Lenin's care and affection.

. . . .

The 'optical illusion' about Lenin is the sole instance of subjectivist thinking in the *History*. Otherwise it is as an objective thinker that Trotsky presents the events. To be sure, only an actor and an eye-witness could feel as intimately as he does the inwardness, the colour, and the flavour of every fact and scene. But as historian he stands above himself as actor and eye-witness. What is said of Caesar—that as author he was only the shadow of the commander and politician— cannot be said of Trotsky. He submits his work to the most exacting tests and supports the narrative by the most rigorous testimony, which as a rule he draws from enemies rather than friends. He never refers to his own authority; and only very rarely does he introduce himself as a *dramatis persona*. He devotes, for instance, only one brief dry sentence to his assumption of the Presidency of the Petrograd Soviet, which was one of the great scenes and momentous events of the time.[2] It is perhaps a defect of the *History* that if one tried to deduce from it alone just how important was Trotsky's role in the revolution, one would form a wrong idea. Trotsky looms incomparably larger, in 1917, on every page of *Pravda*, in every anti-Bolshevik newspaper, and in the records of the Soviets and the party than he does in his own pages. His silhouette is the only almost empty spot on his vast and animated canvas.

. . . .

Hazlitt held that oratorical genius and literary greatness are incompatible. Yet Trotsky who had in such full measure the orator's quickness of perception, spontaneous eloquence, and

[1] *Diary in Exile*, pp. 130–1.   [2] *History*, vol. II, p. 347.

responsiveness to audience, possessed in the same degree the habits of deep and sustained reflection, the indifference to ephemeral satisfaction, and the 'patience of soul' indispensable to the true writer. Lunacharsky, himself a most prominent speaker, describes Trotsky as 'the first orator of his time' and his writing as 'congealed speech'. 'He is literary even in his oratory and oratorical even in literature.'[1] This opinion applies well to Trotsky's early writings; and Lunacharsky expressed it in 1923 before Trotsky the writer rose to his full height. In *My Life* and in the *History* the rhetorical element is sternly disciplined by the needs of narrative and interpretation, and the prose has an epic rhythm. It is still 'congealed speech' in the sense in which all narrative is.

For decades Trotsky's major works have been read only in translations. As the man was exiled, so his literary genius was banished into foreign languages. He found skilful and devoted translators in Max Eastman, Alexandra Ramm, and Maurice Parijanine who acquainted the European and American public with his major works. Yet something of his spirit and style is missing from any translation, although Trotsky, having absorbed so much of the European literary tradition, is the most cosmopolitan of Russian writers. But it was from his native sources that he drank most deeply, imbibing the vigour, subtlety, colour, and humour of the Russian tongue. He is, in his generation, the greatest master of Russian prose. To the English ear his style may sometimes seem to suffer from that 'too muchness' in which Coleridge saw the flaw of even the best German, or continental, style. This is a matter of taste and accepted stylistical standards, which vary not only from nation to nation, but within the same nation from epoch to epoch. Emotional vigour and strong, repetitive emphasis belong to the style of a revolutionary era, when speaker and writer expound to great masses of people ideas over which a life and death struggle is being waged; and, of course, the raised voices in which people communicate on a battlefield or in a revolution are unbearable at the quiet fireplace of the Englishman's castle. However, *My Life* and the *History* are free from 'too muchness'. Here Trotsky exercises a classical economy of expression. Here he is an 'objective word maker',

[1] Lunacharsky, op. cit.

striving for the utmost precision in nuance of meaning or mood—a heavy worker in the field of letters. He moulds his work with a watchful eye on the structure of the whole and the proportions of the parts, with an unflagging sense of artistic unity. So closely does he weave his theoretical argument into the narrative that try to disentangle them and the fabric loses texture and pattern. He knows when to contract and when to expand his story as very few narrators know. Yet it is not by arbitrary design that he expands or contracts it: the pace and cadences are attuned to the pulse of events. The whole has the torrential flow proper to a presentation of revolution. But for long stretches he keeps his rhythms even and regular, until, when they approach a climax, they rise and grow, passionate and tempestuous, so that the Red Guards' assault on the Winter Palace, the sirens of the battleships on the Neva, the final cut and thrust between the parties in the Soviet, the collapse of a social order, and the triumph of revolution are reproduced with symphonic effect.[1] And in all this grand sweep his *Sachlichkeit* is never lost—his originality lies in the combination of classical grandeur and sober modernity.

Over his pages he strews dazzling similes and metaphors; these spring spontaneously from his imagination, but he never loses control over them. His imagery is as precise, conceptually, as it is vivid. He uses metaphor with a definite purpose: to accelerate thought; to illumine a situation; or to clip tightly together two or more threads of ideas. The image may flash out in a single sentence; it may shape itself more slowly over the length of a passage; or it may grow in a chapter like a plant, push up a shoot first, blossom forth a few pages later, and come to fruition before the end of the chapter. Note, for instance, the use of metaphor in a passage describing the beginning of the February revolution: the scene is a demonstration of 2,500 Petrograd workers which, in a narrow place, runs into a detachment of Cossacks, 'those age-old subduers and punishers' of popular revolt:

Cutting their way with the breasts of their horses, the officers first charged through the crowd. Behind them, filling the whole width of the Prospect, galloped the Cossacks. Decisive moment! But the horsemen, cautiously, in a long ribbon, rode through the

[1] Compare, e.g., pp. 301, 305, 313, 315–16, 377–8 in the *History*, vol. III.

corridor just made by the officers. 'Some of them smiled,' Kayurov recalls, 'and one of them gave the workers a good wink.' This wink was not without meaning. The workers were emboldened with a friendly, not hostile, kind of assurance, and slightly infected the Cossacks with it. The one who winked found imitators. In spite of renewed efforts from the officers, the Cossacks, without openly breaking discipline, failed to force the crowd to disperse, but flowed through it in streams. This was repeated three or four times and brought the two sides ever closer together. Individual Cossacks began to reply to the workers' questions and even to enter into momentary conversations with them. Of discipline there remained but a thin transparent shell that threatened to break through any second. The officers hastened to separate their patrol from the workers, and, abandoning the idea of dispersing them, lined the Cossacks out across the street as a barrier to prevent the demonstrators from getting to the centre. But even this did not help: standing stock-still in perfect discipline, the Cossacks did not hinder the workers from 'diving' under their horses. The revolution does not choose its paths: it made its first steps towards victory under the belly of a Cossack's horse.[1]

The generalizing image of the revolution diving under the belly of a Cossack's horse emerges naturally from the descriptive passage: it illumines all the novelty, hopefulness, and uncertainty of the situation. We feel that this time the workers will not be trampled upon though their positon is not yet quite secure. But turn another twenty pages, which narrate the progress of the rising, and the metaphor reappears in modified form, as a reminder of the distance the revolution has travelled:

One after another came the joyful reports of victories. Our own armoured cars have appeared! With red flags flying, they are spreading terror through the districts to all who have not yet submitted. Now it will no longer be necessary to crawl under the belly of a Cossack's horse. The revolution is standing up to its full height.[2]

Not less characteristic is a different kind of image in which the writer depicts a peculiar scene with such intensity that the scene itself grows into a haunting symbol. He describes the antagonism between officers and men in the disintegrating Tsarist army:

[1] *History*, vol. I, pp. 122–3.   [2] Ibid., p. 143.

The blind struggle had its ebbs and flows. The officers would try to adapt themselves; the soldiers would again begin to bide their time. But during this temporary relief, during these days and weeks of truce, the social hatred which was decomposing the army of the old régime would become more and more intense. Oftener and oftener it would flash out in a kind of heat lightning. In Moscow, in one of the amphitheatres, a meeting of invalids was called, soldiers and officers together. An orator-cripple began to cast aspersions on the officers. A noise of protest arose, a stamping of shoes, canes, crutches. 'And how long ago were you, Mr. Officer, insulting the soldiers with lashes and fists?' These wounded, shell-shocked, mutilated people stood like two walls, one facing the other. Crippled soldiers against crippled officers, the majority against the minority, crutches against crutches. That nightmare scene in the amphi-theatre foreshadowed the ferocity of the coming civil war.[1]

This sternly realistic reportage is all terse passion. The scene is rendered in six clipped and harsh sentences. A few words transfer us into the amphitheatre and hit our ears with the 'stamping of shoes, canes, crutches'. A commonplace simile stresses the uncommonness of the spectacle: the cripples stand 'like two walls, one facing the other'. How much tragic pathos is condensed in these few and apparently artless sentences.

Sarcasm, irony, and humour pervade all his writings. He has turned against the established order not only from indig-nation and theoretical conviction but also from a sense of its absurdity. In the midst of the most tense and merciless struggle his eye catches the grotesque or comic incident. He is struck, and struck for ever afresh, by men's weakmindedness, mean-ness, and hypocrisy. In *My Life* he recollects how in New York, early in 1917, Russo-American socialists reacted to his forecast that the Russian Revolution would end in the over-throw of bourgeois rule as well as of Tsardom:

Almost everyone I talked with took my words as a joke. At a special meeting of 'worthy and most worthy' Russian Social Democrats I gave a lecture in which I argued that the proletarian party would inevitably assume power in the next phase of the Russian revolution. The effect was like that of a stone thrown into a puddle alive with puffed up and phlegmatic frogs. Dr. Ingerman did not hesitate to point out that I was ignorant of the

[1] *History*, vol. I, p. 273.

rudiments of political arithmetic and that it was not worth while wasting five minutes to refute my nonsensical dreams.[1]

It is with this kind of amused disdain that Trotsky most often laughs at his adversaries. His laughter is not kindly, except on rare occasions, or in recollections of childhood and youth when he could still laugh disinterestedly. Later, he is too much absorbed in too bitter a struggle; and he derides men and institutions in order to turn the people against them. 'What!' he says in effect, 'Are we going to allow those puffed up and phlegmatic frogs to have it all their way and to manage our human affairs for us?' His satire was to make the oppressed and the downtrodden look down upon the mighty in their seats; and the mighty squirmed under the lash. Like Lessing (in Heine's famous portrait), he not only cuts off the head of his enemy, but 'is malicious enough to lift it from the ground and show the public that it is quite empty'. Never does he cut off so many heads, and show them to have been empty, as when he revisits, with Clio, the great battlefield of October.

[1] *Moya Zhizn*, vol. I, p. 315.

CHAPTER IV

'Enemy of the People'

'FOR the very reason that it fell to my lot to take part in great events, my past now cuts me off from chances of action', Trotsky remarks in his Diary. 'I am reduced to interpreting events and trying to foresee their future course.'[1] This appears to be the only such observation he made about himself; and it expresses more than he probably intended to say. To judge from the context, what he had in mind was that his ostracism made it impossible for him to engage in any large-scale political activity. In truth, his past 'cut him off from chances of action' in another and deeper sense as well. His ideas and methods and his political character belonged to an epoch towards which the present, the period of his banishment, was hostile; and because of this they did not have their impact. His ideas and methods were those of classical Marxism and were bound up with the prospect of revolution in the 'advanced' capitalist West. His political character had been formed in the atmosphere of revolution from below and proletarian demo- cracy, in which Russian and international Marxism had been nurtured. Yet in the period between the two world wars, despite the intense class struggles, international revolution stagnated. The staying power of western capitalism proved far greater than classical Marxism had expected; and it was further enhanced as Social Democratic reformism and Stalin- ism disarmed the labour movement, politically and morally. Only in the aftermath of the Second World War was inter- national revolution to resume its course; but then its main arena was to be in the underdeveloped East, and its forms, and partly also its content, were to be very different from those predicted by classical Marxism. To eastern Europe revolution was to be brought, in the main, 'from above and from outside' —by conquest and occupation; while in China it was to rise not as a proletarian democracy, spreading from the cities to

[1] *Trotsky's Diary in Exile*, p. 21.

the country, but as a gigantic *jacquerie* conquering the cities from the country and only subsequently passing from the 'bourgeois democratic' to the socialist phase. In any case, the years of Trotsky's exile were, from the Marxist viewpoint, a time out of joint, an historic hiatus; and the ground crumbled under the champion of classical socialist revolution. In the stormy events of the nineteen-thirties, especially in those outside the U.S.S.R., Trotsky was essentially the great outsider.

Yet his past, which had 'cut him off from chances of action', did not allow him to be inactive either: the man of October, the founder of the Red Army, and the erstwhile inspirer of the Communist International could not possibly reconcile himself to the outsider's role. It was not that such a part would have been incompatible with his Marxist outlook. Marx and Engels themselves were, for long periods, detached from 'practical' politics, engaged in fundamental theoretical work, and content to 'interpret' events—they were in a sense outsiders. Not they but Lassalle led the first socialist mass movement in Germany; not they but Proudhon and Blanqui inspired French socialism; and their influence on the British labour movement was remote and less than skin deep. They did not take their own philosophical postulate about 'unity of theory and practice' so narrowly as to feel obliged to engage in *formal* political activity at all times.[1] When they had no chance to build their

---

[1] In February 1851, after the defeat of revolution in Europe had become clear, Engels wrote to Marx: 'Now at last we have again . . . the opportunity to show that we need no popularity, no "support" from any party in any country, and that our position is altogether independent of such trifles. . . . Really we should not even complain when these *petits grands hommes* [the leaders of the various socialist parties and sects] are afraid of us; since so many years we have behaved as if rag, tag, and bobtail were our party, whereas we had in fact no party at all, and the people whom we counted as belonging to our party, at least officially, *sous reserve de les appeler des bêtes incorrigibles entre nous*, did not grasp even the rudiments of our problems.' 'From now on we are responsible only for ourselves; and when the moment comes when these gentlemen need us we shall be in a position to dictate our terms. Till then we have at least peace. To be sure, with this goes some loneliness. . . . [Yet] how can people like ourselves, who shun any official post like the plague, fit into a "party" . . . i.e. into a band of asses who swear by us because they think we are of their sort. . . . At the next occasion we can and must take this attitude: we hold no official position in the state, and as long as possible no official party post either, no seats in committees, &c., no responsibility for asses [but instead we exercise] merciless criticism of all and enjoy a cheerfulness of which the plotting of all the blockheads cannot deprive us. . . . The main thing for the

party and fight for power, they withdrew into the realm of ideas. The work they did there was historically, but not immediately, of the utmost practical importance, for, steeped in rich experience of social struggle, it pointed to future action. As to Trotsky, neither his character nor his circumstances permitted him to resign from formal political activity. He would not and could not contract out of the day-to-day struggle. The time of his banishment was not an uneventful political interval like the decades after 1848, when Marx wrote *Das Kapital*; it was an era of world-wide social battles and catastrophes, from which a man of Trotsky's record could not stand apart. Nor was he for a moment free to withdraw from his ceaseless and ferocious duel with Stalin. His past drove him to action as pitilessly as it cut him off from the prospect of action.

All his behaviour in exile is marked by this conflict between the necessity and the impossibility of action. He senses the conflict, but is never clearly conscious of it. Even when he glimpses the impossibility, he sees it as extraneous, temporary, and resulting merely from persecution and physical isolation. This unawareness of his deeper predicament gives him the strength to struggle on against odds perhaps more fearful than any historic figure has ever faced. Necessity impels him to formal political activity. Yet he recoils again and again, not in his conscious mind, which is ever hopeful, but in his involuntary moods and instinctive reflexes. His will wrestles with these moods and never succumbs. But this is a fierce, desperate, and exhausting collision.

During the Prinkipo years sheer physical isolation made his dilemma less pressing. He fretted and longed to come closer to a scene of political action, convinced that this would enable him to intervene effectively. In the meantime he had no choice but to plunge into literary historical work. He withdrew, though not completely, into the realm of theoretical ideas, where his enduring strength now lay. This is why the four Prinkipo years were his most creative period in exile. His emergence from Prinkipo was bound to heighten and sharpen

moment is that we should have the possibility to publish what we write. . . . either in quarterlies or in bulky volumes. . . . What will be left of all the prattle-tattle, in which this entire émigré mob may indulge at your expense, once you have come out in reply with your economic treatise?' Marx-Engels *Briefwechsel*, vol. I, pp. 179–82.

his dilemma. Not only was he presently to experience the full blast of that implacable hostility from which seclusion had partly shielded him. Closeness to a scene of political action was to excite in him all that passion for action, in which his weakness now lay. He was to discover or rediscover that the current of events was passing him by; yet he would exert himself to turn it. In the eight years left to him he was to produce no single work as weighty and enduring as his *History* or even his autobiography, although his hand never dropped the pen. He left Prinkipo planning to write a History of the Civil War which, because of his unique authority, would have been as important as the *History of the Revolution*, and perhaps even more illuminating. He started a large-scale biography of Lenin, which, as he confided to Max Eastman and Victor Gollancz, he expected to be 'the major work of my life' and the occasion for a comprehensive, 'positive and critical', exposition of the philosophy of dialectical materialism.[1] He did not carry out these and other plans, partly because wanderings and persecution did not allow him to concentrate, but mainly because he sacrificed them to his formal political activity, to his untiring labour for the Fourth International.

Thus his whole existence was torn between the necessity and the impossibility of action. Just now, at the moment of departure from Prinkipo, he had a foreboding of the gravity of the conflict. He was leaving in high spirits, full of hope and great expectations, yet with a chilling dread in the innermost recesses of his being.

.   .   .   .

With Natalya, Max Shachtman, and three secretaries, van Heijenoort, Klement, and Sara Weber, he sailed from Prinkipo on board a slow Italian boat *Bulgaria*, on 17 July 1933. The voyage to Marseilles took a full week. Once again all precautions taken to keep the move secret failed. As on the trip to Denmark, he travelled under his wife's name and did his best to remain inconspicuous; but when the boat called at the port of Piraeus, many eager reporters were already

---

[1] Trotsky to Eastman, 6 November 1933, and to Gollancz, 28 September 1933. *The Archives*, Closed Section. In his letter to Gollancz Trotsky wrote that he would like Arthur Ransome to edit the English edition of this work.

waiting for him. He told them that his journey was 'strictly private'; that he and his wife would devote the next few months to medical treatment; and he refused to be drawn into any political statement: 'Our journey has no right to engage public attention, especially now when the world is occupied by infinitely more important questions.' But the Press once again watched suspiciously and speculated on his purpose. There was a rumour that he was going to France on Stalin's initiative to meet Litvinov, the Soviet Commisssar of Foreign Affairs, and to discuss the terms of his return to Russia. So widespread and persistent was the rumour that the *Vossische Zeitung*, a serious German paper, asked him whether it was true, and the Soviet Telegraph Agency issued an official denial.[1]

He spent most of his time *en route* in his cabin, working out his ideas about the Fourth International. He wrote an article, 'One Cannot Remain in One "International" with Stalin . . . and Co.' (He also reviewed, briefly and warmly, a novel just published by one of his young Italian followers, Ignazio Silone's *Fontamara*.[2]) After a few industriously spent days, he fell ill as the boat was nearing France: a severe attack of lumbago laid him low.[3] 'It was very hot', Natalya recollects, 'the pain tormented him . . . he was unable to get up. We called the ship's doctor. The steamer was approaching its destination. We were afraid of disembarking.' His pain, which made even breathing difficult, was somewhat relieved when a good way outside Marseilles, the ship was suddenly stopped and the French police ordered him and Natalya into a small tug, while his secretaries were to go on to Marseilles. He was uneasy at being separated from the secretaries and was about to protest when he noticed Lyova and Raymond Molinier

---

[1] Trotsky suspected that the *Vossische Zeitung* (already Nazified) had made the inquiry on Hitler's orders; and that Stalin hastened to reassure Hitler that he was not thinking of any reconciliation with the man who had suggested that the Soviet Government should reply to Hitler's seizure of power with the mobilization of the Red Army. See the note 'Stalin reassures Hitler' of 19 July 1933 in *The Archives*.

[2] *The Archives*, B.O., nos. 36–37, 1933.

[3] According to *Black's Medical Dictionary* (p. 731), 'an attack of lumbago may be due not to any disease of the muscles of the back, but to emotional disturbances which literally prevent the individual from standing up to the stresses and strains of life.'

waiting for him in the tugboat. He descended slowly, gasping with pain. It was Lyova who had arranged that he should be taken off in order to remove him from the public gaze and to escape the swarm of reporters, who were waiting at the harbour, and among whom G.P.U. agents were sure to be planted. Unobtrusively Trotsky landed at Cassis, near Marseilles, where an officer of the *Sûreté Génerale* handed him an official paper revoking the order under which he had, in 1916, been expelled from France 'for ever'. 'It is a long time', Trotsky noted, 'since I acknowledged the receipt of any official document with so much pleasure.'[1]

The pleasure was at once somewhat spoiled by the outcry of right-wing newspapers against his admission.[2] Ironically, on the day of his arrival, 24 July, *Humanité* also protested against the annulment of the 1916 expulsion order—an order issued at the instigation of Count Isvolsky, the Tsar's last Ambassador, as reprisal for Trotsky's anti-war activity. *L'Humanité* also published a resolution of the French Politbureau, calling upon the whole Communist party to keep a watch on Trotsky's movements. Lyova's fears and precautions proved well justified. From Cassis, accompanied by a few young French Trotskyists, they drove towards Bordeaux, then northwards to St. Palais, near Royan, on the Atlantic coast, where Molinier had rented a villa. Meanwhile, the secretaries disembarked at Marseilles, unloaded Trotsky's library, archives, and luggage, despatched these to Paris, and went there themselves. G.P.U. sleuths concluded from this that Trotsky too had gone to Paris—on this guess Vyshinsky was to base, during the Moscow trials four years later, an essential part of his allegations about Trotsky's terroristic activities in France.

Trotsky's party travelled slowly towards Royan, and because of Trotsky's persistent pain, stopped at a village inn in the Gironde department—at night Lyova and a young Frenchman stood guard at Trotsky's doors. Only next afternoon did they reach St. Palais. On arrival Trotsky went to bed with high fever. But within an hour he had to dress and leave the house in a hurry—a fire had broken out, the rooms were full of smoke; the verandah, the garden, and the fences stood in

---

[1] *The Archives.*
[2] See, e.g., *Le Matin* and *Le Journal*, of 24, 25, and 26 July 1933.

flames. There was something symbolic in this opening incident: more than once during Trotsky's stay in France the ground would catch fire beneath his feet and he would have to rush out and take to the road. But the mishap at St. Palais was quite accidental; the summer was exceedingly hot; and not a few woods and houses were ablaze. The accident might have become embarrassing if Trotsky's identity had been found out; he was under an obligation to keep his incognito. Outside the villa, a crowd had gathered; and to avoid being recognized, he rushed across the road, hid in Molinier's car at the roadside, and there waited until his wife, son, and friends, helped by a change of wind, extinguished the fire. People approached him; but he pretended to be an American tourist, speaking hardly any French; and he noticed with relief that his accent had not given him away. Next day the local paper reporting the event mentioned an 'elderly American couple' who had moved into the villa just before the fire broke out.

He stayed at St. Palais from 25 July to 1 October, remaining all this time indoors, mostly in bed. His health, according to Natalya, deteriorated every time there was something the matter; he suffered from insomnia, headaches, and fevers. 'He could not raise himself up to have a look at the garden or to get out to the beach, and he postponed this "undertaking" from day to day.' When he was a little better, he received visitors; but he tired quickly and spent long hours on a couch indoors or a deckchair in the garden. Visitors were to recall that he could not sustain a conversation for longer than fifteen or twenty minutes and that he perspired profusely and almost fainted, so that some of them stayed at St. Palais for a few days in order to have several short talks with him.[1]

Yet, during the two months at St. Palais, he received no fewer than fifty callers. Among them were, apart from French and other Trotskyists: Jenny Lee (Aneurin Bevan's wife) and A. C. Smith of the British Independent Labour party; Jacob Walcher and Paul Frölich, formerly leaders of the German Communist party, then of the Sozialistische Arbeiterpartei;

---

[1] See Natalya Sedova's statement before the Dewey Commission (of 1 March 1937) and depositions by Klement and 'Erde' (of 31 March 1937) in *The Archives*, Closed Section.

Maring-Sneevliet, once Comintern representative in Indonesia and China, now member of the Dutch Parliament and leader of an Independent Socialist party; Paul-Henri Spaak, future General Secretary of the North Atlantic Treaty Organization, at this time leader of Belgium's Socialist Youth and something of a disciple of Trotsky's, over-awed by the master and diligently yet apprehensively submissive; Ruth Fischer; Carlo Rosselli, the eminent Italian anti-fascist; André Malraux; and others.

Most visitors called in connexion with a conference, convened in Paris at the end of August, of parties and groups interested in the idea of a new International. Trotsky, unable to attend the conference, was active in its preparation, wrote 'Theses' and resolutions for it, and took a close interest in the details of organization. He hoped to win over many of those who stood outside the established Internationals. But of the fourteen small parties and groups represented at the conference only three, the German Sozialistische Arbeiterpartei and two Dutch groups, joined the Trotskyists in working for the Fourth International. All the others were frightened by the fierceness of Trotsky's opposition to both reformism and Stalinism; even the three who joined did so with reservations; and they did not form an International but merely a preliminary organization. Outwardly, Trotsky was pleased with this start, and saw in it an event as significant as the Zimmerwald Conference had been in its time.[1]

Yet he could not fail to sense how feeble a start it really was; and this certainly contributed to his despondency. Of his mood in these weeks we find an intimate expression in his correspondence with Natalya, who early in September left for Paris to consult doctors. Their letters, sad and tender, show him forlorn and morally dependent on her in a way he could hardly ever have been in any of the earlier, more active periods of his life. Her stay in Paris reminded him of the far-off years when they had lived there together; and he had a poignant feeling of declining strength and of advancing age. A day or two after her departure he wrote to her: 'How painfully I long to see your old picture, our common picture, showing us

---

[1] B.O. nos. 36–37, 1933. Trotsky signed some of the resolutions and 'theses', he wrote for the conference with the pen-name 'G. Gurov'.

when we were so young. . . . You are in Paris . . . the day you left . . . I was unwell. . . . I went into your room and touched your things. . . .' Again and again, he strained to recapture the image of their youth, and complained of sleeplessness, lassitude, and loss of memory, 'caused by the sufferings of recent years'. But he reassured her that he felt his intellectual powers to be unimpaired and that he was well looked after by a good doctor, a comrade who had come from Paris and was staying with him. 'Dearest, dearest mine', he wrote on 11 September, 'it was quieter on Prinkipo. Already the recent past seems better than it was. Yet we looked forward with so much hope to our stay in France. Is this definitely old age already? Or is it only a temporary, though all too sharp a decline, from which I shall still rally? We shall see. Yesterday two elderly workers and a schoolmaster came to see me. Naville was also here. . . . I felt weary; there was little relevance in the talk. But I watched with curiosity the elderly provincial workers.'[1]

A week later he recovered somewhat and described to Natalya how, still bedridden, he had received a group of followers and had argued with them vigorously; and how Lyova, having seen them off, had come back, embraced him over the blanket, kissed him, and whispered: 'I love you, Father'—this filial affection and admiration moved him after years of separation. But a few days later he wrote again that he felt very old among the young men who visited him, and that at night he awoke and 'like an abandoned child' called for Natalya—'did not Goethe say that old age catches us by surprise and finds us children?' 'How sad you are', Natalya answered. 'You have never been like that. . . . I see you pale, weary, doleful—this is terribly depressing. This is quite unlike you. . . . You are making superhuman demands on yourself and speak of old age when one should be amazed at how much you are still able to shoulder.' He quailed inwardly before the impossibility of his task; and the visits, the talks, mostly turning round in circles, and the intrigues of tiny cliques, could hardly raise his spirits.[2]

By the beginning of October his health had improved; and to get a complete rest he went with Natalya to Bagnères de

---

[1] *The Archives*, Closed Section.   [2] Ibid.

Bigorre in the Pyrenees, where they spent three weeks, made trips and visited Lourdes, which amused and irritated him as a monument to human credulity. He came to himself and longed to be back at work. From Bagnères he wrote to Gollancz, who had been urging him to get on with the *Lenin*, that he would now concentrate on this book and put aside his plan for the *History of the Red Army*.[1]

Thus three months passed from his landing in France. The protests against his admission had subsided; he had managed to keep his incognito; his whereabouts were unknown to the Press; and few even of the friends and well-wishers who called at St. Palais knew his exact address—so cautiously had Lyova arranged their visits. The Stalinists were unable to trace him and to stage their planned demonstrations against his presence. A Trotskyist sympathizer, who was still a party member, came to Royan to watch what was going on in the cells there and, if need be, to give a warning to St. Palais; but the local Stalinists had no inkling that Trotsky was in the neighbourhood. The government, reassured by his discretion, lifted some of the restrictions on his freedom of movement and permitted him to stay anywhere, except in Paris and the Seine department. And so on 1 November he moved to Barbizon, the little town near Paris which has given its name to a famous school of painting. He lived there in a house outside the town, in a small park on the edge of the Fontainebleau forest, well hidden from prying eyes, guarded by sentries and watchdogs. He kept in close touch with his followers in Paris—messengers regularly carried the correspondence to and fro; and in the winter, escorted by a bodyguard, he made two or three trips to the capital. At Barbizon he hoped to work undisturbed on the *Lenin* at least for a year.

There seemed to be no trace of his recent lassitude. He resumed his customary routine: at six in the morning, while everyone in the house was still asleep, he was at work; and, pausing only for breakfast, he went on until noon. After lunch and an hour's rest he was at work again; at 4 p.m. he, Natalya, and the secretaries took tea standing; then everyone was back at his job till supper. In the evenings the members of the

---

1 Letter to Gollancz of 25 October 1933. In reply to these assurances Gollancz offered Trotsky an advance of £1,500 on the *Life of Lenin*. Ibid.

household and visitors formed a debating circle, over which he presided, of course. He resumed solid research and literary work: he assembled materials for the *Lenin*; delved into the Ulyanovs' family background and Lenin's childhood and adolescence, studied the Russia of the eighteen-seventies and eighteen-eighties, and the formative phases of Lenin's intellectual growth, the topics that fill the first and the only completed part of the biography. Preparing to deal with Lenin's philosophical writings, and conscious of gaps in his own knowledge, he went back to the classics of logic and dialectics, Aristotle and Descartes, but especially to Hegel. He did not allow other projects to tempt him away. About this time Harold Laski urged him to write a book 'Where is America Going?', somewhat on the pattern of *Where is Britain Going?* 'I know no one', wrote the mentor of the British Labour party, himself renowned as an authority on American constitutional history and politics, 'I know no one whose book on this subject would be more interesting to the Anglo-American public.'[1] But Trotsky was not to be diverted.

More than ever he now followed *con amore* French politics and letters. For relaxation he wrote or rewrote character sketches of Briand, Millerand, Poincaré, Herriot; and he reviewed quite a few French novels. Of these minor writings, his essay on Céline's *Voyage au bout de la nuit* and Poincaré's *Memoirs* deserve to be briefly summarized.[2] The occasion for it was Céline's debut with *Voyage*. 'Céline has entered great literature as others enter their homes', Trotsky said, praising the writer's indifference to respectability, his wide experience, fine ear and daring idiom. 'He has shaken up the vocabulary of French literature' and brought back into it words long banished by academic purism. Rooted in a rich tradition, deriving from Rabelais, he had written the *Voyage* 'as if he had been the first to use French words'. He also defied the conventionalities of the French bourgeoisie, of which Poincaré was a perfect embodiment. The juxtaposition of Céline and Poincaré was suggested to Trotsky by the opening scene in *Voyage* which depicts Poincaré inaugurating a dog show. The 'incorruptible

[1] Laski to Trotsky, 15 November 1933, ibid.
[2] The essay is dated 10 May 1933, but Trotsky was still working on it after he had moved to France.

notary of the French bourgeoisie' and patron saint of the
Third Republic had 'not a single individual note of his own'—
everything in him was conventional and imitative; his personality, as it appeared in his speeches and memoirs, was like
a 'barbed-wire skeleton wrapped up in paper flowers and
golden tinsel'. 'I am a bourgeois, and nothing bourgeois is
alien to me', Poincaré might have said. His rapacity, displayed
in exacting reparations from defeated Germany, and his
hypocrisy, 'so absolute that it became a kind of sincerity', were
dressed up as traditional French rationalism. Yet the logic and
clarté of bourgeois France stood to that high philosophical tradition 'as medieval scholasticism stood to Aristotle': 'it viewed
the world not in the three dimensions of reality but in the
two dimensions of documents'. The famous French sense of
proportion was in Poincaré a 'sense of the small proportions'.
The French bourgeoisie had 'inherited from its ancestors a
wardrobe rich in historic costumes', which it used to cover its
stubborn conservatism; and next to rationalism, the 'religion
of patriotism' was to it what religion was to the Anglo-Saxon
middle classes. 'The free-thinking French bourgeois', for whom
Poincaré spoke, 'projected into his own nation all the attributes
that other people vest in Father, Son and Holy Ghost'; France
is to him the Holy Virgin. 'The liturgy of patriotism is an
inevitable part of the political ritual.'

Céline's merit was that he exposed and rejected these sanc-
tities. He depicted a way of life in which murder for small profit
was not the rare exception or the excess which conventional
morality pretended it to be, but almost a natural occurrence.
Yet, an innovator in style rather than in ideas, Céline was
himself a bourgeois, weary, despairing, and 'so disgusted by his
own image in the mirror that he smashes the glass until his
hands bleed'. With only his intense hatred of the lie and his disbelief in any truth, Céline would not be able to write another book
like the *Voyage*, Trotsky concluded—if no radical change were to
occur in him he would sink into darkness. (Presently Céline was
indeed captivated and carried away by the tide of Nazism.)

Trotsky's remarks on Malraux are also noteworthy, for he
was one of the first, if not the first, reviewer of *La Condition
Humaine* which he hailed as the revelation of a great and
original talent. He urged a New York publisher to bring out

an American edition of the book and recommended it in these terms: 'Only a great superhuman purpose for which man is ready to pay with his life gives meaning to personal existence. This is the final import of the novel which is free from philosophical didacticism and remains from beginning to end a true work of art.'[1] In an earlier review, however, he spoke of the streak of 'cheap Machiavellianism' in Malraux, who was fascinated not so much by revolution and its genuine fighters as by pseudo-revolutionary adventurers and 'bureaucratic super-men' seeking to dominate and boss the working class. Fascination with this kind of 'superman' as we now know, was to make it so easy for Malraux to become associated first with Stalinism and then with Gaullism. At this time, however, he was still trying to reconcile his Stalinist inclinations with sympathy and admiration for Trotsky.[2]

. . . .

At Barbizon Trotsky was able to take a close look at his western European followers, especially the French ones; and he tried to go beyond the narrow confines of his faction in recruiting adherents for the Fourth International. He set great store by the accession of Ruth Fischer and Maslov who were émigrés in France; he often received Fischer at Barbizon and, to the annoyance of the German Trotskyists, introduced her as a member to the International Secretariat. He wrote an enthusiastic preface to a brochure by Maria Reese, formerly a communist member of the Reichstag, who exposed the confusion and panic in which the German party had collapsed in 1933 and announced her adherence to Trotskyism. Shortly thereafter, however, Reese deserted the Trotskyists, returned to Germany, and declared for National Socialism.[3] The

[1] Trotsky to Simon and Schuster, New York, 9 November 1933. *The Archives*, Closed Section.

[2] Trotsky first wrote about Malraux in 1931. *B.O.*, nos. 21–22. Some time after Trotsky's arrival in France Malraux was a member of a *Comité pour contribuer à la securité de Leon Trotzky*. The Committee made a money collection which was to cover the expenses of providing Trotsky with a bodyguard; and in an appeal signed *inter alia* by Malraux it addressed itself '*à tous ceux qui refusent de livrer un proscrit dont toute la vie a été au service de l'avenement d'une société meuilleure aux balles de la reaction*'. (Among those who lent support to the appeal was Romain Rolland, who later, however, also justified Stalin's purges.) Quoted from *Les Humbles*, May–June 1934.

[3] *B.O.*, nos. 38–39, 1934.

recruitment of new adherents was hard going. The few groups that had agreed to work together for the new International were at loggerheads. Some old Trotskyists like Nin and his friends broke away to form an independent party, the P.O.U.M. of Catalonia. In France all the Trotskyist groups aggregated about a hundred members at the most, and *Verité* had a circulation of less than 3,000. Rosmer kept aloof: 'During the two years of Trotsky's stay in France', he says, 'we never saw each other. Probably he was waiting for me to make the first step towards him, and I waited for a first step from him.'[1] By now Trotsky was discovering that it was not without reason that Rosmer had refused to associate with Raymond Molinier; he himself was taken aback by Molinier's 'political irresponsibility', even though the Molinier family was very helpful to him during his French peregrinations. He was also irked by Naville's 'arrogance' and 'lack of revolutionary spirit and initiative'.[2] He spent many an hour in discussions with Simone Weil, a 'Trotskyist' at this time, but found her a 'muddlehead', 'without any understanding for working-class politics and Marxism'—in later years she gained fame as a philosophical convert to Catholicism and mystic. The impression which most of his French intellectual adherents made on him is well conveyed in his letter to Victor Serge, written two years later, where he describes them as 'Philistines': 'I have been even in their homes and have felt the smell of their petty bourgeois life—my nose has not deceived me.' All he could count on were a few fervent and young workers and students; yet even these lacked political knowledge and experience and vegetated outside the labour movement. 'We must look for roads to the workers,' he concluded, 'and in the process must avoid ex-revolutionaries and even push them discourteously aside. . . .'[3]

This was the time of the Stavisky affair, the scandal which

[1] This tallies well with what Trotsky wrote later (on 29 April 1936) to V. Serge: 'Rosmer, disagreeing with me over a secondary issue, got over-excited. . . . Because of this we did not meet during [my] stay in France; but our respect and sympathy for both of them, Alfred and Marguerite, are as great as ever. Rosmer is a man on whom one can always count in a difficulty.' *The Archives*, Closed Section.

[2] Trotsky to Lyova, 27 December 1935, *The Archives*, Closed Section. He characterized another one of his French followers, David Rousset, as 'a mixture of opportunism and anarchism'. Ibid.

[3] Trotsky to Serge, 30 July 1936, ibid.

revealed shocking corruption in the Third Republic, its Ministers, Deputies, police chiefs, and Press. The Parliamentary mainstay of the Republic, the Radical party, was deeply involved; and the government was nearly choking in the fumes of the affair. Fascist and quasi-fascist leagues, especially the *Croix de Feu*, or the Cagoûlards, led by Colonel de la Rocque, battened on the popular indignation and threatened to overthrow the parliamentary régime. On 6 February 1934 they staged a semi-insurrection and with the cry '*Daladier au poteau*' assailed the Chamber of Deputies. The *coup* failed, however; and within a week it provoked a General Strike of the workers of Paris, in which socialists and communists spontaneously formed a united front, the first time for years. This happened just when the Comintern was abandoning its 'ultra-left' tactics; and the united front of 12 February set a precedent. In July socialists and communists reached a formal agreement to 'defend jointly the Republic against every fascist attack'. The Radical party did not join them yet —the Popular Front, which was to include it, was to come into existence only the following year. But a new chapter was opened: the Daladier government had been saved by the united front and was increasingly dependent on its support; the political balance of France had shifted; there was an upsurge of energy among the workers and a revival of class struggle.

In these circumstances, Trotsky held, it was all the more urgent that his followers should find themselves within the mass movement. As they could not go back to the Communist party, which slandered them and persecuted them mercilessly, he advised them to join the S.F.I.O., the Socialist party, which, directed by Leon Blum, still held sway over the majority of workers. (The S.F.I.O. was not yet the party of the white-collar man and the *petite bourgeoisie* which it was to become after the Second World War.) Trotsky advised his followers to join that party, not in order to accept its ideas, but on the contrary in order to defy reformism within its own stronghold and to 'carry their revolutionary programme to the masses'. The S.F.I.O. was not a centralized body but a federation of various groups and factions openly competing for influence: in such an organization it should have been possible for the

Trotskyists to convert people to the idea of the Fourth International. This was the 'French turn', which all Trotskyist groups were debating in 1934–5—eventually Trotsky advised nearly all of them to follow a similar course in their own countries, i.e. to join, as distinct groups, the Social Democratic parties.[1]

In this way he implicitly acknowledged that his scheme for the new International was unreal; the 'French turn' was a desperate attempt to salvage it. It could not succeed. Trotskyism could not appeal, except episodically, to the rank and file of a Social Democratic party; it went too strongly against their habits of thought and deep-rooted reformist tradition. Trotsky could not defeat Blum's influence on Blum's native ground, which was what he indirectly undertook to do. His followers entered the S.F.I.O. as a tiny group without authority or prestige, proclaiming in advance their enmity to the party's established leaders and accepted tenets. They made a few converts among the young, but soon ran up against a wall of hostility. Yet the 'French turn' removed the Trotskyists even further from the mass of communists and provided grist to Stalinist propaganda. To the communist rank and file the claim that they had joined the S.F.I.O. only in order to 'give battle to reformism' sounded like a flimsy pretext. Communists saw the Social Democrats making for a time political capital out of the Trotskyists' adherence; and they heard the latter inveighing against Stalinism from Social Democratic platforms. Their old distrust of Trotskyism turned into a blind hatred of the 'renegades and traitors'. True enough, presently they did indeed see the Trotskyists assailing the Social Democratic leaders and their policies and being expelled from the S.F.I.O. But this happened during the Popular Front; and the Communist party applauded and even instigated the expulsion. All the same, the 'French turn' helped to change the antipathy the ordinary French communist felt for Trotskyism into an intense animosity; and even if the difference was only one of a nuance, it was not unimportant: it was by such imperceptible

---

[1] 'Entrism' is the term by which the Trotskyists described and discussed this move even thirty years later, when they still kept entering, leaving, and re-entering other parties, splitting and dispersing their own ranks in the process, and still 'building the Fourth International'.

gradations that the mood of western communists was being worked up to that furious abhorrence of Trotskyism in which they were to receive the Great Purges.

. . . .

Six months had not yet passed since Trotsky's arrival at Barbizon when the comparative peace in which he lived there was suddenly destroyed. He had again preserved his incognito and concealed his whereabouts so well that even his friends did not know where he was and corresponded with him at a cover address. Not a single letter of his was ever posted from Barbizon; a secretary acted as messenger and carried letters between Barbizon and Paris. A trivial incident undid all those prudent arrangements. One evening in April the police stopped Trotsky's messenger for a minor traffic offence. Puzzled by his vague answers and foreign accent—the messenger was Klement, a German political émigré—the police stumbled on the discovery that Trotsky was at Barbizon. As the head-quarters of the *Sûreté* had cautiously withheld this information from them, the local gendarmes, flushed with their feat of detection, broke the news with all drums beating. The local *Procureur*, followed by a platoon of gendarmerie and by re-porters from Paris, came to interrogate Trotsky. The right-wing Press at once renewed its clamour, and *Humanité* once again vied with it. The government was frightened. The fascist leagues had already attacked it for granting Trotsky asylum: this, they screamed, was one of the crimes of the 'rotten and degenerate' régime, the true face of which had been shown up in the Stavisky affair. From Berlin, Goebbels' Ministry of Propaganda spread the tale that Trotsky was preparing a communist insurrection. The *petite bourgeoisie*, terrified by the slump, incensed against the Third Republic, and fed by sensational headlines about Trotsky's mysterious doings, readily believed that the 'ogre of Europe' was after them. *Humanité* maintained that he was conspiring against the French national interest. To appease the hostile clamour the Ministry announced that it was about to expel Trotsky, and it served him an order of expulsion. It did not, however, enforce the order, because no country was prepared to accept him.

T

On 16 or 17 April the police instructed him to move out of Barbizon. His house was besieged by crowds; an attack either from the Cagoûlards or from the Stalinists was to be feared. He shaved off his beard, did what else he could to make himself unrecognizable, and slipped out of the house. He went to Paris and stayed for a few days with his son in a poor student's garret. But Paris was out of bounds for him and too dangerous; and so, leaving Natalya, he took to the road again. With Henri Molinier and van Heijenoort he drove southwards without any definite destination. He was still to stay in France for another fourteen months; but he had either to lead a vagrant life or to take shelter in a remote village in the Alps; and all the time he had to hide his high and all too conspicuous head.

Followed by a police detective, he moved from place to place and from hotel to hotel until he arrived at Chamonix. Almost at once a local newspaper came out with the hot news. 'Apparently the police suspected', he noted, 'that I had some intentions concerning Switzerland or Italy, and gave me away. . . .' He had to move on. The police forbade him to remain in the frontier area and ordered him to find a refuge in a small town or village situated at least 300 km. from Paris. At Chamonix, Natalya rejoined him; and, while Molinier or van Heijenoort looked for a new dwelling-place, they had to be put up in a pension. To enter a pension was 'a very complicated operation', because he could not present himself under his name, and the police would not allow him to use any cover name. He introduced himself finally as Monsieur Sedov, a French citizen of foreign extraction; and to obtain complete privacy he and Natalya pretended to be in deep mourning and took meals in their room. Van Heijenoort posing as a nephew kept an eye on the surroundings. Tragi-comically, the pension turned out to be the centre for local royalists and fascists with whom the 'loyal republican' agent of the *Sûreté*, who continued to escort Trotsky, engaged in acrimonious discussions at table. 'After each meal our "nephew" would tell us about these Molièresque scenes; and half an hour of merry though suppressed laughter (we were, of course, in mourning) repaid us at least partially for the discomfort of our existence. On Sunday Natalya and I went out "to Mass", really for a walk. This heightened our prestige in the house.' From this pension they moved to a cottage in the country. But when the local Prefect learned the address, he wrung his hands: 'You have chosen the most inappropriate place! That is a hotbed of clericalism. The *maire* is a personal enemy of mine.' Having rented the cottage for a few months and being 'bankrupt' by now, Trotsky refused to depart, until another indiscretion in a local newspaper compelled him to leave in a hurry.[1]

After nearly three months of such wanderings he came at last, early in July, to Domesne, near Grenoble, where he and Natalya stayed with a Monsieur Beau, the village teacher. There they remained for nearly eleven months, in complete isolation, without a secretary or a bodyguard. Only two or three visitors, who had come especially from abroad called at Domesne. Once in several weeks a secretary would arrive from Paris; and every now and then a few schoolmasters from the neighbourhood visited Monsieur Beau, and then his two tenants joined them in discussing local school affairs. 'Our life here differs very little from imprisonment', Trotsky noted. 'We are shut up in our house and yard and meet people no more often than we would at visiting hours in a prison . . . we have acquired a radio, but such things probably exist even in some penitentiaries. . . .' Even their daily walks reminded them of taking exercise in a prison yard: they skirted the village to avoid people and could not go far without straying into a neighbouring hamlet. Mail from Paris arrived only twice a month. In democratic France they had far less freedom than they had had on Prinkipo and even at Alma Ata.[2]

He worked less than usual and less fruitfully, and made almost no progress with the *Lenin*. In October he wrote under the title *Où va la France?* a pamphlet about French politics on the eve of the Popular Front. The pamphlet contained many brilliant passages, but it failed to answer, or rather gave the wrong answer, to the question posed in its title. He viewed the French scene through the same prism through which he had viewed the German scene; yet the prism through which he had seen Hitler's advent so clearly blurred his view of the French

[1] *Diary in Exile*, p. 104.
[2] Ibid., pp. 37, 92, and *passim*; also Pierre Naville, *Trotsky Vivant*.

prospects. Once again he diagnosed, rightly, a crisis of bourgeois democracy; but once again he saw, mistakenly, the lower middle classes 'running amok', producing a dynamic fascist mass movement, and confronting the working class with their violence. The February *coup* of the *Croix de Feu* seemed to lend some colour to this view. But Colonel de la Rocque was not to be the French Hitler; nor was the French *petite bourgeoisie* to engender a movement like National Socialism, either because the Popular Front forestalled it, or because its outlook and traditions were different from those of the German *Klein-bürgertum*. It was one of the peculiarities of French political history in the nineteen-thirties, the nineteen-forties and the nineteen-fifties that attempts to launch fascist mass movements were repeatedly made and repeatedly failed. When the Third Republic collapsed in 1940, it did so under the blast of German invasion; even then not a native fascism but Pétain's sclerotic dictatorship doddered over its ruins. Eighteen years later the Fourth Republic too succumbed to a military *coup*. The French reaction against bourgeois democracy took, as it had done in the nineteenth century, a quasi- or pseudo-Bonapartist form, resulting in the 'rule of the sabre', the methods and impact of which were very different from those of totalitarian fascism.[1]

From his premises Trotsky expounded his ideas about the strategy and tactics of the French labour movement. He criticized the united front, as Thorez and Blum practised it, on the grounds that its action was confined to parliamentary manœuvres and electoral alliances; and that it did not seek to arouse the workers to an extra-parliamentary struggle against fascism, a struggle which might have opened up the prospect of socialist revolution as well. He poured out his sarcasm on

[1] Trotsky was, in his time, the only political theorist to produce a precise definition of fascism. Yet on some occasions he applied it rather imprecisely. He saw the imminence of fascism in France; and he insisted on labelling Pilsudski's pseudo-Bonapartist dictatorship over Poland as fascist, although Pilsudski did not rule in a totalitarian fashion and had to put up with the existence of a multi-party system. On the other hand, Trotsky described, rather unconvincingly, the ephemeral governments of Schleicher and Papen, and also Doumergue's feeble government of 1934, as Bonapartist. (Only in 1940 did he at last describe the Petain régime as pseudo-Bonapartist rather than fascist.) I argued on these points with Trotsky in the nineteen-thirties; but the issue is perhaps of too little historical significance and too involved to be taken up here.

the Comintern which had denounced him for urging German socialists and communists to bar jointly Hitler's road to power and which had now, without turning a hair, adopted the united front only to pervert it into a tactic of evasion, 'parliamentary cretinism', and opportunism. Ironically, it was Thorez who now urged Blum to extend their alliance to the Radicals in order to 'associate the *petite bourgeoisie* with the anti-fascist struggle of the working class'. This—the Popular Front—Trotsky argued, would not associate the lower middle classes with the workers but would open up a chasm between them, because the lower middle classes were turning their backs on the Radicals, their traditional party. He appealed to communists and socialists to form workers' militias and prepare to fight fascism arms in hand, if need be; and he repeated these views in another pamphlet, *Encore une fois: Où va la France?*, written in March 1935.

The eventual failure of the Popular Front was to justify most of Trotsky's criticisms. For the moment, however, the joint socialist-communist action succeeded in throwing back the fascist leagues, which never recovered from their defeat; and the Popular Front undeniably aroused the working class for a time and gave a tremendous impulse to its movement. Only subsequently was the policy of the Popular Front to break the energy of the workers, to estrange the *petite bourgeoisie*, and thus to throw the country into the mood of reaction and prostration, in which the outbreak of the Second World War found it. But in 1934–5, as the danger of fascism had receded, Trotsky's call for extra-Parliamentary action and for workers' militias sounded out of season, and evoked no response. Watching from his retreat in the Alps the first manœuvres of the Popular Front, he noted in his Diary that 'this order has hopelessly undermined itself. It will collapse with a stench.'[1] Only a few years lay between the triumphs of the Popular Front and the great stench of the 1940 collapse.

.   .   .   .

Up to the end of 1932 Trotsky was still in contact with his followers in the Soviet Union, and received letters and bulletins from many penal settlements and prisons. Written in Russian,

[1] *Diary in Exile*, p. 48.

French, and German, mostly on rough wrapping sheets, sometimes even on cigarette paper, and dealing with political and
theoretical matters or bringing personal greetings, this correspondence was despatched with incredible ingenuity: once,
for instance, there landed on Trotsky's desk a matchbox inside
which he found a whole political treatise penned in the tiniest
of handwritings. This correspondence, preserved in his Archives, brought to Prinkipo the breath of Siberian and sub-Polar
winds, the smell of dungeons, the echoes of savage struggles,
the cries of doomed and despairing men, but also some lucid
thoughts and unbroken hopes. As long as it went on reaching
him, he felt the throbbing of Soviet reality. Gradually, however,
the correspondence dwindled to a trickle; and even before he
left Prinkipo it ceased altogether.

In France he had no contact at all with the Opposition in
the Soviet Union. Its silence, made even deeper by the capitulators' unending recantations, was on his mind when he stated
that the Russian movement had lost the power of revolutionary
initiative and that only a new International could regain it.
In February 1934, while he was still at Barbizon, the news of
Rakovsky's capitulation reached him. It may well be imagined
how this affected him. Rakovsky had been closer to him as
'friend, fighter, and thinker' than any other associate; despite
his age, he had, unbroken by persecution, held out against
Stalin after nearly all the other leaders of the Opposition
had surrendered; and in the prisons and places of deportation
his moral authority had been second only to Trotsky's. In
almost every issue of the *Bulletin* Trotsky had published something by Rakovsky or about him: an article, a letter, an extract
from an old speech, or a protest against his persecution. After
every defeat of the Opposition and after every series of capitulations, he had pointed to Rakovsky as the shining example and
as proof that the Opposition was alive. Rakovsky's defection
filled him therefore with immense sadness; it marked for him
the passing of an epoch. 'Rakovsky', he wrote, 'was virtually
my last contact with the old revolutionary generation. After
his capitulation there is nobody left. . . .'[1] Was it weariness,
he wondered, that had at last overcome the old fighter? Or
was he, as he stated, guided by the conviction that when the

[1] Op. cit., pp. 41, 53; also *B.O.*, no. 40, 1934.

Soviet Union was threatened by the Third Reich, he too had to 'rally behind Stalin'? In any case, Stalin's triumph could not be more complete. And in the next few months the re-conciliation between Stalin and his many repentant opponents seemed more genuine than ever, although the party still ceaselessly expelled 'disloyal elements' from its ranks.

Then suddenly, before the end of the year, this appearance of reconciliation was exploded. On 1 December, Sergei Kirov, who had nine years earlier replaced Zinoviev as the head of the Leningrad organization and in the Politbureau, was assassinated. The first official version claimed that a body of White Guard conspirators stood behind Nikolaev, the assassin; and that a Latvian consul had pulled the wires—there was no question of any inner party opposition being involved. A second version, however, described the assassin as a follower of Zinoviev and Kamenev and made no mention of White Guards. Nikolaev and fourteen other young men, all Komsomoltsy, were executed. Zinoviev and Kamenev were expelled from the party for the third time; they were imprisoned, and awaited trial by a court martial. Press and radio linked Trotsky with Zinoviev and Kamenev and assailed him as the real instigator. A mass terror was let loose against 'Kirov's assassins', Trots-kyists, Zinovievists, and disgruntled Stalinists; many thousands were deported to concentration camps. Finally, several high officers of the Leningrad G.P.U. were charged with 'neglect of duty' and sentenced, with surprising mildness, to two or three years.

At the cottage in the Alps Trotsky, glued to his wireless set and listening to transmissions from Moscow, followed the unfolding of the plot and recorded his comments.[1] In the clamour rising from Moscow he discerned at once a prelude to events far vaster and more sinister than the Kirov affair. He was convinced that Zinoviev and Kamenev had not been implicated in the attempt on Kirov—old Marxists that they were, nothing could have been more unlike them than a cloak-and-dagger action, which hit an individual holder of office without changing the system. He had no doubt that Stalin was using the assassination as a pretext for a new assault

[1] *B.O.*, no. 41 (the issue for January 1935) was entirely taken up by Trotsky's current commentary on the Kirov affair.

on the Opposition. On 30 December, a fortnight before the
news about the trial of the chiefs of the Leningrad G.P.U. was
broadcast, Trotsky asserted, on the internal evidence of the
official announcements, that the G.P.U. had known about the
preparations of the attempt and had, for their own reasons,
condoned them. What were those reasons? Nikolaev had been
one of the Komsomoltsy who had grown up after the suppression of the oppositions and, who, disillusioned, robbed of
any way of expressing themselves legally, and uninhibited by
Marxist tradition, sought to protest with bomb and revolver.
Not the Opposition, Trotsky asserted, but the ruling group was
responsible for this. The G.P.U. had known about Nikolaev's
intentions and had used him as their pawn. What aims did
they pursue? Nikolaev had allegedly confessed that the Latvian consul had urged him to enter into contact with Trotsky
and to write a letter to him. The 'consul', Trotsky pointed
out, had been acting for the G.P.U., who had planned to
'discover' Nikolaev's plot only after they could produce
'evidence' that he was in correspondence with Trotsky. As
long as they had not obtained that 'evidence', they left
Nikolaev at large, and were confident that they were able to
watch him closely and direct all his moves. They miscalcu-
lated: Nikolaev aimed his revolver at Kirov before the G.P.U.
had achieved their purpose. Hence the contradictions between
the various official versions; hence the secrecy in which
Nikolaev's trial had been held; and hence the trial of the
G.P.U. officers for 'neglect of duty' and the mildness of their
sentences.

Trotsky concluded that the G.P.U., having failed to obtain
false evidence against him from Nikolaev, would try to get it
from—Zinoviev and Kamenev. Meanwhile, Zinoviev and
Kamenev had been sentenced to ten and five years prison
respectively, but had been allowed to state in public that they
had had no dealings with Nikolaev and that they could be
held co-responsible only indirectly, in so far as their criticisms
of Stalin in bygone years might have influenced the terrorist.
The court accepted their plea; and Trotsky concluded that
behind the scenes a bargain was being struck between Stalin
and Zinoviev and Kamenev: Stalin must have promised to
rehabilitate them if they agreed to denounce Trotsky as the

leader of a terroristic conspiracy. 'As far as I can judge . . .', Trotsky wrote, 'the strategy which Stalin displayed around Kirov's corpse has brought him no laurels': the incongruities of the affair had given rise to comment and to rumour which placed the odium on Stalin and his entourage. 'Precisely because of this Stalin can neither halt nor retreat. He must cover up the failure of this amalgam by new amalgams which must be conceived on a much wider scale, on a world scale, and more . . . successfully.'[1] Dissecting the Kirov affair, Trotsky predicted the great trials, which were indeed to be conceived 'on a world scale' with Hitler, instead of a mere Latvian consul, being cast by Stalin as Trotsky's ally.

. . . .

The Kirov affair at once affected the fortunes of Trotsky's family. His two sons-in-law, Nevelson and Volkov, deported since 1928, were arrested, and without trial the terms of their deportation or internment were prolonged. His first wife— now over sixty—was expelled from Leningrad, first to Tobolsk and then to a remote settlement in the Omsk province. His three grandchildren who had been under her care were now put up with an old aunt and were at fate's mercy. 'I receive letters from the little ones', Alexandra wrote to Lyova, 'but I do not have a clear idea of their life. My sister is probably having a hard time . . . although she keeps reassuring me. My health is so-so, there is no doctor here so that I must keep well.'[2] This time the terror also hit Sergei, Trotsky's youngest son who, we remember, was a scientist, shunned politics, and avoided contact with his father. In all the years since 1929 he had been writing only to his mother, confining himself to such matters as his health and his progress in academic work, and inquiring about the family's well-being—there had never been even the slightest political allusion in his letters and postcards. Just a few days after Kirov's assassination he wrote again to his mother about his professional work, describing the variety of the subjects on which he lectured at the Higher Technological Institute in Moscow, the effort this demanded

---

[1] See 'Letter to American Friends' of 26 January 1935 in *The Archives*; and *B.O.*, no. 42, February 1935.
[2] *Diary in Exile*, p. 79.

of him, and so on. Only in the closing lines he hinted that 'something unpleasant is brewing, so far it has taken the form of rumours, but how all this is going to end I do not know'. A week later, on 12 December, he wrote again about his academic work, and concluded alarmingly: 'My general situation is very grave, graver than one could imagine.'[1] Was it possible, the parents wondered in anguish, that the G.P.U. would seize Sergei as a hostage? For many a week they lived in expectation of another letter from him. None came. An old friend of the family, the widow of L. S. Klyachko, domiciled in Vienna, visited Moscow and inquired about Sergei, with the result that she was ordered to leave the country at once, without any explanation.

For weeks and months, through many a sleepless night, his parents' thoughts went out to Sergei. They were tormented by uncertainty. Perhaps his trouble was of a personal and private nature, not political? Perhaps the G.P.U. had only expelled him from Moscow but not imprisoned him? Surely they must realize that he was not politically minded at all? Could they have imprisoned him without Stalin's knowledge? Natalya asked, as if she entertained a faint, unspoken hope that perhaps an appeal to Stalin would help. No, Trotsky replied, only on Stalin's order could they have imprisoned him—only Stalin could have contrived such an act of vengeance. Would they try to extract from Sergei a confession with accusations against his own father? But of what use could these be to Stalin? Would their falsity not be obvious? Yet for what other reason could they have seized him? Would they torture him? Would he break down?[2]

For days and nights on end his parents were haunted by the image of their son facing his inquisitors. They feared that in his political innocence he would not be able to take the blows. They saw him bewildered and crushed; and they reproached themselves for not having insisted that he should go with them into exile. But could they try and tear him away from his academic preoccupations and routine when they themselves did not know what awaited them? It was different with Lyova, whose mind and passions were completely engaged in the political struggle. They remembered Zina whom they

---

[1] *The Archives*, Closed Section.   [2] *Diary in Exile*, pp. 61–72.

had been unable to save after she had joined them abroad. They recollected Sergei's jolly childhood, his reaction against his father and elder brother, his distaste for politics, his restless yet gay adolescence, and finally his serious and dedicated concentration on science. No, they could not have asked him to become involved in his father's affairs. But was he thinking now that they had abandoned and forgotten him? They searched Russian newspapers to see whether there was any mention of him. In the mounting avalanche of abuse against the 'dregs of Zinovievists, Trotskyists, former Princes, Counts and gendarmes' they came across names of relatives and friends; but there was dead silence about Sergei. Stalin, Trotsky noted, 'is clever enough to realize that even today I would not change places with him. . . . But if revenge on a higher [moral-political] plane has not succeeded—and clearly it will not succeed—it is still possible [for Stalin] to reward himself by striking at people close to me'.[1]

The feeling that Stalin had laid hands on the son because he could not reach the father gave Trotsky a sense of guilt. In his Diary, between entries about Sergei, he tells, seemingly out of context, the story of the execution of the Tsar and the Tsar's family. In his anxiety over Sergei falling a victim to his conflict with Stalin, he evidently thought also about those other innocent children, the Tsar's, on whom the sins of the father had been visited. He records in the Diary that he had no part in the decision about the Tsar's execution—the decision having been Lenin's primarily—and that he was taken aback when he first learnt about the fate of the Tsar's family. He does not recall this, however, to dissociate himself from Lenin or to exculpate himself. Seventeen years after the event he defends Lenin's decision as necessary and taken in the interest of the revolution. In the midst of civil war, he says, the Bolsheviks could not leave the White Armies with a 'live banner to rally around'; and after the Tsar's death any one of his children might have served them as the rallying symbol. The Tsar's children 'fell victim to that principle which constitutes the axis of Monarchy: dynastic succession'. The unspoken conclusion of this digression is clear enough: even if one granted Stalin the right to exterminate his adversaries—

[1] Op. cit., pp. 66–67.

and Trotsky was, of course, far from granting him that—Stalin still had not a shred of justification for persecuting the children of his adversaries. Sergei was not bound to his father by any principle of dynastic succession. Immediately after this digression Trotsky notes: 'No news about Seriozha, and perhaps there won't be any for a long time. Long waiting has blunted the anxiety of the first days.'[1]

Yet the anxiety began to tell on Trotsky. He was depressed. He brooded again over his advancing age and death. He was not yet fifty-five, but repeatedly he recalled Lenin's, or rather Turgeniev's, dictum: 'Do you know what is the greatest vice? To be more than fifty-five years old.' With a hint of envy he remarked: 'But Lenin did not live long enough to develop this vice.' 'My condition is not encouraging. The attacks of illness have become more frequent, the symptoms are more acute, my resistance is obviously getting weaker.' 'Of course, the curve may yet take a temporary turn upwards. But in general I have a feeling that liquidation is approaching.' With a clear prescience of what was to come he observed that Stalin 'would now give a great deal to be able to retract the decision to deport me. He will unquestionably resort to a terroristic act . . . in two cases . . .: if there is a threat of war, or if his own position deteriorates greatly. Of course, there could also be a third case, and a fourth . . . we shall see. And if we don't, then others will.' He began to think of suicide, and reflected that he should commit it if and when his physical strength gave out and he could no longer continue his struggle. Perhaps, it occurred to him, in this way he might save Sergei? But these were fleeting thoughts. Although his energy was sapped, he was still to show astonishing vitality and vigour in years to come, when events were to confront him with their challenge even more directly. Meanwhile, he was experiencing something as ordinary and human as the crisis of middle age; he succumbed to bouts of hypochondria and to the weariness of prolonged isolation and passivity.[2]

He was now at his nadir. The ambitious plans and sanguine hopes with which he had left Turkey were in the doldrums. His great campaign against the Stalinist surrender to Hitler had brought him no political rewards. Stalinism was even

[1] Op. cit., p. 82.   [2] Ibid., pp. 51, 109 and *passim*.

exploiting this surrender to make fresh political capital: playing on the fear of Nazism, it ingratiated itself with the European left. Trotsky sensed, although he could not admit it even to himself, that the Fourth International was stillborn. He could neither escape his circumstances nor make peace with them. And so he found some solace in exalted reflections on his 'historic mission' in founding the Fourth International. It was in this context that he contemplated what would have been the course of the Russian Revolution without Lenin and himself and that he asserted that his work for the new International was 'indispensable' in a sense in which even his work in the October insurrection and the Civil War was not. 'There is no arrogance in this claim at all', he noted. 'The collapse of the two Internationals has posed a problem which none of the leaders of these Internationals is at all equipped to solve. . . . There is now no one except me to carry out the mission of arming a new generation with a revolutionary method over the heads of the leaders of the Second and the Third Internationals. And . . . the worst vice is to be more than fifty-five years old! I need at least about five more years of uninterrupted work to ensure the succession', that is to form an International capable of leading the working class to revolution.[1]

At his nadir he challenged fate, which was to grant him exactly 'five more years' yet was not to allow him to 'ensure the succession'.

.   .   .   .

In all the years of their life together—now thirty-three—Trotsky and Natalya had never been as alone as they were during these eleven months at Domesne. Solitude and suffering drew them even closer to each other. In tragic hours, he said, he was 'always amazed at the reserves of her character'. Their love had survived triumph and defeat; and the afterglow of their past happiness broke through even the gloom of these days. Her face was becoming wrinkled and tense with worry and anxiety, and he thought with pain of her bright and gaily defiant youth. 'Today on our walk we went up a hill. Natalya got tired and unexpectedly sat down, quite pale, on the dry leaves. . . . Even now she still walks beautifully, with-

[1] Ibid., p. 54.

out fatigue and her gait is quite youthful, like her whole figure. But for the last two months her heart has been playing up now and then. She works too much. . . . [she] sat down all of a sudden —she obviously just *could not* go any further—and smiled apologetically. What a pang of pity I felt for her youth. . . .' She bore her lot with quiet fortitude, and her life was wholly absorbed in his. Every storm that passed over him shook her; every current of his emotion permeated her being as well; and every reflex of his thought was mirrored in her. She had not been to him the kind of political comrade Krupskaya had been to Lenin; for Krupskaya, being childless, had been a political worker in her own right and sat on the party's Central Committee. Natalya was not only less active but less politically minded. 'Even though she is interested in the small daily facts of politics [these are Trotsky's words] she does not usually combine them into one coherent picture.' The loving husband could not express more clearly a doubt about his wife's political judgement. But this was not important: '. . . when politics go deep down and demand a complete response Natalya always finds in her inner music the right note'.[1]

To this, her 'inner music', he often referred; and, incidentally, when he described her in his Diary it was mostly while she was listening to music. Her independent interests were, as always, in the arts; and she had uncommon gifts of insight, observation, and expression, which appear strikingly in her own Diary pages. Her husband's disciples sometimes raised eyebrows at her political remarks, which caused Trotsky to say that 'sensitive people . . . instinctively feel the depth of her nature. Of those who pass her by with indifference or condescension without noticing the forces concealed in her, one can almost always say with certainty that they are superficial and trivial. . . . Philistinism, vulgarity, and cowardice can never be concealed from her, even though she is exceptionally lenient towards all minor human vices'. Of her 'inner forces' there can indeed be no doubt. At the worst moments, when he was almost at the end of his endurance, it was she who raised him back to his feet and revived in him the strength to carry his burden. At Domesne he noted with gratitude that she never reproached him for Sergei's misfortune and that she concealed

[1] Op. cit., pp. 51, 56.

her suffering from him. Only exceptionally did her anguish break out in a remark like this: 'They will not deport Sergei . . . they will torture him in order to get something out of him, and after that they will destroy him.' She hid her feelings in work, housekeeping, helping her husband in his writing, and discussing the French and Russian novels they read together. 'Her voice made me feel a sudden pang . . . slightly hoarse, [it] comes from deep in her chest', he remarked. 'When she suffers it withdraws even deeper, as if her soul were speaking directly. How well I know this voice of tenderness and suffering.' And on one occasion he noticed that for days she had been thinking more about his first wife than about Sergei, saying that Sergei might after all not be in any trouble and fearing that Alexandra, in her old age, would not survive deportation.[1]

In the slender hope that perhaps an appeal to the conscience of the world might save Sergei, Natalya wrote an 'Open Letter' in his defence and published it in the *Bulletin*.[2] She explained Sergei's complete innocence and, doing some violence to her pride, related how his distaste for politics had been caused by his reaction against his father. Had recent developments changed Sergei's attitude and drawn him to the opposition? 'I would be happy for him if I could think so, since under those conditions it would be immeasurably easier for Seriozha to bear the blow. . . .' Unfortunately, this supposition was unreal: she knew from various people that 'during the last few years [he] had been keeping as much out of politics as before. But personally I would not need even this evidence. . . .' The G.P.U. and the university authorities must know this, for they had undoubtedly watched him; and Stalin, 'whose son was a frequent guest in our boys' room', knew it too. She appealed to famous humanitarians and 'friends of the U.S.S.R.', such as Romain Rolland, André Gide, Bernard Shaw, and others, to speak up; she proposed that an international commission should investigate the mass reprisals that followed the Kirov affair. 'The Soviet bureaucracy cannot stand above the public opinion of the working class of the world. As far as the interests of the workers' state are concerned, these would only benefit by a serious examination of

[1] Ibid., pp. 51, 71, 121–2.   [2] *B.O.*, no. 44, July 1935.

its actions. I . . . offer all the necessary information and documents concerning my son. If, after long hesitation I openly raise the question of Sergei, it is not only because he is my son: that reason would be only too sufficient for a mother, but not adequate for . . . political action. But Sergei's case is a completely clear, simple, and indisputable instance of conscious and criminal abuse of power, and a case which can be examined very easily.' The appeal brought no answer.

By a curious coincidence, about the time when Natalya made this appeal, Trotsky was re-reading the autobiography of Protopop Avakuum, a famous and colourful Russian arch-priest and preacher of the Old Belief, who lived in the seven-teenth century, after the Time of Trouble. Avakuum defended 'true' Greek Orthodoxy against Patriarch Nikon, his harsh rival, who had for temporal reasons changed the Church rites and the prayer book; and he exposed the corruption of the ecclesiastical hierarchy, and took up the cause of the oppressed peasants. He was unfrocked, jailed, banished first to Siberia and then to the Mongolian frontier, starved and tortured; but he refused to recant. His family suffered with him, and he, a loving husband and father, wondered for a while whether he should not give up the struggle and save his next of kin. His children died of disease and starvation in exile. It was in Siberia that he wrote his autobiography, a work which made an epoch in Russian literature; and he continued to preach with such effect that his fame as 'hero and martyr for truth' grew in the country. Banished, he was even more dangerous to his enemies than he had been when he stood near the Throne. They brought him back to Moscow and burned him at the stake.[1] Across the chasm of centuries and ideologies Trotsky could not help feeling with a shudder his affinity with this legendary rebel—how much and how little had changed in Russia! And even the spirit of Avakuum's wife stood before him as if embodied in Natalya:

Reflecting on the blows that had fallen to our lot I reminded Natasha the other day of the life of the Archpriest Avakuum. They were stumbling on together in Siberia, the rebellious priest and his faithful spouse. Their feet sunk in the snow, and the poor exhausted

---

[1] A new edition of *Zhizn Protopopa Avakuma*, with an interesting Preface not lacking topical allusions, was brought out in Moscow in 1960.

woman kept falling in the snowdrifts. Avakuum relates: 'And I came up and she, poor soul, began to reproach me, saying "How long, Archpriest, is this suffering to be?" And I said: "Markovna, unto our very death." And she, with a sigh, answered: "So be it, Petrovich, let us be getting on our way."'[1]

And so it was to be with Trotsky and Natalya: the suffering was to be 'unto our very death'.

.   .   .   .

They could not now remain at Domesne much longer. Any political swing to the right, bringing forward the fascist leagues, and any swing to the left, adding strength to the Communist party, threatened to rob Trotsky of his precarious refuge. It was the swing to the left that came. Since the Kirov affair the Stalinist incitement against 'the leader of world counter-revolution' had grown so brutal and venomous that it was all too likely to provoke an act of violence.[2] He could not feel secure even in the remote village in the Alps. He describes how once in these days he and Natalya, alone in their cottage, listened in tense silence to two men who as they approached were singing the Internationale. In past times only a friend could come with that song; now it might be an enemy and an assailant. They felt like those old Narodniks who, two generations earlier, went out into the country to enlighten and emancipate the muzhiks, and were beaten up and lynched by the muzhiks themselves.

The Government could no longer afford to ignore the Stalinist clamour. In May 1935 Laval had gone to Moscow to negotiate the Soviet-French alliance with Stalin, and he returned with that startling declaration by which Stalin pledged his support to Daladier's and Laval's defence policy. The French communist leaders, who had hitherto opposed that policy on principle, at once took up a 'patriotic' line; and the Popular Front took shape. Trotsky had every reason to believe that the Government would presently enforce the

[1] *Diary in Exile*, p. 121.

[2] An article by Jacques Duclos in *Humanité*, in December 1934, spoke of 'Trotsky's hands covered with Kirov's blood'; and the *Secours Rouge International*, the French section of M.O.P.R. (the international organization for the defence of political prisoners and exiles) clamoured for Trotsky's deportation from France.

expulsion order it had served on him the year before; and as no other country was willing to accept him, he feared deportation to a remote French colony, possibly to Madagascar.

In the spring of 1935 he asked for asylum in Norway. An election had just been held there and the Labour party had taken office. This was a Social Democratic party with a difference: it had belonged to the Comintern; and although it had broken with it in 1923, it did not adhere to the Second International. It was only natural to expect that such a party should give Trotsky refuge. Walter Held, a German Trotskyist, living as an émigré in Oslo, approached Olav Schöffle, one of the party's outstanding leaders, who headed its radical wing and was greatly devoted to Trotsky. It took many weeks before an official reply came. Trotsky supposed that the Norwegians had been stung by an article of his which taunted them for abandoning, on the assumption of office, their republican tradition and making peace with their King. Early in June, however, he was informed that they had granted him asylum. On 10 June he left Domesne and went to Paris, where he was to obtain the visa. But there was a hitch: high Norwegian officials, displeased with the government's decision, sought to obstruct it; he did not receive the visa and he had to cancel the arrangements for the voyage. The French police, suspecting that he had used all this as a pretext for descending on Paris, ordered him to leave France at once, within twenty-four or at the most forty-eight hours. He was resigned to returning to Domesne but was not allowed to do so. He proposed to wait for the final answer from Oslo in a private clinic; but the police, imagining that he was playing another trick on them, objected to this too. For a day or two he found refuge in the home of Doctor Rosenthal, a well-known Parisian surgeon. On 12 June he cabled a reproachful message to the Norwegian Prime Minister, saying that he had left his place of residence, relying on the Norwegian promise and now: 'The French government believes that I have deceived it, and demands that I leave France within twenty-four hours. I am sick and my wife is sick. Situation is desperate. I solicit immediate favourable decision.'[1] To make matters worse, he was

[1] From Trotsky's telegram to Nygaardsvold, Prime Minister of Norway (12 June 1935), *The Archives*, Closed Section.

penniless and had to borrow money for the journey. The Norwegians still asked him to secure a French re-entry permit, which he had no chance of obtaining, before they would allow him to come to Norway. At last, thanks to Schöffle's efforts, they granted him the visa, with a residence permit for six months only. He parted hurriedly from his French followers: 'I was seeing numerous Parisian comrades. The worthy doctor's apartment had unexpectedly been transformed into the headquarters of the Bolshevik-Leninist group. There were meetings going on in all the rooms, the telephone was ringing, more and more new friends kept arriving.'[1] He described the scene in a manner calling back to one's mind the moment of his deportation from Moscow in 1928. But this description is something of a pastiche: the farewells in Moscow had concluded one great epoch of his struggle and opened another; the farewells in Paris concluded and opened nothing.

He once again wrote, as he had done before his expulsion from France in 1916, an 'Open Letter' to the French workers. He told them that during his stay he had been condemned to silence. 'The most "democratic" Ministers like the most reactionary ones see their task in defending capitalist slavery. I belong to a revolutionary party which sees its task in the overthrow of capitalism.' He lashed out at the Stalinists: 'Two years ago *Humanité* reported every day that "the fascist Daladier has called the social-fascist Trotsky to France in order to organize with his assistance military intervention against the Soviets. . . ." Today the same gentlemen have formed . . . an anti-fascist "Popular Front" with the "fascist" Daladier. They have ceased to talk . . . about any French imperialist intervention against the U.S.S.R. Now they see the guarantee of peace in the alliance of French capital with the Soviet bureacracy and . . . say that Trotsky's policy serves not Herriot and Daladier but Hitler. . . .' He concluded vehemently that Stalinism was a 'festering sore' on the labour movement, which should be burned out with 'red hot iron', and that the workers should reassemble under the banner of Marx and Lenin. 'I am leaving with a deep love for the French people and ineradicable faith in the future of the working class. Sooner or later they will render me the hospitality that the bourgeoisie

[1] *Diary in Exile*, pp. 125–6.

refuses me.'[1] After two dismal and wasted years he was leaving France never to return.

.   .   .   .

The tale of Trotsky's sojourn in Norway reads like a large variation on Ibsen's *Enemy of the People*. Ibsen presents the drama of Doctor Stockman, revered for his nobility by all his fellow-citizens, until he threatens to destroy their prosperity by disclosing the truth about the poisoned source of their wealth. Then his own brother, the town Mayor, and his own 'radical' friends, turn against him with cold and murderous fury. We are now in Ibsen country once again. It does not greatly matter that this time the Enemy of the People is a fugitive from abroad; that he speaks not about the contaminated conduit pipes of a Norwegian resort but about a revolution that has been perverted. The drama and the stage are essentially the same; and so are the family traits of the actors, especially of the sons and grandsons of Ibsen's pseudo-radicals—even their *People's Messenger* is still there, as of old, changing sides overnight and manipulating public opinion. In the crowd we may also discern one or two descendants of the modest and courageous captain Horster, who stood up for the Enemy of the People. Only the times are changed; the forces in action are far more potent; and the conflict more cruel.

From the outset the auguries were not promising. Not only had the Norwegians been niggardly in granting Trotsky asylum; they placed him under restrictions not very different from those under which he had lived in France, and they reserved the right to fix the place of his residence at some distance from the capital. No sooner had he disembarked on 18 June than the National Farmers' Union protested against his admission; and on 22 June the Storting was already debating the protest. This had no immediate sequel, but it was clear that the Opposition would use his presence to embarrass the government. The conservative bourgeoisie was scared of the 'ogre'; it was impossible to find lodgings for him; no householder dared to accept him as tenant. The government asked him to pledge himself to refrain from political activity. He did this, on the understanding that what was demanded from him

[1] *B.O.*, no. 44, July 1935.

was that he should not interfere in Norway's domestic affairs. The government was to claim later that it had asked him to refrain from any political activity, a demand to which no political exile can normally submit or is asked to submit. The circumstance that he was thus treated by men still thinking of themselves as schismatics from official communism underlined the meanness of their behaviour.

However, on his arrival, the chiefs of the government and the Labour party made a great show of generosity. 'The working class of this country and all right thinking and un-prejudiced people'—this is how their newspaper *Arbeider-bladet* welcomed him—'will be delighted with the government's decision. The right of asylum must not be a dead letter but a reality. The Norwegian people feel . . . honoured by Trotsky's presence in their country.' Without going into the pros and cons of his dispute with Stalin, on which they held no definite views, they denied Stalin the right to 'persecute and banish a man like Leon Trotsky whose name will stand together with Lenin's in the history of the Russian revolution. Now that, despite his great and imperishable services, he has been exiled from his own country, any democratic nation must consider it a welcome duty to offer him refuge. . . .'[1] Martin Tranmael, the party's founder and leader, came out with personal greetings. Various ministers intimated that the terms of Trotsky's admission, the six months' limit and the restrictions on his freedom of movement, were so many formalities. The government asked Konrad Knudsen, a Socialist editor, to assist in settling Trotsky; and Knudsen, seeing that it was impossible to lease a house, invited him and Natalya to move into his own home.[2]

Presently three party chiefs, Tranmael, Trygve Lie, the Minister of Justice, and the Editor of *Arbeiderbladet* paid Trotsky a formal visit. The meeting was rather awkward. The Norwegians reminded Trotsky that in 1921 they had been in Moscow and negotiated with him, Lenin, and Zinoviev the terms of their adherence to the Comintern; but before they

[1] H. Krog, *Meninger*, p. 220; (I am obliged to Mr. Krog and to Mr. N. K. Dahl for the English translations of passages from this book and of other Norwegian documents quoted later); *Diary in Exile*, pp. 128–9.
[2] This is how Konrad Knudsen has related the facts to me.

proceeded further Trygve Lie wished to make sure that Trotsky was aware of his obligation to refrain from political activity. He answered that he had not the slightest intention of meddling in Norwegian affairs—Trygve Lie later maintained that he, Lie, had demanded there and then that Trotsky should abstain from all action 'hostile towards any friendly government'. An eye-witness recollects that 'Trotsky refused to be drawn into any political discussion with us and talked only about the weather'. But the visitors, having gone through with the official part of the business, were eager to change over to a tone of comradeship, to talk politics, and to bask in the greatness of the man to whom they had given refuge. They begged him to give *Arbeiderbladet* a long and exhaustive interview on the major issues of world politics. According to the same eye-witness, he replied frigidly that the Minister of Justice had just forbidden him to indulge in any form of political activity. His interlocutors shrugged and laughed off the prohibition as a piece of make-believe which they had to go through *pro forma* in order to appease their parliamentary opponents; and the Minister of Justice reassured Trotsky that by expressing his opinions he would in no way offend against the terms of his residence. The Minister himself then turned into an eager journalistic interviewer; and Trotsky answered his questions at length, availing himself of the opportunity to denounce Stalin's policy and the terror unleashed since Kirov's assassination. On 26 July *Arbeiderbladet* published the interview, with much editorial flourish, leaving readers in no doubt that the Minister of Justice himself had been instrumental in making the benefit of Trotsky's views available to them. Thus the 'misunderstandings' of the first days seemed dispelled. The party in office treated Trotsky as an illustrious guest rather than as a refugee on sufferance. Parliamentarians and journalists vied with one another in paying him respects; and for a time nothing bestowed greater distinction on a person in Oslo's leftish circles than the ability to boast of having been received by the great exile.

Before the end of June, Trotsky and Natalya were installed in Knudsen's home at Vexhall, a village near Honnefoss, about thirty miles north of Oslo. Amid the quiet and peace of the country, sharing in the domesticities of a modest, warm-hearted and fairly large family they could recover from recent harassments. Knudsen was a moderate, suave Social Democrat, very remote from Trotskyism—it was from sheer sensitiveness and from defiance of philistinism that he had invited the man of October under his roof. By tacit agreement they never touched on their political differences. And so 'during his entire stay with us', these are Knudsen's words, 'we were not troubled even once by the slightest misunderstanding. Trotsky was too much concentrated on his work to waste time in fruitless discussion. He worked very hard. I have never met anyone as precise, punctual, and pedantic in his habits. When he was not ill, he used to get up at 5.20 or 5.30 in the morning, go down to the pantry, take a little food, and set to work. He did it all very quietly, on tiptoe, so as not to disturb anyone. I have no words to describe his tact and consideration for all who lived in our house. Natalya's behaviour was the same; we nicknamed her affectionately "the little lady of the big house". Their needs were quite incredibly modest.'[1]

For the first time since 1917 Trotsky did not have to live under the protection of a 'comradely bodyguard' or under police surveillance and incognito. The yard gate was wide open day and night, with village folk straying in for amiable chats. Occasionally, visitors came from abroad, German refugees living in Scandinavia, Frenchmen, Belgians, and Americans. Among the Americans was Harold Isaacs who had just returned from China, after a stay of several years, and was a source of valuable information on that country and its communist movement. (He was just writing a book, *The Tragedy of the Chinese Revolution*, to which Trotsky was to contribute a preface.) Shachtman and Muste, the well-known American socialist who had joined the Trotskyists, also came to Vexhall. The French arrived several times with their disputes and quarrels, asking Trotsky to act as umpire. They could not agree whether they should leave the S.F.I.O. and reconstitute themselves as an independent party. Raymond Molinier had set up his own paper, *La Commune*, advocating disaffiliation. This brought the quarrel into the open and at last led Trotsky to break with Molinier. The incident would

[1] I am quoting again Knudsen's account and also his Preface to the Norwegian edition of Trotsky's *My Life*.

not be worth mentioning had it not been for the fact that the feud was to go on for years and to become grotesquely intertwined with the fortunes of Trotsky's family. Amid all this and while his correspondence with his followers, which he could not well carry on from France, grew to enormous bulk, Trotsky began writing a new book, *The Revolution Betrayed*.[1]

Towards the end of the summer, however, on 19 September, he entered the Municipal Hospital of Oslo because of the persistence of his fevers and a general debility. In the stillness of his ward he gave himself to melancholy musings. 'It is nearly twenty years', he wrote, 'since I lay on a bunk in a Madrid prison and wondered in amazement what on earth had brought me there. I remember I burst into a fit of laughter . . . and laughed and laughed until I fell asleep. Now once again I am wondering in amazement what on earth has brought me here, into an Oslo hospital?'[2] A Bible on his bedside table sent his mind wondering farther back, to a prison cell in Odessa where thirty-seven years before he had been learning foreign languages from a multilingual copy of the Bible. 'Unfortunately, I cannot promise that this new encounter with the old and so familiar book will help in saving my soul. But reading the Gospel in Norwegian may help me to learn the language of the country which has shown me hospitality, and its literature which I have . . . loved since my early years.' After many tests and examinations he left hospital, his soul not saved and his body not restored to health. He spent the greater part of December in bed—this, he said later, was 'the worst month of my life'.

His recovery was impeded by old and new anxieties and worries. He was depressed by the futility of his 'organizational' work. He was irritated by the French Trotskyists, who did not cease to pester him with their quarrels; and he wrote to Lyova: 'It is absolutely necessary that I should get at least four weeks' "leave" and should not be approached with any letters from the Sections. . . . Otherwise it will be impossible for me to recover my capacity for work. These disgusting

[1] The enormous growth of Trotsky's correspondence with his French, German, Belgian, Dutch, Austrian, American, Greek, and other followers is reflected in the files of *The Archives*, Closed Section. Harold Isaacs' report on China, ibid.

[2] Quoted from Trotsky's Preface to the Norwegian edition of *My Life*; and from *The Archives*.

trivia (*eckelhafter Kleinkramm*) not only rob me of the ability to occupy myself with more serious affairs, but give me insomnia, fever, etc. . . . I request you to be quite ruthless about this. Then I may perhaps be at your disposal again, say, by 1 February.'[1] In the following weeks and months, however, he repeatedly reproached Lyova for harassing him with the '*eckelhafte Kleinkrämmerei*' and vented his 'despair' at the 'silly intrigues' of the 'French cliques'.[2] His correspondence shows all too clearly that things were no better in most of the other Sections of the would-be Fourth International. And there was the anguish over events in Russia and the uncertainty about Sergei. Indirect inquiries in Moscow brought forth an official explanation that Sergei was not imprisoned but 'placed under police surveillance' to keep him from communicating with his father. But when Natalya tried to transmit a small money order to Sergei's wife in Moscow, this was returned to a bank in Oslo with the note that the addressee was unknown. On top of all this, he was troubled by lack of money. Publishers' advances had just enabled him to cover the expenses of settling in Norway and to pay off a debt to Henri Molinier, which he was anxious to do before he broke with the Molinier set. In what bad straits he was can be seen from a letter to Harold Isaacs which he wrote from the Oslo hospital on 29 September, begging for help in a 'financial catastrophe': He had to pay 10 Krones per day in hospital, and had only 100 Krones left.[3]

Just before Christmas he went with Knudsen and a few young Norwegians to the wild rocky country north of Honnefoss, hoping that a few days of physical activity in the open air might improve his health. The time of this trip should be noted—a year later, at Radek's and Pyatakov's trial, Vyshinsky was to claim that at this time Pyatakov paid Trotsky a secret visit; and Pyatakov himself was to confess that he had come to Oslo by plane from Berlin and had gone by car straight from the airfield to meet Trotsky. These allegations were refuted by Norwegian authorities, who ascertained that no German plane had landed at Oslo airfield at the end of December

[1] The letter, dated 27 December 1935, was addressed to Lyova and also, it seems, to another member of the International Secretariat. *The Archives*, Closed Section.

[2] Letters of 14 January and 22 March 1936, ibid.

[3] Correspondence with Harold Isaacs, ibid.

1935 and for several months before and after that; and Trotsky's companions proved that no one could have come by car to the place where they stayed with Trotsky. 'The winter was extremely severe; the roadless country was completely submerged by snowdrifts, and gripped by Arctic ice. We remember this well, because once during the trip Trotsky was trapped by snow and ice. We were on skis, and he was not good at skiing; and so we had to organize a regular rescue operation, and we were very worried.'[1]

Soon after, by one of those abrupt changes in his health which puzzled his doctors, he recovered, and resumed the writing of *The Revolution Betrayed*. This kept him busy for the next six months until he completed the book.

.   .   .   .

The Revolution Betrayed occupies a special place in Trotsky's literary work. It is the last book he managed to complete and, in a sense, his political testament. In it he gave his final analysis of Soviet society and a survey of its history up to the middle of the Stalin era. His most complex book, it combines all the weakness and the strength of his thought. It contains many new and original reflections on socialism, on the difficulties with which proletarian revolution has to grapple, and on the role of a bureaucracy in a workers' state. He also surveyed the international position of the Soviet Union before the Second World War and tried to pierce the future with daring and partly erroneous forecasts. The book is a profound theoretical treatise and a tract for the time; a creative restatement of classical Marxist views; and the manifesto of the 'new Trotskyism' calling for revolution in the Soviet Union. Trotsky appears here in all his capacities: as detached and rigorously objective thinker; as leader of a defeated Opposition; and as passionate pamphleteer and polemicist. The polemicist's contribution forms the more esoteric part of the work and tends to overshadow the objective and analytical argument. Because of the wealth of its ideas and its imaginative force, this has been one of the seminal books of this century, as instructive as confusing, and destined to be put to adventitious use more often

---

[1] This is what Mr. and Mrs. N. K. Dahl relate, who accompanied Trotsky on this trip. See also *The Case of Leon Trotsky*, pp. 204–23.

than any other piece of political writing. Even its title was to become one of the shibboleths of our time.

The Revolution Betrayed was Trotsky's critical reaction to a crucial moment of the Stalin era. Official Moscow had just proclaimed that the Soviet Union had already achieved socialism—until recently it had contented itself with the more modest claim that only 'the foundations of socialism' had been laid. What emboldened Stalin to proclaim nothing less than the advent of socialism was the progress of industrialization, the first superficial signs of the consolidation of collective farming, and the nation's fresh relief at having left behind the famines and massacres of the early nineteen-thirties. A new Constitution, 'the most democratic in the world', was to be the epitome of the new epoch: it nominally abolished discrimination against members of the former possessing classes, and introduced general and equal franchise for all. This presupposed that the proletarian dictatorship no longer needed any special constitutional guarantees, because a virtually classless society had come into being. Yet while it gave all citizens the equal right to vote, the Constitution deprived all of the right to choose for whom to vote, and unlike previous Soviet Constitutions, it formally consecrated the single party system. That system and the monolithic party, the propagandists maintained, conformed to the very nature of a socialist community, which was not torn by any conflict of class interests, whereas any multi-party system reflected the inherent antagonisms of bourgeois society.

Yet this was also a time of growing inequality, when discrepancies between high and low earnings widened rapidly, when 'socialist competition' degenerated into a wild scramble for privileges and necessities of life, when Stakhanovism carried that scramble to every factory bench and coal seam in the country, and when the contrast between the affluence of the few and the pauperism of the many took on most offensive forms. Stalin, conducting a ferocious drive against the 'levellers', placed himself at the head of the *nouveaux riches*, whetted their appetites, ridiculed the faint scruples that inhibited them, and glorified the new inequality as an accomplishment of socialism. A new hierarchical organization was taking shape. It was elaborately graded, with ranks, titles, and prerogatives sharply differentiated, and with every little rung on all the multiple steep ladders of authority marked out with bizarre precision. Nowhere was this reversal from earlier 'proletarian democratic' ways to the new authoritarianism as pronounced as in the armed forces, where the ranks and distinctions of Tsarist times were reintroduced. Amid the celebrations of the advent of socialism there was thus the flavour of something like Restoration in the air. The educational system and the nation's spiritual life were deeply affected. The progressive school reforms of the nineteen-twenties, which had aroused the admiration of many foreign educationists, were decried as ultra-left aberrations; and a heavy, increasingly nationalist traditionalism and an old-fashioned paternalistic discipline invaded classrooms and lecture halls, stifling the spirit of the young generation. The bureaucratic tutelage over science, literature, and the arts grew unbearably tyrannical. In every field the state exercised absolute power provocatively and brazenly, glorifying itself as the supreme guardian of society. And the autocratic bearer of power was exalted as Father of the Peoples, fount of all wisdom, benefactor of mankind, and demiurge of socialism.

Trotsky set out to refute Stalin's claims; and he did this by confronting the realities of Stalinism with the classical Marxist conception of socialism. He pointed out that the predominance of social forms of ownership did not yet constitute socialism, even though it was its essential condition. Socialism presupposed an economy of abundance; it could not be founded on the want and poverty that prevailed in the Soviet Union and that led to the recrudescence of glaring inequality. Stalin had invoked Marx's dictum about the two stages of communism, a lower one where society would reward its members 'each according to his work', and the higher where it would reward them 'each according to his needs'—it was at the lower stage, Stalin declared, that the Soviet Union found itself. Trotsky pointed out that Stalin was abusing the authority of Marx in order to justify the inequality he was promoting. While it was true that Marx had foreseen that inequality would persist in the early phase of socialism, it would not have occurred to him that it would grow, and even grow by leaps and bounds, as it did under Stalin's rule. Soviet society was still only half-way between capitalism and socialism. It could advance or slide back; and only to the extent to which it overcame inequality would it advance. The growth of inequality indicated backsliding. '

The orgies of Stalinist absolutism were part and parcel of the same retrograde trend. Lenin had, in his *State and Revolution*, wrested from oblivion the Marxian notion of the 'withering away of the state' and made of it the household idea of Bolshevism; and Trotsky now defended the idea against Stalinist manipulation. He insisted that socialism was inconceivable without the withering away of the state. It was from class conflict that the state had arisen; and it existed as an instrument of class domination. Even in its lower phase socialism meant the disappearance of class antagonisms and of political coercion—only the purely administrative functions of the state 'the management of things, not of men' were to survive under socialism. Lenin had imagined the proletarian dictatorship as a 'semi-state' only, modelled on the Commune of Paris, whose officials would be elected and deposed by vote and paid workers' wages, so that they should not form a bureaucracy estranged from the people. In backward and isolated Russia this scheme had proved unworkable. All the same, the advance towards socialism must be measured by the degree to which the coercive power of the state was on the decline. Massive political persecution and the glorification of the state in themselves refuted the Stalinist claim about the achievement of socialism. Stalin argued that the state could not wither away in a single country; to Trotsky this was only an indirect admission that socialism could not be achieved in a single country either. But it was not the 'capitalist encirclement' that was the chief reason for the increased power of the state, for the Stalinist terror aimed primarily at 'domestic enemies', i.e. at communist opposition.

To the non-Marxist much of this critique must seem 'doctrinaire'. To the Marxist it was vital because it stripped Stalinism of 'ideological' pretensions and dissociated Marxism from Stalin's practices. Trotsky sought to establish for the Marxist school of thought a position, from which it could disclaim the moral liabilities which Stalinism was creating for it, and from which it could declare that its ideas were no

more responsible for Stalin's reign of terror than the Ten
Commandments and the Sermon on the Mount had been for
the Holy Inquisition. Nor is the significance of this argument
only moral and historical, for it still has a profound bearing on
communist thinking. The notion, which Khrushchev has
expounded in the late nineteen-fifties and early nineteen-sixties,
that the Soviet Union is passing from socialism to communism
is predicated on the Stalinist claim about the achievement of
socialism in the nineteen-thirties, and is just as unreal as that
claim. Seen from Trotsky's standpoint Soviet society is, as yet,
despite its immense strides forward, very far from having
achieved socialism. As all the thinking of Soviet ideologues,
economists, sociologists, philosophers, and historians is still
entangled in the canon about the completion of socialism, and
is moving within a circle of fictions construed around that
canon, the application of Trotsky's criteria to present Soviet
reality would entail a revision of the legacy of Stalinism far
more thoroughgoing than that undertaken in the Soviet Union
in the first decade after Stalin.

The Revolution Betrayed is Trotsky's classical indictment of
bureaucracy. Once again, in the 'conflict between the ordinary
working woman and the bureaucrat who has seized her by
the throat' he 'sided with the working woman'. He saw the
mainspring of Stalinism in the defence of privilege, which alone
gave a certain unity to all the disparate aspects of Stalin's
policy, connecting its 'Thermidorian' spirit with its diplomacy
and the debasement of the Comintern. The ruling group
shielded the interests of an acquisitive minority against popular
discontent at home and the shocks of revolutionary class struggle
abroad. Trotsky analysed the social composition of the managerial groups, of the party machine, of the civil servants and of
the officer corps, who between them formed 12 to 15 per cent.
of the population, a massive stratum, conscious of its weight,
rendered conservative by privilege, and straining with all its
might to preserve the national and the international *status quo*.

Not content with indicting the bureaucracy, Trotsky considered again how and why it had achieved its power in the
Soviet Union and whether its predominance was not inherent
in socialist revolution at large. He went beyond his earlier
answers and threw into bolder relief the objective causes for

the recrudescence of inequality amid all the 'want and poverty' in the Soviet Union. But he also stated with emphasis that some of these factors would recur in every socialist revolution, for none would be able to abolish inequality immediately. Even the United States, the wealthiest industrial nation, did not yet produce enough to be able to reward labour 'according to needs'; it still suffered from a relative scarcity which would compel it, under communist government, to maintain differential wages and salaries. Consequently, tensions and social conflicts would persist, although they would be much milder than in an underdeveloped country. And so 'the tendencies of bureaucratism . . . would everywhere show themselves even after a proletarian revolution'.[1] Marx and Lenin had been aware of this. Marx had spoken of 'bourgeois law', safeguarding unequal distribution of goods, as being 'inevitable in the first phase of communist society'. Lenin had described the Soviet republic as being in some respects a 'bourgeois state without the bourgeoisie', even if it were governed in the spirit of proletarian democracy. But only the experience of the Stalin era had revealed the full dimensions of the problem and allowed real insight into the contradictions of post-capitalist society. A revolutionary government had to maintain inequality and had to struggle against it; and it had to do both for the sake of socialism. It had to provide incentives to technicians, skilled workers, and administrators in order to ensure the proper functioning and the rapid expansion of the economy; yet it had also to aim at the reduction and the eventual abolition of privileges.

Ultimately, this contradiction could be resolved only by an increase in social wealth, surpassing all that mankind had hitherto dreamt of, and by the attainment of so high and universal a level of education that the gulf between manual labour and intellectual work would vanish. In the meantime before these conditions are fulfilled, the revolutionary state assumes 'directly and from the very beginning a dual character': it is socialist in so far as it defends social property in the means of production; and it is bourgeois in so far as it directs an unequal, differential distribution of goods among the members of society. The clear formulation of this contradiction and

[1] *The Revolution Betrayed*, pp. 57–59.

duality as inherent in the transition to socialism is one of Trotsky's important contributions to the Marxist thought of his time.[1]

Returning to the analysis of Soviet society he admitted that Lenin and he had not foreseen that a 'bourgeois state without a bourgeoisie' would prove inconsistent with genuine Soviet democracy; and that the state could not 'wither away' as long as there was 'the iron necessity' for it to foster and support a privileged minority. The destruction of Soviet democracy was thus due not merely to Stalin's conspiracy, which was the subjective aspect of a wider objective process. He went on to say that the Stalinist government had preserved the 'dual character' inherent in any revolutionary government; but that the bourgeois element in it had gained immense weight and power at the expense of the socialist element. The bureaucracy was by its very nature 'the planter and protector of inequality'; it acted like a policeman who during an acute shortage of goods 'keeps order' while crowds queue up at food-shops—when food is abundant there are no queues and the policeman becomes superfluous. Yet 'nobody who has wealth to distribute ever omits himself. Thus out of a social necessity there has developed an organ which has far outgrown its socially necessary function, and has become an independent factor and therewith the source of great danger for the whole social organism. . . . The poverty and cultural backwardness of the masses have again become incarnate in the malignant figure of the ruler with the great club in his hand.'[2]

Had the bourgeois element in the Soviet state acquired enough force to destroy the socialist element? Trotsky asked. Once again he firmly rejected the view that the bureaucracy was a 'new class' or that the Soviet masses were exploited by 'state capitalism'. State capitalism without a capitalist class was to the Marxist a contradiction in terms. As for the bureaucracy, it lacked the social homogeneity of any class which owed its place in society to the ownership and the command of the means of production. The exercise of mere managerial functions had not turned the directors of the Soviet industry and state into such a class, even though they treated both state

[1] See in particular op. cit., chapter II: 'Socialism and the State.'
[2] Op. cit., p. 111.

and industry as if these were their private domains. The inequality which Stalinism promoted was still confined to the
sphere of private consumption. The privileged groups were
not permitted to appropriate means of production. Unlike
any exploiting class, they could not accumulate wealth in the
form that would give them command over the labour of others
and enable them to appropriate more and more wealth. Even
their privileges and power were bound up with the national
ownership of productive resources; and so they had to defend
that ownership and thereby to perform a function which, from
the socialist viewpoint, was necessary and progressive, though
they performed it at an exorbitant cost to society.

But the social balance of the Stalinist state, Trotsky went on,
was unstable. In the long run either the socialist element or the
bourgeois one must prevail. The continuous growth of inequality was a danger signal. The managerial groups would
not indefinitely content themselves with consumer privileges.
Sooner or later they would seek to form themselves into a
new possessing class by expropriating the state and becoming
the shareholding owners of trusts and concerns. 'One may argue
that the big bureaucrat cares little what are the prevailing
forms of property, provided only that they guarantee him the
necessary income. This argument ignores not only the stability
of the bureaucrat's own rights, but also the question of his
descendants. . . . Privileges have only half their worth if they
cannot be transmitted to one's children. But the right of testament is inseparable from the right of property. It is not enough
to be director of the trust; it is necessary to be a stockholder.
The victory of the bureaucracy in this decisive sphere would
mean its conversion into a new possessing class.' Stalin, Trotsky
pointed out, could not preside over this 'conversion'; his régime
was based on national ownership and a planned economy.
Turning into a new bourgeoisie, the bureaucracy would therefore necessarily come into conflict with Stalinism; and Stalin,
by encouraging its acquisitiveness, was unwittingly under-
mining not only his own rule, but all the conquests of the
revolution. So close did this danger appear to Trotsky that
he had no hesitation in stating that the 1936 Constitution
'creates the political premises for the birth of a new possessing
class'. As in the nineteen-twenties so in the nineteen-thirties, he

w

considered the bureaucracy, or a section of it, as the potential agent of a capitalist restoration; but while earlier he saw it as an auxiliary of the kulaks and the N.E.P. men, now, after the 'liquidation' of those classes, he regarded it as an independent agent.[1]

This view appears altogether erroneous in retrospect. Far from laying its hands on and appropriating the means of production, the Soviet bureaucracy was, in the coming decades, to remain the guardian of public ownership. It should be remarked, however, that Trotsky spoke of the bureaucracy's metamorphosis into a new bourgeoisie as of one of several possibilities; he was careful to point out that the potentiality should not be mistaken for actuality. He dealt, as he emphasized, with an unprecedented, complex, and enigmatic phenomenon, at a time when the Stalinist anti-egalitarianism and reaction against early Bolshevism were at the highest pitch. The theorist could take nothing for granted; he could not rule out the possibility that these trends might release powerful and independent forces utterly inimical to socialism. Stalin, representing an ambiguous combination of 'Leninist orthodoxy' with a revulsion against revolutionary principle, did indeed appear at times to lead Russia to the very brink of Restoration. That he could not cross that brink Trotsky had no doubt. He feared that others might cross it, even if over Stalin's body.[2]

The same fear, however, haunted Stalin as well; and this was why he raged against his own bureaucracy and, on the pretext of fighting Trotskyism and Bukharinism, decimated it in each of the successive purges. It was one of the effects of the purges that they prevented the managerial groups from consolidation as a social stratum. Stalin whetted their acquisitive instincts and wrung their necks. This was one of the most obscure, least discussed and yet important consequences of the permanent terror. While on the one hand the terror annihilated the old Bolshevik cadres and cowed the working class and the peasantry, it kept, on the other, the whole of the bureaucracy in a state of flux, renewing permanently its composition, and not allowing it to grow out of a protoplasmic or amoeboid condition, to form a compact and articulate body

[1] Op. cit., pp. 240, 257, and *passim*.   [2] Ibid., pp. 236–7.

with a socio-political identity of its own. In such circumstances the managerial groups could not become a new possessing class, even if they wanted to—they could not start accumulating capital on their own account while they were hovering between their offices and the concentration camps. Just as he had 'liquidated' the kulaks, so Stalin was constantly 'liquidating' the embryo of the new bourgeoisie; and in this he once again acted, in his own barbaric autocratic manner, from Trotsky's tacitly accepted premiss. In any case, the bureaucratic would-be bourgeoisie was no mere figment of Trotsky's imagination. But he patently exaggerated its vitality and capacity for self-realization, just as he had exaggerated the power of the kulaks; and he underrated once again Stalin's cunning, tenacity, and ruthlessness. The manner in which Stalin both promoted and repressed the bourgeois element in the state was utterly alien and even incomprehensible to Trotsky, who, as always, thought that only a conscious and active working class could check the anti-socialist tendencies of the state.

Yet Trotsky also realized that the Soviet workers were unwilling to rise against the bureaucracy, for even if they were hostile to it 'in their vast majority', they feared 'lest in throwing out the bureaucracy they would open the way for a capitalist restoration. . . .' The workers felt that for the time being 'the bureaucracy continues to fulfil a necessary function' as the 'watchman' guarding some of *their* conquests. 'They will inevitably drive out the dishonest, impudent, and unreliable watchman as soon as they see another possibility.' What a paradox this was! The same social group which might turn into a new possessing class and destroy the revolution was to some extent the revolution's protector. Trotsky knew that 'doctrinaires would not be satisfied' with his appraisal of the situation: 'They would like categorical formulas: yes—yes, and no—no'; and, of course, sociological analysis would be simple 'if social phenomena had always a finished character'. But he refused to force realities into any neat scheme and to give 'for the sake of logical completeness' 'a finished definition to an unfinished process'. Confronted by a completely new and 'dynamic social formation', the theorist could produce only working hypotheses and let events test them.[1]

[1] Ibid., pp. 241–2, 269–70.

Events disproved the hypothesis about the transformation of the bureaucracy into a new possessing class already in the nineteen-thirties; but even more so during and after the Second World War. Then the needs of national defence and the destruction of the bourgeois order in eastern Europe and China powerfully reinforced the nationalized structure of the Soviet economy. The Stalinist state, by promoting or assisting for its own reasons revolution in eastern Europe and Asia, created formidable counter-checks to its own bourgeois tendencies. The post-war industrialization, the immense expansion of the Soviet working class, the growth of mass education, and the reviving self-assurance of the workers tended to subdue the bourgeois element in the state; and after Stalin's death the bureaucracy was compelled to make concession after concession to the egalitarianism of the masses. To be sure, the tension between the bourgeois and the socialist elements of the state continued; and, being inherent in the structure of any post-capitalist society, it was bound to persist for a very long time to come. The managers, the administrators, the technicians, and the skilled workers remained privileged groups. But the gulf between them and the great mass of the toilers was narrowing in the middle and late nineteen-fifties and the early nineteen-sixties; and so the balance between the contradictory elements in the state was very different from what it had been when Trotsky wrote *The Revolution Betrayed*. Trotsky himself anticipated such a development:

> Two opposite tendencies are growing up out of the depth of the Soviet régime. To the extent that, in contrast to a decaying capitalism, [that régime] develops the productive forces, it is preparing the economic basis of socialism. To the extent that, for the benefit of an upper stratum, it carries to more and more extreme expression bourgeois norms of distribution, it is preparing a capitalist restoration. This contrast between forms of property and norms of distribution cannot grow indefinitely. Either the bourgeois norms must in one form or another spread to the means of production, or the norms of distribution must be brought into correspondence with the socialist property system.[1]

It is this latter course that events were to take twenty and twenty-five years later, when Stalin's successors began grudg-

[1] Op. cit., pp. 231–2.

ingly yet unmistakably to bring the norms of distribution into closer correspondence with the socialist property system. Trotsky's hypothesis about the rise of a new possessing class appears therefore unduly pessimistic, even though it reflected a situation in which the balance was strongly and dangerously weighted against the socialist elements. Yet, despite the 'pessimism', Trotsky's analysis of the dynamic contradictions of the post-revolutionary state still offers the best clue to the subsequent social evolution.

It was against a 'greedy, mendacious, and cynical caste of rulers', against the germ of a new possessing class, that Trotsky formulated his programme of a 'political revolution' in the U.S.S.R. 'There is no peaceful outcome . . .' he wrote. 'The Soviet bureaucracy will not give up its positions without a fight . . . no devil has ever yet voluntarily cut off his own claws.' 'The proletariat of a backward country was fated to accomplish the first socialist revolution. For this historic privilege it must, according to all the evidence, pay with a second supplementary revolution—against bureaucratic absolutism.' He preached 'a political, not a social revolution', a revolution, that is, which would overthrow the Stalinist system of government, but would not change the existing property relations.[1]

This was a completely new prospect: Marxists had never imagined that after a socialist revolution they would have to call upon the workers to rise again, for they had taken it for granted that a workers' state could be only a proletarian democracy. History had now demonstrated that this was not so; and that, just as the bourgeois order had developed various forms of government, monarchical and republican, constitutional and autocratic, so the workers' state could exist in various political forms, ranging from a bureaucratic absolutism to government by democratic Soviets. And just as the French bourgeoisie had to 'supplement' the social revolution of 1789–93 by the political revolutions of 1830 and 1848, in which ruling groups and methods of government were changed but not the economic structure of society—so, Trotsky argued, the working class too had to 'supplement' the October Revolution. The bourgeoisie had acted consistently within its class

[1] Op. cit., pp. 271–2.

interest when it asserted itself against its own absolutist rulers; and the working class would also act legitimately in freeing its own state from a despotic stranglehold. A political revolution of this kind had, of course, nothing to do with terroristic acts: 'Individual terror is a weapon of impatient and despairing individuals, belonging most frequently to the young generation of the bureaucracy itself.' For Marxists it was axiomatic that they could carry out the revolution only with the open support of the majority of the workers. It was therefore not with a call for any immediate action that Trotsky came out, for as long as the workers saw in the bureaucracy the 'watchman of their conquests', they would not rise against it. Trotsky advanced the idea, not the slogan, of a revolution; he offered a long-term orientation for the struggle against Stalinism, not guidance for direct action.

This is how he formulated the programme of the revolution:

It is not a question of substituting one ruling clique for another, but of changing the very methods of administering the economy and guiding the culture of the country. Bureaucratic autocracy must give place to Soviet democracy. A restoration of the right of criticism and genuine freedom of elections is the necessary condition for the further development of the country. This assumes a revival of freedom of Soviet parties, beginning with the party of Bolsheviks, and a renascence of the trade unions. The bringing of democracy into industry means a radical revision of plans in the interests of the toilers. Free discussion of economic problems will decrease the overhead expense of bureaucratic mistakes and zigzags. Expensive playthings—Palaces of the Soviets, new theatres, showy Metro subways—will be abandoned in favour of workers' dwellings. 'Bourgeois norms of distribution' will be confined within the limits of strict necessity, and, in step with the growth of social wealth, will give way to socialist equality. Ranks will be immediately abolished. The tinsel of decorations will go into the melting pot. Youth will receive the opportunity to breathe freely, criticise, make mistakes, and grow up. Science and art will be freed of their chains. And, finally, foreign policy will return to the traditions of revolutionary internationalism.[1]

He reiterated here all the familiar desiderata of the period when he still stood for reform. Only in one point did he make a new departure—namely, in his demand for 'genuine freedom

[1] Op. cit., p. 273.

of elections'. On this point, however, he was confronted with a dilemma: he had discarded the principle of the single party; but he did not advocate unqualified freedom of parties. Going back to a pre-1921 formula, he spoke of a 'revival of freedom of *Soviet* parties', that is of the parties that 'stood on the ground of the October Revolution'. But who was to determine which were and which were not 'Soviet parties'? Should the Mensheviks, for instance, be allowed to benefit from the 'revived' freedom? He left these questions in suspense, no doubt because he held that they could not be resolved in advance, regardless of circumstances. He was similarly cautious in discussing equality: he did not speak of any 'abolition' of 'bourgeois norms of distribution'—these were to be maintained, but only 'within the limits of strict necessity'; and dispensed with gradually, 'with the growth of social wealth'. The political revolution was thus to leave some privileges to managers, administrators, technicians, and skilled workers. As he himself sometimes, in polemical utterances, spoke loosely of the 'overthrow' or 'abolition' of bureaucracy, this gloss put the problem in a more realistic perspective. What he envisaged on calm reflection was a drastic curtailment, not the obliteration, of bureaucratic and managerial privilege.

Over a quarter of a century after its formulation, this programme has remained relevant; and most of its ideas have reappeared in the post-Stalinist movement of reform. Yet the question must be asked whether in insisting on the necessity of a political revolution in the U.S.S.R. Trotsky had not taken too dogmatic a view of the prospect and, against his own advice, given 'too finished a definition to an unfinished process'. From the tenor of *The Revolution Betrayed* it is clear that he saw no chance of any reform from above; and there was indeed no chance of it in his lifetime and for the rest of the Stalin era. But during that time there was no chance in the Soviet Union of any political revolution either. This was a period of deadlock: it was impossible either to cut or to untie the Gordian knots of Stalinism. Any programme of change, whether revolutionary or reformist, was illusory. This could not prevent a fighter like Trotsky from searching for a way out. But he was searching within a vicious circle, which only world-shaking events began to breach many years later. And when that happened the Soviet Union moved away from Stalinism through reform from above in the first instance. What forced the reform was precisely the factors on which Trotsky had banked: economic progress, the cultural rise of the masses, and the end of Soviet isolation. The break with Stalinism could only be piecemeal, because at the end of the Stalin era there existed and could exist no political force capable and willing to act in a revolutionary manner. Moreover, throughout the first decade after Stalin there did not emerge 'from below' any autonomous and articulate mass movement even for reform. Since Stalinism had become an anachronism, nationally and internationally, and a break with it had become an historic necessity for the Soviet Union, the ruling group itself had to take the initiative of the break. Thus, by an irony of history Stalin's epigones began the liquidation of Stalinism and thereby carried out, *malgré eux mêmes*, parts of Trotsky's political testament.[1]

But can they continue this work and complete it? Or is a political revolution still necessary? On the face of it, the chances of revolution are still as slender as they were in Trotsky's days, whereas the possibilities of reform are far more real. The conditions for any revolution, as Lenin once put it, are (*a*) that the rulers should not be able to go on ruling as they used to; (*b*) that the ruled, in their misery, despair, and fury, should refuse to go on living as before; and (*c*) that there should exist a revolutionary party determined and able to seize its chance. These conditions are not likely to materialize in a country with a vital and expanding economy and with rising standards of living, when the masses, having unprece-

---

[1] I underlined this circumstance in my book *Russia After Stalin* (1953) and in many articles published just at the end of the Stalin era. The American Trotskyists then devoted a whole issue of their theoretical organ *The Fourth International* (Winter 1954) to the theme: '*Trotsky or Deutscher*'; and James P. Cannon, their leader, vehemently denounced me as a 'revisionist' and as 'the Bernstein of Trotskyism'. My sin was that I forecast that in the next few years there would be no chance for a 'political revolution' in the U.S.S.R. and that a period of 'reform from above' was opening. (This was indeed to be the chief political characteristic of the first decade after Stalin.) I based my argument, *inter alia*, on the fact that the extermination of all oppositions, especially of the Trotskyist Opposition, had left Soviet society politically amorphous, inarticulate, and incapable of 'initiative from below'. It was paradoxical that Trotskyists in the West should have been so utterly unaware of this consequence of the extermination of the Trotskyists (and other anti-Stalinist Bolsheviks) in the U.S.S.R.

dented access to education, see before them prospects of continuous social advance. In such a nation any conflict between popular aspirations and the selfishness of a ruling group, a conflict under which Soviet society is still labouring, is more likely to give rise to pressure for continuous reform than to lead to a revolutionary explosion. History may therefore yet vindicate the Trotsky who had for twelve or thirteen years struggled for reform rather than the Trotsky who, in his last five years, preached revolution.

This, however, can be only a tentative conclusion. The problem of a bureaucracy in a workers' state is indeed so new and complex that it allows little or no certitude. We cannot determine in advance how far a bureaucracy can go in yielding up privileges; what strength and effectiveness popular pressure for reform can acquire under a single party system; and whether a 'monolithic' régime can gradually dissolve and transform itself into one allowing freedom of expression and association on a socialist basis. How far do the social tensions inherent in 'primitive socialist accumulation' soften or abate as the accumulation loses its primitive, forcible, and antagonistic character? To what extent does the rise in popular well-being and education resolve antagonisms between the bureaucracy and the people? Only experience, in which there may be more surprises than are dreamt of in any philosophy, can provide the answer. At any rate, the present writer prefers to leave the final judgement on Trotsky's idea of a political revolution to a historian of the next generation.

.   .   .   .

Mention should be made here of the revision, which Trotsky carried out in *The Revolution Betrayed*, of his conception of the Soviet Thermidor. We have described earlier the passions and the turbulence which this abstruse historical analogy had aroused in the Bolshevik party in the nineteen-twenties; and we have said that this was a case of *le mort saisit le vif*.[1] About ten years later we find Trotsky, under a Norwegian village roof, still wrestling with the French phantom of 1794. We remember that as long as he stood for reform in the Soviet Union, he rejected the view, originally held by the Workers'

[1] *The Prophet Unarmed*, pp. 312–14.

Opposition, that the Russian Revolution had already declined into the Thermidorian or post-Thermidorian phase. Thermidor, he argued, was the danger with which Stalin's policy was fraught, but not yet an accomplished fact. He still defended this attitude against friend and foe alike in the first years of his banishment. But having decided that the Opposition must become an independent party and that political revolution was inevitable in the Soviet Union, he thought again and stated that the Soviet Union had long since been living in the post-Thermidorian epoch.[1]

He admitted that the historical analogy had done more to obfuscate minds than to enlighten them; yet he went on elaborating it. He and his friends, he argued, had committed a mistake in thinking that Thermidor amounted to a counter-revolution and restoration; and having so defined it, they had been right in insisting that no Thermidor had occurred in Russia. But the definition was wrong and unhistoric: the original Thermidor had not been a counter-revolution, but only 'a phase of reaction *within* the revolution'. The Thermidorians had not destroyed the social basis of the French Revolution, the new bourgeois property relations, that had taken shape in 1789–93; but they had on that basis set up their anti-popular rule and set the stage for the Consulate and the Empire. The comparable development in the Soviet Union occurred as early as 1923, when Stalin suppressed the Left Opposition and established his anti-proletarian régime on the social foundations of the October Revolution. With the calendar of the French Revolution before his eyes all the time, Trotsky went on to say that Stalin's rule having assumed a Bonapartist character, the Soviet Union was living under its Consulate. Within this perspective the danger of restoration appeared all too real—in France twenty years had passed between Thermidor and the return of the Bourbons; and Trotsky's call for a new revolution and a return to Soviet democracy echoed the cry raised by the Conspiracy of Equals for a return to the First Republic.

---

[1] Trotsky first 'revised' his Thermidor analogy in an essay 'The Workers State, Thermidor, and Bonapartism', written during the latter part of his stay in France and published in *B.O.*, no. 43 in April 1935. That essay contained in a nutshell the argument of *The Revolution Betrayed*.

Thus, Trotsky involved himself deeper and deeper in that 'summoning up of the ghosts of the past' which Marx had seen as a peculiar feature of bourgeois revolutions. The English Puritans had conjured up the prophets of the Old Testament; and the Jacobins the heroes and the virtues of Republican Rome. In doing so, Marx said, they did not just 'parody the past', but 'genuinely strove to rediscover the spirit of revolution'.[1] Marx was confident that a socialist revolution would not need to borrow its costumes from the past because it would have a clear awareness of its own character and purpose. And indeed, in 1917 the Bolsheviks did not dress up in such costumes and had no use for the pageantry and the symbols of earlier revolutions. In later years, however, they derived from Jacobinism all their nightmares and fears, the nightmares of the *épurations* and the fears of Thermidor; and they magnified these by their own actions and in their own imagination. They did so not from sheer imitativeness, but because they were struggling with similar predicaments and sought to master them differently. They consulted the gloomy experiences of the past in order to avoid their repetition. And although it is true that the Bolsheviks did not escape the horrors of a fratricidal struggle in their midst, yet they did manage to avoid the whole fatal cycle through which Jacobinism had moved to its doom and through which the French Revolution was driven to its end. The fear of Thermidor that haunted the Bolsheviks was a reflex of self-defence and self-preservation. But the reflex often worked irrationally. Trotsky now admitted that for more than ten years the Opposition had raised the alarm about Thermidor without perceiving clearly the meaning of the precedent Thermidor represented. Was he himself more clear about it now?

The original Thermidor was one of the most involved, many-faceted, and enigmatic events in modern history; and this accounts partly for the confusion about it. The Thermidorians overthrew Robespierre after a series of internecine Jacobin struggles, in the course of which Robespierre, leading the centre of his party, had destroyed its right and left wings, the Dantonists and Hebertists. The end of his rule marked the downfall of his faction and of the Jacobin party at large.

[1] *The Eighteenth Brumaire of Louis Bonaparte.*

Soon after Thermidor the Jacobin Club was disbanded and ceased to exist. The Thermidorians replaced Robespierre's 'reign of terror' by the rule of 'law and order' and inflicted final defeat on the plebs of Paris, which had suffered many reverses even earlier. They abolished the quasi-egalitarian distribution of food, which Robespierre had maintained by fixing 'maximum' prices. Henceforth, the bourgeoisie was free to trade profitably, to amass fortunes, and to gain the social dominance which it was to preserve even under the Empire. Thus, against the background of ebbing revolutionary energies and of disillusionment and apathy in the masses, the revolutionary régime passed from the popular to the anti-popular phase.

It is enough to outline briefly these various aspects of Thermidor to see where Trotsky was wrong in his assertion that Russia had gone through her Thermidor in 1923. The defeat of the Opposition in that year was not in any sense an event comparable to the collapse and dissolution of the Jacobin party; it corresponded rather to the defeat of the left Jacobins which had taken place well before Thermidor. While Trotsky was writing *The Revolution Betrayed* the Soviet Union was on the eve of the great purge trials—in France the *épurations* were part and parcel of the Jacobin period; only after Robespierre's downfall was the guillotine brought to a halt. Thermidor was in fact an explosion of despair with the permanent purge; and most of the Thermidorians were ex-Dantonists and ex-Hebertists who had survived the slaughter of their factions. The Russian analogy to this would have been a successful coup against Stalin carried out, after the trials of 1936–8, by remnants of the Bukharinist and Trotskyist oppositions.

Another difference is even more important: Thermidor brought to a close the revolutionary transformation of French society and the upheaval in property. In the Soviet Union these did not come to a halt with Stalin's ascendancy. On the contrary, the most violent upheaval, collectivization of farming, was carried out under his rule. And it was surely not 'law and order', even in a most anti-popular form, that prevailed either in 1923, or at any time during the Stalin era. What the early nineteen-twenties had in common with the Thermidorian period was the ebbing away of the popular revolutionary energies and the disillusionment and apathy of the masses. It was against such a background that Robespierre had sought to keep the rump of the Jacobin Party in power and failed; and that Stalin struggled to preserve the dictatorship of the Bolshevik rump (i.e. of his own faction) and succeeded.

Admittedly, there was a strong Thermidorian flavour about Stalin's anti-egalitarianism. But that was not absent from Lenin's N.E.P. either. Curiously, when in 1921 the Mensheviks described N.E.P. as the 'Soviet Thermidor', neither Lenin nor Trotsky protested. On the contrary, they congratulated themselves on having carried out something like Thermidor peacefully, without breaking up their own party and losing power. 'It was not they [the Mensheviks]', Trotsky wrote in 1921, 'but we ourselves who formulated this diagnosis. And, what is more important, the concessions to the Thermidorian mood and tendencies of the petty bourgeoisie, necessary for the purpose of maintaining the power of the proletariat, were made by the Communist party without effecting a break in the system and without quitting the helm.'[1] Stalin also made the most far-reaching 'concessions to the Thermidorian moods and tendencies' of his bureaucracy and managerial groups, 'without effecting a break in the system and without quitting the helm'. In any case, an historical analogy which led Trotsky, in 1921, almost to boast that he and Lenin had carried out a semi-Thermidor, then to deny that any Soviet Thermidor had occurred, and finally, in 1935, to maintain that the Soviet Union had for twelve years lived under a Thermidor, without Trotsky himself noticing it—such an analogy did indeed serve more to obfuscate minds than to enlighten them.

The historically far more justified charge that Trotsky could have levelled against Stalin was that he instituted a reign of terror like Robespierre's, and that he had monstrously outdone Robespierre. However, Trotsky's own past and the Bolshevik tradition did not allow him to say this. It will be remembered that in 1903–4, when he first dissociated himself from Bolshevism, he levelled the accusation of Jacobinism against Lenin; and in reply Lenin proudly identified himself as the 'proletarian Jacobin' of the twentieth century.[2] The

[1] Trotsky, *Between Red and White*, p. 77. (Trotsky concluded the writing of this book in February 1922.)
[2] See *The Prophet Armed*, pp. 91–97.

two men were thinking of two different Robespierres: Lenin of the one who had secured the triumph of the revolution against the Gironde, Trotsky of the one who had sent his own comrades to the guillotine. Not only in Lenin's eyes, but in those of most western Marxists, the Conductor of the Purges had, after a century, receded behind the great Incorruptible hallowed in the Pantheon of the Revolution. Trotsky the Bolshevik regretted that he had ever raised the charge of Robespierrism against Lenin; and he was wary of throwing it at Stalin. Having in the meantime accepted the Bolshevik glorification of Jacobinism, he virtually identified himself with Robespierre; and this led him to see his enemies as Thermidorians, which they were not. True, his alarms did much to rouse all Bolsheviks, including the Stalinists, to vigilance. Moreover, something of the Thermidorian mood still survives in the Soviet Union; and it can be found (together with the 'bourgeois element' and 'bourgeois norms of distribution') in any workers' state. All the same, we who have seen, in the nineteen-forties and nineteen-fifties, the Russian Revolution in its full Protean power, by far surpasssing the French Revolution in scale and momentum—we can only wonder over the strange *quid pro quo* through which the Thermidorian phantom strayed on to the Russian scene and kept itself there for a whole historic epoch.

.   .   .   .

The pessimism, real and apparent, underlying *The Revolution Betrayed* shows itself also in those pages where Trotsky tried to anticipate the impact of the Second World War on the Soviet Union. He noted that the new social system had provided 'national defence with advantages of which the old Russia could not dream'; that in a planned economy it was relatively easy to switch from civilian to military production and 'to focus on the interests of defence even in building and equipping new factories'. He underlined the progress of the Soviet armed forces in all modern weapons and stated that 'the correlation between the living and mechanical forces of the Red Army may be considered by and large as on a level with the best armies of the West'.[1] This was not, in 1936, a view

[1] *The Revolution Betrayed*, pp. 196–7.

generally accepted by western military experts; and the emphasis with which Trotsky expressed it was undoubtedly calculated to impress the Governments and the General Staffs of the western powers. But he saw the weakness of the Soviet defences in the Thermidorian spirit of its officer corps, in the army's rigidly hierarchical structure which was replacing its revolutionary-democratic organization, and above all in Stalin's foreign policy. He argued that Stalin, having first neglected the danger from the Third Reich, was now, to counter it, relying mainly on alliances with western bourgeois Governments, on the League of Nations, and on 'collective security', for the sake of which he would in case of war refrain from making any genuinely revolutionary, appeal to the armed workers and peasants of the belligerent nations.

'Can we . . .' Trotsky asked, 'expect that the Soviet Union will come out of the approaching great war without defeat? To this frankly posed question we will answer as frankly: if the war should remain only a war, the defeat of the Soviet Union would be inevitable. In a technical, economic, and military sense, imperialism is incomparably stronger. If it is not paralysed by revolution in the West, imperialism will sweep away the régime which issued from the October Revolution.'[1] Divided though the West was against itself, it would eventually unite 'in order to block the military victory of the Soviet Union'. Well before the Munich crisis, Trotsky observed that France was already treating her alliance with the Soviet Union as a 'scrap of paper' and she would continue to do so, no matter how much Stalin tried to secure the alliance through the Popular Front. Only if Stalin were to yield further to French, British and American economic and political pressures, would the alliance assume reality; but even then the allies would take advantage of the Soviet Union's wartime difficulties and seek to sap the socialist foundations of its economy and exact far-reaching concessions to capitalism. At the same time the peasantry's individualism, stirred up by war, would threaten to disrupt collective farming. These external and domestic pressures, Trotsky concluded, would bring the danger of counter-revolution and restoration closer to Russia. The situation was not hopeless, however, because the war

[1] Op. cit., p. 216.

would also bring revolution closer to Europe; and so, on balance, 'the Soviet régime would have more stability than the régimes of its probable enemies'. 'The Polish bourgeoisie' could only 'hasten the war and find in it . . . certain death'; and 'Hitler has far less chance than had Wilhelm II of carrying a war to victory'. Trotsky's confidence in European revolution was as strong as was his despondency about the prospects of the Soviet Union in the absence of such a revolution:

The danger of war and defeat of the Soviet Union is a reality, but the revolution is also a reality. If the revolution does not prevent war, then war will help the revolution. Second births are commonly easier than first. In the new war it will not be necessary to wait a whole two years and a half for the first insurrection [as it was after 1914]. Once it is begun, moreover, the revolution will not this time stop half way. The fate of the Soviet Union will be decided in the long run not on the maps of the General Staffs, but on the map of the class struggle. Only the European proletariat, implacably opposing its bourgeoisie . . . can protect the Soviet Union from destruction, or from an 'allied' stab in the back. Even a military defeat of the Soviet Union would be only a short episode, if there were to be a victory of the proletariat in other countries. And, on the other hand, no military victory can save the inheritance of the October revolution if imperialism holds out in the rest of the world. . . . Without the Red Army the Soviet Union would be crushed and dismembered like China. Only its stubborn and heroic resistance to the future capitalist enemy can create favourable conditions for the development of the class struggle in the imperialist camp. The Red Army is thus a factor of immense significance. But this does not mean that it is the sole historic factor.

It is not under the banner of the *status quo* [which Stalin's diplomacy defended in the nineteen-thirties] that the European workers and the colonial peoples can rise. . . . The task of the European proletariat is not the perpetuation of boundaries, but, on the contrary, their revolutionary abolition, not [the preservation of] the *status quo* but a socialist United States of Europe.[1]

The outcome of the Second World War was to be far less clear cut than this alternative; and nothing would be easier than to compile from *The Revolution Betrayed* a list of Trotsky's errors in prognostication. Yet each of his errors contains important elements of truth and follows from premises which

[1] Op. cit., pp. 219–20.

retain validity; and so more can still be learned from his mistakes than from the correct platitudes of most political writers. Trotsky is in this respect not unlike Marx: his thought is 'algebraically' correct, even when his 'arithmetical' conclusions are wrong. Where his forecasts were erroneous, they were so because too often he viewed the Second World War in terms of the first; but his general insights into the relationship between war and revolution were deep and are still essential to an understanding of the revolutionary aftermath of the Second World War.[1]

The Revolution Betrayed has exercised its influence in a strange, often self-defeating, manner—*pro captu lectoris*. It was published in May 1937, right in the middle of the slaughter of the Old Bolsheviks, just after the trial of Radek, Pyatakov, and Sokolnikov and on the eve of the execution of Tukhachevsky and the other Generals. The volleys of Stalin's execution squads gave a peculiar resonance to the title of the book: it came as a desperate and piercing cry of protest. Focusing all of Trotsky's tragic invective, it suggested that the October Revolution had suffered its last and irretrievable débâcle and that Trotsky and his followers had abandoned all allegiance to the Soviet Union. Thus, the 'revolution betrayed' became a startling, memorable, yet vacuous slogan; and for a long time the title page of the book made a stronger impression than the book itself; often it closed minds to Trotsky's complex and subtle argument. His speculations about the possible emergence of a new possessing class caught readers' attention to the exclusion of his qualifying clauses and counter-balancing ideas. Quite a few of his disciples saw actuality where he saw mere potentiality. The very brilliance of his controversial style helped to produce this distorting response, for it tempted hosts of lesser writers to imitate the master's invective, which was so much easier to do than to enter critically into his thought. Not only did *The Revolution Betrayed* become the Bible of latter-day Trotskyist sects and chapels, whose members piously mumbled its verses long after Trotsky's death. The effect of the book was felt more widely, in the literature of disillusionment produced by western ex-communists in the nineteen-forties and nineteen-fifties. Some of them lived on

---

[1] The argument is further developed in the Postscript to this volume.

mere crumbs, and not the best ones, from Trotsky's rich table; and they gained a reputation for originality by serving these up in their own brands of sauce. James Burnham, a Trotskyist in the nineteen-thirties, based his *Managerial Revolution* on a few fragments of Trotsky's theory torn out of context.[1] *The Revolution Betrayed* re-echoes through the early writings of Ignazio Silone and Arthur Koestler. George Orwell was strongly impressed by it. The fragments of 'The Book', which take up so many pages in his *1984* were intended to paraphrase *The Revolution Betrayed* just as Emmanuel Goldstein, Big Brother's enigmatic antagonist, is modelled on Trotsky. And last but not least, in the nineteen-forties, and nineteen-fifties, many of the intellectually ambitious 'Sovietologists' and propagandists of the cold war drew, directly or indirectly, their arguments and catch phrases from this source.[2]

Despite the adventitious use made of it, *The Revolution Betrayed* remains a classic of Marxist literature. But this is Trotsky's most difficult book; and only the reader who approaches it with discrimination, without accepting or rejecting it *in toto*, can benefit from it. Goethe once said of Lessing that, being the greatest thinker of his generation, his influence on contemporaries was only slight and partly even harmful, because only an intelligence equal to Lessing's could absorb the full complexity of his thought; he therefore swayed the mind of Germany only indirectly and posthumously. This is also true of the author of *The Revolution Betrayed*, and accounts for the distorted and distorting influence of this book in the West. In our time, however, its ideas are already in the air in the U.S.S.R., where Trotsky's writings are still banned. The Soviet Jourdains who nowadays unknowingly speak his prose are legion: they are to be found in universities, factories, literary clubs, Komsomol cells, and even in the ruling circles. To give only a few random illustrations: Trotsky's verdict that the Stalin era 'will go down in the history of artistic creation pre-

---

[1] See further, pp. 471–5.

[2] In 1961 an American Government agency brought out a pamphlet under the title *The Revolution Betrayed*, the purpose of which was to justify the American campaign against Cuba. The man whom the State Department, the Pentagon, ex-owners of Cuban sugar plantations, and some 'radicals' denounced as traitor to the revolution was Fidel Castro. An American sponsored invasion of Cuba aimed presumably at restoring to the Cuban revolution its pristine purity.

eminently as an epoch of mediocrities, laureates, and toadies' has come to be generally accepted. Who does not now agree with him that under Stalinism 'the literary schools were strangled one after the other' and that

The process of extermination took place in all ideological spheres, and it took place more decisively since it was more than half unconscious. The present ruling stratum considers itself called not only to control spiritual creation politically, but also to prescribe its roads of development. The method of command-without-appeal extends in like measure to the concentration camps, to scientific agriculture, and to music. The central organ of the party prints anonymous directive editorials having the character of military orders, in architecture, literature, dramatic arts, the ballet, to say nothing of philosophy, natural science, and history. The bureaucracy superstitiously fears whatever does not serve it directly, as well as whatever it does not understand.[1]

If fortunately not all of this is any longer true, much of it still is; and as critic of the legacy of Stalinism the dead Trotsky still speaks more powerfully than all the living 'de-Stalinizers':

The school and the social life of the student are saturated with formalism and hypocrisy. The children have learned to sit through innumerable deadly dull meetings, with their inevitable honorary presidium, their chants in honour of the dear leaders, their pre-digested debates in which, quite in the manner of their elders, they say one thing and think another. . . . The more thoughtful teachers and children's writers, in spite of the enforced optimism, cannot always conceal their horror in the presence of this spirit of repression, falsity, and boredom. . . . Independent character, like independent thought, cannot develop without criticism. The Soviet youth, however, are simply denied the elementary opportunity to exchange thoughts, make mistakes, and try out and correct mistakes, their own as well as others'. All questions. . . . are decided for them. Theirs only to carry out the decision and sing the glory of those who made it. . . . This explains the fact that out of the millions upon millions of communist youth there has not emerged a single major figure.

In throwing themselves into engineering, science, literature, sport, or chess playing, the young people are, so to speak, winning their spurs for future great action. In all these spheres they compete with the badly prepared older generation, and often equal and beat

[1] Trotsky, *The Revolution Betrayed*, p. 173.

them. But at every contact with politics they burn their fingers.

And how alive still is the prophetic anger, faith, and vision which inspired words like these:[1]

> . . . the actual establishment of a socialist society can and will be achieved, not by these humiliating measures of a backward capitalism, to which the Soviet government is resorting, but by methods more worthy of a liberated humanity—and above all not under the whip of a bureaucracy. For this very whip is the most disgusting inheritance of the old world. It will have to be broken into pieces and burned at a public bonfire before one can speak of socialism without a blush of shame.

.   .   .   .

The months during which Trotsky wrote *The Revolution Betrayed* were, despite intense work, a respite. Life at Vexhall was uneventful and tranquil. The daily routine was rarely interrupted for visitors or for an outing in the bare and rocky countryside to the north. Once a week the Trotskys and Knudsens went to the cinema at Honnefoss to view an old and faded American film. So well did Trotsky progress with his work that, having concluded *The Revolution Betrayed*, he looked forward to taking up the *Lenin* at once. He had found, so it seemed, the security of a real asylum at last. Yet now and then a small cloud showed itself. Elections were due in the autumn; and already in the summer a small pro-Nazi party, the National Sammling, had begun to attack the government for jeopardizing peace and prosperity by harbouring Trotsky. The party's leader was Major Quisling, who a few years later, under German occupation, was to become head of a puppet government, and whose name then became the by-word for 'collaboration' with the occupant. At this time, however, his following was small and it belonged to the lunatic fringe; little notice was therefore taken. More disturbing were the attacks of *Arbeideren*, the communist paper. Although it too had few readers, it voiced the views of the Soviet Embassy, when it charged Trotsky with using Norway as 'a base for terroristic activities directed against the Soviet Union and its leaders, above all against the greatest leader of the world proletariat in our time—Stalin. . . .' 'How long', the paper asked,

[1] Op. cit., p. 125.

'will the Norwegian workers tolerate this? What has the Central Bureau of the Norwegian Labour Party to say? What has the Norwegian Government to say?' This was the first time it was alleged that Trotsky 'was using Norway as a base for terroristic activities'—the charge was to be taken up by Vyshinsky a few months later.

The Labour party firmly rejected the allegation. 'What is the purpose of this?' Schöffle replied. 'To make the Norwegian workers believe a lie . . . and to compel the Labour government to place Trotsky under arrest? Well, gentlemen, neither will happen. You will not so easily make fools either of Norwegian workers or of the Norwegian Labour government. . . .' Other spokesmen for the party in office replied in the same vein.[1]

The Norwegian police nevertheless kept Trotsky under surveillance and regularly reported not only their own findings but communications received from the Belgian and the French police to the Minister of Justice. A Sherlock Holmes in Brussels had discovered that Trotsky was the actual inspirer and leader of the Fourth International; and at Oslo police headquarters cautious minds inquired whether that disquieting piece of information was correct. The French police confirmed it and expressed concern over the comings and goings of Trotsky's secretaries, all agents of the Fourth International. The Norwegian Ministers could only be amused by this feat of detection—a little earlier they themselves, or some among them, might even have been inclined to join the subversive organization. All the same, to appease his police, the Minister of Justice ordered the deportation of Jan Fraenkel, one of Trotsky's secretaries. His place, however, was soon taken by Erwin Wolf, who stayed at Vexhall for about a year unmolested and married Knudsen's daughter. To avoid needless irritation, Trotsky asked his followers to delete his name from the list of the 'International Executive' of their organization; and he published articles on internal Trotskyist affairs anonymously or under a pen-name.[2] He refused to give interviews to

[1] *Arbeideren*, 12 December and *Soerlandet*, 16 December 1935.

[2] 'Crux' was the pen-name Trotsky used most often at this time; he also conducted part of his correspondence with his adherents in Paris and Amsterdam in code. The key to the code is preserved in *The Archives*, Closed Section. See also Krog, op. cit., pp. 245–6

foreign newspapers. And so scrupulously did he avoid even the
slightest involvement in Norwegian politics that when Knudsen,
who stood for Parliament, invited him to attend his election
meetings as spectator, Trotsky refused; he used to accompany
Knudsen and wait for him outside, in his car, until the meeting
was over.[1] The police dutifully reported to the Minister that
Trotsky's behaviour was in this respect irreproachable. 'We
knew, of course, that Trotsky continued to write his commentaries on international affairs', says Koht, the Foreign
Minister, 'but we considered it our duty to respect his right
to do so under the democratic principle of asylum.'[2] The
government was so satisfied that it twice prolonged Trotsky's
residence permit automatically, without raising any question.

Nevertheless, when in the summer of 1936 Koht went on a
mission to Moscow and was ostentatiously fêted there, Trotsky
awaited his return with misgivings. 'They are bargaining over
my head in the Kremlin', he said to Knudsen. 'Do you believe',
Knudsen asked with shocked incredulity, 'that we, the Norwegian Labour party, are ready to sell your head?' 'No',
Trotsky replied, sparing the feelings of his host, 'but I believe
that Stalin is ready to buy it.'[3] According to Koht himself,
he had gone to Moscow only on a courtesy visit: having previously been in Warsaw as guest of the Polish Government,
he had been anxious to avoid giving Moscow the impression
that he had 'ganged up' with the Poles. During his visit, he
says, the question of Trotsky's asylum was never raised—only
once in Geneva, at a session of the League of Nations, had
Litvinov blandly alluded to it in a private talk.[4] Koht's testimony may well be accepted: Stalin would hardly have bargained over Trotsky's head with Koht, a gentle and somewhat
unworldly scholar-diplomat—for that he had to find a much
tougher character.

Trotsky's suspicion arose out of the stupendous growth of

[1] This is Knudsen's account given to me.

[2] Professor Koht made this statement early in 1937; and he emphatically
repeated it to me during my visit in Oslo in 1956.

[3] Trotsky, *Stalin's Verbrechen; The Case of Leon Trotsky*, p. 33; Knudsen's statement to the writer.

[4] This is how Koht explained his motives to me (adding that he had long
been in contact with Moscow's academic circles in connexion with his research on
the early history of Russo-Norwegian relations).

the anti-Trotskyist terror in the Soviet Union. He had recently received first-hand accounts about this from three followers who had come straight from Soviet prisons and concentration camps. They were: A. Tarov, a Russian worker and old Bolshevik; Anton Ciliga, former member of the Politbureau of the Yugoslav Communist party; and Victor Serge, to whose role in the Russian Opposition we have frequently referred.[1] Serge owed his freedom to Romain Rolland's personal intervention with Stalin; Ciliga was released at the instance of western European friends; and Tarov had secretly crossed the frontier. Tarov related that, impressed by the rise of Nazism, he had been ready to make his peace with Stalinism and had negotiated with the G.P.U. over the terms of his capitulation. 'Do you agree or do you not', they asked him, 'that Trotsky is the chief of the vanguard of bourgeois counter-revolution?' This was the formula the capitulators were now required to accept. Tarov replied that to his mind 'Trotsky is the man most devoted to the cause of the world proletariat, an unflinching revolutionary, whom I consider my friend and comrade in a common cause'. Throughout many a night he was interrogated and pressed to renounce Trotsky; but he could not bring himself to do that.[2]

All three described the new, cataclysmic violence of the terror: the huge concentration camps set up all over the U.S.S.R.; the pitiless brutality with which the inmates were being treated since Kirov's assassination; and the torture and deceit by which the G.P.U. extracted 'confessions'. For all the severity of his criticisms of Stalin, Trotsky had not been fully aware how far things had gone. Like any political émigré, he had to some extent preserved the image of his country as he had known it, when the terror had been much narrower in scope and milder. The new accounts (and André Gide's just published *Retour de l'U.R.S.S.*) filled him with shame and

---

[1] *The Archives*, Closed Section. Tarov's 'Letter of an Escaped Bolshevik-Leninist' appeared in *B.O.*, no. 45, 1935. Ciliga's account of the Stalinist terror is in *B.O.*, nos. 47, 48, 49. Victor Serge's 'Open Letter to André Gide', exposing the Stalinist régime to Gide, who was then still favourably disposed towards Stalin, appeared in *B.O.*, no. 51, 1936. These issues of the *B.O.* contained an abundance of fresh information from the U.S.S.R. See also the correspondence between Lyova and Serge of April 1936 in *The Archives*, Closed Section.

[2] Tarov, loc. cit.

anger, and confirmed him in his determination to renounce all
'reformist illusions', and to give the sharpest possible expression
to his break with the Comintern.

These reports, it should be added, left hardly any ray of
hope for the Opposition, for while they dwelt on the depravity
of the ruling group and the hatred and contempt which surrounded it, they described also, in the grimmest terms, the total
dispersal and impotence of the Opposition.[1] It must have
been only a bitter consolation for Trotsky to learn how people
like Tarov still defended his honour in the dungeons and
prison camps. These people appeared to be the last Mohicans
of the Opposition. Yet, before the end of 1935, fresh mass
expulsions from the party were announced. On 30 December,
Khrushchev, then Secretary of the Moscow Committee,
stated that in the capital alone 10,000 members had been
expelled; from Leningrad Zhdanov reported the expulsion of
7,000. All over the country at least 40,000 people had been
deprived of membership; many more had been expelled from
the Komsomol; and most were branded as Trotskyists and Zinovievists. Even if only one-half or one-third of this mass had
been genuine oppositionists, their numbers would have been far
greater than the 4–6,000 who had put their signatures to the
Platform of the Joint Opposition in 1927.[2] Was this a new tide?
Trotsky wondered; and, despite Serge's and Ciliga's depressing accounts, he struck an optimistic note:

. . . under the influence of the Stalinist press and its agents (of
the type of Louis Fischer and his like) not merely our enemies but
many of our friends in the West without noticing it have become
accustomed to thinking that if Bolshevik-Leninists still exist in the
U.S.S.R., they do so only as hard labour convicts. No, this is not
so! It is impossible to eradicate the Marxist programme and a great
revolutionary tradition by police methods. . . . If not as a doctrine
then as a mood, a tradition, and a banner, our movement has now
a mass character in the U.S.S.R. and is evidently absorbing new
and fresh forces. Among the 10 to 20,000 'Trotskyists' expelled in

---

[1] Ciliga presently came out with a full-scale description of the situation in his
book, *Au pays du grand mensonge*. Serge in his correspondence with Lyova also
described the disintegration of the Opposition. Such, according to the old Elzin,
an eminent Trotskyist (whom Serge quoted), was the disarray that 'no two comrades can be found to hold the same view—what unites us is the G.P.U.'

[2] *The Prophet Unarmed*, p. 370.

the last months there are no more than a few tens, perhaps a few hundreds . . . of men of the older generation, oppositionists of the 1923–8 vintages. The mass is made up of new recruits. . . . It can be said with confidence that in spite of thirteen years of baiting, slander, and persecution, unsurpassed in wickedness and savagery, in spite of capitulations and defections, more dangerous than persecution, the Fourth International possesses already today its strongest, most numerous, and most hardened branch in the U.S.S.R.[1]

This seemed to contradict Trotsky's earlier resigned statements that no revolutionary initiative could be expected from the Soviet Union, even from his followers. As a 'mood, tradition, and banner' even if not as an organized party, Trotskyism was still as alive as ever. And both Stalin and Trotsky knew that in favourable circumstances 'a mood and a tradition' could easily cohere into a party. Stalin was therefore preparing his final onslaught on Trotskyism. Meanwhile, in the spring and early summer of 1936, there was still an uneasy lull.

In western Europe this was the hey-day of the Popular Front. The parties of the Popular Front had gained an overwhelming electoral victory in France; and this encouraged the workers to raise demands, join trade unions by the million, occupy factories, and stage nationwide strikes and demonstrations. 'The French revolution has begun', Trotsky proclaimed in the title of an article he wrote for the American *Nation*. (The conservative *Le Temps* spoke of '*les grandes manœuvres de la révolution*'.) He pointed to the collapse of the French economy, the sharpening of all class antagonisms, the panic in the possessing classes and their parties, and the impetus of the mass movement. 'The whole working class has begun to move. This gigantic mass will not be halted by words. The struggle is bound to end either with supreme victory or with the most terrible of defeats.' The leaders of the Popular Front courted defeat; they did what they could to subdue the energy and the self-confidence of the workers and to reassure the bourgeoisie. 'The Socialists and Communists had been working with all their strength for a Ministry headed by Herriot, at the worst by Daladier. What have the masses done? They have

[1] *B.O.*, no. 48, 1936.

imposed on them Blum's Ministry. Does this not amount to a direct vote against the policy of the Popular Front?' For the time being counter-revolution lay low, waiting for the storm to blow over and preparing a comeback. 'It would be frivolous to maintain that its calculations are groundless. With the help of Blum, Jouhaux, and Cachin, the counter-revolution may yet achieve its purpose.' For years the Communist party had clamoured *Les Soviets partout*; but now, when it was time to pass from words to deeds, to rally and arm the workers, and to form Workers' Councils, it declared the slogan to be 'untimely'. He also addressed this warning to his own followers: 'The party or group which cannot find a foothold in the present strike movement and establish solid ties with the embattled workers is not worthy of the name of a revolutionary organization.' Not for the first and not for the last time his followers were unable to find the 'foothold'.

On 4 August, having just mailed to his publishers his Preface to *The Revolution Betrayed*, Trotsky left with Knudsen for a holiday, which they intended to spend on a wild and deserted little island in a southern fjord. They travelled by car and on the way Knudsen noticed that a few men, whom he recognized as Quisling's adherents, were pursuing them. At a ferry, however, he managed to put them off the track; and pleased with this, he and Trotsky crossed the fjord, reached the island and settled down for the night in a fisherman's hut.

Next morning they were aroused by an urgent message from Vexhall. During the night Quisling's followers, disguised as policemen, had broken into Knudsen's house and, claiming that they had orders to carry out a search, had tried to force their way into Trotsky's rooms. Knudsen's daughter, suspecting a fraud, resisted them, while her brother alarmed neighbours. The intruders fled, having seized only a few sheets of typescript from a table. Apprehended by the police, they declared that they had planned to break into the house during Trotsky's absence, and that, having tapped Knudsen's telephone, they had known when he and Trotsky would be away. There was no question then of any attempt on Trotsky's life. Their purpose was to obtain evidence of Trotsky's political activity and of his transgression against the terms of his residence in Norway, evidence which Quisling's party intended to use in the elections. The intruders claimed that they had achieved their purpose.

The incident seemed ludicrous. Trotsky was sure that Quisling's men could not possibly have obtained proof of a transgression he had not committed. Nor could they have seized anything of importance from his archives, which Knudsen had, as a precaution, placed in a bank safe before the departure. And so, after a moment of excitement, he and Knudsen went back to climbing the rocks and to fishing. A week later, on 13 or 14 August, a small aircraft landed on the island; and from it emerged the chief of the Norwegian criminal police. He had come on Trygve Lie's orders to interrogate Trotsky in connexion with the forthcoming trial of Quisling's men. The questions concerned the papers the latter had seized at Knudsen's house, a copy of Trotsky's private letter to a French follower and his article 'The French revolution has begun', to which we have just referred. Trotsky answered all questions put to him; and the police officer left to tell the Press that he had found the Nazi charges against Trotsky absolutely groundless.[1]

Early next morning Knudsen listened as usual to the news. The reception was indistinct: there was no electricity on the island, and he had only a small portable wireless set. But what he heard was enough to send him breathless to Trotsky: Moscow had just announced that Zinoviev, Kamenev, and fourteen other defendants would presently stand trial, charged with treason, conspiracy, and attempts at the assassination of Stalin. A long indictment was then broadcast which branded Trotsky as their chief abettor. Knudsen was not sure of the details, but he had no doubt that Zinoviev and Kamenev were accused of terrorism and also of collusion with the Gestapo. Trotsky was dumbfounded. 'Terrorism? Terrorism?', he kept on repeating. 'Well, I can still understand this charge. But Gestapo? Did they say Gestapo? Are you sure of this?' he asked in amazement. 'Yes, this is what they said', Knudsen confirmed. Later in the day they learned that the indictment

[1] Trotsky, *Stalin's Verbrechen*; Krog, op. cit.: *The Archives*; statements by Knudsen and various official Norwegian personalities.

also claimed that it was from Norway that Trotsky was des-
patching terrorists and assassins to the Soviet Union. They
felt as if the rocks of the tranquil island had suddenly errupted
with flames and lava. They rushed back to Vexhall.

On the same day, 15 August, Trotsky refuted the charges,
describing them to the Press as 'the greatest forgery in the
world's political history'. 'Stalin is staging this trial in order
to suppress discontent and opposition. The ruling bureaucracy treats every criticism and every form of opposition as
conspiracy.' The charge that he was using Norway as a base for
terroristic activity, he said, was designed to rob him of asylum
and of the possibility of defending himself. 'I emphatically
assert that since I have been in Norway I have had no connexion with the Soviet Union. I have not received here even
a single letter from there, nor have I written to anyone either
directly or through other persons. My wife and I have not
been able to exchange even a single line with our son, who was
employed as a scientist and has had no political connexion
with us whatsoever.' He proposed that the Norwegian Government should investigate the charges—he was ready to place
before it all relevant papers and materials. And he also
appealed to the labour organizations of all countries for an
impartial and international Commission of Inquiry.[1]

Thus the culmination of the terror, which he had so many
times predicted, had come. It was more hideous and more
menacing than anything he had foreseen. His ears once again
glued to the wireless set, he listened, from 19 to 24 August, to
the accounts of the trial. Hour by hour he absorbed its horror,
as prosecutor, judges, and defendants acted out a spectacle, so
hallucinatory in its masochism and sadism that it seemed to
surpass human imagination. It was clear from the outset that
the heads of the sixteen defendants were at stake, and with them
the heads of Trotsky and Lyova. (In the indictment Lyova
figured as his father's chief assistant.) As the proceedings went
on, it became obvious that the trial could only be the prelude
to the destruction of an entire generation of revolutionaries.
But worst of all was the manner in which the defendants were
dragged through the mud, and made to crawl to their death
amid indescribably nauseating denunciations and self-de-

[1] Quoted from the originals in *The Archives*.

nunciations. Compared with this all the nightmares of the French Revolution, the tumbrils, the guillotine, and the Jacobins' fratricidal struggles, looked now like a drama of almost sober and solemn dignity. Robespierre had put his adversaries in the dock amid thieves and felons and had loaded them with fantastic accusations; but he had not prevented them from defending their honour and dying as fighters. Danton was at least free to exclaim: 'After me it will be your turn, Robespierre!' Stalin hurled his broken adversaries to unfathomable depths of self-humiliation. He made the leaders and thinkers of Bolshevism behave like the wretched medieval women who had to relate to the Inquisition every act of their witchcraft and every detail of their debauchery with the Devil. Here, for instance, is Vyshinsky's dialogue with Kamenev, conducted in the hearing of the whole world:

Vyshinsky: What appraisal should be given to the articles and statements you wrote in which you expressed loyalty to the Party? Was this deception?
Kamenev: No, it was worse than deception.
Vyshinsky: Perfidy?
Kamenev: Worse than that.
Vyshinsky: Worse than deception, worse than perfidy? Then find the word for it. Was it treason?
Kamenev: You have found the word.
Vyshinsky: Defendant Zinoviev, do you confirm this?
Zinoviev: Yes.
Vyshinsky: Treason? Perfidy? Double-dealing?
Zinoviev: Yes.

And this was how Kamenev wound up his *mea culpa*:

Twice my life was spared, but there is a limit to everything, there is a limit to the magnanimity of the proletariat, and that limit we have reached. . . . We are sitting here side by side with the agents of foreign secret police departments. Our weapons were the same, our arms became intertwined, before our fate became intertwined here, in this dock. We have served fascism, we have organized counter-revolution against socialism. Such has been the path we took, and such is the pit of contemptible treachery into which we have fallen.[1]

---

[1] *Sudebnyi Otchet po Delu Trotskistkovo-Zinovievskovo Terroristskovo Tsentra;* the quotations are from the official English version of the proceedings, pp. 68, 169–70.

Zinoviev followed:

> I am guilty of having been organizer, second only to Trotsky, of the Trotskyist-Zinovievist bloc, which set itself the aim of assassinating Stalin, Voroshilov, and other leaders. . . . I plead guilty to having been the principal organizer of the assassination of Kirov. We entered into an alliance with Trotsky. My defective Bolshevism became transformed into anti-Bolshevism and through Trotskyism I arrived at fascism. Trotskyism is a variety of fascism, and Zinovievism is a variety of Trotskyism.[1]

Ivan Smirnov, who had defeated Kolchak in the civil war and had sat by Trotsky's side on the Revolutionary Military Council, stated:

> There is no other path for our country but the one that it is now treading; and there is not, nor can there be, any other leadership than that which history has given us. Trotsky, who sends directions and instructions on terrorism and regards our state as a fascist state, is an enemy. He is on the other side of the barricade.[2]

Mrachkovsky, another one of Trotsky's old companions and also a hero of the civil war, said:

> Why did I take the counter-revolutionary path? My connexion with Trotsky brought me to this. From the time I made that connexion I began to deceive the party, to deceive its leaders.[3]

Bakayev, the intrepid chief of the Leningrad Cheka during the civil war and leader of the Opposition's demonstrations in 1927, confessed:

> The facts revealed before this court show to the whole world that the organizer of this . . . counter-revolutionary terrorist bloc, its moving spirit, is Trotsky. . . . I have staked my head over and over again in the interests of Zinoviev and Kamenev. I am deeply oppressed by the thought that I became an obedient tool in their hands, an agent of counter-revolution, and that I raised my arm against Stalin.[4]

For hours, Vyshinsky, the ex-Menshevik who had climbed on the Bolshevik band-wagon well after the civil war, and was now Prosecutor-General, fumed and raged in a deliberate affectation of hysteria:

[1] Op. cit., p. 170.   [2] Op. cit., pp. 171–2.   [3] Op. cit., p. 165.   [4] Op. cit., p. 168.

These mad dogs of capitalism tried to tear limb from limb the best of the best of our Soviet land. They killed one of the men of the revolution who was most dear to us, that admirable and wonderful man, bright and joyous as the smile on his lips was always bright and joyous, as our new life is bright and joyous. They killed our Kirov, they wounded us close to our very heart. . . . The enemy is cunning, a cunning enemy must not be spared. . . . Our whole people is quivering with indignation; and on behalf of the State Prosecution I am joining my angry and indignant voice to the rumbling voices of millions. . . . I demand that dogs gone mad should be shot, every one of them![1]

After five days filled with coarse vituperation and obscene insults, days during which the prosecution had not submitted a single piece of evidence, the court pronounced a verdict condemning all defendants to death and concluding that:

Lev Davidovich Trotsky and his son Lev Lvovich Sedov . . . convicted . . . as having directly prepared and personally directed the organization in the U.S.S.R. of terroristic acts . . . are subject, in the event of their being discovered on the territory of the U.S.S.R., to immediate arrest and trial by the Military Collegium of the Supreme Court of the U.S.S.R.[2]

Stalin had timed the trial to be staged just after Hitler's march into the Rhineland and shortly after the Popular Front had formed its government in France. In doing so he blackmailed the labour movement and the left intelligentsia of the West, who looked to him as their ally against Hitler. He threatened in effect that if there were any protests against his purges, he would retaliate by breaking up the Popular Front and leaving western Europe alone to face the Third Reich. He was assisted in his purpose by the sombre irrationality of the trial, which confounded people who might have raised their voices against an infamy they understood, but were utterly reluctant to protest against a dark and bloody mystery and thereby to become involved in it.

Oppressive though the trial and the executions were, they aroused in Trotsky all his fighting spirit. He was determined to meet the challenge with all the concentrated power and confidence with which he had once directed the first battles of

[1] Op. cit., pp. 120, 164.　　　[2] Op. cit., p. 180.

the civil war. He had been the chief defendant in the Zinoviev-Kamenev trial; and he knew that there would be further trials, in which he would be made to bear an ever heavier load of ever more stupendous accusations. He fought for his head and honour, for his surviving children, and for the dignity of all the doomed old Bolsheviks who could not defend themselves. He showed up contradictions and absurdities with which the trial was riddled. He strained every nerve to expose its falsehood and to shatter its mystery. He knew that he stood alone against Stalin's huge power and the legions of propagandists who served it. But at least he was free to speak and organize his counteraction; and he was determined to make the fullest use of this. On the second day of the trial he gave an exhaustive interview to *Arbeiderbladet*, which published it next day, 21 August, on the front page (under the title 'Trotsky claims that Moscow's accusations are false'), and left its readers in no doubt about its sympathy with Trotsky's case. He prepared statements for American, British, and French telegraph agencies and for many reporters, who rushed to Oslo. He was in the thick of battle; and time was of its essence: He had to refute Stalin's accusations before the world's amazed and shocked sensitivity was blunted. All that he needed was the freedom to defend himself.

Of that freedom he was suddenly and insidiously robbed; and those who robbed him of it were the men who had just professed friendship for him, had honoured him, and had prided themselves in having given him refuge. On 26 August, a day after the end of the Moscow trial, two senior police officers called on him to tell him, on the orders of the Minister of Justice, that he had offended against the terms of his residence permit. They asked him to sign an undertaking that henceforth he would refrain from interfering 'directly or indirectly, orally and in writing, in political questions current in other countries'; and that as author he would 'strictly limit his activity to historical works and general theoretical observations not directed towards any specific country'.[1] The demand sounded like a mockery. How could he refrain from expressing

---

[1] Norwegian Ministry of Justice and Police, *Storting Report no. 19*, submitted by Mr. Trygve Lie on 18 February 1937; Krog, op. cit.; Trotsky, '*Ich fordere ein Gerichtsverfahren über mich!*' in *The Archives*; and *Stalin's Verbrechen*.

himself on 'questions current in other countries' now, when Stalin had denounced him as Hitler's accomplice and ringleader of a gang of wreckers and assassins? How could he confine himself to 'theoretical observations not directed towards any specific country'? His silence would only lend colour to all the calumny against him which Stalin was drumming into the world's ears. He flatly refused to sign. Thereupon the police put him under house arrest, placed guards at his doors, and forbade him to make any statement for publication.

What accounted for this sudden change in the Norwegian Government's attitude? On 29 August, Yakubovich, the Soviet Ambassador, delivered in Oslo a formal Note demanding Trotsky's expulsion. The Note insisted that Trotsky was using Norway as 'the base for his conspiracy'; it invoked the verdict of Moscow's Supreme Court; and it ended with this slightly veiled threat: 'The Soviet Government wishes to state that the continued granting of asylum to Trotsky . . . will . . . impair friendly relations between the U.S.S.R. and Norway and will violate . . . rules governing international intercourse.'[1] This was three days after Trotsky had been placed under house arrest, a circumstance which enabled Trygve Lie to maintain that it was not because of Soviet intervention that he had taken action against Trotsky. However, the Soviet Ambassador had already asked for Trotsky's expulsion a few days earlier in an oral *démarche*. 'The difficulty', says Koht, 'in establishing the exact date when the Soviet Ambassador first asked that we should deny Trotsky asylum arises from the fact that he did this in an oral communication, of which no record seems to exist. I was away from Oslo at that time, touring my constituency in the far north; and at the Ministry of Foreign Affairs Trygve Lie acted as my deputy.'[2] In fact, the Ambassador saw Trygve Lie shortly after *Arbeiderbladet* had published its interview with Trotsky about the Moscow trial; it is inconceivable that he should not have protested against the publication of the interview in the paper of the ruling party and demanded that Trotsky be denied asylum. Oslo was astir with rumours that he also threatened to cut off trade with

[1] The Note is attached as an appendix to the *Storting Report no. 19. Izvestya* of 30 August 1936 spoke merely of a statement by the Soviet Ambassador.
[2] Koht's statement to the writer.

Norway; and that the shipping companies and the fishing industry were pressing the government not to endanger their interests at a time of slump and unemployment. 'My colleagues in the government', says Koht, 'were afraid of economic reprisals, although the Russians did not say that they would apply them. I did not believe that they would resort to a commercial boycott; and I held that, in any case, our trade with Russia—herring was our main export—was not large enough for us to be afraid. I was therefore opposed to the proposal that we should intern Trotsky; but I was overridden by my colleagues in the Cabinet.'[1]

The Ministers were afraid of a break with Russia and of losing the elections over this issue. And so, although they knew that the allegation that Trotsky was using Norway as a base for terroristic activities was sheer humbug, and although they denied it in their reply to the Soviet Note, they yielded to the pressure. They could not, however, expel Trotsky, because no other country would accept him. Nor could they hand him over to the Soviet Government, which did not ask for extradition, despite the fact that Trotsky had defied Stalin to ask for it. (Such a demand would have necessitated a hearing of the case before a Norwegian court; and this would have given Trotsky the opportunity to refute the charges.) Afraid of incensing Moscow by allowing Trotsky to conduct his defence in public, the Ministers therefore decided to intern him. Democratic conscience and ministerial self-importance, however, did not allow them to admit that they were yielding to threats and that in their own country they could not give shelter to a man of whose innocence they were convinced and whose greatness they had extolled. They had therefore to cast a slur upon his innocence. They did not dare to take up Vyshinsky's accusations, for although they lacked the courage to stand up for the truth, they did not have the audacity to embrace so big a falsehood either. They were small men capable of telling a small lie only. They decided to accuse Trotsky of having abused their confidence by engaging in criticisms of foreign governments and by being involved with the Fourth International, although they admitted that none of these activities was illegal. They now looked for proof of

1 Idem.

his misconduct. But where was it to be found? At Oslo's Court, Quisling's men were flaunting from the dock the few sheets of paper they had managed to snatch at Knudsen's house, the copy of Trotsky's article 'The French Revolution has begun'. Had he not attacked in it the French Popular Front and Blum's Ministry? Was this not an activity 'directed against a friendly government'? There was, however, nothing clandestine or illegal about it: the article had appeared in the American *Nation* and in two small Trotskyist periodicals, *Verité* and *Unser Wort*: and it would be unseemly for Labour Ministers to make use of papers stolen from Trotsky's desk by Quisling's men. The Minister of Justice had in his files the police reports about Trotsky's contacts with the Fourth International. But the government had taken these contacts for granted and had shrugged off the police reports as recently as June, when they gladly prolonged his residence permit. Wherever they turned they could find no respectable motive for denying him asylum.

Yet deny it they had to, even if the legal motivation was to be bungled. As the days passed and Moscow's wrath grew louder and louder, they became more and more terrified to see their Lilliputian interests and reputations involved in a contest of giants; and they cursed the hour when they had allowed the man-mountain to come into their country. He was in their hands, however, and they were free to make him their prisoner. They did this fumblingly, ashamed of turning into Stalin's accomplices. But, to quote a Norwegian writer: 'A guilty conscience and the sense of shame seldom lead a wrongdoer to penitence . . . he must obtain an imaginary justification of his misdeeds. And it is not unusual for the wrongdoer to come to hate his victim.'[1] And the *amour propre* of the Ministers had been so enormously flattered when they acted as hosts to 'Lenin's closest companion' that they grew fretful and irrascible when they became his jailers.

On 28 August, Trotsky appeared, under police escort, at the Court of Oslo to give, for a second time, testimony in the case of Quisling's men. He found himself almost at once put into the position of a defendant rather than of a witness. Quisling's men pleaded that they had exposed his 'disloyal'

[1] Krog, op. cit., p. 220.

behaviour in Norway; and the presiding judge subjected him to close questioning. Had he during his stay in Norway engaged in correspondence with his comrades abroad? Had he offered them political guidance? Had he criticized any foreign government in his articles? Trotsky answered all these questions in the affirmative, although they had no legal bearing on the case before the court, which was whether the men in the dock, in disguising themselves as policemen and breaking into Knudsen's house, had been guilty of fraud and burglary. The judge then declared that Trotsky had, on his own showing, violated the terms on which he had been admitted into the country. Trotsky replied that he had never assumed the obligation to refrain from expressing his views and communicating with his comrades; and that he was prepared to prove there and then that he had not engaged in any illegal or conspiratorial activity. At this point the judge interrupted him and ordered him to leave the witness stand.

Straight from the courtroom the police took him to the Ministry of Justice, where the Minister, surrounded by officials, asked him peremptorily to sign on the spot the following statement:

I, Leon Trotsky, declare that I, my wife, and my secretaries shall not engage, while in Norway, in any political activity directed against any state friendly to Norway. I declare that I will reside at such place as the government may select or approve . . . that I, my wife, and my secretaries, will in no way . . . involve ourselves in political questions current either in Norway or abroad . . . that my activities as author shall be limited to historical works, biographies, and memoirs . . . that [my] . . . writings of a theoretical nature . . . shall not be directed against any government of any foreign state. I further agree that all mail, telegrams, telephone calls, sent or received by myself, my wife, and my secretaries be censored. . . .[1]

Twenty years later eye-witnesses of the scene still remembered the flashes of scorn in Trotsky's eyes and the thunder of his voice as he refused to comply. How, he asked, did the Minister dare to submit to him so shameful a document? Did he really expect a man of his, Trotsky's, record, to sign it? What the Minister asked of him was complete submission and renunciation of any right to express a political opinion. Had he,

[1] *Storting Report no. 19.*

Trotsky, ever been prepared to accept such terms he would not have been in exile now and would not be dependent on Norway's dubious hospitality. Did Trygve Lie believe himself to be so powerful as to obtain from him what Stalin could never obtain? In admitting him to the country the Norwegian Government knew who he was—how then dare they ask that even his theoretical writings should not be directed against any foreign government? Had he ever allowed himself even the most trivial interference in Norwegian affairs—had they the slightest reproach against him on this count? The Minister admitted that they had none. Did they then believe that he was using Norway as a base for terroristic activities? No, the government definitely refused to believe that, Trygve Lie answered. Did they accuse him of conspiratorial or illegal action against any foreign government? No, the Minister replied again, there was no question of any conspiratorial or illegal activity. The government's case against Trotsky was that he had broken his pledge to refrain from any political activity; and his article 'The French Revolution has begun' and his contacts with the Fourth International were proof of this. Trotsky denied having ever given such a pledge. No communist, no socialist can ever commit himself to refrain from all political activity. What sort of idea did the Minister have of socialism and socialist morality? In what respect was the article on France more reprehensible than the interview for *Arbeiderbladet* he, Trotsky, had given to Trygve Lie himself, when Lie assured him that by expressing his political opinion he would not be offending against the terms of the residence permit? And how dare the government base the charge against him on a document supplied by Nazi burglars? Were they allowing a gang of Hitler's stooges to determine their conduct?

At this point Trotsky raised his voice so that it resounded through the halls and corridors of the Ministry: 'This is your first act of surrender to Nazism in your own country. You will pay for this. You think yourselves secure and free to deal with a political exile as you please. But the day is near—remember this!—the day is near when the Nazis will drive you from your country, all of you together with your *Pantoffel-Minister-President*.' Trygve Lie shrugged at this odd piece of soothsaying. Yet after less than four years the same government had indeed to flee from Norway before the Nazi invasion; and as the Ministers and their aged King Haakon stood on the coast, huddled together and waiting anxiously for a boat that was to take them to England, they recalled with awe Trotsky's words as a prophet's curse come true.[1]

After this encounter Trygve Lie put Trotsky in more stringent internment, deported his two secretaries, and placed guards inside Knudsen's house, so as to prevent Trotsky from communicating even with Knudsen. In ordering all this he exceeded his powers, for the Norwegian Constitution did not allow him to deprive of freedom any person not convicted by a court of justice. Many people, including Conservatives, were scandalized and protested; and so, three days after he had ordered Trotsky's arrest, Lie obtained the King's signature for a decree which invested him with extra-constitutional power for this exceptional occasion; and on 2 September he ordered that Trotsky and Natalya be transferred from Knudsen's house to Sundby, in Hurum, to a fjord twenty miles to the south of Oslo, where they were interned in a small house which the Ministry had rented for this purpose. Guarded day and night, they had to share the house with twenty jack-booted, stamping, pipe-smoking, and card-playing policemen. No one was allowed to visit Trotsky, except his Norwegian lawyer— even his French lawyer was not admitted. He was denied the prisoner's normal right to take physical exercise or a short walk outdoors. To obtain a newspaper he had to apply for special permission; and he had to submit all his correspondence to censorship. The censor was a member of Quisling's party; and so was one of the two officers in charge of the guard, Jonas Lie, who was to become chief of police under Quisling's govern-

---

[1] In his War Memoirs Professor Koht thus describes the scene: 'After the meeting [with the King and the German Ambassador] I called together the members of Parliament . . . and explained to them the new German demands. . . . I had no doubt that the government would turn these down . . . that we would be on the run again . . . and would have to get out of the country. I remembered the words Trotsky had said to Trygve Lie . . .: "In a few years you and your government will be political refugees, without home and country, as I am now." We had brushed his words aside, such things had seemed to us utterly impossible. . . . Several times I had to interrupt my speech to keep back my tears.' *Barricade to Barricade* (Norwegian edition), p. 47. Norwegian parliamentarians who witnessed this scene have described it to me in the same terms. One of them maintains that it was King Haakon who reminded Trygve Lie of 'Trotsky's curse'.

ment. 'Trotsky's isolation was so strict', Knudsen recalls, 'that Trygve Lie repeatedly refused me permission to go to Hurum, even after I had become a Member of Parliament. Only after much trouble and delay was I allowed to send Trotsky a wireless set—at first he had been forbidden even to listen to the radio.'[1]

All this was to prevent Trotsky from replying to Stalin's charges. Yet he did not give in. He wrote articles exposing in detail the trial of Zinoviev and Kamenev; and in letters to his followers and to Lyova he instructed them how to conduct a campaign against the purges and how to assemble factual evidence disproving every count of Vyshinsky's indictment. Under protest he submitted the articles and letters to the censor, and then for weeks waited impatiently for answers. None came. The censor confiscated all his writings without informing him. Meantime, Trotsky and Natalya listened day after day to Moscow radio as it thundered·out the accusations and made these re-echo through the world like an apocalyptic cacophony. How many people, Trotsky wondered, were by now beginning to recover from their first shocked astonishment and to give credence to the incredible? Had not the huge clouds of poison-ous dust that had risen from Moscow begun to settle on peoples' minds and harden into a crust? The fact that the Norwegian Government had seen fit to intern him inevitably prejudiced many against him: people reasoned that if he had been quite innocent then surely his friends, the Norwegian socialists, would not have deprived him of freedom. His very silence seemed to cry out against him; and his enemies made the most of it. Barely a fortnight after the internment, Vyshinsky pointed out in *Bolshevik* that Trotsky evidently had nothing to say in self-defence, for otherwise he would have spoken up.[2]

Straining in the trap, Trotsky then tried to sue for defamation two Norwegian editors, a Nazi and a Stalinist who had in their newspapers, *Vrit Volk* and *Arbeideren*, endorsed Vyshinsky's accusations. On 6 October, Puntervold, his Norwegian lawyer, initiated the suit. The court had already issued the sum-monses—the case was to be heard before the end of the month—

---

[1] I am quoting the words that Knudsen spoke to me.
[2] *Bolshevik*, 15 September 1936. Lyova reported this indignantly to his father in a letter of 26 October.

when the government stopped the proceedings. Having interned Trotsky so as to make it impossible for him to answer Stalin, the government could not allow him now to use the court as his forum. Yet in law it could not prevent him from doing that, for even a jailbird had the right to defend himself in court against libel and slander. But Trygve Lie was not to be put off by legal niceties; and just as he had secured the decree which, after the event, sanctioned Trotsky's internment, so, on 29 October, he obtained another 'Provisional Royal Decree' stating that 'an alien interned under the terms of the decree of 31 August 1936 [Trotsky was the only alien interned under that decree] cannot appear as plaintiff before a Norwegian court without the concurrence of the Ministry of Justice'. The Ministry, of course, refused its 'concurrence,' and it forbade the court to hear Trotsky's suit against the two editors.

Trotsky then asked his French lawyer to sue for defamation Stalinist editors in France, Czechoslovakia, Switzerland, Belgium, and Spain, hoping that, even if he were not to be summoned as witness, he would at least be able to state his case through legal representatives. To this, it would seem, the Norwegians could have no objection—they had no legal ground whatsoever for preventing him from defending his reputation before foreign courts. By now, however, the government's zeal for appeasing Stalin knew no limits. 'The Ministry of Justice', Trygve Lie declared, 'after having conferred with the government, has decided that it will oppose Leon Trotsky's attempts to take legal action before a foreign tribunal, while he remains in Norway.'[1] In addition the Minister forbade Trotsky to communicate with any lawyers abroad. Now at last he had trapped and gagged him completely.

'Yesterday I received the official statement forbidding me to sue anyone even abroad', Trotsky informed Gérard Rosenthal, his French lawyer, on 19 November. 'I am refraining from any comment in order to make sure that this letter reaches you.' To Lyova he wrote: 'You must take into account that the Minister of Justice has confiscated all my important letters relating to my personal defence. I am now confronted by

---

[1] *Storting Report no. 19*; Krog, op. cit.; Trotsky's letters to Gerard Rosenthal of 19 and 22 November; and *Stalin's Verbrechen*.

slanderers, burglars, scoundrels . . . and am completely defenceless. You must act on your own initiative and tell all our friends about this.' In the next letter he gave even stronger vent to exasperation. He remarked that *Arbeiderbladet* was just conducting a campaign for the release of Ossietzky, the famous radical writer, from a Nazi concentration camp, but had nothing to say about his own internment in Norway—'Ossietzky at least was not slandered by his jailers'. 'This letter too naturally goes through [the censor's hands], but I have ceased to pay any attention to this. I am writing these words, privately and confidentially, to my son who is pursued by bandits in Paris and whose life may be in danger, while [I am] imprisoned and tied hand and foot. At stake are matters on which . . . [our] physical and moral existence may depend; and I must speak out.'[1]

There was probably something of a *ruse de guerre* in these letters. Trygve Lie claims that Trotsky communicated with his son by illegal means; that he wrote some of his letters in chemical ink; that he communicated stealthily with his followers while he was allowed to visit a dentist in town; and that his followers smuggled letters to him in cakes sent to Hurum; and so on. For once the charges appear to be based on facts, although Natalya, when asked, twenty years later, whether Lie's allegations were true, did not know what to say about them. But political prisoners use such devices to maintain clandestine contact with their comrades; and it would be strange if Trotsky had not used them, when he was subjected to so much violence, trickery, and chicanery.

.   .   .   .

In view of Trotsky's enforced silence, the burden of the first public campaign against the Moscow trials fell on Lyova. Shy, somewhat inarticulate, and accustomed to keep himself in his father's shadow, he was suddenly brought to the fore in this great and grim affair. Vyshinsky had depicted him as a pillar of the 'terroristic conspiracy', and as his father's deputy and chief of staff, who instructed eminent old Bolsheviks about how to conduct their action inside the U.S.S.R.; and the Court's verdict had referred to him in the same

[1] *The Archives*, Closed Section.

terms as to his father. Now he was indeed compelled to act in his father's place. Within a few weeks of the Zinoviev-Kamenev trial he published his *Livre Rouge sur le proces de Moscou*, the first factual refutation of the Stalinist charges, and the first detailed exposure of their incongruities. He produced proof that he had never been with his father in Copenhagen, and that the Hotel Bristol where he was supposed to have met the conspirators was non-existent. He delved into the mystery of the confessions, saying that 'with their self-accusatory statements based on no facts and no evidence, with their literal repetition of the Prosecutor's pronouncements, and with their zeal for self-defamation, the defendants were in effect saying to the world: "Don't believe us, don't you see that all this is a lie, a lie from beginning to end!" '[1]

He was, however, shaken to his depths by the misfortune and the self-humiliation of the old Bolsheviks. He had known them all from his childhood, had played with their sons in the squares and corridors of the Kremlin, and had, as adolescent, looked up to them as the great men of the revolution and his father's friends. With these feelings still alive, he thus defended their honour: '. . . the inner moral strength of Zinoviev and Kamenev was very considerably above the average, though it proved to be insufficient in these quite exceptional circumstances. Hundreds of thousands . . . would not have been able to stand even one-hundredth part of the continuous and monstrous pressure to which Zinoviev, Kamenev, and the other defendants were subjected.' But—'Stalin wants Trotsky's head—this is his main purpose; and he will resort to the most extreme and villainous concoctions to get it. . . . He hates Trotsky as the living embodiment of the ideas and traditions of the October Revolution. . . .' Not content with 'triumphs' at home, the G.P.U. were in fact seeking to exterminate Trotskyism abroad as well. They accused the Spanish Trotskyists of wrecking the Popular Front and trying to assassinate its leaders; and they branded the Polish Trotskyists as agents of the Polish political police, and the German ones as agents of the Gestapo. 'Stalin aims at reducing all political differences in the labour movement to this formula: G.P.U. or Gestapo? Who is not with the G.P.U. is with the

[1] *B.O.*, no. 52–53, October 1936; *Livre Rouge sur le procès de Moscou*.

Gestapo.' 'Today he uses this method mainly in the struggle against Trotskyism, tomorrow he will turn it against other groupings in the working class. . . . Woe, if the world's labour movement proves unable to defend itself against this mortal poison.'[1]

Trotsky describes the relief with which he received at Hurum the first copy of the *Livre Rouge*: 'There are forms of paralysis under which one can see, hear, and understand everything, but is unable to lift a finger in order to turn away a mortal danger. To such political paralysis the Norwegian 'Socialist' Government had subjected us. What an inestimable gift Lyova's book was for us in these conditions. . . . The opening pages, I remember, seemed pallid to me: they reiterated a [familiar] political evaluation. . . . But from the moment the writer began his independent analysis of the trial I became completely absorbed. Every chapter seemed better than the previous one. "Our brave dear Lyova", I and my wife said to each other, "We have a defender!".'[2] In their correspondence, full of pain, anxiety, and tenderness, Lyova described all he was doing to set afoot a campaign against the purges; and he conveyed to his parents every word of sympathy and encouragement he could pick up from their all too few well-wishers.

Yet the horrible spectacle in which he was involved was probably more than Lyova's sensitivity could bear. He was, next to his father, the G.P.U.'s most important target. The feeling that he was being spied upon and that his mail was intercepted by a mysterious hand never left him. He feared that he would be kidnapped. He was lonely, defenceless, and completely dependent on the comradeship of the little band of Trotskyists around him. He found some solace in the friendship of Alfred and Marguerite Rosmer, who had now rallied to his father's defence, forgetting and forgiving all past misunderstandings. But within the narrow circle of his comrades he confided most of all in Mark Zborowski, a young and well-educated man, who had studied medicine and philosophy and who worked in the organization under the pseudonym Étienne, helping to publish the *Bulletin* and sitting on a little Russian Committee supposed to deal with the Opposition in

[1] Ibid.   [2] *B.O.*, no. 64, March 1938.

the U.S.S.R. Being of Polish-Ukrainian origin, Étienne knew Russian and had an intimate feeling for Soviet affairs—this enabled him to render Trotsky many small services and to gain Lyova's confidence.

This well-educated and fervent 'friend', however, was a Stalinist *agent provocateur*. Such was his knack for dissimulation that he never incurred the slightest suspicion on the part of Lyova and Trotsky. And so complete was Lyova's trust in him that he held the key to Lyova's letter-box and collected the mail for him. The mysterious hand that 'intercepted' Lyova's correspondence was Étienne's. He was also in charge of the most confidential files of Trotsky's archives; he kept these in his own home.[1]

A few months before the internment Trotsky asked Lyova to place a section of his archives with the Dutch Institute for Social History. He did this in part because he was pressed for money, and the Institute was willing to pay for the papers he offered them the modest sum of 15,000 (depreciated) French francs; but he was actuated mainly by a fear that the G.P.U. might try to seize his archives and he was anxious to deposit these in safe hands. In the first days of November, Lyova and Étienne delivered a number of files at the Paris branch of the Institute, at 7, rue Michelet—the branch was then under the management of Boris Nikolayevsky, the well-known Menshevik and one time associate of the Marx-Engels Institute in Moscow. The transaction was made tentatively, as an experiment; and the bulk of the archives, including the most confidential papers, remained with Étienne.[2]

No sooner had the files been delivered at the rue Michelet than a burglary was perpetrated there, on the night of 6–7 November; and some of the files were stolen. At once the suspicion arose that this was the G.P.U.'s work. The burglars

---

[1] Étienne (Mark Zborowski) has since made his confession; and in December 1955 he was sentenced by an American court to five years' imprisonment on a charge of perjury. My account of the relations between Étienne and Lyova is based on their correspondence with Trotsky and depositions which each of them made to the French police and magistrate. Étienne's story has been told in the *New Leader* by H. Kasson (on 21 November 1955) and by David J. Dallin (on 19 and 26 March 1956). See also *Hearing before the U.S. Senate Subcommittee on Internal Security*, Part 51, 14–15 February, 1957, pp. 3423–9. See also Isaac Don Levine, *The Mind of an Assassin*.

[2] Lyova's, 'Memoire pour l'Instruction', 19 November 1936. *The Archives*, Closed Section.

had left untouched valuables and money they had found and carried away only Trotsky's papers—who but G.P.U. agents would have done that? The French police were puzzled by the technical skill of the house-breaking; and they decided that this could not have been the work of French criminals but only of a powerful international gang. They interrogated Lyova, who accused the G.P.U. But how and from whom, they asked, did the G.P.U. learn so quickly that the files had been delivered at the rue Michelet? Who had been acquainted with the transaction? Lyova declared that apart from himself only three people had known about it: Nikolayevsky, a certain Madame Estrine, an employee of the Dutch Institute, and Étienne. He vouched for the integrity of all three, although he suspected that Nikolayevsky might have unwittingly, through careless talk, given the G.P.U. a clue. What about Étienne, the police inquired. Étienne, Lyova replied, was absolutely above suspicion: the proof of it was that at the very moment of the burglary he was guarding the most valuable parts of the archives in his own home.[1] Thus the question from whom the G.P.U. had learned about the deposition of the files appeared insoluble.

It turned out that the burglars had seized only Press cuttings and relatively unimportant papers; but no one doubted that, disappointed by the meagre spoils, the G.P.U. would make another and far more serious attempt. To the end of his days Trotsky was to remain almost as much concerned about the safety of his archives as about the safety of his own person. Yet the G.P.U. never made that other, much feared, attempt to seize his papers; and this was another puzzle. In the light of the facts stated here it is clear that they had no need to get hold of the archives, for they could have them, or copies of them, directly from Zborowski. They evidently staged the burglary in Paris as a feint to shield Étienne and enhance Trotsky's and Lyova's trust in him. Nothing, indeed, could more definitely divert all suspicion from him and turn it in other directions than the fact that, while the G.P.U. appeared to do their utmost to seize the archives, Étienne had 'reliably' guarded them in his apartment.

.   .   .   .

[1] Ibid.

At Hurum the months dragged on with leaden monotony; and nothing promised to open or even to loosen the trap in which Trotsky found himself. American followers were trying to obtain for him asylum in Mexico; but it was far from certain that they would succeed; and Trotsky, though anxious to get out of Norway, was reluctant, at so critical a moment, to seek shelter in a country at once so remote and so ill-famed for its cloak-and-dagger politics (where, as Lyova reminded him, 'an assassin is hired for a few dollars'[1]). He still had a glimmer of hope that he would be able to make himself heard even in Norway. On 11 December he was to reappear as witness in the protracted trial of the Quisling men who had broken into his home at Honnefoss; and he reckoned that this time the government would not dare to stop the proceedings. However, the Minister of Justice intervened once again, not indeed to stop the trial, but to order that the case be heard *in camera*. And so, when Trotsky appeared in the witness-box, surrounded by a platoon of police, the public and the reporters were removed from the courtroom. In contrast to what happened during the previous open hearing, now that everything had been done to suppress Trotsky's testimony, the presiding judge treated him with exquisite consideration and courtesy; and for several hours Trotsky pleaded his case, refuting the Stalinist charges, with as much power and gravity as if he were speaking to a world-wide audience. Not once did the President of the Court interrupt him, not even when he attacked the Norwegian Minister of Justice as Stalin's accomplice. It was almost grotesque for Trotsky to make this plea, which was a forensic masterpiece, in the course of an almost trivial trial and in a closed and empty courtroom. But so uncertain was he now of his future and so doubtful whether he would ever have a chance to state his case, that he availed himself of this opportunity to state it, even if only for the record.[2]

.   .   .   .

A few eye-witnesses offer vivid glimpses of Trotsky in internment. Askvik, one of the officers in charge of the guard, decribes, in unpublished memoirs, his calm dignity, pride, and

---

[1] Lyova to his parents, 7 December 1936. *The Archives*, Closed Section.
[2] Trotsky, *Stalin's Verbrechen*, pp. 37ff. (He spoke in German.)

self-discipline. Trotsky, Askvik says, met every restriction on his freedom with a protest and firmly claimed his rights, without ever offending his guards, whom he addressed in correct and fluent Norwegian.[1] Puntervold, the Norwegian lawyer, recalls how intently Trotsky followed the elections—he was worried that Knudsen, who stood for Parliament in a predominantly conservative constituency and was attacked as Trotsky's host, might be defeated. Puntervold was at Hurum when the news came that Knudsen had been elected with an unexpectedly large majority; and he relates that Trotsky, overjoyed, jumped up, seized Natalya in his arms, and danced with her in celebration of his friend's success (which was also something of a slap in the government's face). Knudsen's steadfast loyalty was one of the few consolations left to him in these dreary months, as was also the campaign in Trotsky's defence which Helge Krog, a radical writer, conducted with fire and brilliance in *Dagbladet*, Oslo's Liberal paper.[2]

Twice or perhaps thrice Trygve Lie visited Trotsky at Hurum. He came first on 11 or 13 December to warn Trotsky that he was going to be transferred from Hurum to a more remote and inaccessible place of internment in the north, because the Ministry 'could not afford to go on paying for the large police guard' it had had to keep at Hurum. Trotsky then told Lie that his friends—he mentioned Diego Rivera—intended to take him over to Mexico, and that he would rather go there than allow himself to be moved to the desert of Norway's far north. While they were talking, Lie noticed Ibsen's *Works* on Trotsky's table. 'Are you reading Ibsen here?' Lie asked. 'Yes, I am re-reading his *Works*; Ibsen used to be the love of my youth, and I have gone back to him.' The dialogue which now followed would be worthy of Ibsen himself. Trotsky remarked how relevant the idea of the *Enemy of the People* was to the situation in which he himself and the Minister were

[1] I am obliged for some of these details to Mrs. Askvik, the widow of the police officer. In April 1956, hearing that I was in Oslo, she brought the manuscript of her husband's memoirs to Knudsen, requesting him to submit it to me. It was with surprise that Knudsen, translating to me the relevant passages, learned about Trotsky's command of the Norwegian language. At Honnefoss they had usually conversed in German.

[2] Trotsky, *Stalin's Verbrechen*, pp. 77–78. Every detail of this account is confirmed by other sources.

involved. Lie replied evasively that 'Ibsen could be variously interpreted'. 'No matter how you interpret him', Trotsky said, 'he will always testify against you. Remember Burgomaster Stockman?' The Minister asked whether Trotsky really intended to compare him with the villain of Ibsen's play, who for the sake of authority and vested interests destroys his own brother? 'With Burgomaster Stockman? . . . at best, Mr. Minister, at best', Trotsky replied. 'Your government has all the vices of a bourgeois government without any of its virtues.' Stung by this remark, the Minister inveighed against Trotsky's 'ingratitude', saying that he made 'a silly mistake' when he allowed Trotsky to come to Norway. 'And this silly mistake you now wish to make good by crime?' Trotsky retorted; and opening Ibsen's drama he read out the challenge with which Dr. Stockman meets his villainous brother: 'We shall yet see whether meanness and cowardice are strong enough to close the mouth of a free and honest man.' This brought the conversation to an end. The Minister rose, yet before leaving he turned to his prisoner with an outstretched hand; but Trotsky refused the handshake.[1]

A week later Lie returned to tell Trotsky that Mexico had granted him asylum and that he, Lie, had already chartered a petrol tanker on board which Trotsky and his wife would sail the following day, under the escort of Jonas Lie, the commander of the Hurum police guard. The haste and the details of the deportation aroused Trotsky's forebodings. Why, he asked, was the Minister leaving him only twenty-four hours to prepare for the voyage? Why did he not release him from internment? He demanded that he should be allowed to leave as a free man, to consult friends, to wind up his affairs and collect his papers, to communicate with the Mexican Government, to chose his own route, and to make his own safety arrangements. 'And what', he asked, 'if Stalin knows about your tanker? We may be torpedoed on the high seas and never reach the English Channel.' (He even inquired whether the vessel had any defences.) The Minister refused all his demands, but tried to reassure him about the safety of the voyage, saying that no one knew about the plan except he himself and

---

[1] Trotsky, *Stalin's Verbrechen*, pp. 77–78; and Trotsky's diary pages in *The Archives*, Closed Section; Krog, op. cit.

the shipowner. Trotsky then asked to be allowed to travel via France: surely the French would give him a transit visa now that he had asylum in Mexico? Lie declined this demand too. He was in great haste to get Trotsky out of the country before Parliament assembled to debate the affair. His haste seemed to Trotsky more sinister than it was. 'Of course', he said, 'you are in a position to destroy us physically, but morally you will break your necks just as the German Social Democrats broke theirs on Karl Liebknecht and Rosa Luxemburg.' He repeated his prophecy: 'In three to five years . . . you will all be émigrés . . .'; and once again he turned his back on the Minister, refusing to shake hands.[1]

He had the feeling that he was being driven from one trap into another, and he was not sure what might happen to him and Natalya on the way. While Natalya was packing luggage, he wrote, in chemical ink, an article under the title 'Shame!'. This was to be 'a reply to slanderers', especially to well-known British and French lawyers who had 'vouched' for the legal correctness of the Zinoviev-Kamenev trial. One of these lawyers was a King's Counsel; another was an eminent member of the *Ligue des Droits d'Homme*; and both had praised the Moscow Court for not having sentenced Trotsky to death *in absentia*. 'Whoever knows anything at all about revolutionary history, human psychology and . . . the biographies of the men involved', Trotsky commented, 'will agree . . . that there is a thousand times more ground to assume that [these lawyers] are in Stalin's service than to admit for a minute that Trotsky can be an ally of the Gestapo. . . . All the Leagues of the Rights of Man of the whole solar system will not be able to prove this. . . . I shall give the final answer to the accusers and their lackeys . . . in Mexico, *if ever I arrive there*.' Before starting on the voyage, he wanted to leave this article behind 'as the ship-wrecked sailor leaves a bottle in the sea'.[2] To Lyova he wrote: 'It seems that tomorrow we are going to be sent to Mexico. This then is our last letter from Europe. If something happens to us *en route* or elsewhere you and Sergei are my heirs. *This letter should have testamentary value.* . . . As you know, I have in

[1] Ibid.
[2] The article appeared in the first issue of *B.O.* (nos. 54–55 March 1937) published after Trotsky's deportation from Norway.

z

mind future royalties on my books—apart from these I possess nothing. If you ever meet Sergei . . . tell him that we have never forgotten him and do not forget him for a single mo- ment.'[1] As he was writing these words, his doctor, his lawyer, and his tax collector presented their bills; and to secure their claims they sequestrated his bank account.[2]

On 19 December the petrol tanker, the *Ruth*, sailed from Norway, with Trotsky, Natalya, and their police escort as the only passengers. The deportation was carried out in such secrecy that for several days afterwards police sentries stood outside the internment place at Hurum to give the impression that Trotsky was still there. The seas were rough at first; and in their cabin Trotsky and Natalya read books about Mexico and wondered what the future held in store for them. Then, as the sea calmed down, Trotsky began to write, partly in diary form, his analysis of the trial of Zinoviev and Kamenev, which he included in his book *Stalin's Crimes*. He worked hard for three weeks while the vessel tacked about, changed course, and avoided normal routes. But the world had already learned about the deportation, and Press agencies were anxious to interview Trotsky by radio. The captain of the *Ruth*, however, was ordered from Oslo not to allow him to use the transmitter. On board the empty ship Trotsky and Natalya were still treated as internees; even at their meals they remained flanked by their police escort.

'This was Cain's year', runs the entry in Trotsky's diary dated 31 December 1936. On the next morning the *Ruth* greeted the New Year with her sirens. No one returned the greetings; and there were no Wishes of the Season. Only the fascist police officer came to the dining table, flourishing New Year greetings which he personally had received from his socialist Minister. The world seemed engulfed in absurdity.

By one of those strange coincidences which run through Trotsky's life, it was exactly twenty years since Trotsky had last sailed from Europe, also as an exile expelled from a country which had given him temporary refuge.[3] But in 1917 the world was at war and the seas were infested with sub- marines. Now the world seemed at peace, and no submarines

---

[1] The letter, dated 18 December 1936, was written in French.
[2] *The Archives*, Closed Section.   [3] *The Prophet Armed*, pp. 238–41.

lurked in the ocean. Yet there was an almost warlike tension on board the tanker; and Trotsky noted in his diary that the captain and the crew kept alluding to the G.P.U., but avoided uttering the name, 'as if they were hinting at underwater rocks'.[1] And twenty years before Trotsky had written from his voyage: 'This is the last time I am casting a glance on that old *canaille* Europe'—only to hurry back across the ocean three months later. Now he had indeed cast his last glance on the 'old *canaille*'. But as he departed from Europe, his head and heart were full of its infernal turmoil; and his thoughts hovered over the graves he was leaving behind, the graves of his two daughters, the graves of so many friends and followers, and the graves of so many hopes.

[1] Some parts of this diary are included in *Stalin's Verbrechen*; others have not yet been published and are in *The Archives*.

CHAPTER V

The 'Hell-black Night'

As the *Ruth* sailed into the great oil harbour of Tampico, on 9 January 1937, Trotsky and Natalya were still so apprehensive of what might await them on the Mexican shore that they refused to land unless they were met by friends. The Norwegian police were threatening to disembark them by force when a small boat approached from which a Mexican general, accompanied by officials, emerged, bringing a message of welcome from Lazaro Cardenas, the President of Mexico. (The President had sent his official train to take Trotsky and Natalya from Tampico.) On the pier two American Trotskyists, George Novack and Max Shachtman, waved greetings; and Frida Kahlo, Diego Rivera's wife, was waiting to offer hospitality. The contrast between the warm reception in Mexico and the icy send-off from Norway was so sharp as to seem unreal. Entering the Presidential train Trotsky and Natalya ran into a police guard and again shrank back. 'A fear crossed our minds . . .', she notes, 'that perhaps we were being taken to just another place of captivity.' At a small station near Mexico City Diego Rivera received them with expansive enthusiasm and took them to Coyoacan, a suburb of the capital, and to his Blue House—this was to be their home for the next two years. The place might have been designed to soothe weary nerves: it was spacious, sunlit, covered with a profusion of pictures, full of flowers and objects of Mexican and Indian art. At every step the newcomers found comforting signs of the care with which their Mexican and American friends had prepared the new abode for them, and had thought of their personal protection and provided facilities for work. Thus, the first few days in Mexico brought quite unexpected relief—there was even a touch of the fleeting idyll about them.[1]

---

[1] This is evident from Trotsky's first Mexican letters to Lyova, in which he expresses delight with the country of his new refuge, its climate, and even with its fruit and vegetables.

The country's political climate also offered attractions. The Mexican Revolution was still at its height. Quite recently Cardenas had signed a decree, under which some of the *latifundias* were shared out among the poor peasants; and he was about to nationalize the American and British owned oil and railway companies. Foreign investors, native landlords, and the Catholic Church fought back; and relations between Mexico and the United States were strained. But Cardenas had behind him the peasants and the Confederation of Mexican Workers which had suddenly grown into a great political force.

In admitting Trotsky, at Rivera's request and on promptings from his own entourage, Cardenas had acted from a sense of revolutionary solidarity. He declared that he had not merely granted Trotsky asylum, but invited him to stay in Mexico as the government's guest. From the outset he did his utmost to protect his guest's head against the storms of hatred gathering over it; and he was to go on doing this till the end. However, his own situation was rather delicate. On the one hand, his political enemies soon began to insinuate that Trotsky was the inspirer of his revolutionary policy, and the insinuation found its way into American newspapers.[1] On the other, the Confederation of Mexican Workers, on whose support he depended, was a Stalinist stronghold; its leader Lombardo Toledano and the Communist party protested fiercely against Trotsky's admission and warned the President that they would not rest until that 'chief of the vanguard of the counter-revolution' was expelled. Cardenas was careful to lend no colour to the charge that he was expropriating British and American investors at Trotsky's instigation; and he was even more anxious to calm the Confederation of Mexican Workers. He was himself politically very far from any form of Trotskyism and indeed of communism. The son of poor peasants, he was guided by agrarian radicalism and the empirical experience of his patriotic struggle against foreign predominance. He was

[1] Cardenas later found it necessary to refute the insinuation publicly (*La Prensa*, 12 November 1938); and Trotsky thought of suing an American paper (*The New York Daily News*) which had violently attacked him as Cardenas's evil spirit. He desisted only when he was told by Albert Goldman that he had no legal grounds for action. See Trotsky's correspondence with Goldman in December 1938. *The Archives*, Closed Section.

therefore wary of getting involved in any of the internal conflicts of communism. In these difficult circumstances he repudiated with dignity the Stalinist clamour against Trotsky's admission; but he kept studiously aloof from his 'guest'—they never met in person. He asked Trotsky to pledge himself that he would not interfere in Mexico's domestic affairs. Trotsky gave this pledge at once, but, taught by his bitter Norwegian experience, he was on his guard and explicitly reserved his 'moral right' to reply in public to any accusations or slanders.[1] Cardenas was satisfied with this. It never occurred to him to ask Trotsky to refrain from political activity; and he himself stood up for Trotsky's right to defend himself against Stalinist attacks. In this attitude of aloof but vigilant benevolence he was to persist. Trotsky often expressed his gratitude and, strictly observing his pledge, never ventured to state any opinion on Mexican politics even in private, although his view of Cardenas' policy, which did not go beyond the 'bourgeois stage' of the revolution, must have been critical to some extent.

During his first years in Mexico, Diego Rivera was Trotsky's most devoted friend and guardian. A rebel in politics as well as in art, the great painter had been one of the founders of the Mexican Communist party and a member of its Central Committee since 1922. In November 1927 he witnessed the Trotskyist street demonstrations in Moscow and the expulsion of the Opposition, which gravely disturbed him. Subsequently he broke with the party, and also with David Alfaro Siqueiros, another of Mexico's great painters, his closest friend and political comrade, who sided with Stalin. The dramatic pathos of Trotsky's fate stirred Rivera's imagination: here was a figure of heroic dimensions who might have been destined to take a central place in his epic frescoes—he had indeed put Trotsky and Lenin in the forefront of that famous mural glorifying class struggle and communism with which he had, to the horror of all respectable America, decorated the walls of the Rockefeller Centre in New York. For Rivera it was a moment of rare sublimity when strange fortunes brought his leader and prophet under his roof at Coyoacan.

Trotsky had long admired Rivera's work. He probably first saw his paintings in Paris during the First World War;

1 '*Aux Représentants de la Presse Mexicaine*', 12 January 1937. *The Archives.*

and references to them occur in Trotsky's Alma Ata correspondence of 1928.[1] Rivera's restless search for new artistic expression aptly illustrated Trotsky's own view that the malaise in contemporary painting was rooted in its divorce from architecture and public life, a divorce which was inherent in bourgeois society and could be overcome only by socialism. The striving for the reunion of painting, architecture, and public life animated Rivera's art, in which Renaissance traditions and Goya's and El Greco's influences merged with Indian and Mexican folk art and cubism. This interplay of tradition and innovation suited Trotsky's taste; he was captivated by Rivera's defiant courage and the soaring and passionate imagination with which he brought motifs of the Russian and Mexican revolutions into his monumental murals. Nor could Trotsky help being fascinated and puzzled by Rivera's elemental temperament, somnambulism, and 'Gargantuan size and appetites', which made of him a riotous and roaring prodigy like any of the chimerical figures appearing in his paintings. And in counterpoint, as it were, to Diego there was his wife Frida, herself a painter of delicate melancholy, introspective and symbolist, and a woman of exquisite beauty—she emanated exotic grace and dreaminess as she moved about in long-flowing, richly shaded, and embroidered Mexican robes which concealed a deformed leg. After the dreary months of internment, it was gratifying, even thrilling, for Trotsky and Natalya to find refuge with such friends.

An onlooker with some insight into the characters might have wondered how Trotsky and Rivera would get along and whether a clash between them was not bound to occur. Not satisfied with his artistic eminence, Rivera saw himself also as a political leader. He was not exceptional in this: painters and sculptors played an extraordinarily large part in Mexican politics—most members of the Politbureau of the Communist party were painters. (Political agitation carried out by means of brush and chisel may have appealed to masses of illiterate but artistically sensitive *campesinos* more directly than any other form of agitation.) Yet as a politician Rivera was even less

---

[1] Andres Nin had sent to Trotsky at Alma Ata a volume of reproductions of Rivera's paintings and sculptures, and Trotsky wrote back to thank him for the book and express appreciation of the artist. *The Archives.*

than an amateur; he frequently fell a prey to his restless temperament. However, in Trotsky's presence, at least at the
beginning, he kept his political ambition under control and
modestly assumed the disciple's role. As to Trotsky, he had
always treated the political vagaries of artists with tender
understanding, even those of lesser artists, to whom he was not
indebted for anything. All the more willing was he in Rivera's
case to say that 'genius does what it must'.

Thus Trotsky might have been in a mood to count the
blessings of his new refuge had he not been driven back into
his grim struggle almost at once. He was daily the object of
threats from the local Stalinists and from Moscow. President
Cardenas had to order police guards to be placed outside the
Blue House. Inside, American Trotskyists, who had come to
serve as secretaries and bodyguards, kept watch. Trotsky's
American followers were aiding him unstintingly in organizing
his defence and his campaign against the Moscow trials. They
were few and poor; but they assisted him as well as they could
in re-establishing his contacts with friends and followers all
over the world and in resuming his work. 'What good luck it is',
he wrote to Lyova on 1 February 1937, 'that we have managed
to come to Mexico just before the start of the new trial in
Moscow.'[1]

.   .   .   .

The new trial opened less than a fortnight after his landing
at Tampico. Radek, Pyatakov, Muralov, Sokolnikov, Sere-
briakov, and twelve others took their place in the dock; and
Trotsky was once again the chief defendant *in absentia*. The
accusations piled up ever more incongruously and incredibly.
Vyshinsky now spoke of Trotsky's formal agreement with
Hitler and the Emperor of Japan: in exchange for their aid
in the struggle against Stalin, Trotsky, he maintained, was
working for the military defeat and dismemberment of the
Soviet Union, for he had pledged himself *inter alia* to cede the
Soviet Ukraine to the Third Reich. Meanwhile he was organizing
and directing industrial sabotage in the Soviet Union; catastrophes in coal mines, factories, and on the railways, mass
poisonings of Soviet workers, and repeated attempts on the lives

[1] *The Archives*, Closed Section.

of Stalin and other members of the Politbureau. The defendants echoed the prosecutor and elaborated his charges. One of them, Romm, who had been *Izvestya's* correspondent in France, confessed that he had seen Trotsky in Paris in July 1933 and taken terroristic instructions from him. Pyatakov told the court that he had visited Trotsky in Oslo in December 1935 and there taken orders from him.[1]

'We listened to the radio, we opened the mail and the Moscow newspapers', Natalya writes, 'and we felt that insanity, absurdity, outrage, fraud, and blood were flooding us from all sides here in Mexico as in Norway. . . . With pencil in hand Lev Davidovich, over-tense and overworked, often in fever, yet tireless, lists the forgeries which have grown so numerous that it becomes impossible to refute them.'[2] The trial lasted a week and was followed by executions—Radek and Sokolnikov, however, were sentenced to ten years' imprisonment each.

For Trotsky to refute the accusations was indeed like wrestling and arguing with monsters in a nightmare. The trials were more and more unreal in their horror, and horrible in their unreality. They seemed designed to paralyse every critical thought and to render every argument grotesquely inadequate. Yet even before Trotsky had had the time to marshal his facts and arguments some of the charges were pricked. The Norwegian Ministry of Foreign Affairs investigated the statement that Pyatakov had arrived in Oslo by plane from Berlin in December 1935 and seen Trotsky; and it ascertained that no plane coming from Berlin had landed at Oslo Airport in that month and for many weeks before and after; the Airport Authority issued a statement to that effect. Trotsky then telegraphed these questions to the Moscow tribunal: when exactly, on what day and at what hour, had Pyatakov landed? And where, when, in what circumstances had he, Trotsky, received him? He asked similar questions about his alleged meetings with Romm.[3] The Prosecutor and the judges ignored the questions, knowing full well that if the defendants tried to answer they would involve themselves in glaring contradictions and discredit the show. On 29 January, just before the end of the proceedings, Trotsky once again challenged

---

[1] *Sudebnyi Otchet po Delu Anti-Sovietskovo Trotskistskovo Tsentra.*
[2] In V. Serge, *Vie et Mort de Leon Trotsky*, p. 258.   [3] *B.O.*, nos. 54–5, 1937.

Stalin to demand his extradition. In an appeal to the League
of Nations he declared himself ready to submit his case to a
Commission on Political Terrorism which the League was
supposed to set up on Soviet initiative—he had already made
one such appeal from Norway. The League kept silent; and
Stalin once again disregarded the demand for extradition.[1] In
another effort to come to grips with his accusers Trotsky
stated in a message to a public meeting in New York:

> . . . I am ready to appear before a public and impartial Com-
> mission of Inquiry with documents, facts, and testimonies . . . and
> to disclose the truth to the very end. I declare: *If this Commission
> decides that I am guilty in the slightest degree of the crimes which Stalin
> imputes to me, I pledge in advance to place myself voluntarily in the hands
> of the executioners of the G.P.U. . . . .* I make this declaration before the
> entire world. I ask the Press to publish my words in the farthest
> corners of our planet. But, if the Commission establishes—do you
> hear me? do you hear me?—that the Moscow trials are a conscious
> and premeditated frame-up, I will not ask my accusers to place
> themselves voluntarily before a firing squad. No, the eternal disgrace
> in the memory of human generations will be sufficient for them!
> Do the accusers in the Kremlin hear me? I throw my defiance in
> their faces, and I await their reply![2]

About this time Trotsky's two sons were finally linked with
him in his ordeal—and here the story turns into the modern
version of the Laocoön legend. Lyova, feeling that he was
pursued by the G.P.U., published in a French newspaper a
statement saying that if he were to die suddenly the world
should know that he had found his death at Stalinist hands—
no other version should deserve credence, for he was in
good health and was not harbouring any thought of suicide.
Sergei had been arrested at Krasnoyarsk in Siberia, according
to the Russian Press, and charged with attempting, on his
father's orders, a mass poisoning of workers in factories. 'Stalin
intends to extract a confession from my own son against me,'
Trotsky noted. 'The G.P.U. will not hesitate to drive Sergei
to insanity and then they will shoot him.' Natalya came out
with another appeal addressed in vain 'To the Conscience of
the World'.[3] 'There were moments', she recollected later, 'when

---

[1] '*Trebovanie Moei Vidachy*', 24 January 1937. *The Archives.*
[2] 'I Stake my Life', Appendix II in *The Revolution Betrayed.*
[3] *B.O.*, nos. 54–55.

L.D. felt crushed' and 'remorseful at still being alive. "Perhaps my death could save Sergei?" he once told me. . . .'[1] She alone knew of those moments. In the eyes of the world Trotsky remained indomitable, unflinching, and possessed of unconquerable energy. He never tired of summoning his followers to action and rousing flagging friends. This, for instance, is what he wrote to Angelica Balabanoff, his old Zimmerwald associate, when he heard that the Moscow trials had plunged her into deep pessimism: 'Indignation, anger, revulsion? Yes, even temporary weariness. All this is human, only too human. But I will not believe that you have succumbed to pessimism. . . . This would be like passively and plaintively taking umbrage at history. How can one do that? History has to be taken as she is; and when she allows herself such extraordinary and filthy outrages, one must fight her back with one's fists.'[2] So he himself fought back.

He undertook to establish his full alibi, to prove that not a single one of the Stalinist charges was or could have been true and to bring to light the political meaning of the gigantic frame-up. This was, in the view of many, an impossible task. He had to retrace all his whereabouts and activities through all the years of his banishment; to assemble evidence from his enormous and partly scattered archives and from newspapers in many languages; to collect testimonials and affidavits from former secretaries and bodyguards and from adherents, some of whom had turned into opponents; and from Ministries, Consulates, police headquarters, travel agencies, landlords, householders, inn-keepers and casual acquaintances in various countries. Yet in a sense this vast and costly undertaking was bound to be useless. Those who wished to know the truth could very well grasp it without such a mass of detailed evidence, while people with indifferent or closed minds were not to be persuaded anyhow. Nor was it likely that posterity would ever require such an accumulation of testimonies in order to be able to form its opinion. Trotsky, the great controversialist might well have been satisfied with exposing the trials on their internal evidence alone, as Lyova and a few

[1] Serge, op. cit., p. 266.
[2] Trotsky to A. Balabanoff, who was then living as exile in New York. Letter of 3 February 1937 in *The Archives*, Closed Section.

friends—and Bernard Shaw—urged him to do[1]. But it was characteristic of the man's relentless meticulousness that once he had resolved to put the whole record straight, he left nothing to chance, allowed not a single relevant incident to remain undocumented and omitted not a single affidavit from the dossier. He behaved as if he were reckoning with the possibility that Stalin's forgery might endure for ages; and for the ages he was preparing a foolproof and indestructible alibi.

This nerve-racking labour took him many months. He put all his strength into it and drove his secretaries and adherents remorselessly, and above all Lyova, who in Paris performed the major part of the work. He brooked no delay, no contradiction, no excuse. At the slightest sign of a let-down he threatened to 'break off all relations' first with Shachtman and then with Naville and to 'denounce their sabotage or even worse', although both men did their best to assist him. In the first letter to Lyova from Mexico he was already venting disappointment at not having received a pile of testimonies he had expected to find on arrival. After a fortnight or so he was bursting with impatience; and every letter to Lyova was bitter with reproach. Why had the papers relating to his Copenhagen trip not yet come? Was this not 'a plain crime'? Why were some testimonies not validated legally by Commissioners of Oaths? Why were the signatures under others illegible? Why were the dates not precise? Why were certain placenames not indicated beyond all possibility of misunderstanding? From week to week his tone grew more scolding and brutal. 'Today I have received your letter . . . with the usual excuses . . . and the usual promises . . .' he wrote to Lyova on 15 February, 'but I have had enough of excuses and have long since ceased to believe in promises!' Lyova's 'slovenliness bordered on treachery'. 'After all the experiences of recent months I must say that I have not yet had a day as black as this one, when I opened your envelope, confident I would find the affidavits in it and instead found only apologies and assurances.' 'It is difficult to say which are the worst blows, those that come from Moscow or those from Paris.'[2] He planned the opening of

[1] For Bernard Shaw's opinion see further pp. 369–70. Lyova expressed his misgivings in a letter to his mother (8 March 1937), *The Archives*, Closed Section.
[2] *The Archives*, Closed Section. Letters of 1 and 15 February 1937.

a counter-trial for the spring; and he was afraid that the dossier would not be ready in time. The Blue House looked almost like a sweat shop in these days, with the secretaries, Trotsky himself, and Natalya translating, copying, and typing endless papers. At the same time he filled pages in American newspapers with his comments, tried to make his views intelligible to the Mexican Press, and arranged for 'Commissions of Inquiry' to be set up in various countries. Obsessed with the importance of what he was doing, suspicious of every hitch, apprehensive of interference by the G.P.U., and despairing perhaps of ever finishing the job, he felt no inhibition about prodding and chiding Lyova, whose life and honour were as much at stake as his own. So indeed might Laocoön have upbraided his sons and goaded them on to strain every nerve in fighting off the giant snakes in whose strangulating coils they had all been caught, father and sons.

Lyova was hurt and wounded in his filial devotion. While Trotsky was interned in Norway he had stepped bravely into the breach. But the incubus with which he was struggling was stronger than himself; and he had looked forward to the day when his father would be released and take the burden on his own broad shoulders. He was distressed now to see that his father was so wrought up and irate. He still doubted the value of the whole undertaking and wrote to Natalya that *Stalin's Crimes*, the short book Trotsky had written *en route* to Mexico, was a far more effective riposte than any 'counter-trial', or the work of any Commissions of Inquiry, could be. Yet, once his father had decided to establish his alibi, Lyova put all his heart into the job. It was not his fault that progress was slow and misunderstandings arose. From Hurum, for instance, Trotsky had directed him to arrange a counter-trial in Switzerland; but in the meantime it had been decided to hold it in America. Lyova, not knowing about this, was still busy with preparations in Switzerland. This brought him a severe rebuke from his father, who threatened to cut off all the money needed and to take all further work out of Lyova's hands, and entrust it to Naville (in whom he had always had so little confidence).[1] The gathering of testimonies was impeded by the animosities of the Trotskyist sects: Lyova had to obtain many statements

[1] Trotsky to Lyova, 24 February, 5 and 16 March 1937, ibid.

from members of the Molinier group whom Trotsky had disowned; and he had to use tact and diplomacy. He was overworked and depressed. He too was engaged in the Press campaign against the trials: his articles appeared now and then in the *Manchester Guardian*. He went on looking after his father's publishing affairs, collected royalties, forwarded them regularly to Mexico, paid parental debts in Norway and France, and brought out the *Bulletin*. Offended by his father's censoriousness, sensing that he was being ensnared by the G.P.U., deeply unhappy in his family life, he began at the age of thirty to suffer from persistent insomnia. He grew weary and exhausted.

As usual, he opened his heart only to his mother: ('Darling Mamochka, I have no doubt that you alone are not cross with me for my silence or for anything else.') But he also met his father's reprimands with this poignant reproach: '. . . I have had to carry out, in very difficult conditions, part of the work which would otherwise have burdened yourself; and I have had to do it without the necessary authority and without the assistance you have; sometimes I do not even have the money to buy postage stamps. I thought that I could count on your support. Instead you are making me your butt and are telling all and sundry about my "criminal carelessness". . . . Even if I bear a share of responsibility for the delay with the Copenhagen documents this does not . . . justify your attitude towards me.'[1] Harassed, and dejected, Lyova confided all the more trustingly in Étienne, whom no one seemed to equal in ingenuity, industry, and devotion to the cause.

Trotsky at first hoped that the counter-trial would be set on a scale appropriate to the provocation, that it would be conducted in such a way as to shake the conscience of the international labour movement. He sought to associate the Second International with it and the so-called Amsterdam International of the Trade Unions. On his advice, Lyova had approached Friedrich Adler, the Secretary of the Second International, who had of his own accord denounced the Moscow purges as 'medieval witch hunts'. Adler did what he could; yet all he achieved was that after much delay the Executive of the International issued a statement condemning

[1] Lyova to Trotsky, 8 March 1937, ibid.

the purges; it refused to take part in any inquiry or counter-trial. So did the Trade Unions' International. Both these organizations, their German and Austrian sections suppressed by Hitler and Dolfuss, were under the thumb of Leon Blum; and as head of the Popular Front government he depended on Stalinist support. Blum was embarrassed even by the International's platonic declaration against the purges; and he used his influence to prevent any further action by his own party and by 'fraternal sections'. And so the western European Social Democrats, usually so eager to defend the 'freedoms and rights of the individual' against communism, preferred this time to observe a diplomatic silence, or even to exculpate Stalin. 'The International', as Trotsky put it, 'boycotted its own Secretary.' This reduced in advance the effectiveness of any counter-trial: without the Socialist parties and the Trade Unions no campaign could engage the attention of the working classes.[1]

Trotsky's adherents then tried to enlist the support of eminent intellectuals of the left. This did not quite suit Trotsky, who often derided the 'peace committees', 'peace congresses', and 'Anti-fascist parades' for which the Stalinists assembled galaxies of literary and academic 'stars'; he despised the showy snobbery of such stage effects, especially when the Comintern substituted them for solid and united action by the labour movement. He reproached his American followers for failing to draw workers into the 'Committees for Trotsky's Defence'; but he had no choice in the matter.[2]

Yet the response of the intelligentsia was also disappointing, for the Stalinists, who in France, Spain, Britain, and the United States exercised a strong influence on them, brought to bear upon the intelligentsia every means of moral pressure to prevent them from lending the slightest support to any protest against the purges. From Moscow, where the flower of Russian literature and art was being exterminated, the voices of Gorky, Sholokhov, and Ehrenburg could be heard, joining in the chorus that filled the air with the cry, 'Shoot the mad dogs!' In the West literary celebrities like Theodore Dreiser, Leon

---

[1] B.O., nos. 56–57, 1937; Lyova's correspondence with F. Adler in 1936. The Archives, Closed Section.

[2] See the Internal Bulletin of the S.W.P. (the American Trotskyist party), March-April 1940. (The question was raised then, in the course of Trotsky's controversy with Shachtman and Burnham.)

Feuchtwanger, Barbusse, and Aragon echoed the cry; and a man like Romain Rolland, the admirer of Ghandi, the enemy of violence, the 'humanitarian conscience' of his generation, used his sweetly evangelical voice to justify the massacre in Russia and extol the master hangman—with such zeal that Trotsky thought of suing him for defamation. Where Gorky and Rolland gave the cue, hosts of minor humanitarians and moralists followed suit with little or no scruple. Their manifestoes and appeals in support of Stalin make strange reading. In the United States, for instance, they declared a boycott on the Commission of Inquiry set up under John Dewey's auspices. They warned 'all men of good will' against assisting the Commission, saying that critics of the Moscow trials were interfering in domestic Soviet affairs, giving aid and comfort to fascism, and 'dealing a blow to the forces of progress'. The manifesto was signed by Theodore Dreiser, Granville Hicks, Corliss Lamont, Max Lerner, Raymond Robins, Anna Louise Strong, Paul Sweezy, Nathaniel West, and many professors and artists, quite a few of whom were to be in the forefront of the anti-communist crusades of the nineteen-forties and nineteen-fifties.[1] Louis Fischer and Walter Duranty, popular experts on Soviet affairs, vouched for Stalin's integrity, Vyshinsky's veracity, and the G.P.U.'s humane methods in obtaining confessions from Zinoviev, Kamenev, Pyatakov, and Radek. Even Bertram D. Wolfe, a member of the Lovestonite Opposition long expelled from the Communist party, still gave Stalin credit for saving the revolution from the Trotskyite-Zinovievite conspiracy.[2] In the Jewish-American Press

---

[1] See, e.g., the manifesto published in *Soviet Russia To-day*, issue of March 1937.

[2] 'Today, whatever his [Trotsky's] subjective intentions may be, and I shall not try to judge them, his objective role is to mobilize labour sentiment against the Soviet Union. He has ordered his followers in France to enter the Socialist International. He has departed ever further from communist fundamentals. . . . He has even come out for civil war in the Soviet Union and thus become an open enemy of the class and land he once served so faithfully.' Thus Bertram D. Wolfe wrote about Trotsky in 1936! ('*Things we want to know*' Workers' Age Publications.) Only when the Great Purges were coming to a close, shortly before Bukharin appeared in the dock, did Wolfe, the 'Communist fundamentalist', express (in *The New Republic* of 24 November 1937) his regret at having given moral support to the purges. This led Trotsky to remark that Wolfe had still a lot of things to learn and un-learn in order to avoid committing grievous mistakes in the future. In later years Wolfe attacked other writers (who had always denounced the Stalinist purges) as 'apologists for Stalin'.

III. Arriving in Mexico. Trotsky and Natalya at Tampico, 1937

IV. Diego Rivera

writers who had hitherto described themselves as 'Trotsky's admirers' turned against him when he spoke of the anti-semitic undertones of the Moscow trials. The editor of one such paper wrote: 'This is the first time that we of the Jewish Press have heard such an accusation. We have been accustomed to look to the Soviet Union as to our only consolation, as far as anti-semitism is concerned. . . . It is unforgivable that Trotsky should raise such groundless charges against Stalin.'[1]

Hypocrisy, bigotry, and the simple-minded fear of aiding Hitler by criticizing Stalin were not the only motives. Some of the intelligentsia saw no point in Trotsky's refutations. Charles A. Beard, America's distinguished historian, held that it was not 'incumbent upon Trotsky to do the impossible, that is to prove a negative by positive evidence. It is incumbent upon his accusers to produce more than confessions, to produce corroborating evidence. . . .'[2] Bernard Shaw also rejected the idea of a counter-trial and wrote: 'I hope Trotsky will not allow himself to be brought before any narrower tribunal than his reading public where his accusers are at his mercy. . . . His pen is a terrific weapon.' A month later he wrote less sympathetically: 'The strength of Trotsky's case was the incredibility of the accusations against him. . . . But Trotsky spoils it all by making exactly the same sort of attacks on Stalin. Now I have spent nearly three hours in Stalin's presence and observed him with keen curiosity, and I find it just as hard to believe that he is a vulgar gangster as that Trotsky is an assassin.'[3] Shaw was, of course, evading the issue, for Trotsky did not make 'exactly

---

[1] B. Z. Goldberg in the New York *Tag* of 26 and 27 January 1937. At this time Trotsky re-formulated his views on the Jewish problem. In an interview with the *Forwärts*, another American-Jewish daily, he admitted that recent experience with anti-semitism in the Third Reich and even in the U.S.S.R. had caused him to give up his old hope for the 'assimilation' of the Jews with the nations among whom they lived. He had arrived at the view that even under socialism the Jewish question would require a 'territorial solution', i.e. that the Jews would need to be settled in their own homeland. He did not believe, however, that this would be in Palestine, that Zionism would be able to solve the problem, or that it could be solved under capitalism. The longer the decaying bourgeois society survives, he argued, the more vicious and barbarous will anti-semitism grow all over the world. *The Archives*, 28 January 1937.

[2] Quoted from *The Case of Leon Trotsky*, p. 464.

[3] G. B. Shaw in letters to the Secretary of the British Committee for the Defence of Leon Trotsky, 20 June and 21 July 1937. Quoted from the Archives of the Committee and the Trotsky *Archives*, Closed Section.

the same sort of attacks on Stalin'. Yet, unlike Rolland, Shaw did not carry his friendship for Stalin to the point of justifying the purges. He saw there a conflict not between right and wrong but between right and right, an historic drama of the kind he had depicted in *St. Joan* (which he had written about the time of the first anathema on Trotsky), a clash between the revolutionary fighting for the future and the established power defending the legitimate interests of the present. André Malraux declared likewise that 'Trotsky is a great moral force in the world, but Stalin has lent dignity to mankind; and just as the Inquisition did not detract from the fundamental dignity of Christianity, so the Moscow trials do not detract from the fundamental dignity [of communism].'[1]

Berthold Brecht's response was similar. He had been in some sympathy with Trotskyism and was shaken by the purges; but he could not bring himself to break with Stalinism. He surrendered to it with a load of doubt on his mind, as the capitulators in Russia had done; and he expressed artistically his and their predicament in *Galileo Galilei*. It was through the prism of the Bolshevik experience that he saw Galileo going down on his knees before the Inquisition and doing this from an 'historic necessity', because of the people's spiritual and political immaturity. The Galileo of his drama is Zinoviev, or Bukharin or Rakovsky dressed up in historical costume. He is haunted by the 'fruitless' martyrdom of Giordano Bruno; that terrible example causes him to surrender to the Inquisition, just as Trotsky's fate caused so many communists to surrender to Stalin. And Brecht's famous duologue: 'Happy is the country that produces such a hero' and 'Unhappy is the people that needs such a hero' epitomizes clearly enough the problem of Trotsky and Stalinist Russia rather than Galileo's quandary in Renaissance Italy.[2]

To Stalin's apologists and to those who washed their hands,

---

[1] Quoted from a summary of Malraux's speech at a banquet given in his honour by the editorial staff of *The Nation*. (The summary was sent to Trotsky by one of his followers and is in *The Archives*, Closed Section.) Malraux had come to the United States with Stalinist backing and was trying to rally support for the International Brigades fighting in Spain. Somewhat earlier Trotsky had attacked him for his attitude towards the purges.

[2] Brecht wrote the original version of *Galileo Galilei* in 1937–8, at the height of the Great Purges.

Trotsky replied with an anger which, however justified, made him look like the proverbial *animal méchant* and gave lukewarm 'defenders of the truth' an excuse for silence. That Sidney and Beatrice Webb refused to join in the protest was not surprising; they had by now become Stalin's admirers. But even men like André Gide and H. G. Wells, whose first impulse was to support the counter-trial, decided in the end to keep aloof. The scope of the campaign thus remained rather narrow; and the various Committees in Defence of Trotsky were composed mostly of declared anti-Stalinists and some anti-communists of long standing; and this restricted still further the effect of their action.

In March 1937 the American, the British, the French, and the Czechoslovak Committees formed a Joint Commission of Inquiry which was to conduct the counter-trial. Its members were: Alfred Rosmer; Otto Rühle, distinguished as the member of the Reichstag, who alone with Karl Liebknecht had voted against war in 1914–15; Wendelin Thomas, also a former communist member of the Reichstag; Carlo Tresca, a well-known anarcho-syndicalist; Suzanne La Follette, a radical, strongly anti-Marxist American writer; Benjamin Stolberg, and John R. Chamberlain, journalists; Edward A. Ross, Professor of Wisconsin University; Carlton Beals, a University lecturer; and Francisco Zamorra, a leftish Latin American writer. Apart from Rosmer, none of the members had ever been associated with Trotsky—most of them were his political opponents. The Commission owed its authority mainly to John Dewey, its chairman, America's leading philosopher and educationist who was also reputed to be a friend of the Soviet Union. John F. Finerty, famous as Counsel of Defence in great American political trials, especially those of Tom Mooney and Sacco and Vanzetti, acted as the Commission's legal counsel.

Trotsky was not at first confident that the Commission would be up to its task. The names of most of its members told him little or nothing; and he had doubts even about its chairman. He wondered whether Dewey, who was nearly eighty, was not too old and too remote from the issues before the Commission. Would he not fall asleep during the hearings? Would he be able to cope with the enormous documentary evidence? And would he as 'friend of the Soviet Union' not be inclined to whitewash Stalin? James Burnham, who was active in organizing the Commission, laid these doubts at rest: 'Dewey is old . . .', he wrote to Trotsky, 'but his mind is still keen and his personal integrity beyond question. It was he, you will recall, who wrote the most searching analysis of the Sacco-Vanzetti case. He will judge the evidence not perhaps as a politician . . . but as a scientist and a logician. He will not sleep during the hearings. . . . It would be a great error to underestimate him. . . . Dewey is, of course, not a Marxist; and all his personal integrity and intelligence does not prevent him from being politically on the fence. In that sense we cannot, obviously, be "quite sure" of him. . . .'[1]

Dewey's accession to the Committee was an act almost heroic. Philosophically he was Trotsky's adversary—they were presently to clash in public controversy over dialectical materialism. For all his radicalism, he stood for the 'American way of life' and parliamentary democracy. As a pragmatist, he was inclined to favour the 'undoctrinaire' and 'practical' Stalin against Trotsky, the 'dogmatic Marxist'. In taking upon himself, at his age, the burden of presiding over the inquiry, he had to break many old associations and give up old friendships. The Stalinists went out of their way to dissuade him. When they failed, they shrank from no obloquy and slander— their mildest aspersion on him was that he had 'fallen for Trotsky' from sheer senility. The *New Republic*, of which he had been a founder and on whose editorial board he had sat for nearly a quarter of a century, turned against him; and he resigned from it. His next of kin implored him not to tarnish the lustre of his name by participating in a shady and shabby business. Intrigue and harassment only hardened his resolve. The fact that so many influences were set in motion to obstruct his action, openly and surreptitiously, was to him an argument in its favour. He put aside work on a treatise *Logic: the Theory of Inquiry*, which he regarded as his *magnum opus*, in order to plunge into the practical experience of this particular inquiry. In the course of weeks and months he pored over the blood-reeking pages of the official reports of the Moscow trials, over Trotsky's voluminous writings and correspondence, and

[1] Burnham to Trotsky, 1 April 1937, *The Archives*, Closed Section.

mountains of other documents. He took notes, compared facts, dates, and allegations, until he was thoroughly versed in all aspects of the case. Again and again he had to resist intimidation and threats. Nothing shook his equanimity or weakened his energy. The Commission was to cross-examine Trotsky as chief witness; and as there was no chance that the American Government would allow him to come to New York, Dewey decided to carry out the investigation in Mexico. He was warned that the Mexican Federation of Workers would not allow the counter-trial to take place; that he and his companions would be met with hostile demonstrations on the frontier; and that they would be mobbed. Unmoved, the old philosopher pursued his course. His mind was open. Though he was convinced that Trotsky's guilt had not been proven in Moscow, he was not yet sure of Trotsky's innocence. Determined not only to maintain strict impartiality, but to make the impartiality evident to all, he never met Trotsky outside the Commission's public sessions, although he 'would have liked to talk to him informally as man to man'.[1]

The Commission opened its hearings on 10 April. It had intended to hold them in a large hall in the centre of Mexico City; but it gave up the idea in order to avoid public disturbance and to save money. The sessions were held at the Blue House, in Trotsky's study. 'The atmosphere was tense. There was a police guard outside . . . visitors were searched for guns and identified by a secretary of Trotsky who was, himself, armed.' The French windows of the room facing the street 'were covered, and behind each of them there were six-foot barricades of cemented brick and sand bags. . . . These brick barricades had been completed the night before.' About fifty people were present, including reporters and photographers. The hearings were conducted in accordance with American judicial procedure. Dewey had invited the Soviet Embassy and the Communist parties of Mexico and the United States to send representatives and take part in the cross examination; but the invitations were ignored.[2]

---

[1] This account is based on what Dewey himself and Dr. Ratner, his secretary, told me in 1950.

[2] *The Case of Leon Trotsky*, appendix III; See also James Farrell in *John Dewey*, (A *Symposium*, edited by Sidney Hook) p. 361.

In a brief inaugural statement Dewey declared that the Commission was neither a court nor a jury but merely an investigating body. 'Our function, is to hear whatever testimony Mr. Trotsky may present to us, to cross-examine him, and to give the results of our investigation to the full Commission of which we are part. . . .' The title of the American Committee for the Defence of Leon Trotsky did not mean that the Committee stood for Trotsky; it acted 'in the American tradition', on the belief that 'no man should be condemned without a chance to defend himself'. Its aim was to secure a fair trial where there was a suspicion that the accused man was denied such a trial. The case was comparable with the cases of Mooney and of Sacco and Vanzetti; but the latter could at least make their pleas before a legally constituted court, whereas Trotsky and his son had twice been declared guilty in their absence by the highest Soviet tribunal; and his repeated demands that the Soviet Government ask for his extradition, which would have brought him automatically before a Norwegian or a Mexican court, had been ignored. 'That he has been condemned without the opportunity to be heard is a matter of utmost concern to the Commission and the conscience of the whole world.' Explaining his own motives for participation, Dewey said that having given his life to social education, he treated his present work as a great social and educational task—'to act otherwise would be to be false to my life's work'.

The proceedings lasted a full week and took up thirteen long sessions. Dewey, Finerty, A. Goldman, Trotsky's lawyer, and others cross-examined Trotsky on every detail of the charges and the evidence. At times the cross-examination almost turned into a political dispute, when some of the examiners insisted on Trotsky's and Lenin's moral co-responsibility for Stalinism, and Trotsky refuted the imputation. There was not a single question into which he refused to go or which he dodged. Despite the controversial interludes, the hearings proceeded calmly and smoothly; they were disturbed only once by the so-called Beals incident.

Carlton Beals, a member of the Commission, repeatedly addressed to Trotsky questions which were more or less irrelevant but showed a marked pro-Stalinist bias and were extremely offensive in form. Trotsky answered composedly and to the point. Towards the end of a long session on 16 April, Beals went into a political argument and maintained that while Stalin, the expounder of socialism in one country, represented the mature statesmanship of Bolshevism, Trotsky was something of an incendiary bent on fomenting world revolution. Trotsky replied that in the Moscow trials he had been described as the fomentor not of revolution but of counter-revolution and as Hitler's ally. Beals then asked him whether he knew Borodin, Stalin's former emissary to China and adviser to Chiang Kai-shek. Trotsky replied that he had not met him personally although he had, of course, known of him. But, Beals asked, had Trotsky not sent Borodin to Mexico in 1919 or 1920 in order to found the Communist party there? The question suggested that Trotsky had lied to the Commission and moreover that he had tried to foment revolution even in the country that was giving him shelter at present. The exchange grew hot. With his Norwegian experience still fresh, Trotsky suspected that the question might have been designed to incite Mexican opinion against him, to rob him of asylum, and to disrupt the counter-trial. He pointed out that he had always set his hopes on world revolution, but had sought to promote it by politically legitimate means, not by staging coups in foreign countries. The allegation that he had sent Borodin to Mexico in 1919–20 was fantastic. At that time, at the height of civil war, he hardly ever left his military train; he had his eyes fixed on the maps of his fronts and had nearly forgotten 'all his world geography'.

Beals emphatically reiterated his allegation and added that Borodin himself had declared that Trotsky had sent him to Mexico, and also that already in 1919 the Soviet Communist party was torn between the statesmen and the fomentors of revolution. 'May I ask the source of this sensational communication?', Trotsky inquired. 'Is it published?' 'It is not published', said Beals. 'I can only give the advice to the Commissioner to say to his informant that he is a liar', Trotsky retorted. 'Thank you, Mr. Trotsky. Mr. Borodin is the liar.' 'Very possible', was Trotsky's laconic reply. Before the end of the hearing he protested against Beals's 'tendentious Stalinist tone'. The incident looked to him more and more sinister.

The Borodin affair had nothing to do with the Moscow trials
and seemed to have been dragged in only to embarrass him
and the Mexican Government. And so at the opening of the
next session he once again denied Beals's assertion and asked
the Commission to throw light on its source. If Beals had his
information directly from Borodin, let him say where and
when he obtained it. If, indirectly, then in what way, through
whom and when did he get it? A probing into these questions
should reveal a design aiming at disrupting the counter-trial.
'If Mr. Beals himself is not consciously and directly involved
in this new intrigue, and I will hope that he is not, he must
hasten to present all the necessary explanations in order to
permit the Commission to unmask the true source of the
intrigue.' As Beals refused to reveal that source, the Commission censured him in private session; and he resigned from
the Commission. The incident had no further sequel.[1]

The results of the cross-examination were summed up by
Trotsky himself in his final plea on 17 April.[2] Showing signs
of strain and fatigue, he asked to be allowed to read his statement sitting. He began by pointing out that either he and nearly
all members of Lenin's Politbureau were traitors to the Soviet
Union and communism, as the accusers in Moscow claimed,
or else Stalin and his Politbureau were forgers. *Tertium non
datur*. It was said that to delve into this question was to interfere with the domestic affairs of the Soviet Union, the Fatherland of the workers of the world. It would be 'a strange
Fatherland' whose affairs the workers were not allowed to
discuss. He himself and his family had been deprived of Soviet
citizenship; they had no choice but to place themselves 'under
the protection of international public opinion'. To those who,
like Charles A. Beard, held that the onus of the proof lay on
Stalin, not on him, and that it was anyhow impossible to
'disprove a negative with positive evidence', he replied that the
legal conception of an alibi presupposed the possibility of
such a disproof and that he was in a position to establish his alibi
and to demonstrate the 'positive fact' that Stalin had organized
'the greatest frame-up in history'.

The juridical examination of the case, however, was 'concerned with the *form* of the frame-up and not with its *essence*',

---

[1] *The Case of Leon Trotsky*, pp. 411–17.   [2] Ibid., pp. 459–585.

which was inseparable from the political background of the purges, the 'totalitarian oppression, to which . . . all are subjected, accused, witnesses, judges, counsel, and even the prosecution itself'. Under such oppression a trial ceases to be a juridical process and becomes a 'play, with the roles prepared in advance. The defendants appear on the scene only after a series of rehearsals which give the director in advance complete assurance that they will not overstep the limits of their roles'. There was no room for any contest between prosecution and defence. The chief actors performed their parts at pistol point. 'The play can be performed well or badly; but that is a question of inquisitorial technique and not of justice.'

In evaluating the accusation one must consider the political record of the defendants. A crime usually arises from the criminal's character or is at least compatible with it. The cross-examination was therefore necessarily concerned with his and the other defendants' work in the Bolshevik party and with their roles in the revolution; and in the light of these the crimes imputed to them were utterly incompatible with their characters. That was why Stalin had to falsify their records. The classical criterion *Cui prodest?* had to be applied here. Was or could the assassination of Kirov be of any advantage to the Opposition? Or was it of advantage to Stalin, whom it provided with a pretext for the extermination of the Opposition? Could the Opposition hope to benefit in any way from acts of sabotage in coal mines, factories, and on the railways? Or did the government, whose insistence on over-hasty industrialization and whose bureaucratic neglect had caused many industrial disasters, seek to exculpate itself by blaming the Opposition for these disasters? Could the Opposition gain anything from an alliance with Hitler and the Mikado? Or was Stalin making political capital out of the defendants' confessions that they were Hitler's allies?

It would have been suicidal folly for the Opposition to commit any of these crimes. The unreality of the accusation accounted for the prosecution's inability to produce any valid evidence. The conspiracy of which Vyshinsky spoke was supposed to have gone on for many years and to have had the widest ramifications in the Soviet Union and abroad. Most of its supposed leaders and participants had all these

years been in the G.P.U.'s hands. Yet the G.P.U. could not
adduce any realistic data or even a single factual piece of evidence for the gigantic conspiracy—only confessions, confessions,
endless confessions. The 'plot had no flesh and blood'. The
men in the dock related not any specific events or actions of
the conspiracy but only their conversations about it—the court
proceedings were a conversation about conversations. The lack
of all psychological verisimilitude and factual content showed
that the spectacle had been enacted on the basis of an especially
prepared 'libretto'. Yet 'a frame-up on such a colossal scale is
too much even for the most powerful police . . . too many
people and circumstances, characteristics and dates, interests
and documents . . . do not fit . . . the ready-made libretto!' 'If
one approaches the question in its artistic aspect, such a task
—the dramatic concordance of hundreds of people and of
innumerable circumstances—would have been too much even
for a Shakespeare. But the G.P.U. does not have Shakespeares
at its beck and call.' As long as they were concocting events
supposed to have taken place inside the U.S.S.R. they could
still maintain a semblance of coherence. Inquisitorial violence
could force defendants and witnesses to be consistent in some
of their fantastic tales. The situation changed when the threads
of the plot had to be extended to foreign countries; and the
G.P.U. had to extend them there in order to implicate him,
the 'public enemy number one'. Abroad, however, the facts,
dates, and circumstances could be verified; and whenever this
was done the story of the conspiracy fell to pieces. Not a single
one of the 'threads' that were supposed to lead to Trotsky had
led to him. It was established that the few defendants, David,
Berman-Yurin, Romm, and Pyatakov, to whom he had
allegedly given terrorist orders (in the presence of his son
or otherwise) had not and could not have seen him (and his
son) at the places and on the dates indicated, because either
he (and his son), or they, were not and could not have been
there. Yet, these contacts disproved, the whole accusation
collapsed because his alleged contacts with Radek (through
Romm) and Pyatakov were crucial to the 'conspiracy'. All
other accusations and testimonies had been based on, or
derived from, Pyatakov's and Radek's confessions that they
had acted as Trotsky's chief agents and as the twin pillars of

the conspiracy. 'All the testimony of the other accused rests upon our own testimony', Radek himself had declared in court; and their own testimony, which centred on the meetings with Trotsky in Paris and Oslo, rested on nothing. 'It is hardly necessary to demolish a building brick by brick once the two basic columns on which it rests are thrown down', Trotsky pointed out; yet he went on demolishing the building 'brick by brick'.

He asked the Commission to consider that his own versions were full of that psychological and historical authenticity which was so conspicuously lacking in Moscow's versions; that the documentation he had placed before the Commission reflected with extraordinary fullness his life and work over many years: and that had he committed any of the crimes, surely his papers would betray him at one point or another. People who easily swallowed camels but strained at gnats were saying that he could have arranged all his archives and all the files of his correspondence so as to camouflage his real designs. Yet for the purpose of camouflage one can compose five, ten, even a hundred documents—not thousands of letters addressed to hundreds of persons, not hundreds of articles and dozens of books. No, he had not 'built a skyscraper to camouflage a dead rat'. If someone had declared, for instance, that Diego Rivera was a secret agent of the Catholic Church, would not any jury investigating the accusation inspect Rivera's frescoes? And would anyone dare to say that the impassioned anti-clericalism evident in those frescoes was mere camouflage? No one can 'pour out his heart's blood and nerves' sap' in works of art, history, and revolutionary politics just in order to deceive the world. How hollow by comparison with his documentation was Vyshinsky's: all it consisted of were Trotsky's letters: two to Mrachkovsky, three to Radek, one to Pyatakov, and one to Muralov—all faked!

But why had the defendants made their confessions? He could hardly be expected to offer precise information about the G.P.U.'s inquisitorial techniques. 'We could not here question Yagoda (he is now himself being questioned by Yezhov) or Yezhov, or Vyshinsky, or Stalin, or . . . their victims, the majority of whom have already been shot.' However, the Commission had before it the affidavits of Russian and

European communists who had themselves been subjected to the G.P.U.'s techniques. It was all too often forgotten that those who made the confessions had not been active opposition leaders but capitulators, who had for years prostrated themselves before Stalin. Their last confessions were the consummation of a long series of surrenders, the conclusion of a truly 'geometric progression of false accusations'. In the course of thirteen years Stalin had with their help erected a 'Babel tower' of slander. A dictator who used terror without inhibition and 'could buy consciences like sacks of potatoes' was well able to perform such a feat. But Stalin himself was terrified by his tower of Babel, for he knew that it must collapse after the first breach in it had been made—and made it would be!

Trotsky ended with an apotheosis of the October Revolution and of communism. Even under Stalin, he said, despite all the horror of the purges, Soviet society still represented the greatest progress in social organization mankind had so far achieved. The blame for the tragic degeneration of Bolshevism lay not on the revolution but on its failure to extend beyond Russia. For the time being Soviet workers were confronted with a choice between Hitler and Stalin. They preferred Stalin; and in this they were right: 'Stalin is better than Hitler.' As long as they saw no other alternative, the workers remained apathetic even in the face of the monstrosities of Stalinist rule. They would shed the apathy the very moment when any prospect appeared abroad of new victories for socialism. 'That is why I do not despair . . . I have patience. Three revolutions made me patient.'

The experience of my life, in which there has been no lack either of success or of failures, has not only not destroyed my faith in the clear, bright future of mankind, but, on the contrary, has given it an indestructible temper. This faith in reason, in truth, in human solidarity, which at the age of eighteen I took with me into the workers' quarters of the provincial Russian town of Nikolayev— this faith I have preserved fully and completely. It has become more mature, but not less ardent.

With these words, and with thanks to the Commission and its chairman, he concluded this *apologia pro vita sua*.

For a long while the Commission sat in silence deeply shaken. Dewey had intended to sum up and close the proceedings in a formal manner; instead he brought the hearings to an end with this single sentence: 'Anything I can say will be an anti-climax.'[1]

The record of the cross-examination is all the more remarkable because of the handicaps Trotsky had imposed upon himself. He often pulled his punches so as not to embarrass the Mexican Government unduly. He sought to explain the many involved issues between himself and Stalin not in his accustomed Marxist idiom, which might have been unintelligible to his audience, but in the language of the pragmatically minded liberal—the difficulty of such a translation can be appreciated only by those who have ever attempted it. Eager for personal contact with his listeners, he conducted his defence not in his native tongue or even in German or French but in English. His vocabulary was limited. His grammar and idiom were shaky. Stripped of the splendours of his mighty eloquence, denying himself the advantages which even the humdrum speaker finds in the use of his native language, he answered impromptu the most varied, complex, and unexpected questions. Day after day, and hearing after hearing, he searched for expression and struggled with the resistance of the language, frequently halting or stumbling into unintentionally comic sentences, and sometimes saying almost the opposite of what he meant to say, or failing to understand questions put to him. It was as if Demosthenes, his stammer uncured and his mouth full of pebbles, had come to court to fight for his life. Thus he recounted the events of his long career, expounded his beliefs, described the many changes in the Soviet régime, analysed the issues that had separated him from Stalin and Bukharin, but also from Zinoviev and Kamenev, portrayed the personalities, and delved into every phase of the terrible contest.

By the end no question had been left unanswered, no important issue blurred, no serious historic event unilluminated. Thirteen years later Dewey, who had spent so much of his life in academic debate and was still as opposed as ever to Trotsky's *Weltanschauung*, recalled with enthusiastic admiration 'the intellectual power with which Trotsky had assembled and

---

[1] Dewey added only a few formal announcements concerning the Commission's further work.

organized the mass of his evidence and argumentation and conveyed to us the meaning of every relevant fact'. The incisiveness of Trotsky's logic got the better of his unwieldy sentences, and the clarity of his ideas shone through all his verbal blunderings. Even his wit did not succumb; it often relieved the gloom of his subject-matter. Above all, the integrity of his case allowed him to overcome all external restraint and constraint. He stood where he stood like truth itself, unkempt and unadorned, unarmoured and unshielded yet magnificent and invincible.

. . . .

It was to be several months before the Dewey Commission got ready with its verdict. Meanwhile, Trotsky had still to supplement the evidence he had laid before it; and he kept the whole household busy. The cross-examination and the work connected with it had worn him out; and he did not recuperate during a brief stay in the country. For the rest of the spring and the summer he suffered again from severe headaches, dizziness, and high blood pressure and complained again about old age that had 'caught him by surprise'. The first echoes of the counter-trial were less than faint.[1] The strains in the family were scarcely diminished. 'Dear Papa', Lyova wrote towards the end of April, 'you continue to subject me to your ostracism . . . it is more than a month since I received any letter from you.' Trotsky, still dissatisfied with the way Lyova was managing the *Bulletin*, had again proposed to transfer it to New York; in reply Lyova calmly pointed out that the paper should remain in Europe where most of its readers were; and he again bitterly complained to his mother about the rough treatment he was getting. In a long and somewhat apologetic letter, Trotsky then tried to smooth matters out.[2] He

---

[1] Both the British and the French Committees for the Defence of Leon Trotsky reported to Coyoacan that the newspapers of their respective countries were ignoring the counter-trial almost completely.

[2] Lyova to Trotsky, 27 April; Trotsky to Lyova, 29 May, 1937. *The Archives*, Closed Section. Over twenty years later Natalya told me that Trotsky had written 'a very long and very cordial letter' to Lyova which 'cleared away all these misunderstandings'. She promised to find that letter, though she feared that it might have been laid astray. She probably had in mind the letter summarized here. This, however, did not quite succeed in 'clearing the misunderstandings'.

explained to Lyova that having lost so many months in Norway before he could prepare for the counter-trial, he was then irritated by further delays and anxious to be able to place his full dossiers before the Dewey Commission; and he was convinced that the delays had been caused by Lyova's reluctance to co-operate with comrades. He advised him to take a rest and steady his nerves: 'great trials are still ahead for both of us'.

The advice was timely enough. Lyova too was suffering from headaches and fevers; and he did not have his father's resilience. 'What is left of my old strength?' he wrote to his mother, hinting that he would presently need 'a small operation'. He lived in poverty, but thought of helping his parents financially by earning a living as a factory worker or gaining an academic scholarship. When Natalya urged him to write for newspapers instead, he replied with a note of frustration: 'Writing ... comes with difficulty to me—I have to read, study, reflect, which requires time. . . . Yet since I have been in emigration I have been burdened almost continually with technical and other chores. I am a beast of burden, nothing else. I do not learn, I do not read. I cannot aspire to do any literary work: I do not have the light touch and the talent that can partly replace knowledge.'[1] This mood of frustration was suffused with tenderness and devotion. When his parents sent him back cheques he had collected from French publishers and forwarded to them, Lyova took only a little for himself and divided the rest among needy comrades or paid it into the organization's funds. He worried lest his father was expending his strength too recklessly and shattering his nerves. Why, he asked Natalya, had they not bought a car in Mexico and organized hunting or fishing trips? Why did L.D. not play croquet of which he used to be so fond? 'My dear darling Mamochka', he wrote in reply to a rather sad letter from her, '. . . think only what might have happened if Stalin had not committed the "mistake" of banishing Papa? Papa would have been dead long ago. . . . Or if I had been allowed to return to the U.S.S.R. in 1929, if Sergei had been active in politics, or if Papa had been in Norway now, or, worse still, in Turkey? Kemal would have handed him over . . . things might have been far, far worse.'[2] These

---

[1] Lyova to his mother, 29–30 June 1937.
[2] Letter of 7 July 1937. *The Archives*, Closed Section.

were poor consolations, of course; yet no better ones were at hand.

About this time there occurred a somewhat tragi-comic incident in the family's intimate life. Amid all these grim events and amid all her anguish, Natalya was troubled by marital jealousy. What caused it exactly is not quite clear: she is most discreet about this even in her letters to her husband, which leave no doubt on one point only—namely that this was the first time she felt she had reason to be jealous. Perhaps a less self-assured woman would have been jealous earlier, for Trotsky's behaviour towards women, in those rare moments when he could notice them, was characterized by a sort of articulate gallantry, not free from male vanity and sensitiveness to female admiration. At any rate, a woman's presence sometimes stimulated him to dashing displays of seductive verve and wit. There was old-fashioned chivalry and artistic finesse in these 'flirtations'; yet they were somewhat at variance with his high seriousness and his almost ascetic life. Natalya, however, was confident enough of his love not to take them amiss. But at Coyoacan she became acutely jealous of someone to whom she referred in her letters only by the initial F. To judge from circumstantial evidence, this may have been Frida Kahlo. Members of the household soon noticed a discord between the two women and a slight cooling off between their husbands. We do not know whether Frida's uncommon delicate beauty and artistry excited in Trotsky more than the normal gallantry or whether Natalya, now fifty-five, fell a prey to the jealousy that often comes with middle age. Enough that a 'crisis' ensued and both Trotsky and Natalya were unhappy and miserable.[1]

In the middle of July he left Coyoacan and with his bodyguard went to the mountains to get physical exercise, to do farmwork on a large estate, and to ride and hunt. Daily, sometimes even twice daily, he wrote to Natalya. He had promised her to say nothing in his letters about her upset but 'could not help breaking the promise': he implored her to 'stop competing with a woman who meant so little' to him while she, Natalya, was all to him. He was full of 'shame and self-hatred' and

---

[1] See their correspondence of July 1937. *The Archives*, Closed Section.

V. Leon Sedov (Lyova)

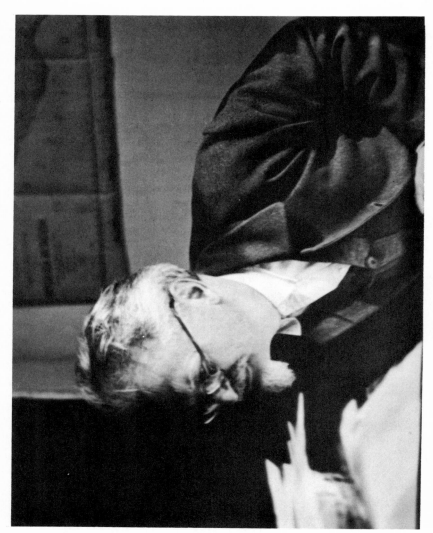

VI. Trotsky at his desk

signed himself 'your old faithful dog'. 'How I love you, Nata, my only one, my eternal one, my faithful one, my love, my *victim*. . . .' 'Ah, if only I could still bring a little joy into your life. As I am writing this, after every two or three lines, I get up, walk about my room, and weep with tears of self-reproach and of gratitude to you; and I weep over old age that has caught us by surprise.' Again and again the note of self-pity, which no stranger and no member of his household could ever detect in him, breaks through in these letters. '. . . I am still living with our yesterdays, with our pangs and memories, and with the torments of my suffering.' Then his resilience and even joy of life come back: 'All will be well, Nata, all will be well—only you must recover and get stronger.' Once he relates to her somewhat teasingly how he 'charmed' a group of men, women, and children—'especially the women'—who had visited him in the mountains. His vitality surges up and he feels a sexual craving for Natalya. He relates to her that he had just re-read the passage in Tolstoy's memoirs where Tolstoy describes how at the age of seventy he would come back from his riding full of desire and lust for his wife—he, Trotsky, at fifty-eight, was returning in the same mood from his strenuous escapades on horseback. In his craving for her he breaks into the slang of sex, and then feels 'abashed at putting such words on paper for the first time in my life' and 'behaving just like a young cadet officer'. And as if to prove the *nihil humanum* . . ., he indulges in an odd marital recrimination. He rakes up a love affair Natalya was supposed to have had as far back as 1918; and he pleads that as he had never made her the slightest reproach and had never even mentioned that affair, she should not be too severe on him, who had not given her any ground for jealousy. In reply she explains the 'affair' of 1918. This was just after she had been appointed Director of the Museums Department in the Commissariat of Education; she did not quite know how to organize her work; and one of her assistants, a comrade who was admittedly 'infatuated' with her, helped her. She was grateful and treated him with sympathy; without, however, reciprocating his feelings or allowing him any intimacy. This gently comic recrimination, in which, after thirty-five years of common life, husband and wife found no other 'infidelity' with which to reproach each

other, reveals in quite unexpected a manner the steadfastness
of their love.[1]

In her letters Natalya appears reserved, somewhat embarrassed by his outbursts, and anxious to bring him back to
himself from his introspective and effusive moods. To his
harping on old age she invariably has this answer: 'One is old
only when one has no prospect ahead' and when one no longer
strives for anything—and this surely could not be true of him!
'Pull yourself together. Get back to work. If only you do this,
your cure will have begun.' Before long she was again mistress
of her emotions; and, though herself ill and tense, she was
pre-occupied with the diseases, fevers, misfortunes, and strivings
of every member of the family, and was calmer and stronger
than any of them. He knew her fortitude and relied on it. In
one of his letters to her there occur these telling words: '*You
will still carry me on your shoulders, Nata, as you have carried me
throughout our life.*'[2]

.   .   .   .

Meanwhile, in the Soviet Union hardly a day passed without
its human hecatomb. Towards the end of May the G.P.U.
announced that they had discovered a conspiracy at the head
of which had stood Marshal Tukhachevsky, the deputy Commissar of Defence, the modernizer and actual Commander-
in-Chief of the Red Army. Outstanding generals Yakir, Uboro-
vich, Kork, Putna, Primakov and others, including Gamarnik,
the chief Political Commissar of the armed forces, were charged
with treason. With the exception of Gamarnik, who committed
suicide, all were executed. Of the four marshals whose signatures appeared under the death sentence, Voroshilov,
Budienny, Blucher, and Yegorov, the last two presently also
faced the firing squad. All these men had risen to their positions
of command while Trotsky was Commissar of War; but most
of them had never belonged to the Opposition and none of
them had been in contact with Trotsky since his banishment.
Yet all were accused of being his and Hitler's accessories and

---

[1] Natalya's 'explanatory' letter is undated; judging from internal evidence, it
was written around 15 July. On 19 July Trotsky answered with two letters. He
also wrote in these days a special diary meant (as he repeatedly underlined) only
for Natalya's eyes.

[2] Trotsky to Natalya on 18 July 1937. *The Archives*, Closed Section.

of working for the military defeat of the Soviet Union and its dismemberment. Their executions were the prelude to a purge which affected 25,000 officers and decapitated the Red Army on the eve of the Second World War. Twenty-five years later, after the formal rehabilitation of Tukhachevsky and most of the other generals, no light has yet been thrown on the background to this purge. According to various anti-Stalinist sources, Tukhachevsky, alarmed by the terror which was sapping the nation's morale and defences, had planned a *coup d'état* in order to overthrow Stalin and break the power of the G.P.U.; but he had done this without any connexion with Trotsky, let alone with Hitler or any foreign power. Trotsky did not believe that there had been any plot, but described Tukhachevsky's fall as a symptom of a conflict between Stalin and the officer corps, a conflict which might place a military *coup* 'on the order of the day'.[1]

By this time the G.P.U. were already rehearsing the 'trial of the twenty-one', casting Rykov, Bukharin, Tomsky, Rakovsky, Krestinsky, and Yagoda, for the main roles. (Of all these, Tomsky alone, by committing suicide, escaped the humiliation of a public trial and confession.) Even before the curtain rose on this spectacle, the terror struck at the Stalinist faction too. Rudzutak, Mezhlauk, Kossior, Chubar, Postyshev, Yenukidze, Okuzhava, Elyava, Chervyakov, and others, members of the Politbureau, party secretaries of Moscow, of the Ukraine, Byelorussia, and Georgia, trade union leaders, heads of the State Planning Commission and of the Supreme Council of the National Economy, nearly all of them Stalinists of long standing, were branded as traitors and foreign spies, and executed. Ordjonikidze, who had been devoted to Stalin for more than thirty years but became troubled in his conscience and began to oppose him, died in mysterious circumstances, or, as some believed, was driven to commit suicide. If the Trotskyists, Zinovievists, and Bukharinists were disgraced publicly, these Stalinists were destroyed secretly without open trials. The havoc which Stalin's rage wrought among them was hidden in obscurity. The terror spread beyond the Bolshevik party and caught many German, Polish, Hungarian, Italian, and Balkan communists who had lived in the Soviet Union as

[1] *B.O.*, nos. 56–57, 1937.

refugees from prisons and concentration camps in their own countries. Then the 'drive against Trotskyism' was carried into foreign lands. In Spain, the G.P.U. established themselves early in the civil war and launched an attack on the P.O.U.M. Andres Nin, the leader of the P.O.U.M., had his differences with Trotsky, who criticized him for his participation in the loyalist government of Catalonia and for adopting a 'timid and semi-Menshevik' attitude in the revolution. Even so, Nin's policy was far too radical and independent for the Stalinism of the Popular Front period; and so he and his party were denigrated as Franco's 'fifth column'; in the end he was kidnapped and assassinated. Whoever dared to protest, exposed himself to G.P.U. vengeance. The witch hunts, the assassinations, and the cynicism with which Stalin used the Spanish revolution demoralized the Republican camp and prepared its defeat. And as if in mockery, Stalin sent none other than Antonov-Ovseenko, the ex-Trotskyist and hero of 1917, to preside over the purge in Catalonia, the stronghold of the P.O.U.M.; then, after Antonov had done his job, he denounced him too as a wrecker and spy and ordered his execution.

In Moscow no one was safe now, not even the inquisitors and the hangmen. After Yagoda's arrest, the G.P.U. and all secret services were purged. Their agents in Europe were lured back to face the usual accusations. As a rule, these agents knew or guessed what awaited them, but as if hypnotized they obeyed the summons—many preferred self-immolation to asylum in any capitalist country. It was therefore a startling event, when Ignaz Reiss, chief of a network of the Soviet secret service in Europe, resigned from his post in protest against the purges. When he made up his mind to do this, he had not even been summoned back to Moscow. Shaken by the purges, he approached Sneevliet, the Dutch Trotskyist parliamentarian (and through him Lyova), in order to warn Trotsky that Stalin had decided to 'liquidate Trotskyism' outside the Soviet Union by the same means he was using to destroy it inside. Reiss described the infernal sadism and blackmail, the long and horrible interrogations, through which the G.P.U. had obtained the confessions in the Moscow trials and the moral torture and confusion in which the old generation of Bolsheviks was meeting its doom; but he also depicted the young communists who refused to submit and still filled prison yards and execution places with the cry: 'Long live Trotsky!'[1]

On 18 July, Reiss addressed a message from Paris to the Central Committee in Moscow announcing his break with Stalinism and 'adherence to the Fourth International'. 'The day is not far', he stated, 'when international socialism will sit in judgement over all the crimes committed in the last ten years. Nothing will be forgotten, nothing forgiven. . . . The "leader of genius, Father of the peoples, Sun of Socialism" will have to give account for all his deeds.' 'I am returning to you the Order of the Red Banner awarded to me in 1928. To wear it . . . would be beneath my dignity.'[2]

Six weeks later, on 4 September, Reiss was found dead, his body bullet-ridden, on a Swiss road near Lausanne. The G.P.U. had known about his decision even before he handed his letter of resignation to an official of the Soviet Embassy in Paris. Knowing the disgust which the purges had aroused even among his former colleagues in the secret service, he had hoped to persuade some of them to follow his example. With this purpose in mind he had arranged to meet in Lausanne Gertrud Schildbach, a Soviet agent resident in Italy, who had been a friend of his for close on twenty years. They met; she pretended sympathy; and after their first talk, she lured him to another meeting on the outskirts of Lausanne. There the G.P.U. had laid a trap for him.

The Swiss and French police soon brought to light some of the circumstances. Using clues found in an abandoned blood-stained car and in luggage left in hotels, they established the identity of the assassins. These, it turned out, had been members of the Society for the Repatriation of Russian Émigrés in Paris, a society sponsored by the Soviet Embassy.

The police ascertained that the gang which had killed Reiss had long kept a watch on Lyova also. A woman in whose name the blood-stained car was hired, had been detailed to shadow him. (He recalled that a year earlier she had followed him to the south of France, where he had gone for a short rest, had installed herself in his *pension* and had occasionally

---

[1] '*Zapiski Ignatya Reissa*', B.O., nos. 60–61, December 1937.
[2] I. Reiss, '*Pismo v Ts.K. V.K.P.*' in B.O., nos. 58–59, September-October 1937.

urged him with strange insistence to go out with her on sailing trips.) Further investigation disclosed that the same gang had laid a trap for Lyova in Mulhouse, near the Swiss border, in January 1937, when he was planning to go there in order to discuss with a Swiss lawyer a lawsuit against Swiss Stalinists. He avoided the trap because ill-health prevented him from making the journey; but the gang had gone on shadowing him throughout the first half of the year and he sensed it. In July and August he was puzzled to notice that the watch on him had almost ceased—evidently his pursuers were then busy keeping track of Reiss. Now they could be expected to return to their old hunt.[1]

Lyova was startled to learn from the interrogation how quickly and accurately the G.P.U. agents were as a rule informed about all his plans and moves. By whom? And who had informed them about Reiss's intentions? Some Trotsky-ists already wondered whether an *agent provocateur* was not to be found among Lyova's closest friends; and suspicion turned on Étienne (who had quite recently worked for the Society for the Repatriation of Russian Émigrés). Sneevliet's distrust of Étienne was so strong that after Reiss had approached him, he at first refused to put him in touch with the Trotskyist centre in Paris, fearing that this would be dangerous.[2] Lyova, however, refused to countenance any suspicion of his 'best and most reliable comrade'.

With the feeling that a mysterious noose was tightening around his neck, Lyova wrote Reiss's obituary for the *Bulletin*.[3] 'The "Father of the Peoples" and his Yezhovs know all too well how many potential Reisses there are around. . . . Stalin's designs will be defeated. . . . No one can bring history to a halt with a gun. Stalinism is doomed; it is rotting and dis-integrating before our eyes. The day is near when its stinking corpse will be thrown into the sewers of history.' Yet Reiss's fate deterred potential imitators. In the next few weeks only two of these came forward: Walter Krivitsky, another senior agent of the secret service, and Alexander Barmin, Soviet

[1] See Lyova's cable to Coyoacan of 16 September and his letters to Trotsky of 4 and 12 October 1937. *The Archives*, Closed Section. See also N. Markin's (Lyova's pen-name) account of the assassination in *B.O.*, nos. 58–59.

[2] In addition to Lyova's letters quoted above see his letter of 7 August 1937.

[3] *B.O.*, nos. 58–59.

chargé d'affaires in Athens. They too, having broken with their government, sought contact with Trotsky, whose adherents they had never been, because, as Krivitsky put it, Trotsky was 'surrounded by an aureole' even in the eyes of the G.P.U. men assigned to the struggle against Trotskyism.[1] These were strange converts. Krivitsky feared that Trotsky and his adherents would distrust and despise him as one who had spent so many years in Stalin's service. He was therefore anxious to justify his past at the very moment he was breaking with it. Reiss's widow accused him of complicity in the assassination of her husband. He bowed his head and confessed that he was not blameless.[2] He was eager to wipe out his guilt by revealing the truth about the purges; yet he was also anxious to guard the many secrets in his possession which had a bearing on Soviet military security. Lyova listened to his tortured confidences with some distaste. But he considered it his duty to transmit the information to his father and also to help, comfort, and as far as possible protect any Soviet citizen who broke with Stalinism. Trotsky, on his part, urged Krivitsky and Barmin, for their own safety and for the sake of political clarity, to come out against Stalin unequivocally and in broad daylight; he was uneasy about their contortions and impatient with Lyova's indulgence. This led to renewed altercations between father and son.[3]

Meanwhile, the presence of an *agent provocateur* in Lyova's circle caused more and more suspicion and confusion. Krivitsky had confirmed Reiss's warnings about the forthcoming assassinations of Trotskyists and said that the G.P.U. had their 'eyes and ears' inside the Trotskyist centre in Paris. He was, however, unable to identify the *agent provocateur*, and he cast a suspicion on Victor Serge, of all people. The G.P.U., he said, would not have released Serge and allowed him to leave the Soviet Union unless they were sure that he would spy for them on the Trotskyists. No one was, of course, less suited to act such a part than Serge. He was one of Trotsky's early adherents, a gifted and generous, though politically

[1] *B.O.*, nos. 60–61, December 1937.
[2] Lyova to Trotsky on 19 November 1937 and Trotsky to Lyova on 22 January 1938. *The Archives*, Closed Section.
[3] See Trotsky, '*Tragicheskii Urok*', *B.O.* ibid.; and Lyova's letters of 16 and 19 November and 17 December 1937.

ingenuous, man of letters. The worst that might be said of
him was that he had a foible for vainglorious chatter and that
this was a grave fault in a member of an organization which
had to guard its secrets from the G.P.U. In any case suspicion began to cling indiscriminately to anyone, even to
Lyova himself, while the actual *agent provocateur* went on
collecting and reading Trotsky's mail, shared all of Lyova's
secrets, and used his wiles to keep his own reputation clear by
casting distrust upon others.[1]

The French police, continuing to investigate the Reiss case,
discovered that one of the gang of assassins had applied for a
Mexican visa and had supplied himself with detailed plans of
Mexico City. Lyova at once conveyed the warning to Coyoacan.
The police also took a grave view of the danger to Lyova's life
and assigned a special guard to him.[2] One of his comrades
—almost certainly Klement ('Adolf')—took Lyova's plight so
much to heart that he wrote to Trotsky and Natalya begging
them to ask Lyova to leave France at once and join them in
Mexico. Lyova, he warned them, was ill, nearly exhausted,
exposed to constant danger, yet convinced that he was 'irreplaceable' in Paris and that he must 'remain at his post'.
This was not so, however, for his comrades could replace him;
and if he were to stay on in Paris, he would be 'quite helpless
against the G.P.U.' At the very least his parents should ask
him to come over to Mexico for a time, to rest and convalesce
there. 'He is able, brave, and energetic; and we must save him.'[3]

This touching solicitude did not have the effect it should have
had. Trotsky was well aware, of course, that Lyova's life was
in jeopardy. He had urged him unceasingly to be prudent and
avoid any contact with people 'on whom the G.P.U. might
have a hold', especially with nostalgic Russian émigrés. Just
before the Reiss affair he had written: 'If an attempt is made
on your life or mine, Stalin will be blamed, but he has nothing
to lose, in honour anyhow.' Yet he discouraged the idea of
Lyova's move from France. When Lyova insisted that he was
'irreplaceable in Paris' and assured him that to protect himself

[1] In addition to Lyova's letters just referred to, Étienne's own correspondence
with Trotsky (*The Archives*, Closed Section) reveals all these details.

[2] Lyova's letters to Trotsky of 1 and 5 November 1937.

[3] The letter, written in German and dated 5 November 1937, is signed only with
the initial A. (which probably stands for Adolf). *The Archives*, Closed Section.

he was assuming an incognito (as Trotsky had done at Bar-bizon), Trotsky wrote back that Lyova would gain nothing by leaving France: the United States was not likely to admit him and Mexico would offer him even less security than France. He did not wish his son to shut himself up in the Coyoacan 'semi-prison'; and the discords between father and son perhaps made both of them reluctant to contemplate reunion. Trotsky's final letter about this ended with these spare and tense sentences: '*Voilà, mon petit*, this is what I can tell you. It isn't much. But . . . it's all. . . . You ought to keep now whatever you can cash from the publishers. You will need it all. *Je t'embrasse. Ton Vieux.*'[1] There was in this letter (of which Trotsky was to think with bitter self-reproach a few months later) something of the message sent out to a fighter holding out in a doomed forward position beyond all succour. Yet Trotsky had some ground for thinking that Mexico would offer Lyova even less security than France. Many G.P.U. agents, often disguised as refugees from Spain, had just installed themselves in Mexico; and the clamour for Trotsky's expulsion was growing more and more strident. Before the turn of the year the walls of Mexico City were covered with posters accusing him of conspiring with reactionary generals to overthrow President Cardenas and to establish a fascist dictatorship in Mexico. There was no saying whither the vilification might lead.

The gloom of these months was only momentarily relieved in September, when the Dewey Commission concluded the counter-trial and pronounced the verdict. This stated pointedly: 'On the basis of all evidence . . . we find that the [Moscow] trials of August 1936 and January 1937 were frame-ups . . . we find Leon Trotsky and Leon Sedov not guilty.'[2] Trotsky received this verdict with joy. Yet its effect was small, if not negligible. Dewey's voice commanded some attention in the United States; but it was ignored in Europe, where opinion was preoccupied with the critical events of the year, the last year before Munich, and with the vicissitudes of the French Popular Front and the Spanish Civil War. Trotsky was again

---

[1] Letter of 18 November 1937, ibid.

[2] *Not Guilty!* (Report of the Commission of Inquiry into the Charges made against Leon Trotsky in the Moscow Trials). See also Trotsky's letter to Lyova of 21 January 1938.

disappointed; and when the *Bulletin* which was to carry the verdict was slow in appearing, he was so irritated that he chided Lyova for 'this crime' and 'political blindness'. 'I am utterly dissatisfied', he wrote to him on 21 January 1938, 'with the way the *Bulletin* is conducted and I must pose anew the question of its transfer to New York.'

By this time Lyova's strength had ebbed away. He had lived, as Serge puts it, 'an infernal life'. He endured poverty and personal frustrations more easily than blows to his faith and pride. To quote Serge again: 'More than once, lingering until dawn in the streets of Montparnasse, we tried together to unravel the tangle of the Moscow trials. Every now and then, stopping under a street lamp, one of us would exclaim: "We are in a labyrinth of sheer madness!" '[1] Overworked, penniless, and anxious about his father, Lyova lived permanently in this labyrinth. He went on echoing his father's arguments, denunciations, and hopes. But with each of the trials something snapped in him. His brightest memories of childhood and adolescence had been bound up with the men in the dock: Kamenev was his uncle; Bukharin almost an affectionate playmate; Rakovsky, Smirnov, Muralov, and so many others— elder friends and comrades, all ardently admired for their revolutionary virtues and courage. He brooded over their degradation and could not reconcile himself to it. How had it been possible to break every one of them and make them crawl through so much mud and blood? Would at least one of them not stand up in the dock, abjure his confession, and tear in shreds all the false and terrible accusations? In vain Lyova waited for this to happen. He was shocked and pained when Lenin's widow was reported to have come out in support of the trials. For the *n*th time he repeated that the Stalinist bureaucracy, aspiring to become a new possessing class, had finally betrayed the revolution. But even this interpretation failed to account for all the blood and fury. Yes, this was the labyrinth of sheer madness—would even his father's clear-sighted genius be able to find the way out?

Sickness of heart, despair, fever, insomnia. Reluctant to leave his 'post', he delayed an operation for appendicitis, despite recurrent sharp attacks. He ate little, was unnerved,

[1] Serge, *Mémoires d'un Révolutionnaire*, p. 375.

and moved about droopingly. Yet in the first days of February he at last brought out the *Bulletin* with the verdict of the Dewey Commission; he joyfully reported this to Coyoacan, enclosing the proofs; and outlined his plans for further work, without giving any hint about his health. This was the last letter he wrote to his parents.

On 8 February he was still working, but ate nothing the whole day and spent much time with Étienne. In the evening he had another attack, the worst of all. He could not delay the operation any longer; and he wrote a letter, which he sealed and handed to his wife, telling her to open it only if some 'accident' happened to him. He talked to Étienne again, and wished to see no one else. They agreed that he must not enter a French hospital and register under his own name; for, if he did, the G.P.U. would easily discover his whereabouts. He was to go to a small private clinic run by some Russian émigré doctors; he was to present himself as Monsieur Martin, a French engineer; and he was to speak only French. No French comrade, however, was to know where he was or to visit him. Having agreed on all this, Étienne ordered the ambulance.[1]

Even on the face of it this was an incredibly absurd arrangement. Russian émigrés were the last people among whom Lyova could hope to pass for a Frenchman. He was all too likely to lapse into his native tongue in fever or under an anaesthetic. It was preposterous that in the whole of Paris the only hospital or clinic found for him should be one staffed by the very people whom, since Reiss's assassination, he had avoided like the plague. Yet he agreed at once to go, although, when his wife and Étienne took him there, he was neither delirious nor unconscious. Evidently, his critical sense and instinct of self-preservation were blunted.

He was operated on that same evening. In the next few days he seemed to recover well and rapidly. Apart from his wife, only Étienne came to see him. The visits cheered him up: they talked about politics and matters of organization; and he invariably begged Étienne to come back as soon as possible. When some French Trotskyists wished to see him, Étienne

[1] See depositions of Madame Estrine, Elsa Reiss, Rous, Jeanne Martin, and of Étienne himself made during the police investigation; also the Report of the Police Prefecture in *The Archives*, Closed Section.

told them, with the appropriate air of mystery, that they could not do so and that the address must be kept secret even from them if it was to remain a secret to the G.P.U. When one of the French comrades was startled by this excess of caution, Étienne promised to talk the matter over with Lyova; but no one was admitted to the patient's bedside. Four days passed. Then, all of a sudden, the patient suffered a grave relapse. He was seized by attacks of pain and lost consciousness. On the night of 13 February he was seen wandering half-naked and delirious through corridors and wards, which for some reason were unattended and unguarded. He was raving in Russian. Next morning his surgeon was so surprised by his state that he asked Jeanne whether her husband might not have attempted to take his own life—had he not been recently in a suicidal mood? Jeanne denied this, burst into tears, and said that the G.P.U. must have poisoned him. Another operation was carried out urgently, but it brought no improvement. The patient suffered terrible agony, and the doctors tried to save him by repeated blood transfusions. It was in vain. On 16 February 1938 he died at the age of thirty-two.

Did he, as his widow claimed, die at the hands of the G.P.U.? Much of the circumstantial evidence suggests that this was the case. In the Moscow trials he had been branded as his father's most active assistant, indeed, as the chief of staff of the Trotskyist-Zinovievist conspiracy. 'The youngster is working well; without him the Old Man would have found the going much harder', it had often been said at G.P.U. headquarters in Moscow, according to Reiss's and Krivitsky's testimony. It was in the G.P.U.'s interest to deprive Trotsky of his help, especially as this was sure to gratify Stalin's vengefulness. The G.P.U. had a most reliable informer and agent at his side who had brought him to the spot where he was to meet his death. The G.P.U. had every reason to hope that once Lyova was out of the way, their agent would take his place at the Russian 'section' of the Trotskyist organization and establish direct contact with Trotsky. At the clinic not only doctors and nurses but even cooks and porters were Russian émigrés, some of them members of the Repatriation Society. Nothing would have been easier for the G.P.U. than to find an agent among them, who would somehow administer poison to the patient. With so many murders on their conscience, would the G.P.U. have scrupled over this one?

But there is no certainty. An inquest held at Jeanne's demand yielded no proof of foul play. Police and doctors emphatically denied poisoning or any other attempt on Lyova's life; they attributed his death to a post-operational complication ('intestinal occlusion'), heart failure, and low powers of resistance. An eminent doctor who was also a friend of the Trotsky family accepted their opinion. On the other hand, Trotsky and his daughter-in-law asked a number of pertinent questions which remained unclarified. Was it by sheer accident that Lyova found himself in the Russian clinic? (Trotsky did not know that no sooner had Étienne called the ambulance than he informed the G.P.U. as Étienne himself has since confessed.) The staff of the clinic maintained that they had been unaware of Lyova's identity and nationality. But eye-witnesses maintain that they had heard him raving and even arguing about politics in Russian. Why had Lyova's surgeon been inclined to attribute the deterioration of his state to a suicidal attempt rather than to any natural cause? According to Lyova's widow, that surgeon lapsed into terrified silence as soon as the scandal blew up; and he took cover behind his duty to guard his professional secrets. It was in vain that Jeanne tried to bring these obscure circumstances to the notice of the examining magistrate; and that Trotsky pointed out that the routine inquest took no acount of the G.P.U.'s 'perfected and recondite' techniques of assassination. Did the French police, as Trotsky surmised, hush up the matter in order to cover their own inefficiency? Or were there, within the Popular Front, powerful political influences at work to prevent a thorough investigation? Nothing was left for the family but to demand a new inquest.[1]

·   ·   ·   ·   ·   ·   ·   ·   ·   ·

When the news reached Mexico, Trotsky was away from Coyoacan. A few days earlier Rivera had noticed unknown

[1] The depositions, eye-witness accounts, doctors' testimonials, and Trotsky's correspondence are quoted partly from *The Archives*, Closed Section, and partly from Lev Sedov's Papers which Jeanne Martin has transmitted to me through the courtesy of Pierre Frank.

people prowling around the Blue House and spying on its inhabitants from an observation post in the neighbourhood. He was alarmed; and he arranged that Trotsky should move out and stay for some time with Antonio Hidalgo, an old revolutionary and Rivera's friend, at Chapultepec Park. There, on 16 February, Trotsky was working on his essay, *Their Morals and Ours*, when evening papers announced Lyova's death. Rivera, when he read the news, telephoned to Paris hoping for a denial, and then went to Trotsky at Chapultepec Park. Trotsky refused to believe it, exploded with anger, and showed the door to Rivera; but then went back with him to Coyoacan to break the news to Natalya. 'I was just . . . sorting out old pictures, photographs of our children', she writes. 'The bell rang, and I was surprised to see Leon Davidovich coming in. I went out to meet him. He entered, his head bowed as I had never seen it, his face ash-grey and his whole look suddenly age-worn. "What has happened?" I asked in alarm. "Are you ill?" He answered in a low voice: "Lyova is ill. Our little Lyova . . .".'[1]

For many days he and Natalya remained closed in his room, petrified with pain, and unable to see secretaries, receive friends, or answer condolences. 'None spake a word unto him; for they saw that his grief was very great.' When after eight days he emerged, his eyes were swollen, his beard overgrown, and he could not bring out his voice. Several weeks later he wrote to Jeanne: 'Natalya . . . is not yet capable of answering you. She is reading and re-reading your letters and weeping, weeping. When I manage to free myself from work . . . I weep with her.'[2] Mingled with his grief was compunction for the harsh rebukes he had not spared his son this last year and the advice he had given him to stay on in Paris. This was the third time he was mourning a child; and each time there was greater remorse in the mourning. After Nina's death, in 1928, he reproached himself for not having done enough to comfort her and not even having written to her in her last weeks. Zina was estranged from him when she killed herself; and now Lyova had met his doom at the post where he had urged

---

[1] Natalya Sedova, 'Father and Son' in *Fourth International* (August 1941) and in *Vie et Mort de Leon Trotsky*.

[2] Trotsky to Jeanne Martin on 10 March 1938, *The Archives*, Closed Section.

him to hold out. With none of his children had he shared so much of his life and struggle as with Lyova; and no other loss had left him so desolate.

In these days of mourning he wrote Lyova's obituary, a threnody unique in world literature.[1] 'Now as together with Lev Sedov's mother I write these lines . . . we cannot yet believe it. Not only because he was our son, faithful, devoted, and loving . . . but because like no one else he had entered our life and grown into it with all his roots. . . .'

The old generation with whom . . . we once embarked upon the road of revolution . . . has been swept off the stage. What Tsarist deportations, prisons, and katorga, what the privations of life in exile, what civil war, and what illness had not done, Stalin, the worst scourge of the revolution, has accomplished in these last few years. . . . The better part of the middle generation, those . . . whom the year 1917 awakened and who received their training in twenty-four armies on the revolutionary front, have also been exterminated. The best part of the young generation, Lyova's contemporaries, . . . has also been trampled down and crushed. . . . In these years of exile we have made many new friends, some of whom have become . . . like members of our family. But we first met all of them . . . when we were already approaching old age. Lyova alone knew us when we were young; he participated in our life from the moment he acquired self-awareness. Remaining young, he became almost like our contemporary. . . .

Simply and tenderly he recollected Lyova's short life, depicting the child, scuffling with his father's jailers, bringing food parcels and books to the prison, making friends with revolutionary sailors, and hiding under a bench in the Soviet Government's conference hall so as to see 'how Lenin directed the revolution'. He portrayed the adolescent, who during the 'great and hungry years' of civil war would bring home, in the sleeves of a tattered jacket, a fresh roll given him by baker's apprentices, among whom he worked as a political agitator; and who, detesting bureaucratic privilege, refused to travel with his father by motor car and left the parental home in the Kremlin for a proletarian students' hostel and, joining voluntary workers' teams, swept the snow off Moscow's

[1] *'Lev Sedov, Syn, Drug, Borets'* In *B.O.*, no. 64, March 1938. (English pamphlet edition: *Leon Sedoff, Son, Friend, Fighter*.)

streets, unloaded bread and timber from trains, repaired locomotives, and 'liquidated' illiteracy. He recalled the young man, the oppositionist, who 'without a moment's hesitation' left his wife and child to go with his parents into banishment; who, at Alma Ata, where they lived surrounded by the G.P.U., assured his father's contact with the outside world, and went out, sometimes in the dead of night, in rain or snow storm, to meet a comrade clandestinely in woods outside the town, in a crowded bazaar, in a library, or even in a public bath. 'Each time he would return animated and happy, with a bellicose little fire in his eyes and with a treasured trophy under his coat.' 'How well he understood people—he knew many more Oppositionists than I did . . . his revolutionary instinct allowed him, without hesitation, to tell the genuine from the false. . . . His mother's eyes—and she knew the son better than I did—shone with pride.'

Here the father's feelings of remorse found their outlet. He mentioned his exacting demands on Lyova and explained these apologetically by his, Trotsky's, 'pedantic habits of work' and his inclination to demand the utmost from those who were closest to him—and who had been closer than Lyova? It might appear that 'our relationship was marked by a certain severity and estrangement. But underneath there . . . lived a deep, a burning mutual attachment, springing from something immeasurably greater than blood kinship—from commonly held views, from shared sympathies and hates, from joys and sufferings experienced together, and from great hopes cherished in common.' Some saw Lyova as merely 'a great father's little son'. But they were mistaken as were those who for a long time thought in this way of Karl Liebknecht; only circumstances had not allowed Lyova to rise to his full stature. Here comes a perhaps over-generous acknowledgement of Lyova's share in his father's literary work: 'On almost all my books written since the year 1929 his name should in justice have figured next to mine.' With what relief and joy his parents, in their Norwegian internment, had received a copy of Lyova's *Livre Rouge*, 'the first crushing retort to the slanderers in the Kremlin'. How right the G.P.U. men were who had said that 'without the youngster the old man would have found the going much harder'—and how much harder it would be now!

He again contemplated the ordeals which this 'very sensitive and delicate being' had had to endure: the endless hail of lies and calumnies; the long series of desertions and surrenders of former friends and comrades; Zina's suicide; and finally the trials which 'deeply shook his moral organism'. Whatever the truth about the direct cause of Lyova's death, whether he died exhausted by these ordeals or whether the G.P.U. had poisoned him, in either case 'it was *they* [and their master] that were guilty of his death'.

The great lament ended on the note on which it had begun:

His mother, who was closer to him than anyone in the world, and I, as we are living through these terrible hours, recall his image feature by feature; we refuse to believe that he is no more and we weep because it is impossible not to believe. . . . He was part of us, our young part. . . . Together with our boy has died everything that still remained young in us. . . . Your mother and I never thought, never expected, that fate would lay this task on us . . . that we should have to write your obituary. . . . But we have not been able to save you.

It was almost certain by now that Sergei had also perished, although there was no official information about this—and none was to be available even twenty-five years later. From a political prisoner, however, who early in 1937 shared a cell with him in Moscow's Butyrki we have this account:[1] for several months in 1936 the G.P.U. pressed Sergei to renounce publicly his father and all his father stood for. Sergei refused, was sentenced to five years' labour in a concentration camp, and was deported to Vorkuta. There towards the end of the year the Trotskyists were being assembled from many other camps. It was there, behind the barbed wire, that Sergei first came into close contact with them; and although he refused to consider himself a Trotskyist even now, he spoke with deep gratitude and respect of his father's adherents, especially those who had held out without any surrender for nearly ten years. He took part in a hunger strike which they proclaimed and which lasted more than three months; and he was near death.[2]

[1] This account of Sergei's behaviour in prison comes from Mr. Joseph Berger, who, having helped to found the Communist Party of Palestine and having served on the Middle Eastern division of the Comintern, spent twenty-three years in Stalin's prisons and concentration camps. He was freed and rehabilitated in 1956.

[2] See later, p. 416.

At the beginning of 1937 he was brought back to Moscow for yet another interrogation (it was then that the prisoner from whom we draw our account met him). He did not hope to be freed or to get any relief, for he was convinced that all his father's followers—and he with them—would be exterminated. Yet he behaved with stoic equanimity, drawing strength from his intellectual and moral resources. 'Discussing the G.P.U.'s methods of interrogation, he expressed the opinion that any educated man . . . should be equal to them; he pointed out that a century earlier Balzac had described all these tricks and techniques very accurately and that they were still almost exactly the same. . . . He faced the future with complete calm and would under no circumstances make any statement that would implicate in the slightest degree either himself or anyone else.' He evidently stuck it out to the end, for if he had not—if the G.P.U. had succeeded in wresting any confession from him —they would have broadcast the fact all over the world. He guessed that his parents must fear that he, their 'non-political' son, might lack the conviction and courage necessary to endure his lot; and 'he regretted most of all that no one would ever be able to tell them, especially his mother, about the change that had occurred in him, for he did not believe that anyone of all those whom he had met since his imprisonment would live to tell the story'. The author of this account soon lost sight of Sergei, but heard of his execution from other prisoners. Much later, in 1939, a message of dubious trustworthiness, which reached Trotsky through an American journalist, claimed that Sergei had still been alive late in 1938; but after that nothing more was heard about him.[1]

.   .   .   .

Of Trotsky's offspring, only Seva, Zina's son who was now twelve years old, was left alive outside the U.S.S.R. Nothing was or is known of what happened to Trotsky's other grandchildren. Seva had been brought up by Lyova and Jeanne, who, herself childless, had been a mother to him and had become passionately and obsessively attached to him. In his first letter after Lyova's death, Trotsky invited her to come with the child to Mexico. 'I love you greatly, Jeanne', he wrote,

[1] *The Archives*, Closed Section.

'and for Natalya you are not only . . . a daughter loved tenderly
and discreetly as only Natalya can love, but also part of Lyova,
of what is left of his most intimate life. . . .' They both wished
nothing more than that she and Seva should live with them in
Mexico. But if this was not Jeanne's wish, let her at least visit
them; 'and if you think that it would be too difficult for you
now to separate yourself from Seva, we shall understand your
feelings.'[1]

Here, however, the sorrowful tale shades off into the grotesque, and becomes entangled in the squabbles of the Trotskyist sects in Paris. Lyova and Jeanne had belonged to two
different groups, he to the 'orthodox Trotskyists', she to the
Molinier set. It says much for his tact and dignity that in the
letter he left in lieu of a will, he declared that despite this
difference (and despite, one may add, their unhappy marital
life) he held her in the highest esteem and had unreserved
confidence in her. Yet the furious competition of the rival
sects did not spare even Lyova's dead body; it fastened on the
little orphan; and it involved Trotsky himself in a preposterous
situation.[2] Jeanne, pressing desperately for a new inquest on
Lyova, authorized a lawyer who was a member of the Molinier
set to represent the family interest *vis-à-vis* the French magistrates and police. The 'orthodox Trotskyists' (and Gérard
Rosenthal, who was Trotsky's lawyer) denied Jeanne the right
to do this and maintained that Lyova's parents alone were
entitled to speak for the family. The conflicting claims only
made it easier for the police and the magistrates to ignore the
demand for the inquest.[3]

Another rumpus broke out over Trotsky's archives. Since
Lyova's death these had been in Jeanne's possession and so,
indirectly, in the hands of the Molinier group. Trotsky asked
that the archives be returned to him through one of his 'orthodox' French followers. Jeanne refused to hand them over. The
relations between her and Lyova's parents cooled off abruptly
and even grew hostile. Trotsky eventually recovered the
archives, but not before he sent an American follower of his
to Paris to collect them. Despite repeated urgings, Jeanne

---

[1] Trotsky to Jeanne, 10 March 1938. *The Archives*, Closed Section.
[2] Trotsky to Rosmer, Jeanne, Rous, and Camille (Klement) on 12 March.
[3] See Trotsky's correspondence with G. Rosenthal in *The Archives*, Closed Section.

refused to come to Mexico or to send the child there. She was neurotic; her mind was quite unsettled now; and she would not agree to part with her ward even temporarily. The rival factions kicked up a row over this too; and, much though Trotsky tried to conciliate his daughter-in-law, they rendered any agreement impossible. Whether because after the loss of all his children Trotsky was more than eager to recover his grandchild, the only one he could recover, or because he was afraid of leaving the orphan, as he put it, under the care of '*un ésprit très ombrageux et malheureusement déséquilibré*', or for both these reasons, he decided to go to law. An unseemly litigation followed, which dragged on for a year, providing grist for sensational newspapers and sectarian sheets.[1] In her despair at losing the child, Jeanne sought to invalidate Trotsky's claim by asserting that he had never legalized either his first or his second marriage; and Trotsky had to prove that this was not true. Even under this provocation he expressed (in a letter to the court) his understanding of Jeanne's emotional predicament; recognized her moral, though not her legal, right to the child; and renewed the invitation to her, offering to pay the cost of her journey to Mexico. He even declared himself willing to consider returning Seva to her, but not before he had had the chance to see him.[2] Twice the court adjudicated in Trotsky's favour and appointed trustees to ensure that the orphan was given back to the grandfather; but Jeanne refused to comply, took the boy away from Paris, and hid him. Only after a long search and a 'winter expedition' to the Vosges, did Marguerite Rosmer trace the child and wrest him from his aunt's hands. This was not the end yet, for Jeanne's friends made an attempt to abduct the child; and it was not until October 1939 that the Rosmers at last brought him to Coyoacan.

[1] Trotsky to the Rosmers on 19 September and to G. Rosenthal on 27 October 1938; Rosenthal to Trotsky, 7 October 1938. *Paris-Soir*, among other papers, reported the affair on the 26 March 1939; and Molinier's *Vérité* (no. 4, April 1939) devoted a special supplement to it under the title: '*Tous les moyens sont bons*', presenting Jeanne's case and attacking Trotsky for laying claims to the child.

[2] Trotsky's statement to M. Hamel, the official trustee, of 7 February 1939; reports of court proceedings, legal communications, and correspondence between Trotsky, his secretaries, and his attorney, especially letters of 22, 27, 29 March and 17 and 29 April 1939. *The Archives*, Closed Section. Also Trotsky's letter to the Rosmers of 26 May 1939.

In a pathetic letter Trotsky tried to explain to Seva why he insisted on his coming to Mexico. As he avoided making any derogatory remark about Jeanne, he could not give the child his main reason, and so the explanation was awkward and unconvincing:

Mon petit Seva . . . uncle Leon is no more, and we should keep in direct touch with each other, my dear boy. I do not know where your father is or whether he is still alive. In his last letter to me, written over four years ago, he asked insistently whether you had not forgotten the Russian language. Although your father is a very intelligent and educated man, he does not speak foreign languages. It would be a terrible blow to him if finding you one day he were unable to communicate with you. The same goes for your sister. You may imagine what a sad reunion this would be if you could not talk with your little sister in your native tongue. . . . You are a big boy now, and so I want to talk to you also about something else that is of great importance, the ideas that were and are common to your mother and father, to your Uncle Leon, and to me and Natalya. I greatly desire to explain to you personally the high value of these ideas and purposes, for the sake of which our family . . . has suffered and is suffering so much. I bear full responsibility for you, my grandson, before myself, before your father if he is alive, and before yourself.

And in words which were strangely stiff and out of place in a letter to a child, he concluded: 'That is why my decision about your journey is irrevocable.'[1]

.   .   .   .

Meanwhile the G.P.U. continued to weave their intrigue. Étienne had no difficulty in taking Lyova's place in the Trotskyist organization in Paris: he now published the *Bulletin*, was Trotsky's most important correspondent in Europe, and kept in touch with new refugees from the Stalinist terror who sought contact with Trotsky. The 'Russian section' of the organization had only three or four members in Paris, of whom none was as well versed in Soviet affairs and as educated and industrious as Étienne. From Lyova's letters Trotsky knew that Lyova had regarded him as his most intimate and reliable friend; and the *agent provocateur* now did what he could to

[1] The letter dated 19 September 1938 is in *The Archives*, Closed Section.

confirm this opinion of himself. Playing on Trotsky's paternal
grief and sensibilities, he sought to arouse Trotsky's distrust
of people who were in his, Étienne's, way. Within a week of
Lyova's death, he wrote to Trotsky with all due indignation
that Sneevliet was spreading the 'slanderous rumour' that
Lyova had been responsible for Reiss's death; and with
seeming casualness he reminded Trotsky of Lyova's complete
trust in him, Étienne, who had all the time held the key to
Lyova's letter box and collected all his mail.[1] Trotsky, who had
his political differences with Sneevliet, replied with an angry
outburst against the 'slanderer'.[2] The *agent provocateur* was,
of course, the model of an orthodox Trotskyist, never dissenting
from the 'Old Man', yet never appearing as a contemptible
yes-man either. Careful to give abundant yet not over-ostenta-
tious proof of devotion, he inquired with touching concern about
the Old Man's health and well-being, addressing, however,
such questions not to Trotsky himself but to one of the secretaries. With Trotsky directly he discussed political questions and
the contents of the *Bulletin*, which now appeared more regularly
than it had done for a long time. He asked Trotsky for a
commemorative article on Reiss, which, he said, he was
anxious to publish on the anniversary of Reiss's death. He saw
to it that the paper should come out with a proper tribute to
Lyova, too, on his first anniversary. He gave Trotsky notice
that the *Bulletin* was going to come out with an article, 'Trotsky's
Life in Danger', exposing the activities of G.P.U. agents in
Mexico. He supplied Trotsky with data and quotations drawn
from files of old Russian newspapers and from other, not easily
accessible, publications, data and quotations which Trotsky
needed for his *Stalin*. In a word, he made himself indispensable,
almost as indispensable as Lyova had been. And all the time
he unobtrusively added fuel to the feud between the sects and
the quarrel between Trotsky and Jeanne, until Trotsky refused
to support Jeanne's application for a new official inquiry into
the circumstances of Lyova's death. Étienne himself did what

---

[1] Étienne corresponded with Trotsky sometimes in his own name, sometimes on
behalf of the 'International Secretariat', and sometimes as editor of the *Bulletin
Oppozitsii*. See, e.g., letter of International Secretariat to Trotsky of 22 February
1938 (signed by Étienne and Paulsen).

[2] Trotsky to the International Secretariat, 12 March 1938. *The Archives*, Closed
Section.

he could to obstruct the inquiry: presenting himself to the French police as 'Leon Sedov's closest friend', he dismissed any suspicion of foul play, saying that Lyova's death had been due to the feeble resistance of his constitution.[1]

The *agent provocateur* was also at the centre of the preparations which the Trotskyists were making for the 'foundation congress' of the Fourth International. In the very middle of the preparations, on 13 July 1938, Rudolf Klement, the German émigré who had been Trotsky's secretary at Barbizon and was the secretary of the would-be International, vanished mysteriously from his home in Paris. About a fortnight later Trotsky received a letter, ostensibly written and signed by Klement, but posted from New York, which denounced Trotsky's alliance with Hitler, collaboration with the Gestapo, &c. Having repeated the usual Stalinist accusations, the writer announced his break with Trotsky. (Several French Trotskyists received copies of this letter which had been posted at Perpignan.) The letter contained so many incongruities and blunders, which Klement could not possibly have committed, that Trotsky at once concluded that it was plain forgery or that Klement had written it under duress, while a G.P.U. man pointed a revolver at him. 'Let Klement, if he is still alive, come forward and state before the judiciary, the police, or any impartial commission everything he knows. One can foretell that the G.P.U. will in no case let him out of their hands.'[2] Shortly thereafter Klement's body, horribly mutilated, was found washed ashore by the waters of the Seine. The gang that had assassinated Reiss had evidently killed him too; and one of the killers had assumed in Klement's name the attitude of a 'disillusioned follower' breaking with Trotsky—two years later Trotsky's assassin was to adopt the same pose.

Why had the G.P.U. picked Klement? He had not been outstanding among the Trotskyists for any special ability;

[1] Étienne's deposition during the police interrogation quoted above; his letter to Trotsky of 6 June 1938; and Trotsky's letter to 'Comrades Lola and Étienne', 17 February 1939. Van Heijenoort, Trotsky's secretary, wrote to Naville on 29 April 1938: 'J'ai reçu des lettres de divers amis de France (*Étienne en particulier*) s'inquietant de la situation du Vieux et demandant des informations'. My italics. *The Archives*, Closed Section. See also *B.O.*, nos. 66–67, 68–69, and 70, 1938, and no. 74, 1939.

[2] *B.O.*, nos. 68–69, 1938.

but he had been a modest and selfless worker who kept his eyes wide open to what was going on in the organization. It was, we think, he who had urged Trotsky and Natalya to ask Lyova to leave France. Had he recently come into possession of some important G.P.U. secret? Had he been on the track of their *agent provocateur*, perhaps about to unmask him? This, Trotsky guessed, would plausibly explain why the G.P.U. pounced and why they killed him in so vindictive and cruel a manner.[1]

By this time Sneevliet's suspicion of Étienne had hardened into a certainty; and both he and Serge voiced it openly. The *agent provocateur* was so brazen as to ask Trotsky what to do about it. Trotsky replied that he should at once challenge his accusers to lay their charges before a competent commission: 'Comrade Étienne should take this step; and the sooner, the more categorically and firmly he does it the better.' Trotsky could give no other advice: in such cases it was customary and obligatory for a man who had come under suspicion to ask for an investigation and for a chance to clear his honour. But Trotsky himself did not believe the accusation.[2]

To make the strange tale stranger, another warning reached Trotsky within a month. It came from a senior officer of the G.P.U., now a refugee in the United States. The author of the warning was, however, so afraid of the G.P.U. that he refused to disclose his identity and pretended to be an old American Jew of Russian origin conveying the message to Trotsky from a relative of his, a G.P.U. officer who had fled to Japan. The correspondent begged Trotsky to beware of a dangerous stool-pigeon in Paris, who was called 'Mark'. He did not know 'Mark's' family name, but gave so detailed and accurate a description of Étienne's person, background, and relations with Lyova, that Trotsky could have no doubt to whom he was referring. The writer was amazed at the credulity and carelessness of the Trotskyists in Paris, whose suspicions had not been aroused even by the fact (which he claimed was well known) that 'Mark' had worked in the notorious Society for the Repatriation of Russian Émigrés;

[1] *B.O.*, nos. 68–69, 1938.

[2] Trotsky to Étienne on 2 December 1938—the letter was formally addressed to the editors of the *Bulletin*. *The Archives*, Closed Section. Referring to Sneevliet's accusations, Trotsky put 'accusations' in quotation marks.

and he assured Trotsky that if only they watched the stool-pigeon they would find that secretly he was still meeting officials of the Soviet Embassy. Whether 'Mark' was guilty of Lyova's death, the correspondent did not know; but he feared that 'what was on the agenda now' was Trotsky's assassination, which was to be carried out either by 'Mark' himself or by some Spaniard posing as a Trotskyist. This was a weighty warning. 'The main thing, Lev Davidovich', the correspondent urged, 'is that you should be on your guard. Distrust any man or woman whom this *agent provocateur* may send or recommend to you.'[1]

Trotsky did not leave the warning altogether unheeded. Through a note in a Trotskyist paper, he asked the correspondent to get in touch with his followers in New York. The correspondent, afraid of revealing himself to them, tried to speak to Trotsky over the telephone from New York, but failed to contact him. The apparent lack of response on the correspondent's part and the strange form of his warning made Trotsky doubt his trustworthiness. Nevertheless, a small commission was formed at Coyoacan to investigate the matter; but it found no substance in the charges against Étienne. Trotsky wondered whether the denunciation was not a G.P.U. hoax, designed to discredit the man who appeared to be the most efficient and devoted of his assistants, who spoke and wrote Russian, was thoroughly versed in Soviet affairs, and edited the *Bulletin*. All too many accusations had already been bandied about in the small Trotskyist circle in Paris anyhow; and if all of these were to be taken seriously, there would be no end to the chasing of *agents provocateurs*. He knew all too well what a curse stool-pigeons were in any organization; but he also knew that constant suspicion and witch-hunting could be even worse. He decided not to lend ear to any accusation unless it was unequivocally presented and substantiated. He preferred to take the gravest risks and to expose himself to

---

[1] Both my wife and I found the letter among Trotsky's papers (in the 'open' section of *The Archives*) early in 1950, and copied it *in extenso*. Since then Alexander Orlov, a former G.P.U. officer, has claimed authorship of this letter. (See his deposition in *Hearings* of the U.S. Senate Committee on the Judiciary, sub-committee dealing with the scope of Soviet activity in the U.S. Part 51, pp. 3423-9.) For Orlov's role in the G.P.U., especially during the purges in Spain, see Jesus Hernandez, *La Grande Trahison*, and Orlov's own *Stalin's Crimes*.

extreme danger rather than to infect and demoralize his followers with distrust and scares. And so the *agent provocateur* went on acting as his factotum in Paris until the outbreak of the war.[1]

.   .   .   .

Within a fortnight of Lyova's death, Bukharin, Rykov, Rakovsky, Krestinsky, and Yagoda appeared in the dock in Moscow. It might have seemed that in the previous trials the macabre imagination of the stage producer had reached the limit. But those trials looked almost like essays in moderate realism compared with the new fantasmagoria. Once again Prosecutor and defendants denounced Trotsky as the chief of the conspiracy, which this time included the Bukharinists, who had been his deadly enemies. Lyova loomed as his father's accomplice even larger than in the earlier indictments. After a feeble attempt to deny the accusations, Krestinsky confessed that he had repeatedly conspired with Trotsky personally and Lyova in Berlin and various European resorts; that he had contacted Lyova with General von Seeckt, chief of the *Reichswehr*; and had paid out two million goldmarks, nearly a million dollars, and various other sums to finance the conspiracy. Trotsky and the defendants were now depicted as the agents not only of Hitler and the Mikado, but of British Military Intelligence as well, and even of the Polish *Deuxième Bureau*. To the familiar tales about attempts on the lives of Stalin, Voroshilov, and Kaganovich, and about railway catastrophes, colliery explosions, and mass poisonings of workers, were added stories about the assassination of Gorky, Menzhinsky, Kuibyshev, and even of Sverdlov, who had died in 1919— all these Trotsky had on his conscience. With each confession, the conspiracy not only grew in scope and swelled beyond the

[1] Mrs. Lilia Dallin (the 'comrade Lola' of the Russian Section of the International Secretariat in Paris in the late nineteen-thirties) testified in the U.S. that when she came to Coyoacan in the summer of 1939, Trotsky showed her the letter warning him against Étienne. 'I felt a bit uncomfortable because the details were very unpleasant. . . . I said: "That is certainly a dirty job of the N.K.V.D., who want to deprive you of your few collaborators . . ." and . . . he [Trotsky] had another letter from another unnamed correspondent telling him that a woman, meaning me, is coming to visit him and will poison him. So we both decided . . . that . . . it was a hoax of the N.K.V.D. . . . And the first thing I did [on returning to Paris] was to tell Étienne about it. . . . I trusted him'. Étienne had a 'hearty laugh' about the matter. Isaac Don Levine, *The Mind of an Assassin*, p. 60.

bounds of reason; it also extended back in time, to the very first weeks of the Soviet régime, and even to earlier periods. Like ghosts, Kamkov and Karelin, once leaders of the left Social Revolutionaries, came to the courtroom to testify that in 1918, when they staged their anti-Bolshevik insurrection, they had acted in secret accord with Bukharin, who was out to assassinate Lenin. Yagoda, who had for ten years been in charge of the persecution of the Trotskyists, had deported them *en masse*, had introduced torture in prisons and concentration camps, and had prepared the trial of Zinoviev and Kamenev, now claimed to have been all this time a mere tool in Trotsky's hands. Alongside former members of the Politbureau or the Central Committee and Ministers and Ambassadors, a group of distinguished doctors sat in the dock. One of them, Doctor Levin, a septuagenarian, had been Lenin's and Stalin's personal physician since the revolution; and he was charged with having, on Yagoda's orders, poisoned Gorky and Kuiby-shev. For many hours in the course of several sessions the doctors related how they plied their poisonous trade within the walls of the Kremlin, describing all manner of sadistic procedures in which they had allegedly indulged.[1]

Trotsky compared this trial with the Rasputin affair, for the trial, he said, reeked with 'the same rot and decay of an autocracy'. Perhaps nothing shows more tellingly than this comparison how his mind boggled at the spectacle. The Rasputin affair had, of course, been a puny and almost innocuous incident, compared with any of these trials; and the trials can hardly be said to have hastened Stalin's downfall even though they were to cover his memory with shame and disgrace. Yet Trotsky found no more adequate precedent or parallel, because none existed. Stalin had in a sense surpassed all historical experience and imagination: he set a new scale to the terror and imparted to it a new dimension. As the trials proceeded, any rational reaction to them became more and more helpless. Trotsky went on exposing the absurdities of the case, elaborating methodically his alibi, and proving that neither he nor Lyova could have conspired with any of the defendants, let alone with General von Seeckt, at the places and the dates indicated.

[1] *Sudebnyi Otchet po Delu anti-Sovietskovo i Pravo-Trotskistskovo Bloka.*

In this criminal activity [he commented] Prime Ministers, Ministers, Generals, Marshals, and Ambassadors, appear invariably to have taken their orders from a single quarter—not from their official leader but from a banished man. A wink from Trotsky was enough for veterans of the revolution to become Hitler's and the Mikado's agents. On Trotsky's 'instructions', transmitted through the first and best Tass correspondent, the leaders of industry, agriculture, and transport were destroying the nation's productive resources and shattering its civilization. On an order from the 'enemy of the people', sent out from Norway or Mexico, railwaymen wrecked military transports in the Far East and highly respectable doctors poisoned their patients in the Kremlin. This is the astounding picture . . . drawn by Vyshinsky. . . . But here a difficulty arises. In a totalitarian régime it is the apparatus [i.e. the party and state machine] that exercises the dictatorship. If my underlings have occupied all the crucial positions in the apparatus, how is it that Stalin is in the Kremlin and that I am in exile?[1]

He referred to the international setting and the consequences of the trials: Hitler's troops had just marched triumphantly into Austria and were getting ready for further conquests:

Is Stalin still chuckling behind the scenes? Has this unforeseen turn of events not yet taken his breath away? True, he is separated from the world by a wall of ignorance and of servility. True, he is accustomed to think that world opinion is nothing and the G.P.U. is everything. But the threatening and multiplying symptoms must be visible even to him. The working masses of the world are seized by acute anxiety. . . . Fascism is gaining victory after victory and finding its chief aid . . . in Stalinism. Terrible military perils knock at all the doors of the Soviet Union. And Stalin has chosen this moment to shatter the army and trample over the nation. . . . Even this Tiflis impostor . . . must find it harder to chuckle. An immense hatred is growing around him; a terrible resentment is suspended over his head. . . .

It is quite possible that a régime which exterminates . . . the nation's best brains may eventually provoke a genuinely terroristic opposition. What is more: it would be contrary to all laws of history if [it did not do so]. . . . But this terrorism of despair and revenge is alien to the adherents of the Fourth International. . . . Individual revenge . . . would be all too little for us. What political and moral satisfaction indeed could the working class derive from the assassination of Cain-Djugashvili, whom any new bureaucratic 'genius'

[1] *B.O.*, no. 65, 1938.

would replace without difficulty? In so far as Stalin's personal fortunes can be of any interest to us at all, we should only wish that he should survive the crumbling of his own system; and that is not very far off.

He forecast 'another trial, a genuine one', at which the workers will sit in judgement over Stalin and his accomplices. 'No words will then be found in the human language to defend this most malignant of all the Cains that can be found in history. . . . The monuments he has erected to himself will be pulled down or taken into museums and placed there in chambers of totalitarian horrors. And the victorious working class will revise all the trials, public and secret, and will erect monuments to the unfortunate victims of Stalinist villainy and infamy on the squares of a liberated Soviet Union.'[1]

Again, this prophecy was to come true, but not for many years. In the meantime, the purges, by their scale and force, acted like an immense natural cataclysm, against which all human reaction was vain. The terror crushed brains, broke wills, and flattened all resistance. The immense hatreds and resentments of which Trotsky spoke were there; but they were pressed deep down, where they were to remain stored for the future; at present and for the rest of the Stalin era they could find no outlet. All those—the Trotskyists in the first instance —in whom such emotions were allied with a political conscious-ness and who had ideas and programmes of action to offer— all such people were being exterminated systematically and pitilessly.

For over ten years Stalin had kept the Trotskyists behind bars and barbed wire, and subjecting them to inhuman per-secution, demoralized many of them, divided them, and almost succeeded in cutting them off from society. By 1934 Trotskyism seemed to have been stamped out completely. Yet two or three years later Stalin was more afraid of it than ever. Para-doxically, the great purges and mass deportations that had followed the assassination of Kirov gave fresh life to Trotsky-ism. With tens and even hundreds of thousands of newly-banished people around them, the Trotskyists were no longer isolated. They were rejoined by the mass of capitulators, who

[1] Ibid. The leading article.

ruefully reflected that things might have never come to the present pass if they had held out with the Trotskyists. Oppositionists of younger age groups, Komsomoltsy who first turned against Stalin long after Trotskyism had been defeated, 'deviationists' of every possible variety, ordinary workers deported for trivial offences against labour discipline, and malcontents and grumblers who began to think politically only behind barbed wire—all these formed an immense new audience for the Trotskyist veterans.[1] The régime in the concentration camps was more and more cruel: the inmates had to slave ten or twelve hours a day; they starved; and they wasted away amid disease and indescribable squalor. Yet the camps were once again becoming schools and training grounds of the opposition, with the Trotskyists as the unrivalled tutors. It was they who were at the head of the deportees in nearly all the strikes and hunger strikes, who confronted the administration with demands for improvements in camp conditions, and who by their defiant, often heroic behaviour, inspired others to hold out. Tightly organized, self-disciplined, and politically well informed, they were the real élite of that huge segment of the nation that had been cast behind the barbed wire.

Stalin realized that he would achieve nothing by further persecution. It was hardly possible to add to the torment and the oppression, which had only surrounded the Trotskyists with the halo of martyrdom. They were a menace to him as long as they were alive; and with war and its hazards approaching, the potential threat might become actual. We have seen that since he had first seized power he had to reconquer it over and over again. He now decided to rid himself of the necessity to go on reconquering it; he was out to ensure it once for all and against all hazards. There was only one way in which he could achieve this: by the wholesale extermination of his opponents; above all, of the Trotskyists. The Moscow trials had been

[1] M. Fainsod in *Smolensk Under Soviet Rule* quotes, from captured G.P.U. documents, cases when even in 1936-7, at the height of the purges, workers who were asked who should be considered an exemplary Bolshevik replied: Trotsky (and/or Zinoviev); and when school children, at a meeting called to commemorate Kirov, proposed that Trotsky be included in the honorary praesidium (p. 302 and *passim*). Trotskyism had not been especially popular in the Smolensk province; and such cases were more frequent in other parts of the country. All culprits, often even the children, were deported as 'Trotskyists.' For a description of the orgy of the denunciation of 'Trotskyists' in the Smolensk area, see ibid., pp. 232-7.

staged to justify this design, the main part of which was now carried out, not in the limelight of the courtrooms, but in the dungeons and camps of the East and far North.

.   .   .   .

An eye-witness, an ex-inmate of the great Vorkuta camp but not a Trotskyist himself, thus describes the last activities of the Trotskyists and their annihilation.[1] There were, he says, in his camp alone about a thousand old Trotskyists, calling themselves 'Bolsheviks-Leninists'. Roughly five hundred of these worked at the Vorkuta colliery. In all the camps of the Pechora province there were several thousands of 'orthodox Trotskyists', who 'had been in deportation since 1927' and 'remained true to their political ideas and leaders till the end'. The writer probably includes former capitulators among the 'orthodox Trotskyists', for otherwise his estimate of their number would appear greatly exaggerated.[2] 'Apart from these genuine Trotskyists', he goes on to say, 'there were about this time more than one hundred thousand inmates of the camps in Vorkuta and elsewhere, who as party members or Komsomoltsy had joined the Trotskyist Opposition and had then, at various times and for various reasons, . . . been forced to "recant and admit their mistakes" and to leave the ranks of the Opposition.' Many deportees, who had never been party members, also regarded themselves as Trotskyists. These numbers again must include oppositionists of every possible shade, even some of Rykov's and Bukharin's adherents, and newcomers of the young and youngest age groups, as our eye-witness himself indicates.

'All the same', he remarks, 'the Trotskyists proper, the followers of L. D. Trotsky, were the most numerous group.' Among their leaders he lists V. V. Kossior, Posnansky, Vladimir Ivanov, and other authentic Trotskyists of long standing. 'They arrived at the colliery in the summer of 1936 and were put up . . . in two large shanties. They refused categorically to work in the pits. They worked only at the pitheads and for not more than eight hours a day, not ten or twelve hours, as the

---

[1] This report ('*Trotskisty na Vorkute*'), signed M.B., appeared in the émigré-Menshevik *Sotsialisticheskii Vestnik*, nos. 10–11, 1961.

[2] Compare Chapter I, p. 81.

regulations required and as all other inmates laboured. They ignored the camp regulations ostentatiously and in an organized manner. Most of them had spent about ten years in isolation, first in jails, then in camps on the Solovky Islands, and finally at Vorkuta. The Trotskyists were the only groups of political prisoners who openly criticized the Stalinist 'general line' and openly and in an organized manner resisted the jailers.' They still proclaimed, as Trotsky did abroad, that in case of war they would defend the Soviet Union unconditionally, but seek to overthrow Stalin's government; and even 'ultra-lefts', like Sapronov's adherents, shared this attitude, though with reservations.

In the autumn of 1936, after the trial of Zinoviev and Kamenev, the Trotskyists arranged camp meetings and demonstrations in honour of their executed comrades and leaders. Shortly after, on 27 October, they began a hunger strike— this was the strike in which, according to the account quoted earlier, Sergei, Trotsky's younger son, took part. The Trotskyists of all the Pechora camps joined in and the strike lasted 132 days. The strikers protested against their transfer from previous places of deportation and their penalization without open trial. They demanded an eight-hour working day, the same food for all inmates (regardless of whether they fulfilled production norms or not), separation of political and criminal prisoners, and the removal of invalids, women, and old people from sub-Polar regions to areas with a milder climate. The decision to strike was taken at an open meeting. Sick and old-age prisoners were exempted; 'but the latter categorically rejected the exemption'. In almost every barrack non-Trotskyists responded to the call, but only 'in the shanties of the Trotskyists was the strike complete'.

The administration, afraid that the action might spread, transferred the Trotskyists to some half-ruined and deserted huts twenty-five miles away from the camp. Of a total of 1,000 strikers several died and only two broke down; but those two were not Trotskyists. In March 1937, on orders from Moscow, the camp administration yielded on all points; and the strike came to an end. In the next few months, before the Yezhov terror reached its height, the Trotskyists benefited from the rights they had won; and this raised the spirits of all other deportees so much that many of them looked forward to the twentieth anniversary of the October Revolution, hoping that a partial amnesty would be promulgated. But presently the terror came back with fresh fury. The food ration was reduced to 400 gr. of bread a day. The G.P.U. armed criminal prisoners with clubs and incited them against the Oppositionists. There were indiscriminate shootings; and all political prisoners were isolated in a camp within the camp, surrounded by barbed wire, and guarded by a hundred heavily armed soldiers, day and night.

One morning, towards the end of March 1938, twenty-five men, mostly leading Trotskyists, were called out, given a kilogram of bread each, and ordered to collect their belongings and prepare for a march. 'After a warm leave-taking with friends, they left the shanties; there was a roll call and they were marched out. In about fifteen or twenty minutes a volley was suddenly fired about half a kilometre from the shanties, near the steep bank of a little river, the Upper Vorkuta. Then a few disorderly shots were heard, and silence fell. Soon the men of the escort were back, and they passed by the shanties. Everyone understood what march it was the twenty-five had been sent on.'

On the next day no fewer than forty people were called out in this way, given their bread ration, and ordered to get ready. 'Some were so exhausted that they could not walk; they were promised they would be put on carts. With bated breath the people in the shanties listened to the creaking of the snow under the feet of those who were marched away. All sounds had already died down; yet everyone was still listening tensely. After about an hour shots resounded across the tundra.' The crowd in the shanties knew now what awaited them; but after the long hunger strike of the previous year and many more months of freezing and starvation, they had not the strength to resist. 'Throughout April and part of May the executions in the tundra went on. Every day or every other day thirty to forty people would be called out. . . .' Communiqués were broadcast over loudspeakers: 'For counter-revolutionary agitation, sabotage, banditry, refusal to work, and attempts to escape, the following have been executed. . . .' 'Once a large group, about a hundred people, mostly

Trotskyists, were taken out. . . . As they marched away, they sang the Internationale; and hundreds of voices in the shanties joined in the singing.' The eye-witness describes the executions of the families of the Oppositionists—the wife of one Trotskyist walked on her crutches to the execution place. Children were left alive only if they were less than twelve years of age. The massacre went on in all the camps of the Pechora province and lasted until May. At Vorkuta 'only a little over a hundred people were left alive in the huts. About two weeks passed away quietly. Then the survivors were sent back to the colliery, where they were told that Yezhov had been dismissed and that Beria was in charge of the G.P.U.'

By this time hardly any of the authentic Trotskyists or Zino-vievists were left alive. When about two years later hundreds of thousands of new deportees, Poles, Latvians, Lithuanians, and Estonians, arrived in the camps, they found among the old inmates many disgraced Stalinists and even a few Buk-harinists, but no Trotskyists or Zinovievists. An old deportee would tell the story of their extermination in whispers or hints, because nothing was more dangerous even for a wretched deportee than to draw on himself the suspicion of harbouring any sympathy or pity for the Trotskyists.[1]

The terror of the Yezhov period amounted to political genocide: it destroyed the whole species of the anti-Stalinist Bolsheviks. During the remaining fifteen years of Stalin's rule no group was left in Soviet society, not even in the prisons and camps, capable of challenging him. No centre of inde-pendent political thinking had been allowed to survive. A tremendous gap had been torn in the nation's consciousness; its collective memory was shattered; the continuity of its revolutionary traditions was broken; and its capacity to form and crystallize any non-conformist notions was destroyed. The Soviet Union was finally left, not merely in its practical politics, but even in its hidden mental processes, without any alternative to Stalinism. (Such was the amorphousness of the popular mind that even after Stalin's death no anti-Stalinist movement could spring from below, from the depth of the

[1] I am indebted for detailed and perceptive descriptions of life in the Vorkuta camps during the later period to Bernard Singer, the well-known Polish journalist, who was deported there in the early years of the Second World War.

Soviet society; and the reform of the most anachronistic features of the Stalinist régime could be undertaken only from above, by Stalin's former underlings and accomplices.)

While the trials in Moscow were engaging the world's awe-struck attention, the great massacre in the concentration camps passed almost unnoticed. It was carried out in such deep secrecy that it took years for the truth to leak out. Trotsky knew better than anyone that only a small part of the terror revealed itself through the trials; he surmised what was happening in the background. Yet even he could not guess or visualise the whole truth; and had he done so, his mind would hardly have been able to absorb its full enormity and all its implications during the short time left to him. He still assumed that the anti-Stalinist forces would presently come to the fore, articulate and politically effective; and in particular that they would be able to overthrow Stalin in the course of the war and to conduct the war towards a victorious and revolutionary conclusion. He still reckoned on the regeneration of the old Bolshevism to whose wide and deep influence Stalin's ceaseless crusades seemed to be unwitting tributes. He was unaware of the fact that all anti-Stalinist forces had been wiped out; that Trotskyism, Zinovievism, and Bukharinism, all drowned in blood, had, like some Atlantis, vanished from all political horizons; and that he himself was now the sole survivor of Atlantis.

.   .   .   .

Throughout the summer of 1938 Trotsky was busy preparing the 'Draft Programme' and resolutions for the 'foundation congress' of the International. In fact this was a small conference of Trotskyists, held at the home of Alfred Rosmer at Périgny, a village near Paris, on 3 September 1938. Twenty-one delegates were present, claiming to represent the organizations of eleven countries.[1] The conference was overshadowed by the recent assassinations and kidnappings. It elected the three young martyrs: Lyova, Klement, and Erwin Wolf, as

[1] This account is based on the 'Minutes of the World Congress of the Fourth International' held on 3 September 1938. *The Archives*, Closed Section. (An identical copy of the Minutes obtained from former British Trotskyists has been in my possession.) In 1938, I remember, I read a more detailed, critical report of the 'Congress' by its Polish participants.

its honorary Presidents.[1] Along with Klement, the organizing secretary of the conference, reports on Trotskyist work in various countries, the draft of the statutes of the Fourth International, and other documents had vanished. In order to prevent another *coup* by the G.P.U. the conference held only one plenary session, which lasted a whole day without a break; and it refused to admit observers from the Catalonian P.O.U.M. and the French *Parti Socialiste Ouvrier et Paysan*.[2] To assure the 'deepest secrecy' a communiqué issued after the conference spoke of the 'congress held at Lausanne'. At the conference, however, Étienne 'represented' the 'Russian section' of the International. Two 'guests' were also present; one of them was a certain Sylvia Agelof, a Trotskyist from New York, who served as an interpreter. She had come over from the States some time earlier and in Paris met a man calling himself Jacques Mornard, whose mistress she became. He hovered somewhere outside the conference room, pretending to take no interest in the highly secret gathering and waiting only for Sylvia to come out.

Max Shachtman presided over the conference, which during its one-day session voted on the reports of the commissions and on the resolutions most of which had come from Trotsky's pen. The formal agenda was so crowded that it would have kept any normal congress busy for a week. Naville delivered the 'progress report', which was to justify the organizers' decision to proclaim the foundation of the Fourth International. Unwittingly, however, he revealed that the International was little more than a fiction: none of its so-called Executives and International Bureaus had been able to work in the past few years. The 'sections' of the International consisted of a few dozen, or at most, a few hundred members each —this was true even of the American section, the most numerous of all, which claimed a formal membership of 2,500.[3] The

[1] Erwin Wolf, Trotsky's secretary in Norway and Konrad Knudsen's son-in-law, had gone to Spain in 1936 and perished there at the hands of the G.P.U.

[2] Both P.O.U.M. and the *Parti Socialiste Ouvrier et Paysan* (a small, lively French party led by Marceau Pivert) 'sympathized' with Trotskyism, but had their differences with it.

[3] In the *Internal Bulletins* of the American Trotskyists the membership was given as 1,000. Dwight Macdonald says in *Memoirs of a Revolutionist*, p. 17: 'We had about eight hundred members'.

conference, however, remained unshaken in its determination to constitute itself a 'foundation congress', as Trotsky had advised. Only two Polish delegates protested that 'the Polish section as a whole was opposed to the proclamation of the Fourth International'. They pointed out that it was hopeless to try to create a new International while the workers' movement, as a whole, was on the ebb, during 'a period of intense reaction and political depression', and that all previous Internationals had to some extent owed their success to the fact that they had been formed in times of revolutionary upsurge. 'The creation of every one of the earlier Internationals constituted a definite threat to bourgeois rule. . . . This will not be the case with the Fourth International. No significant section of the working class will respond to our manifesto. It is necessary to wait. . . .' The Poles agreed with Trotsky that the Second and the Third Internationals were 'morally dead'; but they warned the conference that it was frivolous to underrate the hold those Internationals had on the allegiance of the working class in many countries; and although the Poles endorsed Trotsky's 'Draft Programme' they appealed again and again to their comrades to refrain from 'making an empty gesture' and 'committing a folly'.[1]

These were weighty objections; and they came from the only Trotskyist group outside the U.S.S.R. which had behind it many years of clandestine revolutionary work and a solid tradition of Marxist thought going back to Rosa Luxemburg. Much of the conference was taken up by rejoinders to the Poles; but no serious attempt was made to refute their argument. Naville declared that the moment was 'uniquely suitable' for the creation of the new International. 'It was essential to put an end to the present indeterminate situation and to have a definite programme, a definitely constituted international leadership, and definitely formed national sections.'

[1] Of the two Polish delegates one, 'Stephen', a young scientist studying in France, had spent several years of his adolescence in a Polish prison for juveniles because of his political activity; and the other 'Karl', an elderly Jewish worker, had spent twelve years in prisons under the Tsar and Pilsudski, had taken part in the October revolution in Moscow, and had fought in the first battles of the civil war in Russia, after which he returned to Poland; was sentenced to death there for his revolutionary activity and escaped while he was being led to his execution. I was the author of the argument against the foundation of the Fourth International which these two delegates advanced at the conference.

Shachtman dismissed the historical arguments of the Poles as 'irrelevant and false' and described them as 'the Mensheviks in our midst', for only Mensheviks could show so poor a grasp of the importance of organization and so little faith in the future of the International. In the vote the conference decided by a majority of nineteen against three to proclaim the Fourth International there and then.

After a hurried and almost unanimous acceptance of all other resolutions, the delegates proceeded to elect an Executive Committee. At this point Étienne, who had been the chief speaker on the 'Russian question', protested that the 'Russian section' had not been allocated a seat. The conference made good this oversight and nominated Trotsky as a 'secret' and Honorary Member of the Executive. As Trotsky could not participate in the Executive's work, the *agent provocateur* was to go on representing the 'Russian section'.

.　　.　　.　　.　　.　　.　　.　　.　　.　　.

Trotsky decided to 'found' the new International at a time when, as the Poles warned him, the act could make no impact. His adherents in the Soviet Union ('the strongest section of the Fourth International') had been exterminated. His following in Europe and Asia was dwindling. In nearly all countries east of the Rhine and south of the Alps the labour movement was crushed. No Marxist organization could engage in systematic clandestine activity under Hitler's rule in Germany, Austria, and presently in Czechoslovakia. In France the Popular Front was crumbling in disappointment and apathy. In Spain the civil war was drawing to an end, with the left self-defeated morally even before it was vanquished militarily. The whole of the European continent was politically prostrate, waiting only for Hitler's armed might to roll over it. Years of Nazi occupation and unbearable oppression and humiliation were needed to force the working classes of some countries back into political activity or into the *Résistance*. But then the workers, at least in France and Italy, turned to the Stalinist parties, which were associated with the Soviet Union, the greatest and since 1941 the most effective force of the *Résistance*. Whatever the changing circumstances, the influence of Trotskyism was bound to remain negligible.

The prospects were no better for it in Asia, even though Asia was full of revolutionary ferment. Trotsky devoted much time and attention to the social and political developments in China, Japan, India, Indo-China, and Indonesia. In all these countries he exercised an influence on small groups of communist intellectuals and workers. But nowhere, with the peculiar exception of Ceylon, were his followers able to form an effective political party. Even in China, where his opposition to Stalin's policy in 1925–7 might have been expected to make the greatest impression, the Fourth International did not possess a section worthy of the name. Trotskyist groups, working clandestinely, under the pressure of terrible poverty, and persecuted by Kuomintang and Stalinists alike, consisted of two dozen men in Shangai, a few dozen in Hongkong, and smaller circles scattered over the central and eastern provinces. Even after Chen Tu-hsiu had embraced Trotskyism they never managed to break out of their isolation. Chen Tu-hsiu spent six years in prison; on his release he was banished to a remote village in the Chungking province and forbidden to engage in politics or publish his writings. He lived in starvation and fear, weighed down by the odium of his responsibility for the defeat of 1927, distrusted even by the Trotskyists, calumniated by the Maoists, surrounded by spies, and threatened with murder by Chiang Kai-shek's police who eventually, in 1943, were to imprison him again and assassinate him. In 1938 and 1939 Trotsky tried desperately to bring him out of China, hoping that 'he could play in the Fourth International a role comparable to that Katayama played in the Third, but . . . with greater advantage to the cause of revolution'. But Chen Tu-hsiu was already breaking down under the strain and sinking into the blackest pessimism. He nevertheless occasionally still surveyed the Chinese scene with great perspicacity and pointed out where and why Trotskyism was failing. In a statement written two months after the proclamation of the Fourth International he explained, for instance, why the revolutionary movement in China must base itself on the peasantry, and not (as Trotsky and he himself had expected) on the urban workers. The Japanese had dismantled industry in China's most advanced provinces; consequently 'the Chinese working class was reduced numerically, materially, and spiritually, to the condition in which it had been thirty or forty years earlier'. It was therefore vain to assume that the revolution could find its main centres in the towns. 'If we do not grasp now what are likely to be the political circumstances of the future and if we do not recognize clearly the weakness of the Chinese proletariat and the condition of its party, we shall be shutting ourselves up in our small holes, shall slumber away our chances, and, taking great pride in ourselves, feed on consolations.' The Trotskyists, he went on, by their sectarian arrogance, their purely negative attitude towards Maoism, and their insensitivity to the needs of the war against Japan were cutting themselves off from political realities. He feared that the proclamation of the Fourth International would merely encourage them in their 'conceits and illusions'; and that the venture would end in bankruptcy. He himself leaned towards reconciliation now with the Kuomintang and now with Maoism; but was unable or unwilling to come to terms with either; and it was as a broken man that he lived out his last tragic years. His warnings and his fate summed up the predicament of Trotskyism in his part of the world.[1]

The only country where Trotskyism stirred a little was the United States. In January 1938, after various splits and mergers, the Socialist Workers' party formed itself, and soon gained the title of 'the strongest section' of the Fourth International. It had to its credit some militant activity in trade unions and industry; and it published regularly two periodicals: *The New International*, a 'theoretical monthly', and *The Militant*. At its head there was a fairly large team of, by American standards, experienced and able leaders, of whom James P. Cannon, Max Shachtman, and James Burnham were the best known.[2] Trotsky was always at the party's call, willing to advise, criticize, praise, prod, and settle disputes and squabbles. Emissaries travelled between New York and Mexico City; and

[1] See Trotsky's letters to 'Comrade Glass' of 5 February and 25 June 1938; and H. Fleetman's account of impressions from a journey to China and meetings with Chinese Trotskyists. (19 February 1940.) Trotsky's abundant correspondence with his Chinese followers testifies to his never-flagging intense interest in prospects of the Chinese revolution. I am quoting Chen Tu-hsiu's views from a long essay of his, written in Szechwan and dated 3 November 1938. *The Archives*, Closed Section.

[2] James P. Cannon, *The History of American Trotskyism*; M. Pablo, 'Vingt ans de la Quatrième Internationale' in *Quatrième Internationale*, 1958–9; and M. Shachtman, 'Twenty-five Years of American Trotskyism' in *The New International*, 1954.

contact was facilitated by the circumstance that the secretaries and bodyguards at the Blue House were nearly all Americans. New York rather than Paris was now the centre of Trotskyism. Even so, the American party too was a feeble shoot planted in a soil from which it could draw all too little nourishment.

Why then, despite such unpropitious auguries, did Trotsky go ahead with the proclamation of the Fourth International?

It was more than five years now since he had decided that it was impossible 'to sit in one International with Stalin, Manuilsky, & Co.'. In these years the Third International had deteriorated so much further and had become so depraved that he was impelled to sever himself and his following from it as sharply and dramatically as possible. Lenin, in his revulsion against the Second International, had once urged the Bolsheviks to throw off the old 'dirty shirt' of Social Democracy and call themselves Communists. Trotsky spoke of the 'syphilis of Stalinism' or of the 'cancer that must be burned out of the labour movement with a hot iron'; and he believed that he was bringing to life an organization that would play a decisive part in the revolutionary class struggles to come.[1]

What is less clear is whether he hoped for success in the near future or whether he was working 'for history', without any such hope. His own statements are contradictory. 'All great movements', he wrote once, referring to the smallness of his following, 'have begun as "splinter groups" of old movements. Christianity was at the beginning a "splinter" of Judaism. Protestantism—a "splinter" of Catholicism, that is of degenerate Christianity. The grouping of Marx and Engels came into being as a "splinter" of the Hegelian left. The Communist International was prepared during the last war by "splinters" of the Social Democratic International. The initiators of all these movements were able to gain mass followings only because they were not afraid of remaining isolated.' A passage like this, for all its historical optimism, suggests that Trotsky did not expect any early and decisive success. On the other hand, the Draft Programme, which he wrote for the International, was not so much a statement of principles as an

[1] *B.O.*, no. 71, November 1938, speech in English on the 'Foundation Congress'.

instruction on tactics, designed for a party up to its ears in trade union struggles and day-to-day politics and striving to gain practical leadership immediately. In a message on the 'foundation congress' he wrote: 'Henceforth the Fourth International is confronted with the task of a mass movement. . . . It is now the only organization which has not merely a clear idea of what are the driving forces of this . . . epoch, but also a full set of day-to-day demands capable of uniting the masses for the revolutionary struggle for power. . . .' And he went on: 'The disproportion between our strength today and our tasks tomorrow is clearer to us than to our critics. But the severe and tragic dialectic of our epoch is working for us. The masses whom [war will] drive to utter despair and indignation will find no other leadership than that which the Fourth International offers them.' In an address to his American followers he exalted the mission of the new International in an almost mystical vein and even more confidently: '. . . in the course of the coming ten years the programme of the Fourth International will gain the adherence of millions, and these revolutionary millions will be able to storm heaven and earth.' In the days of the Munich crisis he stated again that though the Fourth International might be weak at the beginning of the next war, 'each new day will work in our favour. . . . In the very first months of the war a stormy reaction against the fumes of chauvinism will set in among the working masses. Its first victims will be, along with fascism, the parties of the Second and Third Internationals. Their collapse will be the indispensable condition for an open revolutionary movement . . . led . . . by the Fourth International.' To Kingsley Martin, who visited him in 1937, he exclaimed: 'I tell you that in three to five years from now the Fourth International will be a great force in the world.'[1]

[1] *B.O.*, loc. cit. and nos. 66–67 and 71, 1938. The meeting with Kingsley Martin, described by the latter in *The New Statesman* of 10 April 1937, was rather unfriendly because of Martin's concern with 'defending the honour' of his friend D. N. Pritt, King's Counsel and M.P. who was busy justifying the Moscow trials before the British public from the legal viewpoint. The British Editor's sensitivity about Pritt's honour, and insensitivity about the honour of all the defendants in the Moscow trials and of Trotsky himself, may have irritated Trotsky and provoked him into making a rash statement. A rather piquant description of Martin's visit at Coyoacan is given by Trotsky himself in his correspondence with the International Secretariat in Paris.

His expectations were based on the twin premiss that the coming world war would be followed by a revolutionary aftermath similar to that which had followed the first world war, but larger in scope and force; and that the Stalinist parties, like the Social Democratic ones, would use all their strength to stem the tide of revolution. More than ever he saw the advanced industrial countries of the West as the main battlefields of socialism; from their working classes was to come the salutary revolutionary initiative that alone could break the vicious circle—socialism in a single country and bureaucratic absolutism—in which the Russian Revolution was imprisoned. It was almost unthinkable to him that western capitalism, already shattered by the slumps and depressions of the nineteen-thirties, should be able to survive the coming cataclysm. He had no doubt that Hitler would endeavour to unify Europe under German imperialism, and would fail. But Europe needed to be united and only proletarian revolution could unite it and bring into existence the United States of Socialist Europe. Not only Germany, with its Marxist heritage, and France and Italy, with their revolutionary traditions, but even North America would be drawn into the social upheaval. In his Introduction to *Living Thoughts of Karl Marx*, written in 1939, he refuted the Rooseveltian New Deal and all attempts to rejuvenate and reform capitalism as 'reactionary and helpless quackery'; he pointed out how relevant *Das Kapital* was to the problems of the American economy; and he greeted the dawn of a new epoch of Marxism in the United States. In Marxism too 'America will in a few jumps catch up with Europe and outdistance it. Progressive technology and a progressive social structure will pave their own way in the sphere of doctrine. The best theoreticians of Marxism will appear on American soil. Marx will become the mentor of the advanced American worker.'[1]

Trotsky did not overlook the vast potentialities of revolution in the underdeveloped countries, especially in China—he dwelt on these more than any other writer of the nineteen-thirties. But he visualized those prospects as subordinate to the prospect of revolution in the West: 'Once it begins, the socialist revolution will spread from country to country with immeasurably greater force than fascism is spreading now. By the example

[1] *Living Thoughts of Karl Marx*, p. 38.

and with the aid of the advanced nations, the backward nations will also be brought into the mainstream of socialism.' By carrying to an extreme the logic of classical Marxism, which had postulated 'progressive technology and a progressive social structure' as the basis for socialist revolution, he was unwittingly exposing the discrepancy between theory and facts. Had the advanced industrial countries played the part for which classical Marxism had cast them in theory, no country should have been more congenial to Marxism and socialism than the United States. Trotsky did not and could not foresee that in the next few decades the backward nations would form the 'mainstream of socialism'; that the 'advanced West' would seek to contain it or to throw it back; and that the United States in particular, instead of evolving its own ultra-modern version of Marxism, would become the world's greatest and most powerful bulwark against it.[1]

He expected the working classes of the West to rise, as they had risen in 1848, 1871, 1905, and 1917–18. Applying the traditional Marxist conception even to China, he viewed with distrust Mao Tse-tung's 'peasant armies', fearing that, like many such armies in China's history, they might turn into instruments of reaction and come into conflict with the workers, if the latter failed to resume the revolutionary initiative. Despite Chen Tu-hsiu's warnings, he believed that the Chinese working class would recover its political *élan* and reassert itself as the leading force of the revolution. It remained an axiom with him that in all modern class struggle supremacy belongs of necessity to the towns; and the idea of an insurgent movement conquering the cities from the outside—from the countryside—was to him both unreal and retrograde. In West and East alike, he insisted, the revolution would either be proletarian in the true sense or it would not be at all. Least of all could he envisage the situation which was to arise during and after the Second World War, when the course of the class struggle in East and West alike was to be governed, and in a sense distorted, first by the alliance between Stalin's Russia and the West, and then by their world-embracing antagonism.

[1] Whether Trotsky's prognostications about the 'advanced West', especially the United States, will look as unreal towards the end of this century as they did at its middle, must, of course, remain an open question.

From his premisses Trotsky could not but pose the question: who—which party—was going to direct the forthcoming revolutionary struggles? The Second International, he answered, was a rotting prop of the old order. The Third was a tool in Stalin's hands, a tool Stalin would throw away when this suited him or use as a mere bargaining counter in his dealings with the capitalist powers. Stalin and his bureaucracy lived in fear of revolution abroad, a revolution which might arouse the working class of the Soviet Union as well and endanger bureaucratic absolutism and privilege. Thus the workers, as they entered a new epoch of social convulsions, had no revolutionary Marxist party at their head. Lack of leadership had been responsible for the long sequence of débâcles they had suffered in the nineteen-twenties and nineteen-thirties; and without revolutionary leadership they would suffer further and even more catastrophic defeats. If Marxism was not a fallacy, if the working class was the historic agent of socialism, and if Leninism was right in insisting that the workers could not win unless they were led by a 'vanguard', then the protracted 'crisis of leadership' could be resolved only by the formation of a new Communist party and International. In his pre-Bolshevik years, Trotsky, like Rosa Luxemburg and so many other Marxists, had been inclined to rely on the untutored activity of the working class and to neglect the directing and organizing functions of the party—the functions that had been at the centre of Lenin's preoccupations. He had since come to see in this the greatest single mistake he had committed in his long political career; and he was not going now to place his trust once again in the 'spontaneous' flow of the tide of revolution. And when all his reasonings led him to set himself a task, he would not shrink from any difficulties, not even from its apparent hopelessness. 'The Second and the Third Internationals are dead—Long Live the Fourth!' His duty, as he conceived it, was to proclaim this; as for the rest, let the future take care of it.

.   .   .   .

In one milieu, among the radical American intelligentsia, especially in literary circles, Trotskyism was making headway at this time. Under the impact of the great slump, the rise of

Nazism, and the Spanish Civil War, many American intellectuals had been drawn towards the Communist party; but the most critically minded baulked at the Popular Front opportunism which caused the party to court Roosevelt and hail the New Deal; and they were shocked and disgusted by the Moscow trials and the equivocal manœuvres and bizarre rituals of Stalinism. Trotskyism appeared to them as a fresh breeze breaking into the stuffy air of the left and opening new horizons. Men of letters responded to the dramatic pathos of Trotsky's struggle and to his eloquence and literary genius. Trotskyism became something of a vogue which was to leave many marks in American literature. Among the writers, especially critics, affected by it, were Edmund Wilson, Sidney Hook, James T. Farrell, Dwight Macdonald, Charles Malamud, Philip Rahv, James Rorty, Harold Rosenberg, Clement Greenberg, Mary McCarthy, and many, many others.[1]

Partisan Review became the centre of that 'literary Trotskyism'. Edited by Philip Rahv and William Phillips, the paper had been published under the auspices of the John Reed Clubs and, indirectly, of the Communist party. The editors, however, irritated by the party's meddling in literature, uneasy at its political gyrations, and shaken by the Moscow trials, suspended publication. Before the end of 1937 they brought the paper out again, but changed its orientation: *Partisan Review* was to stand for revolutionary socialism and against Stalinism. The editors invited Trotsky to contribute. He refused at first, and treated the venture with reserve. 'It is my general impression', he wrote to Dwight Macdonald, 'that the editors of *Partisan Review* are capable, educated, and intelligent, but have nothing to say.'[2] The leaders of the Socialist Workers party did not like to see his prestige thrown behind the periodical; and he himself wondered just how serious was the *Partisan Review*'s commitment

---

[1] Dwight Macdonald, op. cit., pp. 12–15.

[2] Trotsky to Macdonald on 20 January 1938. The Editors of *Partisan Review* had invited Trotsky to contribute to a Symposium on Marxism in which Harold Laski, Sidney Hook, Ignazio Silone, Edmund Wilson, August Thalheimer, John Strachey, Fenner Brockway and others were to participate. The theme was defined as 'What is alive and what is dead in Marxism?' The fact that *Partisan Review* intended to start its 'new chapter' by questioning the validity of Marxism did not recommend it to Trotsky. See his correspondence with the Editors of *Partisan Review* in *The Archives*, Closed Section. The Editors gave up the idea of the symposium.

to revolutionary socialism. Most of its contributors had known
Marxism and Bolshevism only through the Stalinist distortion—would they not now in their disillusionment with
Stalinism react also against Marxism and Bolshevism? On the
other hand, he reproached the editors with reacting too
feebly against the Moscow trials and attempting to remain on
friendly terms with *New Masses*, *The Nation*, and *The New
Republic*, which either defended the trials or were vague about
them. 'Certain measures', Trotsky wrote to Rahv, 'are necessary for a struggle against incorrect theory, and others for
fighting a cholera epidemic. Stalin is incomparably nearer to
cholera than to a false theory. The struggle must be intense,
truculent, merciless. An element of "fanaticism" . . . is salutary.'[1] Later in the year, as *Partisan Review* grew more outspoken
in its anti-Stalinism, the ice was broken. The moment of the
paper's closest association with Trotsky came when Breton
and Rivera, inspired by Trotsky, published in its pages their
Manifesto for the freedom of art and called for an International
Federation of Revolutionary Writers and Artists to resist
totalitarian encroachments on literature and the arts.[2]

André Breton, the French Surrealist poet, arrived at Coyoacan in February 1938. He had long been one of Trotsky's
ardent admirers; and nothing characterizes better his—but not
only his—feeling towards Trotsky than a letter he wrote him
after the visit to Mexico, on board the ship that was taking
him back to France: '*Très cher Lev Davidovich*. In addressing
you now in this way I am suffering less from lack of confidence
than I did in your presence. I felt so often the desire to address
you thus—I am telling you this so that you should realize of
what inhibition I am the victim whenever I am trying to make
a move towards you and trying it *under your eyes*.' That inhibition came from 'boundless admiration', it was a 'Cordelia
complex' which got hold of him whenever he came face to
face with Trotsky. He succumbed to this inhibition only when
he had to approach the greatest of men: 'You are one of
these . . . the only one alive. . . . I need a long process of

[1] Trotsky to Rahv on 21 March 1938, ibid.

[2] *Partisan Review*, Fall 1938; Trotsky's letters to Rahv of 12 May and 30 July
1938. *The Archives*, Closed Section, contain James Burnham's biting characterization of the personnel of *Partisan Review*, coupled with a gossipy *chronique de scandale*
(Burnham to Trotsky, 12 April 1938).

adjustment to persuade myself that you are not beyond my reach.' (Trotsky's answer to this letter was not less characteristic: 'Your eulogies seem to me so exaggerated that I am becoming a little uneasy about the future of our relations.')[1]

During his stay at Coyoacan, Breton, Trotsky, and Rivera went for long walks and trips into the country, arguing, sometimes heatedly, about politics and art. In France the Surrealists and the Trotskyists (especially Naville, the ex-Surrealist) were at loggerheads. Trotsky's attitude towards Surrealism, however, as towards any artistic innovation, was rather friendly, though not uncritical: he accepted the Surrealists' quasi-Freudian concentration on dream and subconscious experience, but shook his head over a 'strand of mysticism' in the work of Breton and his friends. Remote though these issues were from Trotsky's present preoccupations (Breton's visit coincided with Lyova's death and the Bukharin trial), he nevertheless argued at length with Breton and Rivera about communism and art and the philosophy of Marxism and aesthetics. Out of these discussions emerged the idea of the *Manifesto* to writers and artists and of the International Federation. The *Manifesto*, of which Trotsky was co-author, appeared under Breton's and Rivera's signatures in the *Partisan Review*.[2] Trotsky himself thus commented on the venture in letters to Breton and to the *Review*:

I welcome whole-heartedly [he wrote to Breton] your and Rivera's initiative in founding an International Federation of genuinely revolutionary and genuinely independent artists—and why not add of genuine artists? . . . Our planet is being turned into a filthy and evil-smelling imperialist barrack. The heroes of democracy . . . do all they can to resemble the heroes of fascism . . . and the more ignorant and obtuse a dictator is, the more does he feel destined to direct the development of science, philosophy, and art. The intelligentsia's herd instinct and servility are yet another and not inconsiderable symptom of the decadence of contemporary society.

[1] Breton to Trotsky, 9 August 1938; and Trotsky's answer, 31 August, in *The Archives*, Closed Section. See also Breton *La clé des champs*, pp. 142–54; and *Entretiens*, pp. 118–19 and 187–90; M. Nadeau, *Histoire du Surrealisme*, pp. 242–4.

[2] *Partisan Review*, Fall, 1938. Breton maintains that Rivera contributed only his signature; and that Trotsky was the chief author of the Manifesto, but thought it improper to sign.

The ideas of the *Manifesto* were essentially those which he had expressed in *Literature and Revolution* fifteen years earlier, when he sought to forestall the Stalinist tutelage over literature and the arts. He now attacked the sycophants of Stalinism, 'the Aragons, Ehrenburgs, and other petty tricksters', the 'gentlemen who [like Barbusse] compose with the same enthusiasm biographies of Jesus Christ and of Joseph Stalin', and Malraux, whose 'falsehood' in his latest descriptions of the German and Spanish scenes was 'all the more repulsive because he sought to give it an artistic form'. He saw Malraux's behaviour as 'typical of a whole category, almost of a generation of writers: so many of them tell lies from alleged "friendship" for the October Revolution, as if the revolution needed lies'. The struggle for artistic truth and for the artist's unyielding faithfulness to himself had therefore become a necessary part of the struggle for the ideas of the revolution.

In art man expresses . . . his need for harmony and a full existence . . . which class society denies him. [The quotation is from Trotsky's letter to *Partisan Review*.] That is why there is always implied, a conscious or unconscious, active or passive, optimistic or pessimistic, protest against reality in any genuine artistic creation. . . . Decaying capitalism is incapable of assuring even the minimum conditions necessary for their development to those currents of art which to some extent meet the needs of our epoch. It is superstitiously terrified of any new word. The oppressed masses live their own life. The Bohemian artistic milieu is shut in in its own narrowness. . . . The artistic schools of the last decades, Cubism, Futurism, Dadaism, Surrealism, have superseded each other without any of them coming to fruition. . . . It is impossible to find a way out of this impasse by artistic means alone. This is a crisis of the entire civilization. . . . If contemporary society does not succeed in reconstructing itself, art will inevitably perish as Greek art perished under the ruins of the slave civilization. . . . Hence the function of art in our epoch is determined by its attitude towards the revolution.

But here precisely history has laid a tremendous trap for the arts. A whole generation of the 'left' intelligentsia has . . . turned its eyes eastwards and has tied . . . its fate not so much to the revolutionary working class as to a victorious revolution, which is not the same. In that victorious revolution there is not only the revolution, but also a new privileged stratum . . . [which] has strangled artistic creation with a totalitarian hand. . . . Even under absolute monarchy Court art was based on idealization, but not on falsification, whereas in the Soviet Union official art—and none other exists there—is sharing in the fate of official justice; its purpose is to glorify the 'Leader' and to manufacture officially a heroic myth. . . .

The style of official Soviet painting is being described as 'socialist realism',—the label could have been invented only by a bureaucrat at the head of an Arts Department. The realism consists in imitating provincial daguerrotype pictures of the third quarter of the previous century; the 'socialist' style—in using tricks of affected photography to represent events that have never taken place. One cannot without revulsion and horror read the poems and novels or view the pictures and sculptures, in which officials armed with pen, brush, or chisel, and surveyed by officials, armed with revolvers, glorify the 'great leaders of genius' in whom there is not a spark either of genius or of greatness. The art of the Stalin epoch will remain the most striking expression of the deepest decline of proletarian revolution.

The problem, he pointed out, was not limited to the U.S.S.R.:

Under the pretence of a belated recognition of the October Revolution, the 'left' intelligentsia of the West has gone down on its knees before the Soviet bureaucracy. . . . A new era has opened with all sorts of centres and circles, . . . with the inevitable epistles by Romain Rolland, and with subsidized editions, banquets, and congresses (where it is difficult to draw any line between art and the G.P.U.). Yet, despite its wide sweep, this militarized movement has not brought forth a single artistic work capable of surviving its author and his Kremlin inspirers.

Art, culture, and politics need a new perspective. Without it mankind will not move forward. . . . But a genuinely revolutionary party cannot and will not wish to 'guide' art, let alone take it under command. . . . Only an ignorant and insolent bureaucracy run amok with arbitrary power could conceive such an ambition. . . . Art can be the revolution's great ally only in so far as it remains true to itself.[1]

Despite these rousing appeals, the International Federation of writers and artists never assumed reality. In Europe its

[1] *B.O.*, no. 74, 1939.

call for the defence of artistic freedom was soon drowned in the rumblings of approaching war; and in America the heyday of 'literary Trotskyism' was of short duration. As Trotsky had feared, the intelligentsia's revulsion against Stalinism was turning into a reaction against Marxism at large and Bolshevism.

For the nth time we can follow here the strange cycle through which ran the emotions roused by Trotsky in his intellectual followers. Most of them had turned towards him with an exalted reverence and in most he had evoked the 'Cordelia complex', of which Breton spoke. But gradually they found his way of living and thinking an unbearable moral strain; they found him indeed 'beyond their reach'. Their King Lear, they discovered, was still the hardest of revolutionaries. He was not out to gather around him a retinue of lyrical admirers— he strove to rally fighters to the most impossible of causes. He sought to set his followers, as he himself was set, against every power in the world: against fascism, bourgeois democracy, and Stalinism; against every variety of imperialism, social-patriotism, reformism, and pacifism; and against religion, mysticism, and even secularist rationalism and pragmatism. He required his adherents to 'defend the Soviet Union unconditionally' despite Stalin, and to assail Stalinism with a vehemence matching his own. Himself never yielding an inch from his principles, he would not tolerate yielding in others. He demanded of his adherents unshakeable conviction, utter indifference to public opinion, unflagging readiness for sacrifice, and a burning faith in the proletarian revolution, whose breath he constantly felt (but they did not). In a word, he expected them to be made of the stuff of which he himself was made.

They balked; and their exalted reverence for him gave place first to uneasiness and doubt, or to a weariness which was still mingled with awe, then to opposition, and finally to a covert or frank hostility. One by one the intellectual *Trotskisants* came to abjure first timidly then angrily their erstwhile enthusiasms and to dwell on Trotsky's faults. As nothing fails like failure, they brought up whatever mistakes or fiascos of his, real or imaginary, they could seize on until they came to denounce him as a fanatical and dogma-ridden day-dreamer, or until they decided that there was not much to choose between him and Stalin.

Behind the persistent pattern of these disillusionments and broken friendships there was the growing exasperation of the radical intelligentsia of the West with the experience of the Russian revolution in all its aspects, and with Marxism. This was one of those recurrent processes of political conversion by which the radicals and revolutionaries of one era turn into the middle-of-the-roaders or conservatives and reactionaries of the next—among the literary *Trotskisants* of the nineteen-thirties there were only a very few who would not be found at the head of the propagandist crusaders against communism of the late nineteen-forties and nineteen-fifties. To those crusades they were to bring a familiarity with communism, an acute though one-sided grasp of its vulnerable points, and a passionate hatred which Trotsky had inculcated in them, in the hope that Stalinism, not communism, would be its object. (Of course, many former Stalinists, who had never succumbed to any Trotskyist influence, were also to be prominent in the anti-communist crusades, but more often as vulgar informers than as ideological inspirers.)

The beginnings of this conversion are half hidden in the confusion of a few minor controversies. During the winter of 1937–8 Eastman, Serge, Souvarine, Ciliga, and others raised the question of Trotsky's responsibility for the suppression of the Kronstadt revolt in 1921. The context in which they raised it was an attempt to find out where and when exactly that fatal flaw in Bolshevism had shown itself from which Stalinism took its origin. It had shown itself, they answered, at Kronstadt, in the suppression of the 1921 revolt. That was the decisive turn, the original sin, as it were, that led to the fall of Bolshevism! But was not Trotsky responsible for the suppression of the Kronstadt revolt? Did he not appear in that act as the true precursor of the Stalinist terror? The critics found it all the easier to condemn him, as they had a highly idealized image of the Kronstadt rising and glorified it as the first truly proletarian protest against the 'betrayal of the revolution'. Trotsky replied that their image of Kronstadt was unreal and that if the Bolsheviks had not suppressed the rising they would have opened the floodgates to counter-revolution. He assumed full political responsibility for the Politbureau's decision about this, a decision he had supported, and denied only the allegation that he had personally directed the attack on Kronstadt.[1]

The polemic was full of a strange and unreasonable passion. There was no need to accept Trotsky's version to see that his critics greatly inflated the importance of the Kronstadt rising, detaching it, as it were, from the historic flux and the many cross-currents of events. Kronstadt as the prelude to Stalinism overshadowed in their eyes the fundamental factors that favoured Stalinism such as the defeats of communism in the West, the poverty and isolation of the Soviet Union, the weariness of its working masses, the conflicts between town and country, the 'logic' of the single party system, and so on. And such at times was the venom of the discussion over the relatively distant and ambiguous episode that Trotsky remarked: 'One would think that the Kronstadt revolt occurred not seventeen years ago but only yesterday.' What angered him was that his supposed well-wishers should have chosen to heckle him about Kronstadt right in the middle of his campaign against the Moscow trials. Moreover, while he was denouncing the present executions of the wives and children of the anti-Stalinists, Serge and Souvarine blamed him for the shooting of hostages during the Civil War. Did not this 'hue and cry' aid Stalin? And did they not see the moral and political difference between his use of violence in civil war and Stalin's present terror? Or were they denying the Bolshevik government of 1918–21 the right to defend itself and impose discipline?

---

[1] *B.O.*, no. 70, 1938. In a letter to Lyova (19 November 1937) Trotsky relates that, when the issue came before the Politbureau, he was for attacking Kronstadt while Stalin was against it, saying that the rebels, if left to themselves, would surrender within two or three weeks. Curiously, in his public polemics against Stalin (and in his biography of Stalin) Trotsky never mentioned this fact, although he usually made the most of any instance of Stalin's political 'softness' or deviation from Lenin's line. Is it that Trotsky somehow felt that in this case 'softness' might redound to Stalin's credit? The debate about Kronstadt went on in *The New International* (Trotsky, 'Hue and cry over Kronstadt', April 1938; Serge, 'Letter to the Editors', February 1939, etc.) and in books (Ciliga's *Au pays du Grand Mensonge* and Serge's *Memoires d'un Revolutionnaire*). One of Trotsky's American secretaries, Bernard Wolfe, who spent a few months at Coyoacan in 1937, has since written a novel, *The Great Prince Died*, the main idea of which is that Trotsky's conscience and life were corroded by his guilt over Kronstadt. Unfortunately, the novel is as crude and cheap artistically as it is unreal historically.

I do not know . . . whether there were any innocent victims [at Kronstadt]. . . . I cannot undertake to decide now, so long after the event, who should have been punished and in what way. . . especially as I have no data at hand. I am ready to admit that civil war is not a school of humane behaviour. Idealists and pacifists have always blamed revolution for 'excesses'. The crux of the matter is that the 'excesses' spring from the very nature of revolution, which is itself an 'excess' of history. Let those who wish to do so reject (in their petty journalistic articles) revolution on this ground. I do not reject it.

The critics accused him of 'Jesuitic' or 'Leninist immorality', that is of holding that the end justifies the means. He replied with his essay *Their Morals and Ours*, an aggressive and eloquent statement on the ethics of communism.[1] The essay begins with a burst of invective against those democrats and anarchists of the 'left' who at a time when reaction triumphs 'exude double their usual amount of moral effluvia, just as other people perspire doubly in fear'; but who preach morality not to the mighty persecutors but to persecuted revolutionaries. He did not indeed accept any absolute principles of morality. Such absolutes had no meaning outside religion. The Popes at least derived them from divine revelation; but whence did his critics, those 'petty secularist priests', draw their eternal moral truths? From 'man's conscience', 'moral nature' and similar concepts which are but metaphysical circumlocutions for divine revelation.

Morality is embedded in history and class struggles and has no immutable substance. It reflects social experience and needs; and so it always must relate means to ends. In a striking passage he 'defended' the Jesuits against their moralistic critics. 'The Jesuitic Order . . . never taught . . . that *any* means, though it be criminal . . . is permissible, if only it leads to the "end". . . . Such a . . . doctrine was malevolently attributed to the Jesuits by Protestant and partly by Catholic adversaries, who had no scruples in choosing the means for the attainment of *their* ends.' Jesuit theologians expounded the truism that the use of

---

[1] Trotsky was concluding the first draft of this essay when Rivera brought him the news of Lyova's death; and he devoted the essay to Lyova's memory. *B.O.*, nos. 68–69, 1938, and *The New International*, June 1938. The essay also appeared as a pamphlet in many languages.

any means, which by itself may be morally indifferent, must be justified or condemned according to the nature of the end it serves. To fire a shot is morally indifferent; to shoot a mad dog threatening a child is a good deed; to shoot to murder is a crime. 'In their practical morals the Jesuits were not at all worse than other priests and monks . . . on the contrary, they were superior to them, at any rate more consistent, courageous, and perspicacious. They represented a militant, closed, strictly centralized and aggressive organization, dangerous not only to enemies but also to allies.' Just like the Bolsheviks, they had had their heroic era and periods of decadence, when from warriors of the Church they turned into bureaucrats, and 'like all good bureaucrats were quite good swindlers'. In the heroic period, however, the Jesuit differs from the average priest as a soldier of the church differs from one who is a merchant in it. 'We have no reason to idealize either of them. But it is altogether unworthy to look upon the fanatic warrior with the eyes of the obtuse and slothful shopkeeper.'

The idea that the end justifies the means, Trotsky argued, is implicit in every conception of morality, not least in that Anglo-Saxon utilitarianism, from the standpoint of which most of the attacks against Jesuitic and Bolshevik 'immorality' are made. In so far as the ideal of 'the greatest possible happiness of the greatest possible number' implies that what is done to achieve that end is moral, that ideal coincides with the 'Jesuitic' notion of ends and means. And all governments, even the most 'humanitarian', who in time of war proclaim it the duty of their armies to exterminate the greatest possible number of the enemy, do they not accept the principle that the end justifies the means? Yet, the end too needs to be justified; and ends and means may change places, for what is seen as an end now may later be the means to a new end. To the Marxist the great end of increasing man's power over nature and abolishing man's power over man is justified; and so is the means to it—socialism; and so is the means to socialism—revolutionary class struggle. Marxist-Leninist morality is indeed governed by the needs of revolution. Does this signify that all means—even lies, betrayal, and murder—may be used if they further the interests of revolution? All means are permissible', Trotsky replied, 'which genuinely lead to mankind's emancipation'; but such is the dialectic of ends and means that certain means *cannot* lead to that end. 'Permissible and obligatory are those and only those means which impart solidarity and unity to revolutionary workers, which fill them with irreconcilable hostility to oppression, . . . which imbue them with the consciousness of their historic tasks, and raise their courage and spirit of self-sacrifice. . . . Consequently, *not* all means are permissible.' He who says that the end justifies the means says also that the end 'rejects' certain means as incompatible with itself. 'A wheat grain must be sown in order that wheat should grow.' Socialism cannot be furthered by fraud, deceit, or the worship of leaders which humiliates the mass; nor can it be imposed upon the workers against their will. As Lassalle put it:

> Show not only the goal; show also the road.
> So inseparably grow goal and road into each other,
> That the one always changes with the other;
> Another road brings another goal into being.

Truthfulness and integrity in dealing with the working masses are essential to revolutionary morality, because any other road is bound to lead to a goal other than socialism. The Bolsheviks, in their heroic period, were 'the most honest political party in the whole of history'. Of course, they deceived their enemies, especially in civil war; but they were truthful with the working people whose confidence they gained to an extent to which no other party had ever gained it. Lenin, who repudiated all ethical absolutes, gave the whole of his life to the cause of the oppressed, was supremely conscientious in ideas and fearless in action, and never showed the slightest attitude of superiority towards the plain worker, the defenceless woman, the child. As to his own, Trotsky's, immorality in decreeing that families of White Guard officers be taken as hostages, he assumed full responsibility for that measure, which was dictated by the necessities of civil war, although to his knowledge not a single one of those hostages had ever been executed. 'Hundreds of thousands of lives would have been saved, if the revolution had from the outset shown less superfluous magnanimity.' He trusted that posterity would judge his behaviour as it judged Lincoln's ruthlessness in the American

Civil War: 'History has different yardsticks for the cruelty of the northerners and that of the southerners. A slave owner who uses cunning and violence to shackle the slave, and a slave who uses cunning and violence to break the chains—only contemptible eunuchs will tell us that they are equal before the court of morality!'

It was a perversion of the truth to blame the October Revolution and 'Bolshevik immorality' for the atrocities of Stalinism. Stalinism was the product not of revolution or Bolshevism but of what had survived of the old society—this accounted for Stalin's pitiless struggle against the old Bolsheviks, a struggle through which the primordial barbarity of Russia was taking revenge on the progressive forces and aspirations that had come to the top in 1917. Moreover, Stalinism was the epitome of all the 'untruths, brutality, and baseness' that made up the mechanics of any class rule and of the state at large. The apologists of class society and of the state, including the defenders of bourgeois democracy, were therefore hardly entitled to feel morally superior: Stalinism was holding up to them their own mirror, even if it was partly a distorting mirror.

Of the many rejoinders to *Their Morals and Ours* John Dewey's deserves to be mentioned here.[1] Dewey accepted Trotsky's view of the relationship between means and ends and of the relative historical character of moral judgements. He agreed also that 'a means can be justified only by its end . . . and the end is justified if it leads to the increase of man's power over nature and the abolition of man's power over man.' But he differed from Trotsky in that he did not see why this end should be pursued mainly or exclusively by means of class struggle—to his mind Trotsky, like all Marxists, treated the class struggle as an end in itself. He detected a 'philosophical contradiction' in Trotsky, who on the one hand asserted that the nature of the end (i.e. of socialism) determines the character of the means and, on the other, deduced the means from 'historical laws of the class struggle' or justified them by reference to such 'laws'. To Dewey the assumption of *fixed laws*, allegedly governing the development of society, was irrelevant. 'The belief that a law of history determines the particular way in

[1] John Dewey, 'Means and Ends' in *New International*, August 1938.

which the struggle is to be carried on certainly seems to tend to a fanatical and even mystical devotion to the use of certain ways of conducting the class struggle, to the exclusion of all other ways. . . . Orthodox Marxism shares with orthodox religionism and . . . traditional idealism the belief that human ends are interwoven into the very texture and structure of existence —a conception inherited presumably from its Hegelian origins.'

Dewey's conclusion became the keynote of nearly all the attacks on Trotsky that presently came from his former disciples and friends—all aimed at the 'Hegelian heritage of Marxism', dialectical materialism, and the 'religious fanaticism' of Bolshevism. Max Eastman, for instance, spoke of the final collapse of the 'dream about socialism'. 'I advocate that we abandon those utopian and absolute ideals.' Not only was Marxism in his eyes now an 'antique religion' or a 'German romantic faith', but it was the progenitor of fascism as well as of Stalinism. 'Do not forget that Stalin was a socialist. Mussolini was a socialist. Hundreds of thousands of the followers of Hitler were socialists or communists. . . .' Sidney Hook like-wise renounced the idea of proletarian dictatorship and finally abandoned Marxism in favour of pragmatic liberalism. So did Edmund Wilson, Benjamin Stolberg, James Rorty, and others.[1]

With forty years of 'ideological' controversy behind him, Trotsky found little new or original in these arguments. They must have reminded him of Tikhomirov's *Why I ceased to be a Revolutionary*, the almost classic statement of recantation by an old Narodnik who left the revolutionary movement to make peace with the established order. Since then in every generation, in every decade, the weary and disillusioned, as they withdrew from the fray or changed sides tried to answer this question. What was new this time was the vehemence of the disillusion-ment: it matched the savage blows that Stalinism was in-flicting on faith and illusion. Never yet had men withdrawn from a revolutionary struggle with so much deep-felt emotion and genuine indignation; and never yet had any cause looked as hopeless as Trotsky's began to look to the professors, authors,

---

[1] Max Eastman, *Marxism, is it Science?* pp. 275–97; Sidney Hook, *Political Power and Personal Freedom*.

and literary critics who were deserting him. They came to feel that by opting for Trotskyism they had needlessly involved themselves in the huge, remote, obscure and dangerous business of the Russian revolution; and that this involvement was bringing them into conflict with the way of life and the climate of ideas which prevailed in their universities, editorial offices, and literary coteries. It was one thing to lend one's name to a Committee for the Defence of Trotsky and to protest against the purges, but quite another to subscribe to the Manifestoes of the Fourth International and to echo Trotsky's call for the conversion of the forthcoming world war into a global civil war. What was galling to Trotsky was to see even such old friends and associates as Eastman and Serge turn their back on him. He emptied the vials of his scorn on them and 'their ilk'; and like another great controversialist, not too fastidious in the choice of his victims, he preserved in his prose—as one preserves insects in amber—the names of quite a few scribblers who would otherwise have been long forgotten. Here is a sample of his polemic—with Souvarine as his target:

Ex-pacifist, ex-communist, ex-Trotskyist, ex-democrato-communist, ex-Marxist . . . almost ex-Souvarine is all the more insolent in his attacks on proletarian revolution . . . the less he knows what he wants. This man loves . . . to collect and file . . . documents, excerpts, quotation marks, and commas; and he has a sharp pen. He once imagined that this equipment would do him for his lifetime. Then he had to learn that it was also necessary to know how to think. . . . In his book on Stalin, despite an abundance of interesting quotations and facts, he himself produced the certificate of his own intellectual poverty. He understands neither revolution nor counter-revolution. He applies the criteria of a petty *raisonneur* to the historic process. . . . The disproportion between the critical bent and the creative impotence of his mind corrodes him like an acid. Hence he is constantly in a state of savage irritation and lacks elementary scruple in appraising ideas, men, and events; and he covers all this by dry moralizing. He is, like all misanthropes and cynics, drawn towards reaction. But has he ever openly broken with Marxism? We have never heard about this. He prefers equivocation; that is his native element. In his review of my pamphlet [*Their Morals and Ours*] he writes: 'Trotsky once again mounts his hobby horse of class struggle.' To the Marxist of yesterday class struggle is already 'Trotsky's hobby horse'. He, Souvarine, prefers to sit astride the dead dog of eternal morality.[1]

In such polemical excursions Trotsky was eagerly accompanied by two of his disciples: James Burnham and Max Shachtman, who sprang fiercely on 'The Intellectuals in Retreat', tearing them to pieces for their 'Stalinophobia' and 'treason to the working class and Marxism'. Before long these disciples too were to desert the master and join 'The Intellectuals in Retreat'.[2]

.   .   .   .

After a friendship which lasted two years, Trotsky and Rivera fell out. The quarrel broke out rather suddenly, just after the manifesto on the freedom of art had appeared in *Partisan Review*. In the summer Trotsky, hoping that Rivera would attend the 'foundation congress' of the Fourth International, had written to the organizers in Paris: 'You should invite him . . . personally . . . and underline that the Fourth International is proud to have in its ranks, him, the greatest artist of our epoch and an indomitable revolutionary. We should be at least as attentive towards Diego Rivera as Marx was towards Freiligrath and Lenin towards Gorky. As an artist he is far superior to Freiligrath and Gorky and he is . . . a genuine revolutionary, whereas Freiligrath was only a petty bourgeois sympathizer and Gorky a somewhat equivocal fellow-traveller.'[3] It was therefore a rude shock to Trotsky when, before the end of the year, Rivera bitterly attacked President Cardenas as 'an accomplice of the Stalinists', and in the Presidential elections backed Cardenas' rival, Almazar, a right-wing general who promised to bring the trade unions to heel and tame the left. Rivera too had caught the 'virus of Stalinophobia' (but such was the whimsicality of his political behaviour that a few years hence he was to return contritely to the Stalinist fold). Trotsky was wary of becoming involved in Mexican politics; and he would in any case have nothing to do with the kind of anti-Stalinism for which Rivera now stood and with his campaign against Cardenas. He tried to dissuade Rivera, but failed. As in the public eye he was extremely closely associated

[1] *B.O.*, nos. 77–78, 1939 and *New International*, August 1939.
[2] *New International*, January 1939.
[3] Trotsky to the International Secretariat in Paris, 12 June 1938.

with the painter, nothing short of an open break with him could free Trotsky of responsibility for his political vagaries. In a special statement Trotsky deplored Rivera's stand in the Presidential elections and declared that henceforth he could not feel any 'moral solidarity' with him or even benefit from his hospitality.[1] However, when the Stalinists attacked Rivera as one who 'sold himself to reaction' Trotsky defended him against the charge of venality and expressed undiminished admiration for the 'genius whose political blunderings could cast no shadow either on his art or on his personal integrity'.[2]

The break with Rivera and the decision to leave the Blue House put Trotsky in a difficult financial situation. His earnings had been greatly reduced, which had not mattered much as long as he did not have to pay for the roof over his head. Now he was compelled to do what he could to raise his earnings; and in the meantime he had to borrow from his friends to be able to run his household.[3] He had undertaken to write a biography of Stalin; but work being frequently interrupted he progressed with it slowly. His publishers, disappointed that his *Lenin* had not been forthcoming, were cautious with advances.[4] He thought of writing a short and popular book that

---

[1] Trotsky's statement to the Mexican Press of 11 January 1939. *The Archives.* See also Van Heijenoort's letter to Breton (11 January 1939) informing him, on Trotsky's instruction, about the breach. Breton, replying to Trotsky on 2 June, refused to take sides in the quarrel between Trotsky and Rivera.

[2] Trotsky's article ('Ignorance is not a Weapon of Revolution') for *Trinchera Aprista*, written on 30 January 1939. *The Archives.*

[3] I am told that a Mexican publisher and bookseller of Russian origin, a descendant of Russian revolutionaries, was Trotsky's creditor on this and other occasions. I have also heard fantastic tales about the 'financial side' of Trotsky's existence in exile. Thus, the Editor of a great American magazine has assured me that Trotsky drew money from a large American bank account which Lenin had opened in his and Trotsky's name during the civil war, when he reckoned with the possibility of a Bolshevik defeat and with the need to resume the revolutionary struggle from abroad. The story would be interesting if it were true. It is not.

[4] *The Archives* (Closed Section) contain Trotsky's correspondence with his publishers, detailed royalty statements, accounts, etc. which give a clear idea of his financial difficulties in the year 1939. Thus, Doubleday had paid him as far back as 1936 an advance of 5,000 dollars on the *Lenin* and now pressed for the manuscript. They had paid him, also in 1936, 1,800 dollars and a smaller amount later on, for *The Revolution Betrayed*; but until 1939 the sales had not covered the advances. Trotsky had signed contracts for *Stalin* with Harpers in New York and Nicholson and Watson in London in the first half of 1938; but before the end of the year Harpers already refused him advances on the ground that he was slow with delivering portions of the MS.

might become a best-seller and free him from journalistic chores; but he could not bring himself to do this. He negotiated with the New York Public Library and the Universities of Harvard and Stanford about the sale of his Archives. Eager to place his papers in safety, he had asked an almost ludicrously low price for them; but the prospective buyers were in no hurry, and the negotiations dragged on for over a year.[1] Even in journalism his stock had slumped badly; and literary agents often found it difficult to place his articles, although he wrote on subjects of burning topicality such as the Munich settlement, the state of the Soviet armed forces, American diplomacy, Japan's role in the coming war, and so on.[2]

Financial difficulties led him to a strange quarrel with *Life* magazine.[3] At the end of September 1939, on Burnham's initiative, one of *Life*'s editors came to Coyoacan, and commissioned him to write a character sketch of Stalin and also an article on Lenin's death. (Trotsky had just concluded the chapter in *Stalin* in which he suggested that Stalin had poisoned Lenin, and he was to present this version in *Life*.) His first article appeared in the magazine on 2 October. Although it contained relatively inoffensive reminiscences, the article raised the ire of pro-Stalinist 'liberals', who flooded *Life* with vituperative protests. *Life* printed some of these to the annoyance of Trotsky, who maintained that the protests had come from 'a G.P.U. factory' in New York, and were defamatory of him. He nevertheless sent in his second article, the one on Lenin's death; but *Life* refused to publish it. Ironically, the objections of the editors were reasonable enough: they found Trotsky's surmise that Stalin had poisoned Lenin unconvincing; and they demanded from him 'less conjecture and more unquestionable facts'. He threatened to sue *Life* for breach of contract; and in a huff submitted the article to the *Saturday Evening Post* and *Collier's*, where he again met with refusals,

[1] Trotsky to Albert Goldman, 11 January 1940. As late as March 1940 Harvard University offered to pay for the archives not more than 6,000 dollars. Eventually, the University bought the archives for 15,000 dollars, a small sum for the 'value' it received.

[2] Among Trotsky's several articles in which American and British Editors found no 'news-value' was one written early in the summer of 1939 and saying that Stalin was about to sign a pact with Hitler.

[3] See Trotsky's letter to J. Burnham of 30 September 1939 and his correspondence with *Life* magazine. *The Archives*, Closed Section.

until *Liberty* finally published it. It is sad to see how much time in his last year the irate and futile correspondence about this matter took. In the end *Life* paid him the fee for the rejected article. This and a few other earnings, he could report to his friends, 'insured' him financially for 'a few months' and allowed him to go on bargaining a little longer about the sale of his Archives.

. . . .

In February or March 1939 he rented a house at the Avenida Viena on the far outskirts of Coyoacan, where the long street grew empty, stony, and dusty, with only a few *campesino* hovels scattered on either side. The house was old and roughly built, but fairly solid and spacious; and it stood in its own grounds, separated by thick walls from the road and the surroundings. No sooner had the Trotskys moved in than a rumour spread that 'the G.P.U. was about to buy up the property'. To forestall this, Trotsky himself purchased it, although he had to borrow money for this his 'first deal in real estate'. In view of the unceasing Stalinist threats of physical violence it was necessary, or so it seemed, to fortify the house. Later on a watch tower was to be erected at the entrance gate; immediately doors were heavily barred, sand bags were put up against walls, and alarm signals were installed. Day and night five policemen were on duty in the street outside; and eight to ten Trotskyists guarded the house inside. The Trotskyists lived in; after a turn of duty at the gate, they worked as secretaries and participated in domestic activities, especially in the regular debates which took place in the evenings—unless the arrival of visitors turned day into debating time.

The visitors were sometimes political refugees from Europe, but more often Americans, radical educationists, liberal professors, journalists, historians, occasionally a few Congressmen or Senators, and, of course, Trotskyists. The debates ranged from dialectics and Surrealism to the condition of the American Negroes, and from military strategy to Indian agriculture or the social problems of Brazil and Peru. Every visitor was a source of fresh knowledge to Trotsky, who listened, interrogated, took notes, argued and questioned again—there seemed to be no limit to his curiosity and capacity to absorb

facts. The men of his bodyguard were uneasy at the unconcern
with which he received strangers, but they could do nothing
about it. Only when his curiosity turned to his immediate
neighbourhood and he peeped into the hovels across the road
to find out how people lived there and 'what they thought of
the land reform', his guards stopped him. They considered it
safer for him to go under their protection on long trips into
the country than to slip past the gate and wander around
outside the house.

The trips into the country had to be undertaken suddenly
and in great secrecy. He usually went by car, accompanied by
Natalya, a friend, and the bodyguard. When they passed
through Mexico City, he had to crouch down in his seat and
cover his face—otherwise a crowd on the pavements would
recognize him and cheer or boo. Just as at Alma Ata and on
Prinkipo these trips were 'military expeditions', with much
marching, climbing, and toiling. Since there was less chance
of fishing and hunting, he developed a new hobby and collected
rare, huge cacti on the rocky pyramid-shaped mountains.
When he was not ill, he still had enormous physical strength,
although with his white head and deeply lined face he sometimes looked prematurely aged. He had also preserved his
military bearing; and the strongest of his bodyguards could not
easily keep pace with him as he climbed up a steep slope with
a load of heavy 'bayonet-bladed' cacti on his back. 'On one
occasion', a secretary relates, 'we accompanied some friends
to Tamazunchale, a distance of about 380 kilometres from
Coyoacan, in hopes of finding a special variety of cactus.
We were unsuccessful, but on the way down, nearer to Mexico
City, L.D. had noticed some *viznaga*s. He decided, despite
the fact that we reached the spot long after dark, to stop and
collect a carful. It was a .balmy night; L.D. was in a cheerful
mood; he moved briskly about the little group, digging cactus
by the light from the headlamps of the cars.'[1] More often his
companions had to follow him in the heat of the blazing sun,
as he climbed among the boulders, his figure, in a blue French
peasant jacket, sharply outlined against the rocks and his

---

[1] Karl Mayer, 'Lev Davidovich' in *Fourth International*, August 1941; Charles
Cornell, 'With Trotsky in Mexico', ibid., August 1944; A. Rosmer in Appendix II
to the French edition of Trotsky's *My Life*.

white thatch of hair torn by the wind. Natalya teasingly called these outings 'days of penal labour'. 'He was in a frenzy', she recollects, 'always the first on the job and the last to leave . . . hypnotically driven by an urge to *complete* the job in hand.'[1]

With time, and with the growing violence of Stalinist threats, even these outings seemed more and more risky; and all of Trotsky's existence was becoming compressed within the walls of his half court half prison. This showed itself even in his manner of taking physical exercise and in his hobbies. He took to planting the most exotic cacti in his garden and to raising chickens and rabbits in his yard. Even in these melancholy chores he remained rigorously methodical: every morning he spent a long while in the yard, feeding the rabbits and chickens (according to 'strictly scientific' formulas), tending them, and scrubbing the coops and hutches. 'When his health was poor', says Natalya, 'the feeding of rabbits was a strain on him; but he could not give it up, for he pitied the little animals.'

·   ·   ·   ·   ·   ·   ·   ·   ·

How remote, how infinitely remote, was now his tumultuous, world-shaking past; and how poignant his and Natalya's loneliness. Very rarely a face or a voice from that past would come back, but only to bring it home to him that nothing bygone could be recaptured or revived. In October 1939 Alfred and Marguerite Rosmer at last came to Coyoacan. They were the Trotskys' only surviving friends of the years of the First World War. They stayed with them at the Avenida Viena, nearly eight months, till the end of May 1940, during which they spent many an hour in intimate talk and reminiscences. Trotsky and Rosmer went over the archives together sorting them out, and pondering old documents. Sometimes they were joined by Otto Rühle, another veteran, who as an exile also lived in Mexico. Rühle, we know, had at the beginning of the First World War distinguished himself as one of the two socialists in the Reichstag—the other was Karl Liebknecht—who voted against the war. He had been one of the founding members of the German Communist party and one of the first dissenters to break with it. In emigration he

---

[1] Natalya Sedova in 'Father and Son', *Fourth International*, August 1941 and in *Vie et Mort de Leon Trotsky*.

devoted himself to a study of Marx and kept aloof from political activity, though he agreed to sit on Dewey's Commission of Inquiry. Since the counter-trial he had become a frequent guest at the Blue House and then at the Avenida Viena; and Trotsky, who respected his scholarship, showed him a warm friendship and helped him as much as he could—together they brought out *The Living Thoughts of Karl Marx*.[1]

In the first days of the war the thoughts of the three men, naturally enough went back to the days when they had all been engaged in the same revolutionary opposition to war, the days of the Zimmerwald movement. Trotsky (the author of the Zimmerwald Manifesto) proposed that they should come out with a new manifesto to assert and to symbolize the continuity of the revolutionary attitude in both world wars. Rosmer was all for it; but as Rühle had his differences with them and anyhow would not allow himself to be tempted into political action, the idea of the 'new Zimmerwald Manifesto' was abandoned. The past was too remote to answer even with an echo.

.   .   .   .

With the Rosmers, Seva had come to Coyoacan; and Trotsky and Natalya hugged the recovered grandchild. It was nearly seven years since they had sent him away from Prinkipo. The child had lived those years in Germany, Austria, and France, had changed guardians, schools, and languages, and had almost forgotten how to speak Russian. His grandfather's huge drama was as if mirrored in the tiny compass of his childhood. He had scarcely left the cradle when his father was torn away from him; and no sooner had he rejoined his mother in Berlin than she killed herself. Then Lyova, who had become father to him, died suddenly and mysteriously; and the child became the object of the family quarrel, was abducted, hidden, and seized again until he was brought to his grandfather, whom he could scarcely remember but whom he had been brought up to adore. And now the bewildered orphan stared

---

[1] Trotsky had advised the American Publishers Longmans, Green and Co. to ask Rühle, who had written a biography of Marx, to be the sole author of this book, assuring them that after Ryazanov Rühle was 'the greatest living Marx scholar'. The Publishers agreed that Rühle should select and edit the Marxian texts, but insisted that Trotsky write the Introduction.

restlessly at the strange and crowded fortress-like house to which he had been brought, a house already marked by death.

Behind the most welcome guests, the Rosmers, an ominous shadow was to creep in, the shadow of Ramon Mercader-'Jacson'. This was the 'friend' of Sylvia Agelof, the American Trotskyist who had attended the foundation conference of the Fourth International at the Rosmers' home. Some claim that it was then or shortly thereafter that 'Jacson' had been introduced to the Rosmers; and that ever since he had unobtrusively sought their company and rendered them, with seeming disinterestedness, many small services and favours. Rosmer emphatically denies this and asserts that he met him only in Mexico; and Rosmer's version is confirmed by 'Jacson' himself.[1] 'Jacson' posed, plausibly enough, as a non-politically minded businessman, sportsman, and *bon viveur*; it was supposedly as an agent of an oil company that he went to Mexico City at the time when the Rosmers arrived there. He kept himself in the background, however, and for many months sought no access to the fortified house at the Avenida Viena. But he was getting ready for his dreadful assignment.

.   .   .   .

Stalin was the only full-scale book, his last, on which Trotsky worked in these years. As posthumously published, the volume is pieced together from seven completed chapters and a mass of diverse fragments, arranged, supplemented, and linked up by an editor, not always in accordance with Trotsky's trend of thought. No wonder that the book lacks the ripeness and balance of Trotsky's other works. But perhaps even if he had lived to give it final shape and to eliminate the many tentative statements and overstatements of its early drafts, the *Stalin* would probably have remained his weakest work.

Trotsky had no awareness at all that he was somehow lowering himself by assuming the role of his rival's and enemy's portraitist. He never found any literary or journalistic work beneath him provided he could carry it out conscientiously. It is said that his publishers pressed him to tackle the biography

[1] See A. Rosmer, 'Une Mise au Point sur L'Assassinat de Léon Trotsky' in *La Révolution Prolétarienne*, Nr. 20, November 1948. See also 'Jacson's' statement in A. Goldman, *The Assassination of Leon Trotsky*, pp. 11, 15 and 25.

of Stalin, and that financial necessity compelled him to yield. This is not quite borne out by the evidence. The publishers were at least as keen, if not more so, on the *Life of Lenin* he had promised to complete.[1] If the need of money played its part in causing him to give priority to *Stalin*, he was nevertheless mainly actuated by a literary-artistic motive. He was eager to reassess Stalin's character in the fresh and fierce light of the purges; and his fascination with this task was stronger than any pride or vanity that might have prevented him becoming Stalin's biographer. His chief character, the Super-Cain now revealed, was to some extent unfamiliar even to him. He scrutinized Stalin's features anew, dug deep into archives, and searched his own memories for those scenes, incidents, and impressions that now seemed to acquire new meanings and new aspects. He delved with unrelenting suspicion into the hidden nooks and crannies of Stalin's career; and everywhere he discovered, or rediscovered, the same villain. Yes, he concluded, the Cain of the Great Purges had been there all the time, concealed in the Politbureau member, in the pre-1917 Bolshevik, in the agitator of 1905, even in the pupil of the Tiflis Seminary and the boy Soso. He drew the sinister, malignant, almost ape-like figure, stealthily making its way to the highest seat of power. The image, rough, lop-sided, sometimes unreal, derives an artistic quality from the force of the passion that animates it. It does present the torso of a terrifying monster.

There is no question that even here Trotsky treats the facts, dates, and quotations with his usual historical conscientiousness. He draws a clear line of distinction between the established facts, the deductions, the guesses, and the hearsay, so that the reader is able to sift the enormous biographical material and form his own opinion. Such indeed is Trotsky's pedantry here that his method of inquiry and exposition is exceptionally repetitive and wearisome. Armed with a formidable array of quotations and documents, he polemicizes at great length against hosts of Stalin's flatterers and courtiers, without realizing what a grotesque honour he pays them by doing so. Nevertheless, in composing the portrait, he uses abundantly and far too often the

[1] See Trotsky's correspondence with Curtis Brown, the Literary Agents. *The Archives*, Closed Section.

material of inference, guess, and hearsay. He picks up any piece of gossip or rumour if only it shows a trait of cruelty or suggests treachery in the young Djugashvili. He gives credence to Stalin's schoolmates and later enemies who in reminiscences about their childhood, written in exile thirty or more years after the events, say that the boy Soso 'had only a sarcastic sneer for the joys and sorrows of his fellows': that 'compassion for people or for animals was foreign to him'; or that from 'his youth the carrying out of vengeful plots became for him the goal that dominated all his efforts'. He cites Stalin's adversaries who depict the youngster and the mature man as almost an *agent provocateur*; and although Trotsky does not accept the accusation, he attaches 'significance' to it as showing what Stalin was held to be capable of by his former comrades![1]

There is no need to go into many examples of this approach. The most striking is, of course, Trotsky's suggestion, mentioned earlier, that Stalin had poisoned Lenin. He relates that in February 1923 Lenin, paralysed and losing speech, wanted to commit suicide and asked Stalin for poison—Stalin himself confided this to Trotsky, and to Zinoviev and Kamenev. He recalls the queer expression Stalin's face bore at that moment; and he makes his accusation on the ground that Lenin's death—a year later—came 'unexpectedly', and that Stalin was just then in so severe a conflict with Lenin that 'he 'must have made up his mind' to hasten Lenin's death. 'Whether Stalin sent the poison to Lenin with the hint that the physicians had left no hope for his recovery or whether he resorted to more direct means, I do not know. But I am firmly convinced that Stalin could not have waited passively when his fate hung by a thread, and the decision depended on a small, very small, motion of his hand.'[2] And here Trotsky presents in a startingly new context the story he had told so many times before, of how Stalin manœuvred to keep him, Trotsky, away from Moscow during Lenin's funeral: 'He might have feared that I would connect Lenin's death with last year's conversation about poison, would ask the doctors whether poisoning was involved, and demand a special autopsy.' He

---

[1] Trotsky, *Stalin*, pp. 11–12, 53, 100, 116, 120 and *passim*.
[2] Op. cit., pp. 372–82.

recalls that on his return to Moscow after the funeral, he found that the physicians 'were at a loss to account' for Lenin's death; and that even two and three years later Zinoviev and Kamenev eschewed all talk about this and answered Trotsky's questions 'in monosyllables and avoiding my eyes'. Yet ·he never states whether he himself had conceived the suspicion or conviction of Stalin's guilt already in 1924 or whether he formed it only during the purges, after Yagoda and the Kremlin doctors had been charged with using poison in their murderous intrigues. If he had felt this conviction or suspicion in 1924, why did he never voice it before 1939? Why did he, even after Lenin's death, describe Stalin as a 'brave and sincere revolutionary' to none other than Max Eastman? Even in this denunciatory biography, Trotsky still expresses the opinion that if Stalin had ever foreseen in what bloody convulsions the inner party struggle would end, he would never have started it.[1] Thus he still treats the Stalin of 1924 as a basically honest though short-sighted man, who would have hardly been capable of poisoning Lenin. Such inconsistencies suggest that in charging Stalin with this particular crime, Trotsky is projecting the experience of the Great Purges back to 1923–4. He concludes that Stalin, the hangman of all of Lenin's disciples was surely capable of killing Lenin as well, and that he did kill him. Yet it is difficult not to wonder whether the 'enigma' of Lenin's death, the suspicion of foul play, the tricks Stalin used to avoid a post-mortem, whether all these parts of the story are not so many transposed circumstances, say, of the story of Lyova's death.

Stalin's personality admittedly confronts any biographer with this difficult problem. His character was undoubtedly a vital element in the purges; and it is the biographer's task to trace the formation of that character and to show how early, at what stages, and to what extent its propensities had revealed themselves. The task, however, is not different from that which the student analysing the life course of a criminal has to solve. The potentiality of the criminal act may be present in the given character early enough; but it must not be presented as an actuality before it has turned into one. To be sure, deep suspiciousness, secretiveness, and a resentful craving for

[1] Op. cit., p. 393, and M. Eastman, *Since Lenin Died*, p. 55.

power reveal themselves in Stalin long before his rise; yet for
many years they are only his secondary characteristics. The
biographer ought to treat them with a sense of proportion and
with an eye to the dynamics of the personality and the all-
important interplay of circumstance and character. Trotsky's
Stalin is implausible to the extent to which he presents the
character as being essentially the same in 1936–8 as in 1924,
and even in 1904. The monster does not form, grow, and emerge
—he is there almost fully-fledged from the outset. Any better
qualities and emotions, such as intellectual ambition and a
degree of sympathy with the oppressed, without which no
young man would ever join a persecuted revolutionary party,
are almost totally absent. Stalin's rise within the party is not
due to merit or achievement; and so his career becomes very
nearly inexplicable. His election to Lenin's Politbureau, his
presence in the Bolshevik inner cabinet, and his appointment
to the post of the General Secretary appear quite fortuitous.
Trotsky himself sums up his approach in a single sentence:
'The process of [Stalin's] rise took place somewhere behind an
impenetrable political curtain. At a certain moment his figure,
in the full panoply of power, suddenly stepped away from the
Kremlin wall. . . .'[1] Yet even from Trotsky's disclosures it is
evident that Stalin did not at all come to the fore in this way:
that he had been, next to Lenin and Trotsky, the most influential man in the party's inner councils at least since 1918;
and that it was not for nothing that Lenin in his will described
Stalin as one of the 'two most able men of the Central Committee'.

As biographer not less than as leader of the Opposition
Trotsky underrates Stalin and the forces and circumstances
favouring him. 'The current official comparisons of Stalin to
Lenin are simply indecent', he rightly remarks. 'If the basis o
comparison is sweep of personality', he then adds, 'it is impossible to place Stalin even alongside Mussolini or Hitler.
However meagre the "ideas" of fascism, both the victorious
leaders of reaction, the Italian and the German, from the
beginning of their respective movements, displayed initiative,
roused the masses to action, pioneered new paths through the
political jungle. Nothing of the kind can be said about Stalin.'

[1] Trotsky, op. cit., p. 336.

These words were written while the U.S.S.R. was entering into the second decade of planned economy; and there was an unreal ring about them even then. They sounded altogether fantastic a few years later when Stalin's role could be viewed against the background of the Second World War and its aftermath. 'In attempting to find an historical parallel to Stalin', Trotsky went on, 'we have to reject not only Cromwell, Robespierre, Napoleon, and Lenin, but even Mussolini and Hitler. [We come] closer to an understanding of Stalin [when we think in terms of] Mustapha Kemal Pasha or perhaps Porfirio Diaz.'[1] Here the lack of historical scale and perspective is striking and disturbing.

What guides Trotsky's pen in passages like these is, of course, his holy anger and disgust with the monstrosities of the Stalin cult. He reduces to less than life-size the autocrat who has puffed himself up to superhuman stature, the self-deified despot. In doing so, Trotsky paves the way, as it were, for those who will many years later pull down Stalin's monuments, evict his body from the Red Square Mausoleum, efface his name from the squares and streets, and even rename Stalingrad Volgograd. With a lucid premonition of all this, Trotsky recalls that Nero too had been deified, but that 'after he perished his statues were smashed and his name was scraped off everything. The vengeance of history is more powerful than the vengeance of the most powerful General Secretary. I venture to think that this is consoling.'[2] About to be struck down by the ultimate act of Stalin's treachery, Trotsky already savours history's coming retribution and his own victory beyond the grave. He prepares that retribution in words weighty enough to serve as texts for posterity's judgement. He treats Stalin as the symbol of an immense vacuum, the product of an epoch in which the morality of the old order has dissolved and that of the new one has not yet formed.

L'état c'est moi is almost a liberal formula by comparison with the actualities of Stalin's totalitarian régime. Louis XIV identified himself with both the state and the Church—but only during the epoch of temporal power. The totalitarian state goes far beyond Caesaro-Papism. . . . Stalin can justly say, unlike le Roi Soleil, la société c'est moi.

[1] Op. cit., p. 413.   [2] Op. cit., p. 383.

And this is how Trotsky conveys in a single epigram the whole tragic tension between Stalin and the old Bolsheviks:

Of Christ's twelve Apostles Judas alone proved to be traitor. But if he had acquired power, he would have represented the other eleven Apostles as traitors, and also all the lesser Apostles whom Luke numbers as seventy.[1]

. . . .

Trotsky's comments on the events leading up to the war and on the prospects of war and revolution could be the subject of a special monograph. In these writings one is struck more strongly than ever by the contrast between his lucid and almost flawless analyses of the strategic-diplomatic elements of the world situation and his blurred vision of the prospects of revolution. He saw the Second World War as being basically a continuation of the First, a prolongation of the struggle of the great imperialist powers for a redivision of the world. At the time of the Munich Crisis he saw 'Hitler's strength (and weakness) in . . . his readiness to use . . . blackmail and bluff and to risk war', whereas the old colonial powers, having nothing to win but much to lose, were frightened of armed conflict. 'Chamberlain would give away all the democracies of the world —and not many are left—for one-tenth of India.' The Munich settlement, in his view, hastened the outbreak of war; and so did Franco's successes in Spain, in so far as they freed the bourgeois governments from the fear of revolution in Europe. Stalin's policy had the same effect: selling out the labour movement, 'as if it were petrol or manganese ore', he too was helping capitalism to regain self-confidence.[2] But it was the attitude of the United States that was decisive, for both Chamberlain and Stalin were afraid of committing themselves against Hitler as long as the United States remained un-committed. Yet as the world's leading imperialist power, in-heriting Britain's place, the United States could not remain

[1] Op. cit., pp. 416, 421.

[2] In an article dated 22 September 1938 (B.O., no. 70) Trotsky wrote: 'One may now be sure that Soviet diplomacy will attempt a rapprochement with Germany. . . .' 'The compromise made over Czechoslovakia's dead body . . . gives Hitler a more convenient base for starting the war. Chamberlain's flights [to Munich] will enter history as [the] symbol of the diplomatic convulsions which a divided, greedy, and helpless imperialist Europe lived through on the eve of the new bloodbath awaiting our planet.' See also B.O., nos. 71, 74, and 75-76.

isolationist; it was vitally interested in stopping the expansion of German and Japanese imperialism; and it would be compelled to join in the Second World War 'much earlier than it had entered the First'. The United States was also destined to play a far more decisive part in the peace-making, for 'if peace is not concluded on the basis of socialism, then the victorious United States will dictate the conditions of peace'.

One may well imagine the thunderous denunciation with which Trotsky met the German-Soviet pact of August 1939: the master of the Great Purges now stood self-exposed as Hitler's accomplice. Ever since 1933 Trotsky had repeated that nothing would suit Stalin better than an accommodation with Hitler. Now, after the decapitation of the Red Army, the fear of his own weakness had driven Stalin into Hitler's arms. 'While Hitler is conducting his military operations, Stalin is acting as his *intendant*', Trotsky remarked in the first days of the war.[1] But Stalin's purpose, he added, was not to help the Third Reich to victory, but to keep the Soviet Union out of the war for as long as possible, and in the meantime to obtain a free hand in the Baltic states and in the Balkans. When Stalin and Hitler, applauded by the Comintern, proceeded to partition Poland, Trotsky commented: 'Poland will resurrect, the Comintern never.' But even in his most vehement assaults on Stalin's lack of principle and cynicism, he did not put all the blame on Stalin. He reiterated that 'the key to the Kremlin's policy is in Washington', and that in order that Stalin should change his course the United States must throw its weight against Hitler. He repeated the same thought during the 'phoney war' in the winter of 1939–40, saying that France and Britain, in avoiding real military collision with Germany, were conducting a sort of 'a military strike' against the United States. From East and West alike Hitler was abetted in the conquest of Europe. The Polish and Czech Governments had already fled to France. 'Who knows', Trotsky wrote on 4 December 1939, many months before the collapse of France, 'whether the French Government, together with the Belgian, Dutch, Polish and Czechoslovak Governments, will not have to seek refuge in Great Britain?' He did not accept 'even for a

---

[1] The article 'Stalin, Hitler's *Intendant*' bears the date '2 September 1939, 3 a.m.' *The Archives.*

moment' the possibility of a Nazi victory; 'but before the hour of Hitler's defeat strikes, many, very many in Europe will be wiped out. Stalin does not want to be among them and so he is wary of detaching himself from Hitler too early'.[1]

When France capitulated and nearly the whole of Europe succumbed to Hitler's armed might, Trotsky stigmatized Stalin and the Comintern for their share in bringing about the catastrophe. 'The Second and Third Internationals . . . have deceived and demoralized the working class. After five years of propaganda for an alliance of the democracies and collective security, and after Stalin's sudden passage into Hitler's camp, the French working class was caught unawares. The war provoked a terrible disorientation, a mood of passive defeatism. . . .' Now the U.S.S.R. was 'on the brink of the abyss'. All Stalin's territorial gains in eastern Europe counted little in comparison with the resources and the power which Hitler had seized and which he would use against the Soviet Union.[2]

Having said all this, Trotsky insisted with the utmost firmness that the Soviet Union remained a workers' state, entitled to be unconditionally defended against all its capitalist enemies, fascist and democratic. He did not even deny Stalin the right to bargain with Hitler, although he himself thought that the Soviet-German Pact had not brought the Soviet Union any significant advantage; he would have preferred a Soviet coalition with the West. But he held that the question with whom the Soviet Union should align itself should be decided solely on grounds of expediency; and that no political or moral principle was involved in the choice, because the western powers no less than the Third Reich fought only for their imperialist interests. What Trotsky repudiated in Stalin's policy was not so much his choice of ally or partner, but his making a virtue of the choice and his proclaiming ideological solidarity with whoever happened to be his partner at the moment. Stalin and Molotov now extolled the German-Soviet friendship 'cemented with blood'; their underlings, conniving in Hitler's atrocities, declared that Poland would never rise again; and their propagandists, like Ulbricht,

[1] 'The twin stars: Hitler–Stalin'; quoted from *The Archives*.
[2] A statement for the Press ('The Kremlin's role in the European catastrophe'), 17 June 1940. Ibid.

turned all their 'anti-imperialist' zeal exclusively against the western powers. This was, Trotsky concluded, how 'Stalinism exercised its counter-revolutionary influence on the international arena'; and this was one more reason why the Soviet workers must overthrow it by force. But he reasserted that even under Stalin's rule, the workers' state remained a reality, which must be protected against any foreign enemy and fought for to the last.[1]

He was well aware that his ideas would again seem paradoxical to many—but was reality not just as paradoxical? Having in collusion with Hitler annexed Poland's eastern marches, Stalin proceeded to expropriate the big landlords there, to divide their estates among the peasants, and to nationalize industry and banking. Anxious to secure military control over the annexed territories, his new 'defensive glacis', he adjusted in every respect their social and political régime to that of the Soviet Union. Thus an act of revolution resulted from Stalin's co-operation and rivalry with the most counter-revolutionary power in the world. At a stroke, Stalin fulfilled the main desiderata which had always figured in every programme of Polish and Ukrainian socialists and communists, the desiderata they themselves had not been able to realize. The social upheaval in the annexed lands was, of course, the work of the Soviet occupation forces, not of the Polish and Ukrainian toilers— it was the first of the long series of revolutions from above which Stalin was to impose upon eastern Europe. And while he was expropriating the possessing classes economically, he expropriated the workers and the peasants politically, depriving them of freedom of expression and association.[2]

Trotsky, contemptuous of Stalin's 'bureaucratic methods' and 'horse-trading with Hitler', acknowledged the 'basically progressive' character of the social changes in Poland's eastern marches. He argued that Stalin overthrew the old order there only because the workers' state was a reality in the Soviet Union—only *that* had stopped him from coming to terms with the Polish landlords and capitalists. In other words, the revolutionary dynamic of the Stalinist state had now overlapped the

[1] 'The U.S.S.R. in War', *New International*, November 1939; articles in subsequent issues of this periodical; and *In Defence of Marxism*.

[2] *New International*, loc. cit.

boundaries of the U.S.S.R. However, in making this assertion Trotsky involved himself in a contradiction. Had he not maintained that Stalinism continued to play a 'dual', progressive and reactionary part, *only* within the Soviet Union, but that its role 'on the international arena' was 'exclusively counter-revolutionary', i.e. directed towards the preservation of the capitalist order? Had this not been Trotsky's chief argument in favour of the creation of the Fourth International? He still held that the wider international influence of Stalinism remained counter-revolutionary; and that the social upheaval on Poland's eastern marches was only a local phenomenon. He pointed out how little the expropriation of landlords and capitalists in the western Ukraine (or later in the Baltic states) weighed against the demoralization by Stalinism of the French workers, the betrayal by it of the Spanish revolution, and the services it had rendered Hitler. Again and again he returned to the disparity of the two facets of Stalinism, the domestic and the foreign; and he sought to explain it by the fact that inside the U.S.S.R. the elements of the workers' state (national ownership, planning, and revolutionary traditions) refracted themselves even through Stalin's bureaucratic despotism and limited Stalin's freedom of movement; whereas in the 'international arena' Stalinism acted without any such inhibition, pursuing only its narrow interests and following freely its opportunistic bent.[1]

The argument, although it contained some truth, could not resolve or even conceal the theoretical and political difficulty which now beset Trotskyism, a difficulty that was to grow immensely with the events of the coming decade. How real indeed was the distinction Trotsky had drawn between the domestic (partly still progressive) and the international (wholly counter-revolutionary) functions of Stalinism? Could any government or ruling group have for any length of time one character at home and quite a different one abroad? If the Soviet body politic preserved the quality of a workers' state, how could this leave unaffected its relationship with the outside world? How could the government of a workers' state be consistently a factor of counter-revolution?

Trotsky and his disciples could deal with this problem in only one of two ways: Either they had to declare that the

[1] Ibid.

Soviet Union had ceased to be a workers' state; that this accounted for the anti-revolutionary direction of Stalin's policies both at home and abroad; and that consequently Marxists had no reason whatsoever to go on 'defending the Soviet Union'. Or else, they had to admit that Stalinism was continuing to act a dual or ambivalent (progressive and reactionary) role both abroad and at home; that this was consistent with the contradictory character of the régime of the U.S.S.R., with the survival of the workers' state within the bureaucratic despotism; and that Marxists could cope with this intricate situation only by opposing Stalinism yet defending the Soviet Union.

Quite a few of Trotsky's disciples tried to find a way out of the predicament by declaring that the Soviet Union was no longer a workers' state, because its bureaucracy formed a new class, exploiting and oppressing the workers and peasants. This idea, we know, had been in the air since 1921, when the Workers' Opposition first voiced it in Moscow; and although Trotsky had always rejected it, the idea never ceased to appeal to some of his followers. In 1929 Rakovsky startled them when he wrote that the Soviet Union had already changed from a proletarian state which was bureaucratically deformed into a bureaucratic state with only a residual proletarian element.[1] Trotsky approvingly quoted the epigram (which underlay some of his reasonings in *The Revolution Betrayed*); but he drew no conclusions from it. Some of his disciples now wondered what could possibly be left of that 'residual proletarian element' after ten years—and what years! Was it not, they asked, preposterous to go on talking about a workers' state? They found encouragement for such a conclusion in some of Trotsky's speculations, hints, and *obiter dicta*. In *The Revolution Betrayed* he had argued that the Soviet managerial groups were preparing to denationalize industry and to become its stockholding owners—in other words, that the Stalinist bureaucracy was incubating a new capitalist class. Years had passed and of such a development there was no sign. Was Trotsky then not mistaken in his conception of Soviet society? He saw the Stalinist bureaucracy hatching out a new bourgeois class and a new capitalism; but was not that bureaucracy itself

[1] See *B.O.*, nos. 15–16, 1930; Correspondence from the U.S.S.R.

the new class hatched out by the October revolution and fully fledged already?

Just before the outbreak of the war an Italian ex-Trotskyist, Bruno Rizzi, answered this question affirmatively in a little noticed but influential book, *La Bureaucratisation du Monde*, published in Paris. Rizzi was the original author of the idea of the 'managerial revolution', which Burnham, Shachtman, Djilas and many others were to expound later in far cruder versions. He based himself on part of Trotsky's argument, as stated in *The Revolution Betrayed*, in order to reject the argument as a whole. The Russian revolution, he maintained, having set out, like the French, to abolish inequality had merely replaced one mode of economic exploitation and political oppression by another. Trotsky, haunted by the phantom of a capitalist restoration in the U.S.S.R., failed to see that 'bureaucratic collectivism' had established itself there as the new form of class domination. He refused to treat the bureaucracy as the 'new class' because it did not own the means of production and did not accumulate profits. But the bureaucracy, Rizzi replied, did own the means of production and did accumulate profits, only it was doing that collectively and not individually, as the old possessing classes had done. 'In Soviet society the exploiters do not appropriate surplus value directly, as the capitalist does when he pockets the dividends of his enterprise; they do it indirectly, through the state, which cashes in the sum total of the national surplus value and then distributes it among its own officials.'[1] *De facto* possession of the means of production, possession *through* the state and possession *of* the state, had taken the place of bourgeois possession *de jure*. The new state of affairs was not, as Trotsky supposed, a bureaucratic interval or a transient phase of reaction, but a new stage in the development of society, even an historically necessary stage. Just as feudalism was followed not by Equality, Liberty, Fraternity, but by capitalism, so capitalism was being followed not by socialism but by bureaucratic collectivism. The Bolsheviks were 'objectively' just as incapable of achieving their ideal as the Jacobins had been of realizing theirs. Socialism was still utopia! The workers inspired by it were once again cheated of the fruits of their revolution.

[1] Bruno, R., *La Bureaucratisation du Monde*.

In so far, Rizzi went on, as bureaucratic collectivism organized society and its economy more efficiently and productively than capitalism had done, or could do it, its triumph marked historic progress. It was therefore bound to supersede capitalism. State control and planning were predominant not only in the Stalinist régime, but also under Hitler, Mussolini, and even under Roosevelt. In different degrees Stalinists, Nazis, and New Dealers were the conscious or unconscious agents of the same new system of exploitation, destined to prevail the world over. As long as bureaucratic collectivism stimulated social productivity, Rizzi concluded, it would be invulnerable. The workers could only do what they had done under early capitalism—struggle to improve their lot and wrest concessions and reforms from their new exploiters. Only after the new system had begun to decay and to retard and shackle social growth, would they be able to resume the fight for socialism successfully. This was a remote prospect, yet it was not unreal: bureaucratic collectivism was the last form of man's domination by man, so close to classless society that bureaucracy, the last exploiting class, refused to acknowledge itself as a possessing class.[1]

Trotsky, knowing that Rizzi had expressed a trend of ideas that was gaining ground among Trotskyists, dealt with his argument in an essay 'The U.S.S.R. in War', written in the middle of September 1939.[2] 'It would be a piece of monstrous nonsense', he began, 'to break with comrades who differ from us in their views about the social nature of the U.S.S.R. as long as we are in agreement about our political tasks.' The argument whether the U.S.S.R. was a workers' state or not was often only a quibble—Rizzi had at least the merit of having 'raised it to the height of historical generalization'. He identified bureaucratic collectivism as the new order of society, essentially the same behind the different façades of Stalinism, Nazism, Fascism, and the New Deal. His equation of Stalinism and Nazism (Trotsky replied) might sound plausible enough in the days of the pact between Hitler and Stalin. That pact, many argued, had merely brought out the kinship of the two régimes, a kinship so evident in their

[1] Bruno, R., op. cit.
[2] New International, November 1939 and In Defence of Marxism, pp. 8–11.

techniques of government; and, in Rizzi's opinion, it was only a matter of time before the Nazi and Fascist (but also the Rooseveltan) state would carry its control of the economy to a logical conclusion and nationalize all industry. Against this, Trotsky asserted that whatever the resemblances between Hitler's and Stalin's methods of government, the economic and social differences were qualitative and not merely quantitative—this was the gulf between their régimes. Neither Hitler nor Roosevelt would or could go beyond 'partial nationalization'—each was only superimposing state intervention upon an essentially capitalist order. Stalin alone exercised control over a truly post-capitalist economy. To be sure, the growth of bureaucracy was evident in various countries and under different régimes. But bureaucratic collectivism as a distinctive social order, if it existed at all, was still confined to a single country; and there it rested upon foundations created by a socialist revolution.

It was therefore rash, Trotsky pointed out, to speak of any 'universal trend', by dint of which bureaucratic collectivism was the real successor to capitalism. If this had been so, then any socialist revolution, even in the most advanced industrial country (or in several such countries), would inevitably usher in something like the Stalinist régime. This was indeed Rizzi's view. Against this Trotsky referred to the empirical evidence which showed how decisively Russia's backwardness, poverty, and isolation had contributed to the ascendancy of Stalinism. The Russian Revolution had deteriorated under the burden of circumstance; and there was no reason to assume that any socialist revolution must, regardless of circumstance, deteriorate likewise. Stalinism was not the norm of the new society, as Rizzi thought, but an historic abnormality; not the final outcome of the revolution, but an aberration from the revolutionary course. Soviet bureaucracy was still a parasitic outgrowth of the working class, as dangerous as such an outgrowth can be; but it was not an independent body. Contrary to Rizzi's view, bureaucratic collectivism did not represent any historic progress—the progress the Soviet Union was making was due to collectivism not to bureaucracy. Stalinism could survive only as long as the Soviet Union was merely borrowing, imitating, and assimilating superior western technology. Once this stage was left behind, the requirements of social life would become more complex; and social initiative would have to reassert itself. A major conflict between bureaucracy and social initiative was therefore looming ahead; and the conflict would be all the deeper, because unlike the French bourgeoisie after the revolution the bureaucracy 'is not the bearer of a new economic system', which could not function without it. On the contrary, in order to function properly the new system would have to free itself from the stranglehold of bureaucracy.

The idea that underlay all the theories about bureaucratic collectivism was that the working class had shown itself incapable of accomplishing the socialist revolution which Marxism had expected it to accomplish. Yet capitalism too had shown itself unable to function and survive. Some form of a collectivist economy was therefore bound to replace it. But as the working class had failed to cope with this task, the bureaucracy was performing it; and not socialist but bureaucratic collectivism was superseding the old order. Trotsky agreed that here was the crux of the controversy.[1] The question whether the Soviet Union was a workers' state or whether its régime was one of bureaucratic collectivism was secondary. All that he himself intended to say when he spoke of the 'workers' state' was that its potentiality and its elements were preserved in the social structure of the Soviet Union—it had not occurred to him to suggest that the Stalinist régime was a workers' state in the ordinary and political sense of the term. One might, on the other hand, speak of 'Soviet' bureaucratic collectivism and still hold that this included the potentiality of the workers' state. What was far more important was whether one held that bureaucratic collectivism had come to stay because the working class was inherently incapable of achieving socialism.

That the record of the labour movement was compounded of failures and disappointments was undeniable. The workers had not been able to bar Mussolini's, Hitler's, and Franco's roads to power; they had allowed themselves to be manœuvred into defeat by Popular Fronts, and they had not prevented two world wars. But how were these failures to be diagnosed? As

[1] Loc. cit.

faults of leadership, faults which could be remedied? Or as the historic bankruptcy of the working class and evidence of its inability to rule and transform society? If the leadership was at fault, the way out was to create a new leadership in new Marxist parties and a new International. But if the working class was at fault, then the Marxist view of capitalist society and socialism must be admitted to have been wrong, for Marxism had proclaimed that socialism would either be the work of the proletariat or it would not be at all. Was Marxism then just another 'ideology' or another form of the false consciousness that causes oppressed classes and their parties to believe that they struggle for their own purposes when in truth they are only promoting the interests of a new, or even of an old, ruling class? Viewed from this angle, the defeat of the pristine Bolshevism would indeed appear to have been of the same order as the defeat of the Jacobins—the result of a collision between Utopia and a new social order—and Stalin's victory would present itself as the triumph of reality over illusion and as a necessary act of historic progress.

Thus at the close of his days Trotsky interrogated himself about the meaning and the purpose of all his life and struggle and indeed of all the struggles of several generations of fighters, communists, and socialists. Was a whole century of revolutionary endeavour crumbling into dust? Again and again he returned to the fact that the workers had not overthrown capitalism anywhere outside Russia. Again and again he surveyed the long and dismal sequence of defeats which the revolution had suffered between the two world wars. And he saw himself driven to the conclusion that if major new failures were to be added to this record, then the whole historic perspective drawn by Marxism would indeed come under question. At this point he indulged in one of those overemphatic and hyperbolic statements which from time to time occur to any great controversialist and man of action, but which taken literally lead to endless confusion. He declared that the final test for the working class, for socialism, and for Marxism was imminent: it was coming with the Second World War. If the war were not to lead to proletarian revolution in the West, then the place of decaying capitalism would indeed be taken not by socialism, but by a new bureaucratic and totalitarian system of exploitation.

And if the working classes of the West were to seize power, but then prove incapable of holding it and surrender it to a privileged bureaucracy, as the Russian workers had done, then it would indeed be necessary to acknowledge that the hopes which Marxism placed in the proletariat had been false. In that case the rise of Stalinism in Russia would also appear in a new light: 'We would be compelled to acknowledge that . . . [Stalinism] was rooted not in the backwardness of the country and not in the imperialist environment, but in the congenital incapacity of the proletariat to become a ruling class. Then it would be necessary to establish in retrospect that . . . the present U.S.S.R. was the precursor of a new and universal system of exploitation. . . . However onerous this . . . perspective may be, if the world proletariat should actually prove incapable of accomplishing its mission . . . nothing else would remain but to recognize openly that the socialist programme, based on the internal contradictions of capitalist society, had petered out as a Utopia.'[1]

Perhaps only Marxists could sense fully the tragic solemnity which these words had in Trotsky's mouth. True, he uttered them for the sake of the argument; but even for the sake of argument he had never yet contemplated the possibility of an utter failure of socialism so closely; he insisted that the final 'test' was a matter of the next few years; and he defined the terms of the test with painful precision. He went on to state: 'It is self-evident that [if the Marxist programme turned out to be impracticable] a new minimum programme would be required—to defend the interests of the slaves of the totalitarian bureaucratic system.' The passage was characteristic of the man: if bureaucratic slavery was all that the future had in store for mankind, then he and his disciples would be on the side of the slaves and not of the new exploiters, however 'historically necessary' the new exploitation might be. Having lived all his life with the conviction that the advent of socialism was a scientifically established certainty and that history was on the side of those who struggled for the emancipation of the ex-ploited and the oppressed, he now entreated his disciples to remain on the side of the exploited and the oppressed, even if history and all scientific certainties were against them. He, at

[1] Loc. cit., and *passim*.

any rate, would be with Spartacus, not with Pompey and the Caesars.

Having explored this dark prospect, he did not, however, resign himself to it. Was there, he asked, sufficient evidence for the view that the working class was incapable of overthrowing capitalism and transforming society? Those who held this view, including some of his disciples, had never seen the working class in revolutionary action. They had watched only the triumphs of fascism, Nazism, and Stalinism; or they had known only bourgeois democracy in decay. All their political experience was indeed compounded of defeat and frustration; no wonder that they had come to doubt the political capacity of the proletariat. But how could he doubt it, he who had seen and led the Russian workers in 1917? 'In these years of world-wide reaction we must proceed from the possibilities which the Russian proletariat revealed in 1917.' The revolutionary intelligence and energy the Russian workers had shown then was surely latent in German, French, British, and American workers as well. The October Revolution was therefore still 'a colossal asset' and 'a priceless pledge for the future'. The subsequent record of defeats must be blamed not on the workers but on their 'conservative and utterly bourgeois leaders'. Such was the 'dialectics of the historic process that the proletariat of Russia, a most backward country . . . has brought forth the most far-sighted and courageous leadership, whereas in Great Britain, the country of the oldest capitalist civilization, the proletariat has even today the most dull-witted and obsequious leaders'. But leaders come and go, the social class remains. Marxists must still work for the renewal of the leadership and must stake everything on the 'organic, deep, irrepressible urge of the toiling masses to tear themselves free from the sanguinary chaos of capitalism. . . .'

He reasserted his Marxist conviction not with the flamboyant optimism of his earlier years, but with a hard-tested and enduring loyalty:

. . . the basic task of our epoch .has not changed for the simple reason that it has not been solved. . . . Marxists do not have the slightest right (if disillusionment and fatigue are not considered 'rights') to draw the conclusion that the proletariat has forfeited its revolutionary possibilities and must renounce all aspirations. . . .

Twenty-five years in the scales of history, when it is a question of most profound changes in economic and cultural systems, weigh less than an hour in a man's life. What good is the individual who, because of setbacks suffered in an hour or a day, renounces a goal he has set for himself on the basis of all the experience . . . of his life?

If this war provokes, as we firmly believe it will, a proletarian revolution, this must inevitably lead to the overthrow of the bureaucracy in the U.S.S.R. and to the regeneration of Soviet democracy on an economic and cultural basis far higher than that of 1918. In that case the question whether the Stalinist bureaucracy was a 'new class' or a malignant growth on the workers' state will be solved . . . it will become clear to everyone that in the world wide process of revolution the Soviet bureaucracy was only an *episodic* relapse.

To 'put a cross' over the Soviet Union because of this 'episodic relapse' and so to lose all historic perspective, would be unpardonable. The Soviet Union—and for the time being the Soviet Union alone—contained within itself the socio-economic framework for a reborn socialist democracy; and this must be defended. 'What do we defend in the Soviet Union? Not the features in which it resembles the capitalist countries, but precisely those in which it differs from them', not privilege and oppression, but the elements of socialism. This attitude 'does not at all mean any *rapprochement* with the Kremlin bureaucracy, any acceptance of its policies or any conciliation with the policies of Stalin's allies. . . . We are not a government party; we are the party of irreconcilable opposition. . . . We realize our tasks . . . exclusively through the education of the workers . . . by explaining to them what they should defend and what they ought to overthrow.'

Turning again to Stalin's moves in eastern Poland, Trotsky pointed out that if Stalin had left private property untouched there then it would have been necessary to reassess thoroughly the nature of the Soviet state. But Stalin acted as Napoleon had done when, having tamed revolution at home, he carried it abroad on bayonets. (Here Trotsky tacitly revised the notion about the 'wholly counter-revolutionary' character of Stalin's foreign policy.) To be sure, this was not the Marxist method of revolution: 'We were and remain against seizures of new territories by the Kremlin. We are for the independence of the Soviet

Ukraine and . . . of Soviet Byelorussia. At the same time, in the provinces of Poland which are occupied by the Red Army, the adherents of the Fourth International must be most active in expropriating the landlords and capitalists, in sharing out the land among the peasants, in creating soviets, workers' councils, &c. In doing so they must preserve their political independence; they must fight in elections for the complete independence of the soviets and the factory committees *vis-à-vis* the bureaucracy; and they must conduct their revolutionary propaganda in a spirit of distrust of the Kremlin and its local agencies.'

Trotsky could not offer his Polish and Ukrainian followers any other advice and remain true to himself; yet they had no chance whatsoever to act on his advice. They were weak; they held lost positions; and the G.P.U. crushed them in no time. They too were caught, as he had been, between the necessity and the impossibility of action.

.　　.　　.　　.　　.　　.　　.　　.　　.　　.

This dispute was to last till the end of May 1940, that is until the armed raid on Trotsky's home. James Burnham, Max Shachtman, and other American Trotskyists, members of the S.W.P., held views similar to Rizzi's, though they were less definite. With the outbreak of war and the Stalin-Hitler Pact these views crystallized rapidly. Early in September 1939 Burnham submitted to the National Committee of the S.W.P. a statement saying that 'it is impossible to regard the Soviet Union as a workers' state in any sense whatever.'[1] Before the end of the month Shachtman tabled a motion branding the Soviet occupation of the western Ukraine and Byelorussia as 'imperialist'; denying that the occupation had any of the progressive consequences of which Trotsky spoke; and urging the party to disavow its pledge to defend the Soviet Union. Burnham, as Professor of Philosophy at New York University, and Shachtman, the party's popular spokesman, exercised a strong influence on the Trotskyist intelligentsia. They had hitherto been committed to oppose war with revolutionary defeatism, if the war was waged by a bourgeois government, even a democratic one; and to defend the Soviet Union no matter to which

[1] See the *Internal Bulletin* of S.W.P. and the *New International* of the last months of 1939; Dwight Macdonald, *Memoirs of A Revolutionist*, pp. 17–19.

imperialist camp it was allied. For men like Burnham and Shachtman it was easy enough to expound such a view theoretically before the outbreak of the war, when it was generally assumed that the Soviet Union would be the ally of the western democracies. But with the Stalin-Hitler pact and the beginning of hostilities much had changed. The national mood, even in the years of American neutrality, was one of cautious sympathy with Britain and France and furious indignation against the German-Soviet Pact. Even Trotskyists found it hard to resist that mood. Burnham and Shachtman could not help feeling that if they went on 'defending' the Soviet Union, they would take upon themselves an unbearable odium. Yet in order to refuse 'defending' it they had, in Marxist terms, to declare that the U.S.S.R. was no longer a workers' state, but just another counter-revolutionary power fighting for imperialist aggrandizement. If Rizzi still argued that bureaucratic collectivism was 'historically necessary' and to some extent progressive, Burnham and Shachtman denied it any such merits. The logic of the argument led them further to deny that there was anything progressive in the Soviet economy. Implicitly or explicitly, they attacked national ownership of industry and national planning, saying that these served as the foundations for bureaucratic collectivism and totalitarian slavery. Gradually every principle of the Marxist-Leninist programme, including dialectics and morality, came again under debate. Burnham, Shachtman, and those who followed them, found themselves rejecting the programme point after point. This was, in fact, a continuation of that 'Retreat of the Intellectuals' which they themselves had just described, when they attacked Eastman, Hook, and others in the pages of *The New International*—only that now the attackers joined in the retreat.

In his criticism of Rizzi, Trotsky had said all that he had to say in this debate. The controversy with Burnham and Shachtman was conducted on far lower levels of political thought and style. The argument was remarkable mainly as an outburst of the disillusionment and pessimism pent-up among Trotsky's followers, and as Trotsky's last stand against them—the finale of all his controversies.[1]

---

[1] Trotsky's most characteristic statements in this debate are collected in *In Defence of Marxism*.

All the issues under debate were brought to a head before the end of the year 1939, when Stalin ordered his armies to attack Finland. Trotsky in his commentaries castigated Stalin's 'stupid and incompetent' conduct of the Finnish War, which had outraged the world and exposed the Red Army to humiliating defeats.[1] He nevertheless insisted that what Stalin was trying to do in Finland was to secure an exposed flank of the Soviet Union against a probable attack from Hitler. This was a legitimate endeavour; and any Soviet government, acting in the circumstances in which Stalin acted (circumstances which were, however, partly of Stalin's making), might well be compelled to protect its frontiers at Finland's expense. The strategic interest of the workers' state must take precedence over Finland's right to self-determination.[2] As Stalin's invasion of Finland was met in the Allied countries by a campaign for 'switching the war', and for armed intervention in favour of Finland, Trotsky called all the more emphatically for the 'defence of the Soviet Union'. This brought an outcry from his erstwhile disciples: 'Has Trotsky become Stalin's apologist?! Does he want us to become Stalin's stooges?!' 'No, Comrade Trotsky, . . .', Burnham replied, 'we will not fight alongside the G.P.U. for the salvation of the counter-revolution in the Kremlin.'[3]

Words like these echoed the language Trotsky himself had used in connexion with the Great Purges, when he called on 'every honest man' to expose the murderous G.P.U. plots, and to 'burn out with iron the cancer of Stalinism,' and when he inveighed against those 'friends of the Soviet Union' who, in the name of the sacrosanct interests of the workers' state, condoned Stalin's crimes. True, even in the heat of the most furious polemics, he had always reiterated that, despite everything, he and his followers would defend unconditionally the U.S.S.R. against all foreign enemies. But quite a few of his followers had treated these declarations as merely his *façon de parler*; and they were

---

[1] Trotsky's commentaries on the Finnish war appeared in American and British newspapers; and he summed up his view in an article 'Stalin After the Finnish Experience' written in March 1940. *The Archives.*

[2] *In Defence of Marxism*, pp. 56–59 and *passim.*

[3] The controversy was conducted within the S.W.P., in its *Internal Bulletin* (which published the resolutions of the Majority and of the Minority in December 1939) and finally in the *New International.*

dismayed to find that he meant what he had said. They charged him with inconsistency, duplicity, even betrayal. They searched his reasonings and arguments for the loose threads that could be found in them; and out of these threads they spun their own theories. Had Trotsky not said that 'internationally' Stalinism was only a factor of reaction and counter-revolution? How could he now dwell on the 'progressive and revolutionary consequences' of Stalinist expansion in eastern Europe? When they spoke about the Soviet Union's 'new class' and bureaucratic collectivism, he reproached them with abandoning Marxism and said that it was preposterous to speak of any new mode of exploitation in a country where the means of production were nationalized. Yet had he himself not declared that if within the next few years socialism were to fail in the West, bureaucratic collectivism would supersede capitalism as the new and universal system of exploitation? If bureaucratic collectivism was conceivable as the new, universal system of exploitation, why was it inconceivable as the national system of the U.S.S.R.? In saying that if the working classes of the West did not overthrow capitalism by the end of the Second World War, Marxism and socialism would be bankrupt—he knocked all his followers on the head.[1] They had watched so many of his prophecies come true that they were not inclined to take this prophecy lightly. The faithful and naïve among his disciples spent the next few years looking for the signs of revolution in the West and having visions of revolution. The sceptics and cynics concluded (at once or somewhat later) that on Trotsky's own showing Marxism and socialism were already bankrupt; and that the epoch of bureaucratic collectivism— had set in. Burnham was the first to dot the i's. He had been a 'good Bolshevik-Leninist', even a 'fierce enemy of American imperialism', as long as he felt that he was riding the tide of history. But having, with Trotsky's unwitting assistance, convinced himself that the managerial class was riding it, he hastened to cast off the ideological ballast of Marxism and to proclaim the advent of the managerial

---

[1] 'Some comrades evidently were surprised', Trotsky remarked, 'that I spoke in my article of the system of "bureaucratic collectivism" as a theoretical possibility. They even discovered in this a complete revision of Marxism.' *In Defence of Marxism*, p. 30.

revolution.[1] Shachtman accepted Burnham's prognosis; but being more strongly attached to Marxism, he viewed the prospect with grief rather than exhilaration; and he tried to fit it in with the wreckage of his earlier beliefs.[2]

In terms of the new Trotskyism which they had culled from *The Revolution Betrayed*, Burnham and Shachtman used fairly strong arguments; and both now claimed to defend Trotskyism against Trotsky himself. 'Then I am not a Trotskyist', the master replied paraphrasing Marx.[3] But to counter their arguments he had to disavow, at least implicitly, his own polemical exaggerations and excesses. 'The comrades are very indignant about the Stalin-Hitler Pact', he said in a letter. 'This is comprehensible. They wish to get revenge on Stalin. Very good. But today we are weak, and we cannot immediately overthrow the Kremlin. Some comrades try then to find a purely verbalistic satisfaction: they take away from the U.S.S.R. the title Workers' State, as Stalin deprives a disgraced functionary of the Order of Lenin. I find it, my dear friend, a bit childish. Marxist sociology and hysteria are absolutely irreconcilable.'[4] After all he had suffered at Stalin's hands, nothing distressed him more than to see the judgement of his own disciples clouded by Stalinophobia; and to his last breath he pleaded with them 'against hysteria' and for 'objective Marxist thinking.'

The American Trotskyists had split into a 'majority' which, led by James P. Cannon, accepted Trotsky's view, and a 'minority' which followed Burnham and Shachtman. Trotsky urged all of them to exercise tact and tolerance; and while he encouraged the 'Cannonites' to conduct the argument against Burnham and Shachtman vigorously, he also warned them that Stalinist agents in their ranks would seek to exacerbate the quarrel; and he advised them to allow the minority to express itself freely and even to act as an organised faction within the S.W.P. 'If someone should propose . . . to expel comrade Burn-

[1] See Burnham's 'Science and Style (A Reply to Comrade Trotsky)' (reprinted as Appendix in Trotsky's *In Defence of Marxism*). 'The Politics of Desperation' in *New International*, March-April 1940, and *The Managerial Revolution*.
[2] Shachtman's 'The Crisis of the American Party—An Open Letter to Trotsky' and 'The U.S.S.R. and the War' also appeared first in the *Internal Bulletin* and then in *New International* of March-April 1940.
[3] Trotsky, op. cit., p. 168.   [4] Ibid., p. 23.

ham', he gave notice, 'I would oppose it energetically.'[1] Even after the minority had held its own National Convention, Trotsky still counselled the majority not to treat this as an excuse for expulsions.

The minority, however, of its own accord constituted itself as a new party and appropriated *The New International*, the 'theoretical monthly' of the S.W.P. Almost at once the new party also split, for Burnham broke with it, declaring that 'of the most important beliefs, which have been associated with the Marxist movement, whether in its reformist, Leninist, Stalinist, or Trotskyist variants, there is virtually none which I accept in its traditional form. I regard these beliefs as either false, or obsolete, or meaningless. . . .' This was a startling confession coming as it did from someone who had been a leading Trotskyist these last years. Only a few weeks earlier Burnham and his friends had felt offended by Trotsky's remarks about his 'unMarxist' way of thinking, 'On the ground of beliefs and interests. . .', Burnham now stated, 'I have for several years had no real place in a Marxist party.'[2] Whether this was true or not, whether the future author of *The Managerial Revolution* was merely trying to make his ideological somersault appear less indecently sudden, or whether he had in fact only posed as a zealous Marxist and Leninist all these years, nothing that Trotsky said against him was even remotely as devastating as was Burnham's present picture of himself. After the event Trotsky was not sorry to lose so dubious a 'disciple' whom he had characterized in private letters with epithets of which 'intellectual snob' is the mildest.[3] He expected others to follow in Burnham's footsteps: 'Dwight Macdonald is not a snob, but a bit stupid. . . . [He] will abandon the party just as Burnham did, but possibly because he is a little lazier, it will come later.' He was, however, truly saddened by the break with Shachtman, for whom he had a soft spot, even though he was often annoyed by his 'clownishness', 'superficiality', &c. Their connexion dated back to Shachtman's visit to Prinkipo early in 1929; and it had become close through many subsequent

---

[1] Ibid., pp. 97, 101, 148 and *passim*.

[2] Burnham's letter of resignation is appended to *In Defence of Marxism*, pp. 207–11.

[3] This epithet appears also in *In Defence of Marxism*, p. 181.

meetings, letters and proofs of Shachtman's devotion. In the present fight between the factions Trotsky, of course, supported Cannon, but on personal grounds he felt much closer to Shachtman. 'If I could do so', he wrote to him at the height of the controversy, 'I would immediately take an aeroplane to New York City in order to discuss with you for forty-eight or seventy-two hours uninterruptedly. I regret very much that you do not feel . . . the need to come here to discuss the questions with me. Or do you? I should be happy.'[1]

The split might be said to have ruined the Fourth International, if so shadowy an organization could be ruined at all. Trotsky trusted that after the exit of the 'petty bourgeois and careerist elements', the S.W.P. would strike deeper roots in the American working class. This was not to happen: the S.W.P. remained a tiny chapel, the members of which were zealously devoted to the letter of Trotsky's teaching, and later to his memory, but which was never able to acquire any political weight; while its rival, Shachtman's group, devoid even of such virtues as may keep the feeblest of sects alive for decades, renounced more and more of its 'Trotskyism' until it crumbled away and vanished.[2] Trotskyist groups in other countries were also affected, for everywhere, but especially in France, quite a few members accepted Burnham's or Shachtman's views.

Thus at his sunset Trotsky watched for the last time the rock he rolled up his dreary mountain rolling down the slope again.

.   .   .   .

On 27 February 1940 Trotsky wrote his testament. He had drafted several brief wills earlier, but he had done so for legal purposes only, to ensure that Natalya and/or Lyova inherited the copyright on his books. The present document was his real last will and testament; every line of it was permeated by his sense of the approaching end. In writing it he supposed, however, that he might die a natural death or commit suicide—he did not think of dying at the hand of an assassin. 'My high (and still rising) blood pressure is deceiving those near me about my

[1] Ibid., p. 64.
[2] Shachtman and his group have since formally and categorically renounced every connexion with Trotskyism and Leninism and joined the Social Democratic group led by Norman Thomas, whose influence on American politics has also been negligible.

actual condition. I am active and able to work. But the end is evidently near.' Yet in the course of the six months which he still had before him, his health, despite the usual ups and downs, was not so bad as to justify this gloomy foreboding. In a postscript, dated 3 March, he repeated: '. . . at present I feel . . . a surge of spiritual energy because of the high blood pressure; but this will not last long.' He suspected that he was in an advanced stage of arteriosclerosis and that his doctors were hiding the truth from him. Evidently, Lenin's last illness and protracted paralysis often came to his mind; and he declared that rather than suffer such agony, he would commit suicide or, to put it more accurately, 'cut short . . . the too slow process of dying'. Yet he hoped that death would come to him suddenly, through a brain hemorrhage, for 'this would be the best possible end I could wish for'.[1]

Unwittingly he modelled his testament to some extent on Lenin's. Both documents consist of the main texts and postscripts added a few days later. In content, however, they reflect all the striking contrast of characters and circumstances. Lenin's will is absolutely impersonal. He gave it the form of a letter to the forthcoming party congress; and he did not say or even hint that he was writing it with his approaching death in mind. Although he too was tormented by the gravest dilemmas, he felt no need to make of his will a *credo*, knowing full well that his principles and beliefs would be taken for granted. His mind was occupied exclusively with the crisis in Bolshevism (which he knew his death would precipitate), and with the means and ways to prevent it. He told the party what he thought about the virtues and the failings of every one of its top leaders; he submitted to it his scheme for the reorganization of the Central Committee; and he advised the Committee to remove Stalin from the post of the General Secretary. To his last breath he remained, in his whole being, the chief of a great movement. Trotsky's testament on the other hand is intensely personal. He states briefly that there is no need for him to refute Stalin's 'stupid and vile slander' for there is 'not a single spot' on his revolutionary honour; and that a new 'revolutionary generation will rehabilitate the political honour' of himself and of thousands of other victims. In a single sentence he thanks friends

---

[1] *The Archives*; and *Trotsky's Diary in Exile*, pp. 139–41.

and followers who kept faith with him in his most difficult hours; but he offers them no advice—the testament contains not a single mention of the Fourth International. About half the text is devoted to Natalya:

> In addition to the happiness of being a fighter for the cause of socialism, fate gave me the happiness of being her husband. During almost forty years of our common life she has remained an inexhaustible source of love, magnanimity, and tenderness. She has undergone great sufferings . . . but I find some comfort in the fact that she has also known days of happiness.

He interrupts this tribute to her with a profession of faith:

> For forty-three years of my conscious life I have been a revolutionary; and for forty-two I have fought under the banner of Marxism. If I were to begin all over again, I would . . . try to avoid making this or that mistake, but the main course of my life would remain unchanged. I shall die a proletarian revolutionary, a Marxist, a dialectical materialist, and consequently an irreconcilable atheist. My faith in the communist future of mankind is not less ardent, indeed it is firmer today, than it was in the days of my youth.

As he penned these lines he looked out of the window, saw Natalya approaching the house, and the sight of her stirred him to conclude with this poetic passage:

> Natasha has just come up to the window from the courtyard and opened it wider so that the air may enter more freely into my room. I can see the bright green strip of grass beneath the wall, and the clear blue sky above the wall, and sunlight everywhere. Life is beautiful. Let the future generations cleanse it of all evil, oppression, and violence, and enjoy it to the full.

In an addendum he bequeathed to Natalya his literary rights and started another paragraph with the words: 'In case we both die. . . '; but he did not finish the sentence and left a blank. In the postscript of 3 March he went again into the nature of his illness and recorded that he and Natalya had more than once agreed that it was preferable to commit suicide rather than allow old age to turn one into a physical wreck. 'I reserve the right to determine for myself the time of my death. . . . But whatever may be the circumstances . . . I shall die with unshaken faith in the communist future. This faith in man and in his future gives me even now such power of resistance as cannot be given by any religion.'[1]

.   .   .   .

By now Stalin had decided that he could no longer allow Trotsky to live. This may seem strange. What, it might be asked, had he still to fear? Had he not exterminated all of Trotsky's adherents and even their families so that no avenger should rise? And what could Trotsky, from the other end of the world, alone undertake against him? A few years earlier Stalin might have feared that Trotsky could place himself at the head of a new communist movement abroad; but did he not realize now that the Fourth International had come to nothing?

The fact remains that Stalin was not reassured. He could not bring himself to believe that his violence and terror had indeed accomplished all that he wanted, that the old Bolshevik Atlantis had really vanished. He scrutinized the faces of the multitudes that acclaimed him, and he guessed what terrible hatred might be hidden in their adulation. With so many existences destroyed or broken up and with so much discontent and despair all around him, who could say what the unforeseeable shocks of war might not bring? Might not Atlantis somehow re-emerge, with new denizens, but with the old defiance? And even if the Fourth International was quite impotent now, who could say how the cataclysms of war might change the political landscape, what mountains they might not flatten and what hillocks they might not raise into mighty peaks? All the prospects that were so real to Trotsky in his hopes were equally real to Stalin in his fears; and Trotsky alive was their supreme and never-resting agent. He remained the mouthpiece of Atlantis, still uttering all its undying passions and all its battle cries. At every critical turn, when the inglorious Finnish campaign came to an end, when Hitler occupied Norway and Denmark, and when France collapsed, his voice rose from beyond the ocean to thunder on the consequences of these disasters, on Stalin's blunders that had helped to bring them about, and on the mortal perils threatening the Soviet Union. True, his indictments, condemnations, and warnings, did not reach the Soviet people; but they appeared in American, British, and other newspapers; and as the war spread to the East, they

---

[1] *The Archives* and *Trotsky's Diary in Exile*, pp. 139–41.

VII. Two views of the 'little fortress' at Coyoacan

Associated Press

VIII. Trotsky, Natalya, and Seva. At the end of 1939

might, in the turmoil and confusion of military defeats and retreats, penetrate there too.

At the end of April 1940 Trotsky addressed to 'Soviet workers, peasants, soldiers, and sailors' a message entitled 'You are being Deceived'. It is said that a leaflet with the message was smuggled into the U.S.S.R. by sympathetic sailors; but it must be doubted whether the message ever reached its destination.[1] Still, every sentence in it was dynamite. 'Your newspapers', he told the Soviet workers and soldiers, 'are telling you lies in the interest of Cain-Stalin and his depraved commissars, secretaries, and G.P.U.-men.' 'Your bureaucracy is bloodthirsty and ruthless at home, but cowardly vis-à-vis the imperialist powers.' Stalin's infamies were robbing the Soviet Union of sympathy abroad, isolating it, and strengthening its enemies; these infamies were 'the main source of danger to the Soviet Union'. He called the workers and soldiers 'never to surrender to the world bourgeoisie the nationalized industry and the collectivized economy, because upon this foundation they could still build a new and happier society'. 'It is the duty of revolutionaries to defend tooth and nail every position gained by the working class . . . democratic rights, wage scales, and an achievement as colossal as the nationalization of the means of production and a planned economy.' But these 'conquests' of the October revolution would benefit the people only if they proved themselves capable of dealing with the Stalinist bureaucracy as they had once dealt with the Tsarist bureaucracy. No, Stalin could not allow Trotsky's voice to go on summoning insurrection.

Several former G.P.U. officers and foreign communists have since described how the final assault on Trotsky was prepared.[2] At the end of the Spanish Civil War, G.P.U. agencies specialized in the 'liquidation of Trotskyism' were transferred to Mexico. Mexican Stalinists did what they could to whip up

---

[1] I am quoting the text, dated 23 April 1940, from *The Archives*. About this time shortly before the German invasion of Norway, Walter Held, the German Trotskyist, left that country, hoping to reach America via the U.S.S.R. and Japan. En route, however, he vanished without trace. He was almost certainly arrested and executed in the U.S.S.R. It is possible, but not very probable, that he tried to convey Trotsky's message to people in the U.S.S.R.

[2] See, e.g., Budenz, *This is my Story*, pp. 257–63; and Orlov's deposition quoted above.

mass hysteria against 'the traitor sheltering at Coyoacan'. Day in, day out, they accused him not only of plotting against Stalin, but of conspiring, in the interest of the American oil magnates, against Cardenas, and preparing a general strike and a fascist *coup d'état* in Mexico. Even so, at the beginning of the year 1940 Moscow charged the leaders of the Mexican Communist party with adopting 'a conciliatory attitude towards Trotskyism'; and these leaders were demoted. The anti-Trotskyist campaign rose to a new pitch; and a minor blunder committed by Trotsky himself provided grist to his enemies. Just before the end of the year 1939 he agreed to go to the United States and appear as witness before the so-called Dies Committee of the House of Representatives, a body which carried out 'investigations into un-American activities' (and which did this in a manner anticipating the witch-hunts conducted by Senator McCarthy in the nineteen-fifties). Senator Dies, chairman of the Committee, demanded the suppression of the American Communist party on the ground that it was the agency of a foreign power. Trotsky intended to use the Committee as a forum from which he would expose the G.P.U.'s murderous activities directed against himself and his followers. But he made it clear beforehand that he would speak up against the suppression of the Communist party and would call the workers of the world to turn world war into world revolution. Nothing came of the plan, partly because Trotsky's own followers, especially Burnham, strongly objected to it; and partly because the Dies Committee, forewarned about the kind of deposition Trotsky was ready to make, did not wish to hear him; and the American Government refused him the entry visa. Yet whatever the terms on which he had intended to appear before the Committee, the mere fact that he had been willing to do so made it easy for the Stalinists to accuse him of 'intriguing with Dies and the oil magnates against the Mexican people'. On 1 May 1940, 20,000 uniformed communists marched through Mexico City with the slogan 'Out with Trotsky' on their banners. He replied with denials, published his correspondence concerning the Dies Committee, and asked for an official Mexican inquiry into the matter.[1] President Cardenas

---

[1] 'Why did I agree to appear before the Dies Committee?' Trotsky's statements to the Press of 11 and 12 December 1939. *The Archives.*

shrugged off the Stalinist accusations; yet these had made an impression; and Trotsky's well-wishers wondered whether he would not be deprived of asylum, especially if Cardenas were to lose at the forthcoming elections.

. . . .

At this time the assassin already stood at the gate of the house at Avenida Viena. This was the man who had, in the summer of 1938, introduced himself as Jacques Mornard, the son of a Belgian diplomat, to Sylvia Agelof, the American Trotskyist who was present at the founding conference of the Fourth International. What his real name was has not yet been officially established, although it seems quite certain that he was Ramon Mercader, the son of Caridad Mercader, a Spanish communist well known in her country during the civil war *inter alia* for her close connexions with the G.P.U. Mornard's meeting with Sylvia Agelof in Paris was not accidental; it had been carefully prearranged. G.P.U. agents had for some time past watched Sylvia and her sister: both were Trotskyists; and Sylvia's sister travelled occasionally as courier to Coyoacan and did secretarial work for Trotsky. As to Sylvia, she had studied philosophy under Sidney Hook and psychology at Columbia University; she knew Russian, French, and Spanish, and could be especially helpful to the 'Old Man,' who so often complained that he was 'paralysed at work' because of the lack of a Russian secretary. A lonely spinster of rather unattractive appearance, she found herself all of a sudden assiduously courted by the handsome and well-groomed Mornard. She succumbed, and spent with him several absent-minded, dream-like months in France. Now and then she was puzzled by his behaviour. He exhibited so complete a lack of any interest in politics that this seemed to amount to an indolence of mind, quite surprising in the educated 'son of a diplomat'. He had impenetrably obscure connexions in commerce and journalism; even his family background was enigmatic. The stories he told her about himself were odd, even incoherent; and he spent lavishly, as if from a horn of plenty, on feasts and amusements.[1]

In February 1939, Sylvia Agelof returned to the States. In

[1] M. Craipeau, 'J'ai connu l'assassin de Trotsky', *France-Observateur*, 19 May 1960.

September he joined her in New York. Again she was somewhat perplexed by his behaviour. He had given her notice that he would come to the States as the American correspondent of a Belgian newspaper; instead he arrived with a false Canadian passport, assumed the name of Frank Jacson, and said that he had done this in order to avoid military service in Belgium. He claimed that he had never been in New York before; yet he knew his way about the City like someone familiar with the place. But to any puzzled query he had a plausible answer; and as he never dropped his role of playboy and *bon viveur*, he aroused not a shred of political suspicion. The worst she could reproach him with was frivolity and a penchant for fanfaronade. She tried to improve him and to get him interested in Trotskyism; but he invariably met these attempts with a closed mind and a bored face. And so, when soon after his arrival in New York, he told her that he was going to Mexico as sales agent or manager of an import-export firm, she found nothing strange in this; and when he urged her to join him in Mexico, she eagerly consented.

He was in Mexico before the middle of October; she came in January. She went at once to worship at the shrine in the Avenida Viena—she certainly delivered there messages from American Trotskyists. She soon returned to help with secretarial work. 'Jacson' usually drove her to the Avenida Viena in his expensive car; and, when her work was done, he awaited her at the gate. The guards came to know him and often chatted with him. Yet for several months he never ventured into the compound. (He still pretended to have only a condescending smile for Sylvia's political activities; but just to please her he began to show a little more curiosity about them.) At the gate he ran into Alfred and Marguerite Rosmer who presently became familiar with him as the 'obliging young man', 'Sylvia's husband'. He invited them to dinners in Mexico City and took them out into the country on sightseeing trips.

During the hours which he was supposed to work as a business agent, he kept in touch with G.P.U. men from whom he took orders and, it seems, with his mother who, according to several sources, was in Mexico then. Of these contacts of his Sylvia never had the slightest glimpse; he never brought his 'wife' and his mother together. Only sometimes he committed an indiscretion that for a moment put even Sylvia on guard. He gave her the address of his business office; and this turned out to be fictitious. He apologized for the 'mistake' and gave her another address. Sylvia, remembering that he had made a similar 'mistake' in Paris once, was so worried that she asked Marguerite Rosmer, a shrewd and observant person, to investigate the matter. However, the new address was found to be genuine; and even the Rosmers were so convinced that if there was anything slightly *louche* in Mornard-'Jacson's' affairs it had nothing to do with politics, that no one tried to pry into the nature of his 'business office'. (Only much later was it discovered that the same 'office' was used by various local Stalinist big-wigs.) Sylvia was scrupulous enough never to bring 'Jacson' into Trotsky's home—she even told Trotsky that as her husband had come to Mexico on a false passport, his visit might needlessly embarrass Trotsky. And when, in March, she left for New York, she took from 'Jacson' a formal promise that he would never in her absence enter the house in Avenida Viena.

Soon thereafter, however, he did enter. Rosmer had fallen ill, and 'Jacson' was asked to take him to the French hospital in Mexico City; then to bring him back, to buy medicines, &c. While chance thus smoothed his way, he was cautious enough to write to Sylvia and explain apologetically why he had 'broken his promise'. And although he was now becoming more and more familiar with Trotsky's household, three more months were to pass before he would meet Trotsky himself.

It seems that so far 'Jacson' had not yet been assigned the job of assassin. His task was rather to reconnoitre the house, its layout and defences, to ferret out details of Trotsky's daily routine, and to get any other information that might be useful for a massive armed assault which others were to carry out.

The man in charge of this attack was to be David Alfaro Siqueiros, Rivera's former friend, the celebrated painter, communist, and leader of Mexican miners. The year before he had returned from Spain, where he had commanded several brigades during the Civil War—he withdrew from the fighting at the head of only two or three score survivors. That so eminent and even heroic an artist should have agreed or volunteered to become Trotsky's murderer speaks volumes about the morals of Stalinism in these years; but it was, of course, a national habit in Mexico to settle political accounts gun in hand. In Siqueiros art, revolution, and gangsterism were inseparable— he had in himself much of the Latin American buccaneer. In Spain he had entered into a close connexion with the G.P.U. and, some say, with the Mercader family. Yet, despite the zealous services he had rendered, the Communist party had censured him recently for a misdemeanour in his handling of party funds. He was hurt and eager to regain favour by a conspicuous and hazardous act of devotion. He worked out the plan of an armed raid on Trotsky's home, and for its execution he called on men who had fought under him in Spain, and on Mexican miners.[1]

At Avenida Viena everyone had lived in the expectation of such an attack. Reading the local Stalinist papers railing against him, Trotsky remarked: 'People write like this only when they are ready to change the pen for the machine-gun.' True, at the insistence of his American followers, the house had been fortified: heavily barred doors, electrified wires, automatic alarm signals, and machine guns were in the way of would-be assailants. The guards had been increased. Ten Mexican policemen were on duty outside and around the house. Inside, sentries kept watch at the gate, day and night; and four or five men were at the ready in the guards' quarters. No doubt some of the guards, American boys from middle-class homes and just out of college, were little suited for their duty; but this could not be helped: the few workers who were members of the Trotskyist organization could rarely afford to give up jobs, leave families, and come to Coyoacan. Men came and went—after a few months of the monotonous routine a member of the bodyguard easily became jaded, undisciplined, and had to be replaced. It was therefore unavoidable that now and then the sentry at the entrance door should be an inexperienced recruit. Robert Sheldon Harte, who was to be on duty on the night of the Siqueiros raid, had come from New York on 7 April. During his six weeks at Coyoacan his comrades and Trotsky himself found him a warm-hearted and devoted

---

[1] General L. A. S. Salazar, *Murder in Mexico*. (I am obliged for some of the details about Siqueiros's character and background to an American writer who knew him well in the nineteen-thirties.)

but rather gullible and feckless creature.[1] Much later his comrades were to recall that he had struck up a quick friendship with Mornard-'Jacson' and that they had often been seen going out together. Clearly, Trotsky's security depended now on quite a few accidental circumstances. However, even these circumstances were not quite accidental, for they reflected his general situation, the heavy odds against him, and the extreme scantiness and the limitations of his following.

On 23 May, Trotsky worked hard the whole day, went late to bed; and could not fall asleep until he took a sleeping pill. About 4 a.m. a noise like the rattling of machine-guns awakened him. He was tired and drowsy and for a moment he thought that Mexicans outside were celebrating with fireworks one of their uproarious religious or national holidays. But 'the explosions were too close, right here within the room, next to me and overhead. The odour of gunpowder became more acrid, more penetrating . . . we were under attack'.[2] Natalya had already jumped out of bed and shielded him with her body. A moment later, under a hail of bullets, she pushed him down to the floor, into a corner between the bed and the wall; and, pulled by him, she went down herself, again covering him with her body. Silent and motionless, they lay in the darkness, while unseen assailants kept the room under a steady cross-fire coming through windows and doors. Perhaps 200 shots were fired; a hundred fell on and near the beds—over seventy holes were later counted in walls and doors. Natalya raised herself a little; he dragged her down again; and again they lay without stirring, breathed the gunpowder, and wondered what had happened to the guards and the police outside.

Suddenly a high-pitched cry 'Grandpa!' came from behind a wall or door. The attackers had broken into Seva's bedroom. 'The voice of the child', Trotsky said later, 'remains the most tragic recollection of that night.' 'This cry', Natalya recalls, 'chilled us to the marrow.' Then silence fell. 'They have

---

[1] Trotsky relates that shortly after Sheldon's arrival he saw him giving away the key to the front gate of the house to one of the builders working there. Trotsky warned him not to do this and said: 'If you behave like this, you will in case of an attack be its first victim.' Trotsky's statement on Sheldon of 15 July 1940. *The Archives.*

[2] Trotsky's account of the event, written on 8 June 1940, appeared posthumously under the title 'Stalin Seeks my Death' in *Fourth International*, August 1941.

kidnapped him', Trotsky whispered. As in a dream Natalya saw a man's silhouette which was illumined by the flare of an incendiary bomb exploding in the child's room, 'the curve of a helmet, shining buttons, an elongated face'. The man stopped at the doorstep between the Trotskys' bedroom and the child's, as if to check whether there was any sign of life and, though there seemed to be none, he fired another volley at the beds and vanished. The shooting now resounded through the courtyard, and the child's bedroom was on fire. Seva was not there—amid the flames a thin trail of blood could be seen leading out into the patio. 'Then all was silence . . . unbearable silence', Natalya recalled. ' "Where can I hide you safely?" [she was thinking] I was losing my strength from the tension and the hopelessness. Any moment now, they will return to finish him.' Where were all the members of the household, the Rosmers, the secretaries, the guards, the police? Were they all killed? '. . . we felt the stillness of the night, like the stillness of the grave, of death itself. . . . And suddenly there came again the same voice, the voice of our grandson; but this time it came from the patio and sounded quite differently, ringing out like a staccato passage of music, bravely, joyously: "Al-fred! Mar-gue-rite!" It returned us to the living!' Seva had saved himself also by hiding under his bed; and even before the shooting had ceased, thinking that his grandparents were dead, he went out with a wound in his toe, to look for the Rosmers.[1]

Within a few minutes the household assembled in the patio. No one had been killed or seriously wounded. The guards were still so dazed that they had not even checked what had happened to the police outside. Trotsky rushed out into the street and found the sentries disarmed and tied up. Brief, rapid, excited accounts: just before 4 a.m. over twenty men, in police and army uniforms, surprised the sentries and overpowered them without firing a shot. Then the assailants, led by a 'major', approached the gate; and one of them spoke to Robert Sheldon Harte who was on duty. The latter at once opened the gate. The attackers rushed into the courtyard, surprised and terrorized the other guards, placed machine guns behind trees at various points opposite Trotsky's bedroom, took up other positions, and opened fire. They were obviously out to kill

[1] Natalya Sedova in *Vie et Mort de Leon Trotsky*, pp. 309–10.

Trotsky and his family—they aimed not a single shot at anyone else. The raid lasted twenty minutes. Convinced that neither Trotsky nor his wife and grandchild could have survived, the raiders withdrew, throwing incendiary grenades into the house and a powerful bomb (which failed to explode) into the patio. Some left in two cars which belonged to Trotsky and usually stood in the yard ready to depart at a moment's notice, with the ignition keys in the locks. Sheldon disappeared with the raiders. The policemen, who had seen him, maintained that he had put up no resistance, but that two of the raiders had led him out, holding him fast by the arms.

Relief and joy at the 'miraculous escape' were the first emotions; and Trotsky's sense of irony was aroused. It amused him to see that so heavy an attack, so laboriously mounted, should have failed so miserably—only because he, Natalya, and the child had, in their utter helplessness, done the only thing they could do and thrown themselves under their beds! Now Stalin and his agents stood exposed and covered with ridicule! There could be no doubt for whose benefit, at whose instigation, and by whose orders the raid had been carried out. But with the exhilaration and the triumphant irony there mingled some perplexity. How familiar the raiders had been with the layout and the defences of the little fortress—they even knew that they could drive away in their victim's cars! How had Sheldon come to let them in, apparently without hesitation? He was feckless and gullible; but surely before he opened the gate, he must have been approached by someone whom he trusted and whose voice he knew? Who was it? Or had the raiders climbed into the courtyard over the high walls and electrified wires? Why then did they abduct Sheldon (whom they must certainly be going to kill)?

Within half an hour Colonel Salazar, chief of the Mexican Secret Police, was on the spot; and this is how he describes the scene:[1] 'I asked to see Trotsky, who soon arrived accompanied by his wife . . . [he] was in pyjamas, over which he had slipped a dressing-gown. They greeted me with friendliness . . . but they preserved a surprising calm. One might have thought that nothing had happened. . . . Trotsky smiled, with his eyes bright

[1] Salazar, op. cit., pp. 6–10.

and clear behind his tortoiseshell glasses—eyes always keen and piercing—his glance sharp and penetrating, with a jesting, sarcastic, slightly Mephistophelian air. His hair . . . almost white . . . seemed a little untidy, thrown back from his forehead, with stray locks falling at the sides. . . .' There was a 'striking contrast' between Trotsky and Natalya: 'He, energetic and authoritative . . . his features still young, firm; she sweet, calm, and almost resigned.' But both behaved with a coolness and 'perfect self-control' which seemed quite unnatural to the police chief. At once a suspicion crossed his mind: 'Had there really been an attempt on their lives or was it a put-up job?' As he listened to Trotsky, giving him, in his study, 'without the slightest emotion', a full and precise account of what he had just experienced, Salazar reflected again: 'So many attackers, so many fire-arms, even bombs, and nothing has happened to them! It is all very strange!' They went back into the garden which looked, with its lovingly tended cacti, as peaceful as ever; and the officer asked Trotsky whether he suspected anyone as 'the author of the attempt'.

'I most certainly do!' he replied in a very decided tone of voice. 'Come. . . .'
He put his right arm on my shoulder and slowly led me towards the rabbit hutches. . . . He stopped, glanced all round him, [as if] to make sure that we were alone, and, placing his right hand near his mouth, as though wishing to make the confidence more secret, he said in a low voice and with deep conviction:
'The author of the attack is Joseph Stalin, through the medium of the G.P.U.'

Now the officer was sure that Trotsky was pulling his leg. 'I looked at him stupefied. . . . My first suspicion became a certainty. Again I said to myself, "It is a put-up job! There is not the least doubt of it." ' And when Trotsky advised him to interrogate some of the 'most conspicuous' local Stalinists, from whom he might learn a lot about the raid, Salazar concluded that 'the old revolutionary was trying to distract my attention from the real path'. He ordered the arrest first of three domestic servants, a cook, a parlour maid, and a handyman, and then of two of Trotsky's secretaries, Otto Schüssler and Charles Cornell. The turn the investigation now took bred the most sensational rumours. Some said that Diego Rivera had organized the attack and that the raiders had broken into the house with the cry 'Long live Almazar'. (Almazar was the name of the reactionary general whose candidature for the Presidency Rivera backed against Cardenas.) Others maintained that Trotsky or his followers had staged the attack in order to turn suspicion on the Stalinists and discredit them.[1]

Curiously, the chief of the Secret Police felt no hostility towards Trotsky and had no axe to grind. But to the mind of a professional soldier and policeman, unfamiliar with the issues, the personalities, and the atmosphere of the terrible struggle which the raid was to have brought to an end—the whole affair did indeed appear extremely enigmatic. He had just counted seventy-three bullet holes in the wall over Trotsky's bed and the 'miraculous escape' seemed to him all the more mysterious. He observed Trotsky's and Natalya's self-possession and reflected that he, a veteran of Mexico's many civil wars, had never seen anyone behaving with such calm so soon after facing such dangers.[2] The precision and the humour of Trotsky's talk seemed quite out of place and all the more suspect. (Only in the next few months, when his duties brought him to Trotsky very frequently, was he to realize that the man's 'unnatural' calm, courage, and humour were his nature.) On the other hand, the raid was so gross a scandal even by Mexican standards that Salazar found it hard to believe that the Stalinists, Cardenas' supporters (of whom he was no friend), could have been behind it. The behaviour of Trotsky's guards also aroused his distrust: why had they been so strangely passive? Why had none of them been shot at? Salazar was convinced that Sheldon had been in collusion with the raiders and had left with them of his own free will. Trotsky vehemently asserted that Sheldon was their victim, not their accomplice; but he could offer no proof. And there was this grain of truth in Salazar's reasoning: the raid could not have been carried out without the co-operation of someone in Trotsky's entourage or at least of someone in close contact with his household. Who was it? This question

---

[1] Salazar, op. cit., pp. 18–25. Not only the pro-Stalinist newspapers of Mexico but even *The Nation* of New York published a correspondence suggesting that Trotsky himself or members of his household had staged the raid. 'What an infamous reptile breed', Trotsky commented, 'these "radicals" of *The Nation* are.' 18 June 1940. *The Archives*.

[2] Salazar, op. cit., pp. 10–11 and 100.

should now have engaged all their attention and aroused all their vigilance.

A week after the attack, Trotsky, outraged by the suspicions directed towards himself and Rivera, protested to President Cardenas against the arrest of his two secretaries.[1] Referring to what he knew (*inter alia* from Reiss and Krivitsky) about the workings of the G.P.U. in many countries, he demanded that the magistrate or the police interrogate the present and former General Secretaries of the Mexican Communist party and also Siqueiros and Lombardo Toledano. The President ordered the immediate release of Trotsky's secretaries. But for some time yet the investigation followed wrong tracks; and Trotsky was busy refuting imputations made against him, defending his collaborators, and affirming the innocence of Robert Sheldon Harte. 'If Harte', he said, 'had been a G.P.U. agent he could have stabbed me on the quiet', without all the hubbub of a massive and sensational raid. In the meantime the police apprehended several of the raiders, who confirmed that Siqueiros had been their leader; and Siqueiros himself went into hiding.[2] Finally, on 25 June, Salazar's men dug out Sheldon Harte's corpse from the grounds of a little farm outside Mexico City— the farmhouse had been rented by two well-known painters, both Stalinists.

At 4 a.m., the hour when a month and a day earlier the raid took place, Salazar came with this news to Trotsky's house. The guards refused to awaken Trotsky; and so he returned to the farmhouse with one of the guards to identify the body.

We arrived at the foot of the slope at dawn. The soaked earth made the ascent extraordinarily difficult. The body was lying on the stretcher where I had left it, outside the house. . . . Otto . . . immediately recognized his comrade.

'It was daylight when we arrived at San Angel. The corpse was placed in a courtyard. General Nunez arrived shortly afterwards, and gave the order for it to be washed. Then he had the guard

---

[1] Trotsky's letter to President Cardenas of 31 May 1940. *The Archives.*

[2] Salazar, op. cit., p. 184. Arrested on 4 October 1940, (after Trotsky's assassination), Siqueiros did not deny his participation in the May raid, but maintained that the Communist Party had nothing to do with it; and that his purpose was not to kill Trotsky but to produce a 'psychological shock' and to protest against Trotsky's presence. Released on bail, Siqueiros disappeared from Mexico for several years.

strengthened, for the news had spread throughout the town and inquisitive crowds began to arrive. The formalities finished, the magistrate left.

All at once a movement occurred among the crowd.

'Trotsky! Trotsky!'

It was indeed he. Ten o'clock struck. The old Russian exile approached the body. He looked sad and depressed. He stood for a long moment looking at his ex-secretary: his eyes were filled with tears. This man had directed a great revolution, had survived bloody battles, had seen his friends and his family disappearing one by one, had remained unmoved by an attack which had almost cost him not only his, but his wife's and grandson's lives—and now he wept in silence.[1]

The enigma of Harte's role was not definitely solved, however. Salazar still maintained that Harte had been a G.P.U. agent; but that the G.P.U. killed him because they feared that he would fall into the hands of the Mexican police and talk too much. This supposition was partly confirmed by eye-witnesses who said that they had seen Harte moving around the farm-house freely and going out for walks without any guard or escort. Against this Trotsky insisted that this was the eighth of his secretaries to perish and that all that he and his American comrades knew of Harte contradicted Salazar's version.[2] He sent a moving message of condolence to the victim's parents and put up a plaque commemorating 'Bob'—opposite that plaque Trotsky's own tombstone was soon to be raised.

After 24 May the mist of doom hung still and stifling over the 'little fortress' at Avenida Viena. From week to week and from day to day another attack was expected. To Trotsky himself it was a freak of fortune that he was still alive. He would get up in the morning and say to Natalya: 'You see, they did not kill us last night, after all; and you are still dissatisfied.' Once or twice he added pensively: 'Yes, Natasha, we have had a reprieve.'[3] He remained as active and energetic as ever, intervened in every phase of the police investigation, appeared in

[1] Salazar, op. cit., pp. 76–77.

[2] Trotsky listed the following secretaries and assistants of his who had perished as victims of Stalinist vengeance: Glazman, Butov, Blumkin, Sermuks, Posnansky, Klement, and Wolf. Statement of 25 June 1940 in *The Archives*.

[3] Natalya Sedova, 'Tak eto bylo', *B.O.*, no. 85, March 1941; 'How it happened', *Fourth International*, May 1941.

the court, replied to never-ending calumny, commented on such events as the capitulation of France and Molotov's declaration of support to the Third Reich, and went on debating the position of the Negroes in the U.S.A., the tactics of revolutionary defeatism, and so on. A group of American friends, who visited him before the middle of June, implored him to 'go underground', assume an incognito, and allow himself to be smuggled into the United States, where they were confident they could provide him with a safe clandestine retreat. He refused to listen to their entreaties. He could not, he said, skulk for his life and do his work furtively; he had to meet foe and friend in the open—his bare head had to endure the 'hell-black night' to the end.[1] Reluctantly, he yielded to friends and to Mexican authorities who urged that the defences of his house be strengthened by higher concrete walls, new watch towers, armoured doors, and steel shutters on windows. He dutifully inspected the 'fortification works', suggested changes and improvements, but then shrugged with distaste: 'This reminds me of the first jail I was in', he remarked to Joseph Hansen, his secretary. 'The doors make the same sound. . . . This is not a home; it is a medieval prison.' 'One day [Hansen says] he caught me gazing at the new towers. His eyes twinkled in one of those warm, intimate smiles of his. . . . "Highly advanced civilization—that we must still make such constructions." '[2] He was indeed like a man awaiting the fatal day in the condemned cell—only that he was determined to make judicious use of every hour, and his irony and humour did not abandon him.

He went on his last drives into the country over muddy, boulder-strewn roads; and his mind wandered back to Russia's roads in the years of civil war. On this last trip 'he slept much more than usual, as if he were exhausted and this were his first opportunity in a long time to rest. He relaxed in the seat beside me and slept from Cuernavaca almost to Ameccamecca, where the volcanoes, Popocatapetl and Ixtaccihuatl, the Sleeping Woman, gather great fleecy clouds about their white summits . . . we stopped beside an ancient hacienda with towering, strongly buttressed walls. The Old Man regarded the walls with interest: "A fine wall, but medieval. Like our own

[1] This has been related to me by one of those who put the proposal to Trotsky.
[2] Joseph Hansen, 'With Trotsky to the End', Fourth International, October 1940.

prison." '[1] In this description 'medieval', which so often came to his lips, he expressed not merely his repugnance for his own incarceration, but his sense that the world was relapsing from what might have been the age of progress and triumphant humanity into the savage cruelties of the Dark Ages; and that even he, by surrounding himself with turrets, buttresses, and ramparts, was somehow involved in the general backsliding. After the raid friends presented him with a bullet-proof vest; and even as he thanked them he could not hide his displeasure; he put the vest away and suggested that it would best be worn by the sentry on duty in the watch tower. His secretaries repeatedly proposed to search visitors for concealed weapons and objected when he received strangers alone in his study. 'He could not bear having his friends submit to search', says Hansen. 'No doubt he felt that in any case this would be useless and could even give us a false sense of security . . . a G.P.U. agent . . . would find some way of setting at nought what search we could make.' He frowned when any of his bodyguards tried to be present while he talked with visitors, some of whom 'had personal problems [and] would not talk freely in the presence of a guard'.[2]

.   .   .   .

It was on 28 May, a few days after the raid, that the assassin came for the first time face to face with Trotsky. The encounter could not have been more casual. The Rosmers were about to leave Mexico and board a ship at Vera Cruz; and 'Jacson' had offered to take them there in his car, pretending that he had to go to Vera Cruz anyhow, on one of his regular business journeys. He came to fetch them early in the morning and was asked to wait in the courtyard until they were ready. As he entered, he ran into Trotsky, who was still at the hutches feeding the rabbits. Without interrupting his chores, Trotsky shook hands with the visitor. 'Jacson' behaved with exemplary discretion and amiability: he did not stare at the great man, try to engage him in conversation or hang around; he went instead to Seva's room, gave the child a toy glider, and explained its working. At a hint from Trotsky Natalya then asked him to join the family and the Rosmers at the breakfast table.[3]

[1] Hansen, loc. cit.   [2] Hansen, loc. cit.   [3] Ibid.

After his return from Vera Cruz, 'Jacson' did not show himself at Avenida Viena for a fortnight. When he reappeared there on 12 June, he came for a few minutes to say that he was going to New York and leaving his car with the guards so that they might have its use in his absence. He returned to Mexico a month later, but did not call at Avenida Viena for three weeks, until the Trotskys invited him and Sylvia to have tea with them on 29 July. This was his longest visit—it lasted a little over an hour. According to the detailed records kept by the guards, he crossed the gate only ten times between 28 May and 20 August; and he saw Trotsky only twice or thrice. This was enough for him to survey the scene, to take the measure of his victim, and to put the finishing touches to his plan. He could not have behaved more unobtrusively, obligingly, innocuously; he came with a modest bouquet or a box of sweets for Natalya—'gifts from Sylvia'. He offered, as an experienced Alpinist, to accompany Trotsky in climbing mountains; but he did not dwell on the offer or take it up. When he chatted with the guards, he threw out with familiarity the names of well-known Trotskyists of various nationalities so as to give the impression that he was in and of the movement; *en passant* he mentioned his own donations to party funds. In Trotsky's and Natalya's presence, however, he behaved almost bashfully as befitted an outsider who was just being converted into a 'sympathizer'. This was the time of the split among the American Trotskyists. Sylvia had sided with Burnham and Shachtman; but she was as welcome as ever at Avenida Viena—only when she and 'Jacson' were invited to tea there was a lively argument at table. 'Jacson' did not take part, but let it be understood that he was on Trotsky's side, that he agreed that the Soviet Union was a workers' state and had to be defended 'unconditionally'. With the secretaries he was less reserved and he told them of the heated arguments he had about this with Sylvia. Yet he was careful not to appear over-zealous —had Trotsky not warned his followers that *agents provocateurs* in their midst would show *trop de zèle* and seek to exacerbate the quarrel? Well, 'Jacson' did nothing of the sort; he only tried judiciously to bring Sylvia round to the right viewpoint.

Yet even this master dissembler (who during the twenty years of his imprisonment was to foil all investigators, judges,

IX. Trotsky searching for rare cacti. 1940

X. Still arguing—a few days before his assassination

doctors, and psychoanalysts attempting to discover his real identity and his connexions) began to lose nerve as his deadline approached. He returned from New York, where he probably got the final briefing on his assignment, in a brooding mood. Usually robust and gay, he became nervous and gloomy; his complexion was green and pale; his face twitched; his hands trembled. He spent most of his days in bed, silent, shut up in himself, refusing to talk to Sylvia. Then he had fits of gaiety and garrulousness which startled Trotsky's secretaries. He boasted of his Alpinist exploits and of the physical strength which enabled him 'to split a huge ice-block with a single blow of an ice-axe'. At a meal he demonstrated the 'surgical skill' of his hands by carving a chicken with unusual dexterity. (Months later those who witnessed this 'demonstration' recalled that he had also said that he had known Klement well, Klement whose dead body had been found dismembered with such 'surgical skill'.) He talked of the 'financial genius' of his commercial 'boss' and offered to carry out with him some operations on the Stock Exchange in order to help the Fourth International financially. One day, watching with Trotsky and Hansen the 'fortification works' at Avenida Viena, he remarked that these were worthless because 'in the next attack the G.P.U. would use quite a different method'; and asked what method that might be, he answered with a shrug.

Members of the household were to recall these and similar incidents only three and four months later when they realized how ominous they had been. For the time being they saw in them nothing worse than signs of 'Jacson's' erratic temper. Trotsky alone, who knew him so little, became apprehensive. True, even he had defended 'Jacson' rather half-heartedly when someone said with indignation that 'Jacson', during his trip to New York, had not even called at the Trotskyist headquarters there. Well, well, Trotsky replied, Sylvia's husband was, of course, a flippant fellow who might never be of much use as a comrade, but perhaps he would improve—it took all sorts of people to make a party. But 'Jacson's' talk about his 'boss', the 'financial genius', and the Stock Exchange speculations he would undertake for the 'movement', made Trotsky bristle. 'These brief conversations', says Natalya, 'displeased me; Leon Davidovich was also struck by them. "Who is this very rich

'boss'?", he said to me. "One should find out. It may, after all,
be some profiteer of the fascist type perhaps—it might be better
for us not to receive Sylvia's husband any more. . . ." ' He had
broken with Molinier who had also had his 'financial plans';
but he had never had the slightest doubt about Molinier's
political sincerity; and he was quite willing to forgive him his
offences even now. But in 'Jacson' he sensed something sinister—
was he perhaps connected with the fascists? Yet despite this
vague intuition, he would not affront him without verifying
the grounds for the distrust.[1]

On 17 August, 'Jacson' returned, saying that he had written
an article against Burnham and Shachtman (with some references also to the situation in German-occupied France)—
would Trotsky go over the draft and suggest corrections? He
touched cunningly a sensitive chord in his victim, the urge to
instruct and improve comrades and followers. Reluctantly but
dutifully, Trotsky invited 'Jacson' to come with him to the study.
There they remained alone and discussed the article. After only
ten minutes Trotsky came out disturbed and worried. His
suspicion was suddenly heightened; he told Natalya that he had
no wish to see 'Jacson' any more. What upset him was not what
the man had written—a few clumsy and muddled clichés—
but his behaviour. While they were at the writing table and
Trotsky was looking through the article, 'Jacson' seated himself
on the table and there, placed above his host's head, he remained to the end of the interview! And all the time he had his
hat on and clutched his coat to himself! Trotsky was not only
irritated by the visitor's discourtesy; he sensed a fraud again.
He had the feeling that the man was an impostor. He remarked
to Natalya that in his behaviour 'Jacson' was 'quite unlike a
Frenchman',—yet he presented himself as a Belgian brought up
in France. Who was he really? They should find this out.
Natalya was taken aback; it seemed to her that Trotsky 'had
perceived something new about "Jacson", but had not yet
reached, or rather was in no hurry, to reach, any conclusions'.
Yet the implication of what he had said was alarming: if 'Jacson'
was deceiving them about his nationality, why was he doing it?
And was he not deceiving them about other things as well?
About what? These questions must have been on Trotsky's

[1] Natalya Sedova in *Vie et Mort de Leon Trotsky*, p. 319.

mind, for two days later he repeated his observations to Hansen, as if to ascertain whether similar misgivings had occurred to anyone beside himself. However, the assassin moved faster than the victim's intuition and instinct of self-preservation: it was on the day before the attempt on his life that Trotsky confided his vague suspicions to Hansen.[1]

The interview on 17 August was for Jacson his dress-rehearsal. He had enticed Trotsky into the study for a *tête-à-tête*, made him read a manuscript, and placed himself above his head. He had come to the dress-rehearsal with ice-axe, dagger, and pistol concealed in the coat he clutched in his arms. In his pocket he may already have had the letter purporting to explain his motives—the text had been typed out well ahead of time; on the day of the attempt he had only to insert the date and to sign it. In that letter he presented himself as Trotsky's 'devoted follower' who had been ready to give the 'last drop of blood' for him, who had gone to Mexico on instructions from the Fourth International, and for whom meeting Trotsky was 'the realization of a dream.' But in Mexico 'a great disillusionment' awaited him: the man whom he had imagined to be *the* leader of the working class unmasked himself as a criminal counter-revolutionary and urged him 'to go to Russia to organize there a series of attempts against various persons and, in the first place, against Stalin'. He found Trotsky conspiring 'with certain leaders of capitalist countries'—'the consul of a great foreign nation paid him frequent visits'—and conspiring against both the Soviet Union and Mexico.[2] The purpose of the letter was to make even Trotsky's death corroborate all the Stalinist accusations, except that, in view of the pact between Stalin and Hitler, the charge that Trotsky was Hitler's accomplice was replaced by a hint that he was in the service of American imperialism. Even the trick by which Trotsky's 'disillusioned follower' was to confirm the Stalinist charges was not new: the hand that had murdered Klement had written the same 'disclosures' of a 'disillusioned Trotskyist' in Klement's name. To make the concoction even shabbier, 'Jacson' added that Trotsky had urged him to 'desert his wife' because she had joined the Shachtman

---

[1] Natalya Sedova, ibid., and *B.O.*, no. 85, 1941; *Fourth International*, May 1941.
[2] The full text of 'Jacson's' 'Confession' is in Albert Goldman's *The Assassination of Leon Trotsky*, pp. 5–8.

group; but he, 'Jacson', could not live or go to Russia without Sylvia. The forgery was crude, but not too crude for the gullible; and, anyhow, who would find the time and patience to scrutinize it carefully now, during the interval between the capitulation of France and the Battle of Britain, when the existence of so many peoples and the foundations of so many states were shattered?

.   .   .   .

And so the last day, Tuesday, 20 August, had come. Whoever recalled it later remembered the exceptional peace and serenity that prevailed in the house up to the fatal hour. The sun shone brightly. The Old Man emanated calm, confidence, and energy. When he got up at 7 a.m. he turned to his wife not with the grim and by now habitual joke, 'You see, they did not kill us last night', but with an expression of physical well-being. 'It is a long time since I felt so well', he said to her; and added that the sleeping drugs he had taken had a good effect on him. 'It is not the drug that does you good', she replied, 'but sound sleep and complete rest.' 'Yes, of course', he chimed in contentedly. He looked forward to a 'really good day's work', dressed quickly, and 'vigorously walked out into the patio to feed his rabbits'. He had neglected them somewhat, for, on doctor's orders, he had spent the Sunday in bed; so he now tended them diligently for a full two hours. At breakfast he again assured Natalya of his excellent health and mood. He was eager to get back to work on 'my poor book', *Stalin*, which he had to put aside after the May raid in order to give his time to the police investigation and to current polemics. But now he had said all that he had to say about the raid; the investigation was moving in the right direction and he hoped that he would not be bothered with it any more. But before going back to the *Stalin*, he still wanted to write an 'important article', not for the great bourgeois Press but for the little Trotskyist periodicals, and speaking with some excitement about the article he went into the study.

He found the morning mail satisfactory. He had at last placed his archives in safety. A cable from the librarian of Harvard University had just acknowledged their receipt. There had been some uneasiness about these, because of hitches *en route* caused either by the G.P.U. or the F.B.I.; and a couple of days earlier Trotsky had instructed Albert Goldman, his American lawyer and comrade, to take action if the F.B.I. tried to pry into his papers. 'I personally have nothing to hide', he wrote, 'but in my letters many third persons are mentioned.' He had deposited the archives at Harvard on condition that one section of them would remain closed until the year 1980.[1] But the hitch *en route* was evidently not serious; and this matter was now happily settled. In his characteristic English, he wrote a few brief, kindly, and jovial, letters to American Trotskyists.[2] He inquired about the health of one who, after a spell as secretary at Coyoacan, had returned home; he thanked the comrade and his wife for a dictionary of American slang they had sent him and he promised to study it diligently so as to be able to follow his bodyguard's conversation at mealtimes. He sent greetings to two comrades who had been imprisoned for strike activities and were about to be released. And then he settled down to record his last article on a dictaphone.[3]

The tentative shapeless text of the article suggests that his mind was in a ferment and that he was trying to modify an old idea of his or to produce a new one. He had until quite recently expounded 'revolutionary defeatism', as Lenin had done during the First World War, telling the workers that their task was not to defend any imperialist fatherland, be it democratic or fascist, but to turn the war into revolution. But now, after the Nazis had conquered virtually the whole of Europe and while the British and American working classes were reacting to this with militant anti-fascism, he felt that the mere repetition of old formulae was of no use. 'The present war, as we have stated on more than one occasion, is a continuation of the last war. But a continuation is not a repetition [but] a development, a deepening, a sharpening.' Similarly, the continuation of the Leninist policy of 1914–17 should not be mere repetition, but 'development, deepening'. Lenin's revolutionary defeatism had rendered the Bolshevik party immune to the fetishes of bourgeois

---

[1] Trotsky to Goldman, 17 August 1940. *The Archives*, Closed Section. This section of the archives contained his correspondence with his followers. At a time when nearly the whole of Europe was controlled either by the Gestapo or by the G.P.U. he felt it to be his duty to protect his correspondents in this way.

[2] See 'Trotsky's Last Letters' in *Fourth International*, October 1940.

[3] Trotsky's 'Last Article', ibid.

patriotism; but—contrary to a widespread belief—'it could not win the masses who did not want a foreign conqueror'. The Bolsheviks had gained popular support not so much by their 'refusal to defend the bourgeois fatherland' as by the positive aspects of their revolutionary agitation and action. Marxists and Leninists in this war must realize this, he concluded; and he came out against Shachtman's group and the pacifists among the Trotskyists who opposed conscription in the United States. In a letter written a few days earlier he had commented on a Public Opinion Poll which had shown that 70 per cent. of American workers favoured conscription. 'We place ourselves on the same ground as the 70 per cent. of the workers. [We say] you, workers, wish to defend . . . democracy. We . . . wish to go further. However, we are ready to defend democracy with you, only on condition that it should be a real defence, and not a betrayal in the Pétain manner.' In the article his mind wandered between France, humiliated and saddled with a 'treacherous senile Bonapartism', and the vastly different American scene. But he had no time to develop these inchoate thoughts; his voice in the dictaphone was to remain the only trace of his last inconclusive groping in a new direction.

.   .   .   .

At one o'clock Rigault, his Mexican attorney, came to see him to advise him to reply at once to an attack in *El Popular*, Toledano's paper, which had accused him of defaming the Mexican trade unions. Trotsky feared that this would drag him back into arid polemics with the local Stalinists, but he agreed that he had to answer *El Popular* at once; and he put aside the article on revolutionary defeatism 'for a few days'. 'I will take the offensive and will charge them with brazen slander', he said to Natalya. He was defiant, but cheerful; and he assured her once more that he was in excellent form. After a brief siesta he was at his desk again, taking notes from *El Popular*. 'He looked well', says Natalya, 'and was in an even mood all the time.' Somewhat earlier she saw him standing in the patio, bareheaded under the scorching sun; and she hastened to fetch his white cap so as to protect his head. Now from time to time she slightly opened the door to his study 'so as not to disturb him'; and she saw him 'in his usual position, bent over his desk, pen

in hand'. On tiptoe, from behind the door, the modern Niobe
cast her last fond glances at the only beloved being left to her.

Shortly after 5 p.m. he was back at the hutches, feeding the
rabbits. Natalya, stepping out on a balcony, noticed an 'unfamiliar figure' standing next to him. The figure came closer,
took off the hat, and she recognized 'Jacson'. ' "Here he is
again," it flashed through my mind. "Why has he begun to
come so often?" I asked myself.' His appearance deepened her
foreboding. His face was grey-green, his gestures nervous and
jerky, and he pressed his overcoat to his body convulsively. She
remembered suddenly that he had boasted to her that he never
wore a hat and a coat even in winter; and she asked him why
he had the hat and the coat on on so sunny a day. 'It might
rain', he replied; and saying that he was 'frightfully thirsty'
asked for a glass of water. She offered him tea. 'No, no, I dined
too late and I feel the food up here', he pointed at his throat:
'it's choking me.' His mind wandered; he did not seem to catch
the meaning of what was said to him. She asked whether he had
corrected his article and he, clutching his coat with one hand,
showed her several typewritten pages with the other. Pleased
that her husband would not have to strain his eyes over an
illegible manuscript, she went with 'Jacson' towards the hutches.
As they came near, Trotsky turned to her and said in Russian
that 'Jacson' was expecting Sylvia to come and, as they would
both be leaving for New York the next day, Natalya ought
perhaps to invite them to a farewell meal. She answered that
'Jacson' had just refused tea and was not feeling well. 'Lev
Davidovich glanced at him attentively, and said in a tone of
slight reproach: "Your health is poor again, you look ill. That's
no good." '[1] There was a moment of awkward silence. The
strange man stood waiting with the typewritten pages in hand
and Trotsky, having advised him to rewrite the article, felt
obliged to have a look at the result of his fresh effort.

'Lev Davidovich was reluctant to leave the rabbits and was
not at all interested in the article', Natalya relates. 'But controlling himself, he said: "Well, what do you say, shall we go
over your article?" Unhurriedly, he fastened the hutches and
took off his working gloves. . . . He brushed his blue jacket and
slowly, silently, walked with myself and "Jacson" towards the

[1] Natalya Sedova, as above.

house. I accompanied them to the door of L.D.'s study; the door closed and I went into the adjoining room.' As they entered the study, the thought 'this man could kill me' flashed across Trotsky's mind—so at least he told Natalya a few minutes later when he lay bleeding on the floor. However, thoughts like this must have occurred to him sometimes—only to be dismissed—when strangers visited him singly or in groups. He had resolved not to let his existence become cramped by fear and misanthropy; and so now he suppressed this last faint reflex of his self-protective instinct. He went to his desk, sat down, and bent his head over the typescript.

He had just managed to run through the first page, when a terrific blow came down upon his head. 'I had put my raincoat . . . on a piece of furniture', 'Jacson' testifies, 'took out the ice-axe, and, closing my eyes, brought it down on his head with all my strength.' He expected that after this mighty blow his victim would be dead without uttering a sound; and that he himself would walk out and vanish before the deed was discovered. Instead, the victim uttered 'a terrible, piercing cry'—'I shall hear that cry all my life', the assassin says.[1] His skull smashed, his face gored, Trotsky jumped up, hurled at the murderer whatever object was at hand, books, inkpots, even the dictaphone, and then threw himself at him. It had all taken only three or four minutes. The piercing, harrowing cry raised Natalya and the guards to their feet, but it took a few moments for them to realize whence it had come and to rush in its direction. During those moments a furious struggle went on in the study, Trotsky's last struggle. He fought it like a tiger. He grappled with the murderer, bit his hand, and wrenched the ice-axe from him. The murderer was so confounded that he did not strike another blow and did not use pistol or dagger. Then Trotsky, no longer able to stand up, straining all his will not to collapse at his enemy's feet, slowly staggered back. When Natalya rushed in, she found him standing in the doorway, between the dining-room and the balcony, and leaning against the door frame. His face was covered with blood, and through the blood his blue eyes, without the glasses on, shone on her sharper than ever; his arms were hanging limply. ' "What has happened?" I asked. ' "What's happened?" I put my arms

[1] Salazar, op. cit., p. 160.

around him . . . he did not answer at once. For a second I wondered whether something had not fallen on him from the ceiling—repair work was being done in the study—and *why was he standing here*? Calmly, without anger, bitterness, or sorrow he said: "Jacson." He said it as if he wished to say: "Now it has happened." We took a few steps, and slowly, aided by me, he slumped down on to a mat on the floor.'[1]

' "Natasha, I love you." He uttered these words so unexpectedly, so gravely, almost severely that, weak from inner shock, I swayed towards him.' 'No one, no one,' she whispered to him, 'no one, must be allowed to see you without being searched.' Then she carefully placed a cushion under his broken head and a piece of ice on his wound; and she wiped the blood off his forehead and cheeks. 'Seva must be kept out of all this,' he said. He spoke with difficulty, his words were becoming blurred but he seemed unaware of it. 'You know, *in there*'— he turned his eyes towards the door of the study—'I sensed . . . I understood *what he wanted to do* . . . he wanted . . . me . . . once more . . . but I did not let him.' He said this 'calmly, softly, with a breaking voice'; and as if with a note of satisfaction he repeated: 'But I did not let him'. Natalya and Hansen knelt by his sides, opposite each other; and he turned towards Hansen and spoke to him in English, while she 'strained all her attention to catch the meaning of his words, but failed'.

'This is the end,' he said to his secretary in English; and he wanted to find out what exactly had happened. He was convinced that 'Jacson' had fired at him and was incredulous when Hansen told him that he had been hit with an ice-axe and that the wound was superficial. 'No, no, no,' he replied pointing to his heart, 'I feel here that this time they have succeeded.' When he was assured again that the wound was not very dangerous, he smiled faintly with his eyes as if it amused him to see that someone sought to comfort *him* and to conceal the truth from *him*. Most of the time he was pressing Natalya's hands to his lips. 'Take care of Natalya,' he went on in English; 'she has been with me many, many years.' 'We will', Hansen promised. 'The Old Man pressed our hands convulsively, tears suddenly in his eyes. Natalya cried brokenly, bending over him, kissing his wound.'[2]

[1] Natalya Sedova, loc. cit.   [2] Hansen, loc. cit.

Meanwhile, in the study the guards fell upon the assassin, beat him with revolver butts; and his whining and moaning were heard outside. 'Tell the boys not to kill him', Trotsky said, struggling to articulate his words clearly. 'No, no, he must not be killed—he must be made to talk.' The guards related that under the blows 'Jacson' said: 'They have got something on me, they have imprisoned my mother . . . Sylvia has nothing to do with this . . .'; and when they tried to get out of him who had imprisoned his mother, he denied that it was the G.P.U. and said that he had 'nothing to do with the G.P.U.'

When the doctor arrived, Trotsky's left arm and leg were already paralysed. As the stretcher-bearers came—simultaneously with them the police entered—Natalya shrank away: she thought of Lyova's death in the hospital, and she did not want her husband to be moved. He too had no wish to be taken away. Only when Hansen promised that the guards would accompany him, he replied: 'I leave it then to your decision', as if aware that for him 'all the days of making decisions were gone'. While he was being placed on the stretcher, he whispered again: 'I want everything I own to go to Natalya . . . .You will take care of her.'[1]

At the gate the guards, with belated vigilance, stopped the stretcher-bearers; afraid of another attack, they would not allow Trotsky to be taken away unless General Nunez, Chief of Police, came to take charge of the escort. 'I noticed [an ambulance worker relates] that the wounded man's wife had covered her husband with a white shawl. The Señora sobbed and held his bleeding head between her two hands. Señor Trotsky neither spoke nor groaned. We thought that he was dead, but . . . he was still breathing.'[2] They carried him to the ambulance between two lines of police; and as they were about to start, another ambulance arrived to fetch the assassin.

'Through the roaring city, through its vain tumult and din, amidst its garish evening lights, the emergency ambulance sped, winding its way through the traffic and overtaking cars; the sirens were incessantly wailing and the police cordon on motor-cycles whistled shrilly. With unendurable anguish in our hearts and alarm increasing with every minute, we were bearing the wounded man. He was conscious.' His right hand

[1] Hansen, loc. cit.   [2] Salazar, op. cit., pp. 102–3.

described circles in the air as if it could not find a place to rest; then it wandered above the blanket, touched a water basin overhead, and at last found Natalya. She, bending over him, asked how he felt. 'Better now.' Then he motioned Hansen to himself and in a whisper instructed him how to conduct the investigation. 'He is a political assassin . . . a member of the G.P.U. . . . or a fascist. More likely the G.P.U. . . . but possibly aided by Gestapo.' (Almost simultaneously in the other ambulance the assassin was handing to his escort the letter giving his 'motives' and making it clear that the Gestapo had nothing to do at least with this crime.)

A large crowd was already gathered outside the hospital when Trotsky was lifted out of the ambulance. 'There may be enemies among them. . .' Natalya worried. 'Where are our friends? They should surround the stretcher.' A few minutes later he lay on a narrow hospital bed and doctors examined his wound. A nurse started to cut his hair; and he, grinning at Natalya, who stood at the head of the cot, recalled that only the day before they had wanted to send for a barber to cut his hair: 'You see,' he winked, 'the barber has also come.' Then, his eyes almost closed, he turned towards Hansen with the question with which he had turned to him so many times: 'Joe, you . . . have . . . notebook?' He remembered that Hansen did not know Russian and he made a great effort to dictate a message in English. His voice was barely audible, his words blurred. This is what Hansen claims to have taken down: 'I am close to death from the blow of a political assassin . . . struck me down in my room. I struggled with him . . . we . . . entered . . . talk about French statistics . . . he struck me . . . please say to our friends . . . I am sure . . . of victory . . . of Fourth International . . . go forward.' When he started dictating, he evidently still hoped to be able to give his account of the attempt on his life as well as a political message. But suddenly he felt that his life was ebbing; and he cut short the account and hastened to give his followers his last encouragement.

The nurses began to undress him for the operation, cutting with scissors his jacket, shirt, and vest, and unstrapping the watch from his wrist. When they began to remove his last garments, he said to Natalya 'distinctly but very sadly and gravely': 'I do not want them to undress me . . . I want you to undress me.' These were the last words she heard from him. When she finished undressing him, she bent over him and pressed her lips against his. 'He returned the kiss. Again. And again he responded. And once again. This was our final farewell.'[1]

About 7.30 the same evening he fell into a coma. Five surgeons carried out the trepanning of the skull. The wound was two and three-quarter inches deep. The right parietal bone was broken, its splinters embedded in the brain; the meninges were damaged and part of the brain substance was ruptured and destroyed. He 'bore the operation with extraordinary strength' but did not regain consciousness; and he struggled with death for more than twenty-two hours. Natalya 'dry-eyed, hands clenched', watched him day and night, waiting for his awakening. This is the last image she retained of him:

> They lifted him up. His head drooped on to his shoulder. The arms fell just as the arms in Titian's 'Descent from the Cross'. Instead of a crown of thorns the dying man wore a bandage. The features of his face retained their purity and pride. It seemed that any moment now he might still straighten up and become his own master again.[2]

Death followed on 21 August 1940 at 7.25 p.m. The autopsy showed a brain of 'extraordinary dimensions', weighing two pounds and thirteen ounces; and 'the heart too was very large'.[3]

.   .   .   .

On 22 August, in accordance with a Mexican custom, a large funeral cortège marched slowly behind the coffin carrying Trotsky's body, through the main thoroughfares of the city, and also through the working-class suburbs, where ragged, barefoot, silent crowds filled the pavements. American Trotskyists intended to take the body to the United States; but the State Department refused a visa even to the dead. For five days the body lay in state; and about 300,000 men and women filed past, while the streets resounded with the *Grand Corrido de*

---

[1] Natalya Sedova, loc. cit.   [2] Ibid.
[3] Salazar, op. cit., p. 110.

Leon Trotsky, a folk ballad composed by an anonymous bard.[1]

On 27 August the body was cremated; and the ashes were buried in the grounds of the 'little fortress' at Coyoacan. A white rectangular stone was raised over the grave, and a red flag was unfurled above it.

Natalya was to live on in the house for another twenty years; and every morning, as she rose, her eyes turned to the white stone in the courtyard.

---

[1] Salazar, loc. cit. Here are a few lines of the *corrido*, which breathes truly plebeian contempt for the 'sly and cowardly' assassin:

> Murio Trotsky asesinado
> de la noche a la mañana
> porque habian premeditado
> venganza tarde o temporano.
>
> .   .   .
>
> Fué un dia martes por la tarde
> este tragedia fatal,
> Que ha conmovido al pais
> y a todo la Capital.

Pravda announced the event in a few lines, saying that Trotsky had been killed by a 'disillusioned follower'.

Postscript: Victory in Defeat

IN the whole history of the Russian Revolution, and in the history of the labour movement and Marxism no period has been as difficult and sombre as the years of Trotsky's last exile. This was a time when, to paraphrase Marx, 'the idea pressed towards reality' but as reality did not tend towards the idea—a gulf was set between them, a gulf narrower yet deeper than ever. The world was riddled with extraordinary contradictions. Never had capitalism been so close to catastrophe as during the slumps and depressions of the nineteen-thirties; and never had it shown so much savage resilience. Never had the class struggle driven so stormily towards a revolutionary climax and never yet had it been so incapable of rising to it. Never had such vast masses of people been inspired by socialism; and never had they been so helpless and inert. In the whole experience of modern man there had been nothing as sublime and as repulsive as the first Workers' State and the first essay in 'building socialism'. And perhaps never yet had any man lived in so close a communion with the sufferings and the strivings of oppressed humanity and in such utter loneliness as Trotsky lived.

What was the meaning of his work and the moral of his defeat?

Any answer must be tentative, for we still lack the long historical perspective; and our appraisal of Trotsky follows primarily from our judgement on the Russian Revolution. If the view were to be taken that all that the Bolsheviks aimed at—socialism—was no more than a *fata morgana*, that the revolution merely substituted one kind of exploitation and oppression for another, and could not do otherwise, then Trotsky would appear as the high priest of a god that was bound to fail, as Utopia's servant mortally entangled in his dreams and illusions. Even then he would attract the respect and sympathy due to the great utopians and visionaries— he would stand out among them as one of the greatest. Even if it were true that it is man's fate to stagger in pain and blood from defeat to defeat and to throw off one yoke only to bend his neck beneath another—even then man's longings for a different destiny would still, like pillars of fire, relieve the darkness and gloom of the endless desert through which he has been wandering with no promised land beyond. And no one in our age has expressed those longings as vividly and sacrificially as Trotsky.

But has the Russian Revolution been able only to give the people one yoke instead of another? Is this to be its final outcome? Such a view seemed plausible to people who contemplated Stalinism in the last years of Trotsky's life and later. Against them Trotsky asserted his conviction that in the future, after Soviet Society had progressed towards socialism, Stalinism would be seen as merely 'an episodic relapse'. His optimism seemed gratuitous even to his followers. After nearly twenty-five years, however, his forecast may still sound bold, but hardly gratuitous. It is clear that even under Stalinism Soviet society was achieving immense progress in many fields, and that the progress, inseparable from its nationalized and planned economy, was disrupting and eroding Stalinism from the inside. In Trotsky's time it was too early to try to draw a balance of this development—his attempts to do so were not faultless; and the balance is not yet quite clear, even a quarter of a century later. But it is evident that Soviet society has been striving, not without success, to rid itself of the heavy liabilities, and to develop the great assets, it had inherited from the Stalin era. There has been far less poverty in the Soviet Union, far less inequality and far less oppression in the early nineteen-sixties than in the nineteen-thirties or the early nineteen-fifties. The contrast is so striking that it is an anachronism to speak of the 'new totalitarian slavery established by bureaucratic collectivism'. The issues over which Trotsky argued with his disciples in his last controversy are still being debated, not within tiny sects but before a world-wide audience. It is still a matter of argument whether the Soviet bureaucracy is 'a new class' and whether reform or revolution is needed to bring its arbitrary rule to an end. What is beyond question is that the reforms of the first post-Stalin decade, however inadequate and self-contradictory, have greatly mitigated and limited bureaucratic despotism and that fresh currents of popular aspirations are working to transform Soviet society further and more radically.

Even so, Trotsky's belief that one day all the horrors of Stalinism would appear to have been merely 'an episodic relapse' may still outrage contemporary sensitivity. But he applied the grand historical scale to events and to his own fate: 'When it is a question of the profoundest changes in economic and cultural systems, twenty-five years weigh less in history than an hour does in a man's life.' (His inclination to take the long historical view did not blunt his sensitivity to the injustices and cruelties of his time—on the contrary, it sharpened it. He denounced the Stalinist perversion of socialism so passionately because he himself never lost sight of the vista of a truly humane socialist future.) Measured by his historical scale, the progress which Soviet society has achieved since his day is merely a modest, an all too modest, beginning. Yet even this beginning vindicates the revolution and his basic optimism about it, and lifts the dense fog of disillusionment and despair.

Trotsky's huge life and work are an essential element in the experience of the Russian Revolution and, indeed, in the fabric of contemporary civilization. The uniqueness of his fortunes and the extraordinary moral and aesthetic qualities of his endeavour speak for themselves and bear witness to his significance. It cannot be, it would be contrary to all historical sense, that so high an intellectual energy, so prodigious an activity, and so noble a martyrdom should not have their full impact eventually. This is the stuff of which the most sublime and inspiring legends are made—only the Trotsky legend is woven throughout of recorded fact and ascertainable truth. Here no myth is hovering above reality; reality itself rises to the height of myth.

So copious and splendid was Trotsky's career that any part or fraction of it might have sufficed to fill the life of an out-standing historic personality. Had he died at the age of thirty or thirty-five, some time before 1917, he would have taken his place in one line with such Russian thinkers and revolutionaries as Belinsky, Herzen, and Bakunin, as their Marxist descendant and equal. If his life had come to a close in 1921 or later, about the time Lenin died, he would have been remembered as the leader of October, as founder of the Red Army and its

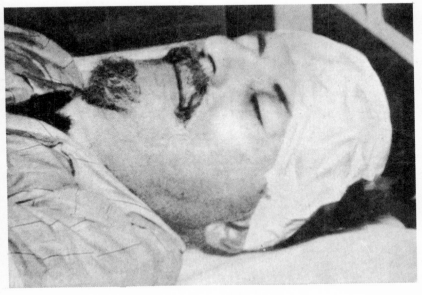

XI. (*upper*) Trotsky's assassin
(*lower*) 'Trotsky is dead'

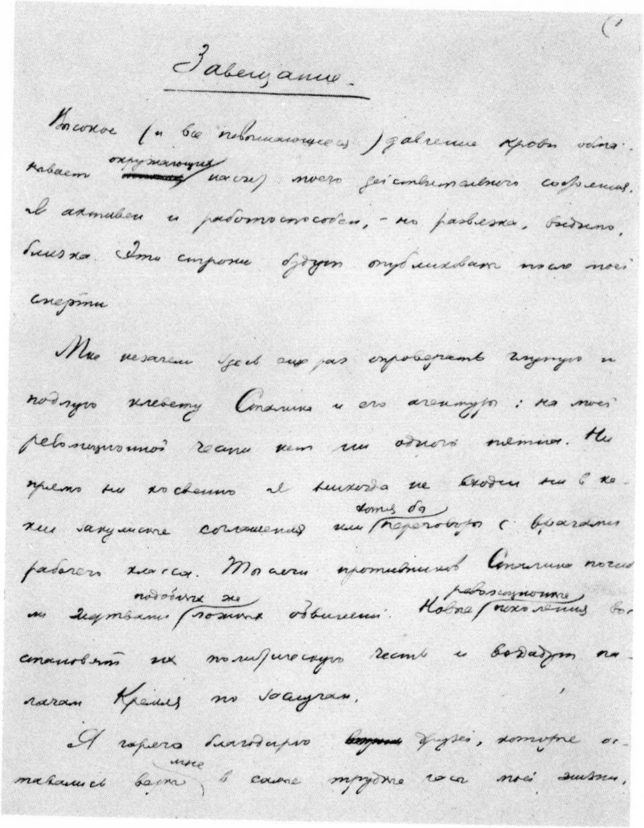

Завещание.

Высокое (и все повышающееся) давление крови обманывает окружающих насчет моего действительного состояния. Я активен и работоспособен, — но развязка, видимо, близка. Эти строки будут опубликованы после моей смерти.

Мне незачем здесь еще раз опровергать глупую и подлую клевету Сталина и его агентуры: на моей революционной чести нет ни одного пятна. Ни прямо ни косвенно я никогда не входил ни в какие закулисные соглашения или хотя бы переговоры с врагами рабочего класса. Тысячи противников Сталина погибли жертвами подобных же ложных обвинений. Новые революционные поколения восстановят их политическую честь и воздадут палачам Кремля по заслугам.

Я горячо благодарю друзей, которые оставались мне верны в самые трудные часы моей жизни,

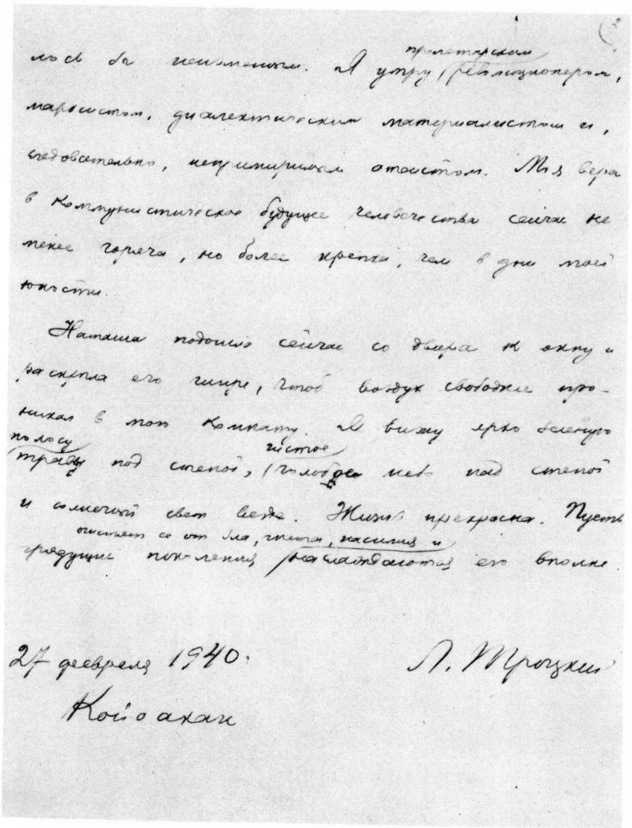

лось бы неизменным. Я умру пролетарским революционером, марксистом, диалектическим материалистом и, следовательно, непримиримым атеистом. Моя вера в коммунистическое будущее человечества сейчас не менее горяча, но более крепка, чем в дни моей юности.

Наташа подошла сейчас со двора к окну и раскрыла его шире, чтоб воздух свободнее проникал в мою комнату. Я вижу ярко зеленую полосу травы под стеной, и чистое голубое небо над стеной, и солнечный свет везде. Жизнь прекрасна. Пусть грядущие поколения очистят ее от зла, гнета и насилия и наслаждаются ею вполне.

27 февраля 1940.

Койоакан

Л. Троцкий

XII. Trotsky's testament. A facsimile.

captain in the Civil War, and as the mentor of the Communist International who spoke to the workers of the world with Marx's power and brilliance and in accents that had not been heard since the *Communist Manifesto*. (It took decades of Stalinist falsification and slander to blur and erase this image of him from the memory of two generations.) The ideas which he expounded and the work which he performed as leader of the Opposition between 1923 and 1929 form the sum and substance of the most momentous and dramatic chapter in the annals of Bolshevism and communism. He came forward as protagonist in the greatest ideological controversy of the century, as intellectual initiator of industrialization and planned economy, and finally as the mouthpiece of all those within the Bolshevik party who resisted the advent of Stalinism. Even if he had not survived beyond the year 1927, he would have left behind a legacy of ideas which could not be destroyed or condemned to lasting oblivion, the legacy for the sake of which many of his followers faced the firing squad with his name on their lips, a legacy to which time is adding relevance and weight and towards which a new Soviet generation is gropingly finding its way.

On top of all this come his ideas, writings, struggles, and wanderings of the period narrated in this volume. We have reviewed critically his fiascos, fallacies, and miscalculations: his fiasco with the Fourth International, his mistakes about the prospects of revolution in the West, his fumblings about reform and revolution in the U.S.S.R., and the contradictions of the 'new Trotskyism' of his last years. We have also surveyed those of his campaigns which are now fully and incontrovertibly vindicated: his magnificently far-sighted, although vain, efforts to arouse the German workers, the international Left, and the Soviet Union to the mortal danger of Hitler's ascendancy; his sustained criticisms of Stalin's hideous abuses of power, not least in the conduct of economic affairs, especially in collectivization; and his final titanic struggle against the Great Purges. Even the epigones of Stalinism, who are still doing all they can to keep Trotsky's ghost at bay, admit by implication that on these great issues he was right—all that after so many years they themselves have been able to do, with all the courage that the dead Stalin has inspired in them, is to echo disparately Trotsky's protests, accusations, and criticisms of Stalin.

It must be emphasized again that to the end Trotsky's strength and weakness alike were rooted in classical Marxism. His defeats epitomized the basic predicament by which classical Marxism was beset as doctrine and movement—the discrepancy and the divorce between the Marxist vision of revolutionary development and the actual course of class struggle and revolution.

Socialist revolution made its first, immense conquests not in the advanced West but in the backward East, in countries, where not the industrial workers but the peasants predominated. Its immediate task was not to establish socialism but to initiate 'primitive socialist accumulation'. In the classical Marxist scheme of things revolution was to occur when the productive forces of the old society had so outgrown its property relations as to burst the old social framework; the revolution was to create new property relations and the new framework for fully grown, advanced, and dynamic productive forces. What happened in fact was that the revolution created the most advanced forms of social organization for the most backward of economies; it set up frameworks of social ownership and planning around underdeveloped and archaic productive forces, and partly around a vacuum. The theoretical Marxist conception of the revolution was thereby turned upside down. The new 'productive relations' being above the existing productive forces were also above the understanding of the great majority of the people; and so the revolutionary government defended and developed them against the will of the majority. Bureaucratic despotism took the place of Soviet democracy. The State, far from withering away, assumed unprecedented, ferocious power. The conflict between the Marxist norm and the reality of revolution came to permeate all the thinking and activity of the ruling party. Stalinism sought to overcome the conflict by perverting or discarding the norm. Trotskyism attempted to preserve the norm or to strike a temporary balance between norm and reality until revolution in the West resolved the conflict and restored harmony between theory and practice. The failures of revolution in the West were epitomized in Trotsky's defeat.

How definite and irrevocable was the defeat? We have seen that as long as Trotsky was alive Stalin never considered him to have been finally vanquished. Stalin's fear was no mere paranoiac obsession. Other leading actors on the political stage shared it. Robert Coulondre, French ambassador to the Third Reich, gives a striking testimony in a description of his last interview with Hitler just before the outbreak of the Second World War. Hitler had boasted of the advantages he had obtained from his pact with Stalin, just concluded; and he drew a grandiose vista of his future military triumph. In reply the French ambassador appealed to his 'reason' and spoke of the social turmoil and the revolutions that might follow a long and terrible war and engulf all belligerent governments. 'You are thinking of yourself as victor . . .', the ambassador said, 'but have you given thought to another possibility—that the victor may be Trotsky?' At this Hitler jumped up (as if he 'had been hit in the pit of the stomach') and screamed that this possibility, the threat of Trotsky's victory, was one more reason why France and Britain should not go to war against the Third Reich. Thus, the master of the Third Reich and the envoy of the Third Republic, in their last manœuvres, during the last hours of peace, sought to intimidate each other, and each other's governments, by invoking the name of the lonely outcast trapped and immured at the far end of the world. 'They are haunted by the spectre of revolution, and they give it a man's name', Trotsky remarked when he read the dialogue.

Were Hitler and the ambassador quite wrong in giving the spectre Trotsky's name? It may be argued that although their fear was well grounded, they should have given the spectre Stalin's name, not Trotsky's—it was, at any rate, Stalin who was to triumph over Hitler. Yet as so often in history so here the underlying realities were far more confused and ambiguous than the surface of events. Stalin's victory over Trotsky concealed a heavy element of defeat while Trotsky's defeat was pregnant with victory.

The central 'ideological' issue between them had been socialism in one country—the question whether the Soviet Union would or could achieve socialism in isolation, on the basis of national self-sufficiency, or whether socialism was conceivable only as an international order of society. The answer events have given is far less clear-cut than were the theoretical arguments, but it comes much closer to Trotsky's view than to Stalin's. Long before the Soviet Union came anywhere near socialism, revolution had spread to other countries. History, it might be said, did not leave the Soviet Union alone long enough to allow a laboratory experiment with socialism in a single country to be carried into any advanced stage, let alone to be completed. In so far as in the struggle between Trotskyism and Stalinism revolutionary international-ism had clashed with Bolshevik isolationism it is certainly not Stalinism that has emerged with flying colours: Bolshevik isolationism has been dead long since. On the other hand, the staying power of the Soviet Union, even in isolation, was far greater than Trotsky sometimes assumed; and, contrary to his expectations, it was not the proletariat of the West that freed the Russian Revolution from isolation. By a feat of history's irony, Stalinism itself *malgré lui-même* broke out of its national shell.

In his last debate Trotsky staked the whole future of Marxism and socialism upon the sequel to the Second World War. Convinced that war must lead to revolution—the classical Marxist revolution—he asserted that if it failed to do so Marx-ism would be refuted, socialism would lose once and for all by default, and the epoch of bureaucratic collectivism would set in. This was, in any case, a rash, dogmatic, and desperate view; historic reality was once again to prove immeasurably more intricate than the theorist's scheme. The war did indeed set in motion a new series of revolutions; yet once again the process did not conform to the classical pattern. The western prole-tariat again failed to storm and conquer the ramparts of the old order; and in eastern Europe it was mainly under the im-pact of Russia's armed power, advancing victoriously to the Elbe, that the old order broke down. The divorce between theory and practice—or between norm and fact—deepened even further.

This was not a fortuitous development. It represented a continuation of the trend which had first announced itself in 1920-1, when the Red Army marched on Warsaw and when it occupied Georgia.[1] With those military acts the revolutionary

[1] See *The Prophet Armed*, pp. 463-77.

cycle which the First World War set in motion had come to a close. At the beginning of that cycle Bolshevism had risen on the crest of a genuine revolution; towards its end the Bolsheviks began to spread revolution by conquest. Then followed the long interval of two decades, during which Bolshevism did not expand. When the next cycle of revolution was set in motion by the Second World War, it started where the first cycle had ended—with revolution by conquest. In military history there exists, as a rule, a continuity between the closing phase of one war and the opening phase of another: the weapons and the ideas about warfare invented and formed towards the end of one armed conflict dominate the first stage of the next conflict. A similar continuity exists also between cycles of revolution. In 1920–1 Bolshevism, straining to break out of its isolation, tried, rather fitfully, to carry revolution abroad on the point of bayonets. Two and three decades later Stalinism, dragged out of its national shell by war, imposed revolution upon the whole of eastern Europe.

Trotsky had expected the second revolutionary cycle to begin in the forms in which the first had begun, with class struggles and proletarian risings, the outcome of which would, in the main depend on the balance of social forces within each major nation and on the quality of national revolutionary leadership. Yet the new cycle started not where the previous one had begun, but where it had ended, not with revolution from below, but with revolution from above, with revolution by conquest. As this could be the work only of a great power applying its pressure in the first instance to its own periphery, the cycle ran its course on the fringes of the Soviet Union. The chief agents of revolution were not the workers of the countries concerned, and their parties, but the Red Army. Success or failure depended not on the balance of social forces within any nation, but mainly on the international balance of power, on diplomatic pacts, alliances, and military campaigns. The struggle and the co-operation of the great powers superimposed themselves upon class struggle, changing and distorting it. All criteria by which Marxists were wont to judge a nation's 'maturity' or 'immaturity' for revolution went by the board. Stalin's pact with Hitler and the division between them of spheres of influence provided the starting-point for the social upheaval in eastern Poland and in the Baltic States. The revolutions in Poland proper, in the Balkan countries, and in eastern Germany were accomplished on the basis of the division of spheres which Stalin, Roosevelt, and Churchill carried out at Teheran and Yalta. By dint of that division the western powers used their influence and force to suppress, with Stalin's connivance, revolution in western Europe (and Greece) regardless of any local balance of social forces. It is probable that had there been no Teheran and Yalta compacts, western rather than eastern Europe would have become the theatre of revolution—especially France and Italy, where the authority of the old ruling classes was in ruins, the working classes were in revolt, and the Communist parties led the bulk of the armed Resistance. Stalin, acting on his diplomatic commitments, prevailed upon the French and Italian Communists to resign themselves to the restoration of capitalism in their countries from the virtual collapse and even to co-operate in the restoration. At the same time Churchill and Roosevelt induced the bourgeois ruling groups of eastern Europe to submit to Russia's preponderance and consequently to surrender to revolution. On both sides of the great divide the international balance of power swamped the class struggle. As in the Napoleonic era, revolution and counter-revolution alike were the by-products of arms and diplomacy.

Trotsky saw only the opening of this great chain of events. He did not realize what it portended. All his habits of thought made it difficult, if not impossible, for him to imagine that for a whole epoch the armies and diplomacies of three powers would be able to impose their will upon all the social classes of old Europe; and that consequently the class struggle, suppressed at the level on which it had been traditionally waged, would be fought at a different level and in different forms, as rivalry between power blocs and as cold war.

From theoretical conviction and political instinct alike Trotsky felt nothing but distaste for revolution by conquest. He had opposed the invasions of Poland and Georgia in 1920–1, when Lenin favoured these ventures. As Commissar of War he had categorically disavowed Tukhachevsky, the early exponent of the neo-Napoleonic method of carrying revolution into foreign countries. Twenty years before the Second World War he had castigated the armed missionary of Bolshevism, saying that 'it were better that a millstone be hanged about his neck and he cast into the sea'. His attitude in 1940 was still the same as in 1920. He still saw in revolution by conquest the most dangerous aberration from the revolutionary road. He was still confident that the workers of the West were impelled by their own circumstances to struggle for power and for socialism and that it would be as criminal on the part of the Soviet Government to try to make the revolution for them as it would be to act directly against their revolutionary interests. He still saw the world pregnant with socialism; he still believed that the pregnancy could not last long; and he feared that any tampering with it would result in abortion. He was not quite wrong: Stalin's armed tampering with revolution has produced many a stillbirth—and many a live monstrosity.

Yet, confronted with revolution by conquest, Trotsky once again found himself in a grave quandary. He was for revolution and against conquests; but when revolution led to conquest or when conquest promoted revolution, he could not press his opposition to it beyond the point of an open and irrevocable breach. He did not press it to that point over Georgia and Poland in 1920–1; and he did not do so over Poland and Finland in 1939–40 either. Had he lived to witness the aftermath of the Second World War, he would have found his dilemma aggravated, huge, insoluble. We need not doubt that he would have denounced Stalin for bargaining away the interests of communism in the West; and also that the logic of his attitude would have compelled him to accept the reality of the revolution in eastern Europe, and, despite all distaste for the Stalinist methods, to recognize the 'Peoples' Democracies' as workers' states. Such an attitude, whatever its merits and integrity, could provide no clue to practical political action; and so Trotsky, the man of practical action, would hardly have found any effective role for himself in the whole post-war drama. There was no room for classical Marxism in this cycle of revolution.

This cycle, however, like the previous one was to end differently from the way it had begun. It culminated in the Chinese revolution which was neither imposed from above nor brought in on the point of foreign bayonets. Mao Tse-tung and his party struggled for power despite Stalin (who in 1945–8, as in 1925–6,

aimed at a deal with the Kuomintang and Chiang Kai-shek);
and having seized power they did not stop at the 'bourgeois
democratic' stages of the upheaval but, obeying the logic of
'permanent revolution', carried it to the anti-bourgeois conclusion. This, the 'Chinese October' was, in a sense, yet another
of Trotsky's posthumous triumphs.

Yet here again 'grey is all theory and evergreen is the tree
of life'. The industrial proletariat was not the driving force of
the upheaval. Mao's peasant armies 'substituted' themselves
for the urban workers and carried the revolution from country
to town. Trotsky had been convinced that, if these armies were
to remain confined to the rural areas for long, they would
become so assimilated with the peasantry as to champion its
individualistic interests against the urban workers, and against
socialism, and become the mainstay of a new reaction. (Had
not rebellious Chinese peasant armies in the past fought jac-
queries and overthrown established dynasties only to replace
them by new dynasties?) This analysis was correct in terms of
classical Marxism, which assumed that a party of socialist
revolution needs not only to 'represent' the urban workers, but
must necessarily live with them and act through them—otherwise it must become socially displaced and express alien class
interests. And it may indeed be that if this revolution had
depended solely on the social alignments within China, Mao's
partisans would have become, during their Yenan period, so
closely assimilated to the peasantry that, despite their communist origin, they would have been unable to bridge the gulf
between jacquerie and proletarian revolution. But the outcome
of the struggle was even in China determined as much by
international as by national factors. Amid the cold war and in
face of hostile American intervention, Mao's party secured its
rule by attaching itself to the Soviet Union and transforming
the social structure of China accordingly. Thus the revolutionary hegemony of the Soviet Union achieved (despite Stalin's
initial obstruction) what otherwise only the Chinese workers
could have achieved—it impelled the Chinese revolution into
an anti-bourgeois and socialist direction. With the Chinese
proletariat almost dispersed and absent from the political stage,
the gravitational pull of the Soviet Union turned Mao's
peasant armies into agents of collectivism.

With this the tide of revolution had moved farther to the east, farther away from the 'advanced' west; and it became once again embedded in a primitive and destitute pre-industrial society. More than ever classical Marxism appeared to be practically irrelevant to the problems of East and West alike. Yet such were the dialectics of the situation that at the same time processes were at work which were in an unexpected manner investing it with fresh validity. Thanks to intensive industrialization the backward East was becoming less and less backward. The Soviet Union emerged as the world's second industrial power, its social structure radically transformed, its large industrial working class striving for a modern way of life, and its standards of living and mass education rising rapidly, if unevenly. The very pre-conditions of socialism which classical Marxism had seen as existing only in the highly industrialized countries of the West were being created and assembled within Soviet society. In relation to the new needs of that society Stalinism, with its amalgamation of Marxism and barbarity, was anachronistic. Its methods of primitive accumulation were too primitive; its anti-egalitarianism was too shocking; its despotism absurd. The traditions of Marxism and of the October Revolution, having survived in a state of hibernation, as it were, began to awaken in the minds of millions and to struggle against bureaucratic privilege, the inertia of Stalinism, and the dead-weight of monolithic dogma. Through the forcible modernization of the structure of society Stalinism had worked towards its own undoing and had prepared the ground for the return of classical Marxism.

The return has been slow and accompanied by confusion and endless ambiguities. The conflict between Stalinism or what was left of it and a renascent socialist consciousness filled the first decade after Stalin.. Had the Trotskyist, Zinovievist, Bukharinist Oppositions survived into the nineteen-fifties, the task of de-Stalinization would have fallen to them; and they would have accomplished it with honour, whole-heartedly and consistently. But as they had all gone down with the old Bolshevik Atlantis, and as de-Stalinization was an inescapable necessity, Stalin's acolytes and accomplices had to tackle the job; and they could not tackle it otherwise than half-heartedly, with trembling hands and minds, never forgetting their own share in Stalin's crimes, and for ever anxious to bring to a halt the shocking disclosures and the reforms they themselves had had to initiate. Of all the ghosts of the past none dogged them as mockingly and menacingly as the ghost of Trotsky, their arch-enemy, to whom each of their disclosures and reforms was an unwitting tribute. Nothing indeed troubled Khrushchev more than the fear that young men, not burdened by responsibility for the horrors of the Stalin era, might become impatient with his evasions and quibblings and proceed to an open vindication of Trotsky.

The open vindication is bound to come in any case, though not perhaps before Stalin's ageing epigones have left the stage. When it does come, it will be more than a long-overdue act of justice towards the memory of a great man. By this act the workers' state will announce that it has at last reached maturity, broken its bureaucratic shackles, and re-embraced the classical Marxism that had been banished with Trotsky.

How all this may affect the rest of the world is a question too large to be discussed in a postscript to a biographical study. Suffice it to say here that if the historic development has already been cancelling out Trotsky's defeat by obliterating the old antithesis between *backward* Russia and the *advanced* West, the antithesis in which his defeat had been rooted, then the regeneration of the Russian revolution must help to obliterate that antithesis to the end. The West, in which a Marxism debased by Mother Russia into Stalinism inspired disgust and fear, will surely respond in quite a different manner to a Marxism cleansed of barbarous accretions; in that Marxism it will have to acknowledge at last its own creation and its own vision of man's destiny. And so history may come full circle

till Hope creates
From its own wreck the thing it contemplates.

Trotsky sometimes compared mankind's progress to the barefooted march of pilgrims who advance towards their shrine by moving only a few steps forward at a time, and then retreat or jump sideways in order to advance and deviate or retreat again; zigzagging thus all the time they approach laboriously their destination. He saw his role in prompting the 'pilgrims' to advance. Mankind, however, when after some progress it succumbs to a stampede, allows those who urge it forward to be abused, vilified, and trampled to death. Only when it has resumed the forward movement, does it pay rueful tribute to the victims, cherish their memory and piously collect their relics; then it is grateful to them for every drop of blood they gave— for it knows that with their blood they nourished the seed of the future.

Bibliography

(See also Bibliography in *The Prophet Armed* and *The Prophet Unarmed*)

AVAKUM, PETROV, *Zhizn Protopopa Avakuma*. Moscow, 1960.

BRECHT, B., *Galileo Galilei*.
BRETON, A., *La clé des champs*. Paris, 1953.
—— *Entretiens*. Paris, 1960.
—— Correspondence with Trotsky in *The Trotsky Archives*, Closed Section.
BUDENZ, L. F., *This is My Story*. New York, 1947.
BULLOCK, A., *Hitler—A Study in Tyranny*. London, 1953.
BURNHAM, J., articles and essays in *New International* and *Internal Bulletin* of the S.W.P. (the American Trotskyist organization).
—— Correspondence with Trotsky in *The Trotsky Archives*, Closed Section.
—— *Managerial Revolution*. New York, 1941.
—— *The Coming Defeat of Communism*. New York, 1950.

CANNON, J. P., *History of American Trotskyism*. New York, 1944.
—— Articles in *Fourth International*, and *Internal Bulletin* of the S.W.P.
—— Correspondence in *The Trotsky Archives*, Closed Section.
CÉLINE, L.-F., *Voyage au bout de la nuit*.
CHEN TU-HSIU, unpublished memoranda, essays, and correspondence in *The Trotsky Archives*, Closed Section.
CHURCHILL, W. S., *Great Contemporaries*. London, 1939.
—— *The Second World War* (vol. iv). London, 1951.
CILIGA, A., *Au pays du grand mensonge*. Paris, 1937.
—— Articles in *Bulletin Oppozitsii* and *Sotsialisticheskii Vestnik*.
COULONDRE, R., *De Staline à Hitler, Souvenir de deux Ambassades*. Paris, 1950. (Trotsky commented on Coulondre's report of his last meeting with Hitler on the basis of Coulondre's article in *Paris-Soir*.)
M. CRAIPEAU, 'J'ai connu l'assassin de Trotsky', *France-Observateur*, 19 May 1960.

DEWEY, JOHN, 'Means and Ends', in *New International*, 1938. (See also *The Case of Leon Trotsky*, and *John Dewey, Philosopher of Science and Freedom, A symposium*, ed. S. Hook.)
DRAPER, TH., *American Communism and Soviet Russia*. New York, 1960.
—— *Roots of American Communism*. New York, 1957.

EASTMAN, MAX, *Since Lenin Died*. London, 1925.
—— *The End of Socialism in Russia*. London, 1937.
—— *Marxism, is it Science?* London, 1941.
—— *Stalin's Russia and the Crisis in Socialism*. London, 1940.
—— *Great Companions*. London, 1959.
—— Correspondence with Trotsky in *The Trotsky Archives*, Closed Section.

ENGELS, F., and MARX, K., *Briefwechsel*. Berlin, 1949–50.

FAINSOD, M., *Smolensk under Soviet Rule*. London, 1959.
FARRELL, J. T., 'Dewey in Mexico' in *John Dewey, A Symposium*, ed. Hook, S.
FREEMAN, J. *An American Testament*. London, 1938.

GIDE, ANDRÉ, *Retour de l'U.R.S.S.* Paris, 1936.
GOLDMAN, A., *The Assassination of Leon Trotsky*. New York, n.d.
GREF, YA., Contributions in *Bulletin Oppozitsii*.
GUÉRIN, D., *Jeunesse du Socialisme Libertaire*. Paris, 1959.
—— *Fascisme et Grand Capital*. Paris, 1936.

HANSEN, J., Reminiscences about Trotsky in *Fourth International*.
HEGEL, G. W. F., *Philosophie der Weltgeschichte*.
HERNANDEZ, J., *La Grande Trahison*. Paris, 1953.
HOOK, S., *The Hero in History*. London, 1945.
—— *Political Power and Personal Freedom*. New York, 1955.
—— ed. *John Dewey, Philosopher of Science and Freedom, A Symposium*. New York, 1950.
LES HUMBLES, Cahiers 5–6, *À Leon Trotsky*. Paris, 1934.

ISAACS, H., *The Tragedy of the Chinese Revolution* (Preface by Trotsky). London, 1938.
—— Reports on China and correspondence with Trotsky in *The Archives*, Closed Section.

KAGANOVICH, L., speeches in Reports of party congresses.
KAROLYI, M., *Memoirs*. London, 1956.
KERENSKY, A., *The Crucifixion of Liberty*. London, 1934.
KHRUSHCHEV, N., *The Dethronement of Stalin*. (*Manchester Guardian* publication June 1956.)
—— Speeches in *22 Syézd K.P.S.S.* Moscow, 1962.
KNUDSEN, K., Preface to Norwegian edition of Trotsky's *My Life*, Oslo.
KOHT, H., *Barricade to Barricade*. (Norwegian edition.) Oslo.
KROG, H., *Meninger*. Oslo, 1947.
KUN, BELA, (ed.) *Kommunisticheskii International v Dokumentakh, 1919–32*. Moscow, 1933.

LENIN, V. I., *Sochinenya*. Moscow, 1941–50.
LEVINE, I. DON, *The Mind of an Assassin*. New York, 1959.
LIE, TRYGVE (On behalf of the Norwegian Ministry of Justice and Police), *Storting Report*, nr. 19 (concerning Trotsky's internment and deportation from Norway), submitted on 18 February 1937.
LUNACHARSKY, A., *Revolutsionnye Siluety*. Moscow, 1923.

MACDONALD, DWIGHT, *Memoirs of a Revolutionist*. New York, 1958.
MALRAUX, A., *La Condition Humaine*.

MANUILSKY, D., *The Communist Parties and the Crisis of Capitalism*. Report at 11 Plenum of Comintern Executive, March–April 1931. London, n.d.
—— Other articles and speeches quoted from Reports of Party Congresses and *Kommunisticheskii International*.
MARX, K., *Das Kapital*.
—— and ENGELS, F., *Das Kommunistische Manifest*.
—— *Der 18 Brumaire des Louis Bonaparte*.
—— and ENGELS, F., *Briefwechsel*. Berlin, 1949–50.
—— *Living Thoughts of Karl Marx* (ed. by L. Trotsky and O. Rühle) London, 1946
MAURIAC, F., *Mémoires Intérieures*. Paris, 1959.
M.B. 'Trotskisty na Vorkute', *Sotsialisticheskii Vestnik*, 1961. (An eyewitness's report on the extermination of the Trotskyists at the Vorkuta concentration camp in 1938.)
MERLEAU-PONTY, M., *Les Aventures de la Dialectique*. Paris, 1955.
—— *Humanisme et terreur*. Paris, 1947.
MILIUKOV, P. N., *Istorya Vtoroi Russkoi Revolutsii*. Sofia, 1921.
MOLINIER, RAYMOND and HENRI, Correspondence with Trotsky and Leon Sedov quoted from *The Trotsky Archives*, Closed Section, and Leon Sedov's Papers.
MOLOTOV, V., speeches and reports in Reports of Party Congresses.

NADEAU, M., *Histoire du Surréalisme*. Paris, 1945.
NAVILLE, P., *Trotsky Vivant*. Paris, 1962.
—— Correspondence in *The Trotsky Archives*, Closed Section.
NIN, A., Correspondence in *The Trotsky Archives*, Closed Section.

ORLOV, A., *The Secret History of Stalin's Crimes*. London, 1953.
ORWELL, G., *Homage to Catalonia*.
—— *1984*.

PABLO, M., 'Vingt Ans de la Quartrième Internationale' in *Quatrième Internationale*, 1958–9.
PARIJANINE, M., 'Léon Trotsky ou la Revolution Bannie' in *Les Humbles*. Paris, 1934.
—— Correspondence in *The Trotsky Archives*, Closed Section.
PAZ, MAURICE and MAGDELEINE, Correspondence with Trotsky in *The Archives*, Closed Section.
PFEMFERT, FRANZ, Correspondence with Trotsky, *The Trotsky Archives*, Closed Section.
PLEKHANOV, G., *Izbrannye Filosofskie Proizvedenya* (vol. ii). Moscow, 1956.
—— *The Role of the Individual in History*. London, 1940.
POPOV, N., *Outline History of the C.P.S.U.(b)* (vols. i-ii). English translation from the 16th Russian edition. London, n.d.
PREOBRAZHENSKY, E., *Novaya Ekonomika*, vol. i, part i. Moscow, 1926.
—— Essays and memoranda (including Manifesto 'Ko Vsem Tovarishcham po Oppozitsii') are quoted from *The Trotsky Archives*.
PRITT, D. N., *The Zinoviev Trial*. London, 1936.

RADEK, K., 'Ot Oppozitsii v Kloaku Kontrrevolutsii' in *Partiya v Borbe z Oppozitsijami*. Moscow, 1936.

—— Articles in *Izvestya* and other Soviet newspapers. His 'Confession' at his trial is in *Sudebnyi Otchet po Delu Antisovietskovo Trotskistskovo Tsentra*. Moscow, 1937.

RAHV, PH., Correspondence with Trotsky. *The Trotsky Archives*, Closed Section.

RAKOVSKY, CH., Essays, articles, and correspondence in *Bulletin Oppozitsii* and *The Trotsky Archives*.

RAMM, A., Correspondence in *The Trotsky Archives*, Closed Section.

REISS, I., 'Letter to Central Committee' and 'Zapiski' in *Bulletin Oppozitsii*, 1937.

R(IZZI), BRUNO, *La Bureaucratisation du Monde*. Paris, 1939.

ROSMER, A., *Moscou sous Lénine*. Paris, 1953.

—— Introduction and Appendixes in Trotsky's *Ma Vie*, Paris, 1953.

—— Articles in Trotskyist periodicals and *La Révolution Proletarienne*.

—— Correspondence with Trotsky in *The Trotsky Archives*, Closed Section.

—— Correspondence with the author.

ROWSE, A. L., *End of an Epoch*. London.

RÜHLE, O. (and L. TROTSKY), *Living Thoughts of Karl Marx*. London, 1946.

SALAZAR, L. A. S. *Murder in Mexico*. London, 1950.

SAYERS, M., and KAHN, A. E., *The Great Conspiracy*. New York, 1947.

SEDOV, LEON, *Livre Rouge sur le procès de Moscou*. Paris, 1936. The Russian text of this appeared simultaneously as a special issue of the *Bulletin Oppozitsii*. Articles and essays in *Bulletin Oppozitsii* (sometimes signed N. Markin), *Manchester Guardian* and other papers.

—— Correspondence with Trotsky, Natalya, and other members of the family, *The Trotsky Archives*, Closed Section.

—— L. Sedov papers transmitted to the author by Jeanne Martin des Paillères.

SEDOVA, NATALYA, (with V. SERGE), *Vie et Mort de Trotsky*.

—— Reminiscences about Trotsky and Lev Sedov in *Bulletin Oppozitsii* and *Fourth International*, 1941.

—— Family correspondence in *The Trotsky Archives*, Closed Section.

—— Correspondence with the author.

—— *Hommage à Natalia Sedova-Trotsky*. (Funeral orations and reminiscences) Paris, 1962.

SERGE, V., (and NATALYA SEDOVA) *Vie et Mort de Trotsky*. Paris, 1951.

—— *Mémoires d'un Révolutionnaire*. Paris, 1951.

—— Articles and letters in *The New International*, and other Trotskyist or near Trotskyist papers. Correspondence in *The Trotsky Archives*, Closed Section.

SHACHTMAN, M., Articles and essays in *New International*, *Militant*, *Internal Bulletin* of S.W.P. etc.

—— Correspondence in *The Trotsky Archives*, Closed Section.

Shaw, G. B., *Saint Joan.*
—— *To a Young Actress.* London, 1960.
—— Correspondence quoted from the Archives of the British Committee for the Defence of Leon Trotsky and from *The Trotsky Archives*, Closed Section.
Shirer, W. L., *The Rise and Fall of the Third Reich.* London, 1960.
Smirnov, Ivan, Memoranda and correspondence quoted from *Bulletin Oppozitsii* and *The Trotsky Archives.*
Sobolevicius-Senin, *alias* Jack Soble, and his brother, Dr. Soblen (*alias* Robert Well), correspondence with Trotsky and Leon Sedov in *The Trotsky Archives*, Closed Section.
Sokolovskaya, (Bronstein) Alexandra, correspondence with Trotsky and Leon Sedov, *The Trotsky Archives*, Closed Section.
Souvarine, B., *Stalin*, London, n.d.
—— Correspondence in *The Trotsky Archives*, Closed Section.
Stalin, J., *Sochinenya* (vols. XII-XIII). Moscow, 1949–51.

Tarov, A., Contributions in *Bulletin Oppozitsii.*
Togliatti (Ercoli), P., Speeches and articles in *Kommunisticheskii Internatsional* and Reports of Comintern Congresses and Conferences.
Thaelmann, E., speeches, reports, and articles quoted from *11 Plenum IKKI, 12 Plenum IKKI, Rote Fahne, Internationale,* and *Kommunistische Internationale* (or the Russian edition of the latter).
Trotsky, L., *Chto i Kak Proizoshlo?* Paris, 1929.
—— *Moya Zhizn*, vols. i-ii. Berlin, 1930. (The English edition, *My Life,* London, 1930; the French, *Ma Vie*, with Introduction and Appendix by Alfred Rosmer, Paris 1953; the German *Mein Leben*, Berlin, 1929.)
—— *The History of the Russian Revolution,* vols. i-iii. Translated by Max Eastman. London, 1932–3.
—— *Écrits, 1928–40.* vols. i-iii, with Introductions by Pierre Frank. Paris, 1955–9.
—— *O Lenine.* Moscow, 1924. (French edition *Lénine.* Paris, 1925), a collection of character sketches about Lenin, not be to confused with the biography of Lenin, of which Trotsky concluded only the first part and which has so far been published only in French as
—— *Vie de Lénine, Jeunesse.* Paris, 1936.
—— *The Third International after Lenin.* New York, 1936.
—— *Nemetskaya Revolutsia i Stalinskaya Burokratiya.* Berlin 1932. (In German: *Was Nun?* Berlin, 1932; in English *What Next?* New York, 1932; French version *Écrits*, vol. iii.)
—— *Edinstvennyi Put'* (in German: *Der einzige Weg*) Berlin, 1932.
—— *Germany, the Key to the International Situation.* London, 1931.
—— *Où va la France?* and *Encore une Fois, Où va la France?* Paris, 1936. (Reproduced in *Écrits*, vol. ii.)
—— *The Revolution Betrayed.* London, 1937.
—— *Permanent Revolution.* Calcutta, 1947.

Trotsky, L. *Problems of the Chinese Revolution*. New York, 1932.
—— *Trotsky's Diary in Exile*. London, 1958.
—— *Stalin's Verbrechen*. Zürich, 1937.
—— *The Real Situation in Russia*. London, n.d.
—— *Stalinskaya Shkola Falsifikatsii*. Berlin, 1932. (The American edition: *The Stalin School of Falsification*. New York, 1937.)
—— *Between Red and White*. London, 1922.
—— *Stalin*. New York, 1946.
—— *Their Morals and Ours*. New York, 1939.
—— *Leon Sedov, Son, Friend, Fighter*. New York, 1938.
—— Articles, essays, treatises, and theses in *Bulletin Oppozitsii*, 1929–40, *New International*, and other Trotskyist periodicals.

The Trotsky Archives, Houghton Library, Harvard University. A description of these was given in the Bibliography in *The Prophet Armed*. Since then *The Archives* have been reorganized. The documents are no longer divided into Sections A, B, and C, but have been rearranged in chronological order. An Index, in two volumes, is available to students of this (the 'Open') part of *The Archives*. All references in *The Prophet Armed* and *The Prophet Unarmed* were to this part of *The Archives*.

What was described in the Bibliography in *The Prophet Armed* as 'Section D' is now described as 'The Closed Section of *The Archives*. It covers only the years 1929–40 and contains Trotsky's correspondence with groups and members of the Fourth International and with other well-wishers and friends, his family correspondence, household papers, correspondence with publishers, documentation prepared for The Dewey Commission, papers of the Fourth International, &c. According to Trotsky's wish, this Section of *The Archives* was not to be opened before the year 1980; but Harvard University gave the author access to it on the basis of a special authorization from Natalya Sedova, Trotsky's widow. (In references to *The Trotsky Archives* at large, the open section of *The Archives* is meant.)

The Closed Section of *The Archives* consists of forty-five boxes, containing 309 folders with documents and correspondence. Thus folders 1–16 contain Trotsky's family correspondence; folders 17–25, his household papers; folders 26–33, correspondence with publishers and literary agents; 34–35, documentation for the Mexican Counter-trial; while Fourth International papers are in folders 36–40. The rest of the material is arranged according to countries, for instance, folders 65–70 contain Trotsky's correspondence concerning China; folders 90–121 refer to France; the German correspondence is in folders 122–6; the British 165–75; folders 214–86 contain correspondence with U.S.A.; folders 287–92 correspondence with the U.S.S.R., and 293–309 letters to and from Soviet citizens exiled from the U.S.S.R. To this Section of *The Archives* some papers were added by Trotsky's widow in 1953.

See also: *The Case of Leon Trotsky*, London, 1937 (Trotsky's depositions and cross-examination before the Dewey Commission in Mexico).
—— *Not Guilty*. Report of the Dewey Commission. London, 1938.

VOLKOV, ZINAIDA (ZINA, Trotsky's daughter), correspondence in *The Trotsky Archives*, Closed Section.

WEBB, SIDNEY and BEATRICE, *Soviet Communism, a New Civilization*. London, 1944.
—— Correspondence with Trotsky. *The Trotsky Archives*, Closed Section.
WEIL, SIMONE, *Oppression et Liberté*. Paris, 1955.
WILSON, E. *To the Finland Station*. London, 1941.
WOLFE, B., *The Great Prince Died*. New York, 1959.
WOLFE, BERTRAM, D., Articles in *Things we Want to Know*. Workers' Age Publications, New York, 1936, and *The New Republic*, 1937.
WOLLENBERG, E., *The Red Army*. London, 1940.

YAROSLAVSKY, E., *Partiya v Borbe z Oppozitsiami*, with contributions by K. Radek, A. Pankratova, and others. Moscow, 1936.
—— *O Noveishei Evolutsii Trotskizma*. Moscow, 1930.
—— *Vcherashny i Zavtrashny Den Trotskizma*. Moscow, 1929.

ZBOROWSKI, MARK, (ÉTIENNE) Correspondence with Trotsky and other documents, concerning his relationship with Leon Sedov are in *The Trotsky Archives*, Closed Section.

The following are the official reports of the Moscow Trials:

Sudebnyi Otchet po Delu Trotskistskovo-Zinovievskovo Terroristskovo Tsentra. Moscow, 1936.
Sudebyni Otchet po Delu Anti-Sovietskovo Trotskistskovo Tsentra. Moscow, 1937.
Sudebnyi Otchet po Delu Anti-Sovietskovo i Pravo-Trotskistskovo Bloka. Moscow, 1938.
(Official English versions or *Reports of Court Proceedings* were published simultaneously by the People's Commissariat of Justice in Moscow:)

The following official Protocols, Verbatim Reports, and Collections of documents have been referred to:

16 Syezd V.K.P.(b). Moscow, 1931.
17 Syezd V.K.P.(b). Moscow, 1934.
20 Syezd K.P.S.S. Moscow, 1956.
22 Syezd K.P.S.S. Moscow, 1962.
11 Plenum IKKI. Moscow, 1932.
12 Plenum IKKI. Moscow, 1933.
Kommunisticheskii Internatsional v Dokumentakh. Moscow, 1933.
K.P.S.S. v Rezolutsyakh, vols. i-ii. Moscow, 1953.
V.K.P.(b) o Profsoyuzakh. Moscow, 1940.
Narodnoe Khozyaistvo S.S.S.R. Moscow, 1959.

Hearing before the Subcommittee to Investigate the Administration of the Internal Security Act, U.S. Senate (14–15 February 1957), part 51. Washington, 1957.

Hearing before the Subcommittee to Investigate the Administration of the Internal Security Act, U.S. Senate (21 November 1957), part 87. Washington, 1958.

Newspapers and periodicals:

Bolshevik, Bulletin Oppozitsii, Izvestya, Pravda, Proletarskaya Revolutsia, Sotsialisticheskii Vestnik, Kommunisticheskii Internatsional. (Stencilled or hand-copied periodicals circulated by Trotskyist deportees and prisoners in the U.S.S.R. are in *The Trotsky Archives.*)

Internationale, Internationale Presse Korrespondenz, Kommunistische Internationale, Rote Fahne, Roter Aufbau, Rundschau Unser Wort, Permanente Revolution, Arbeiterpolitik, Aktion, Berliner Börsenzeitung, Hamburger Nachrichten, Vossische Zeitung.

Militant, New International, Fourth International, Partisan Review, Internal Bulletin Fourth International (International Secretariat), The Times, Manchester Guardian, Daily Express, The Observer, Morning Post, The New Statesman and Nation, The New York Times, The New York American, The New York Daily News, The New York World Telegram, Life, The Nation, The New Republic, The New Leader, Soviet Russia To-day.

New York Tag, and *Vorwärts* (Yiddish—U.S.A.).

Vérité Quatrième Internationale France-Observateur, Intransigeant, Paris-Soir, Le Matin, Le Journal, Le Temps, Humanité, Journal d'Orient.

Politiken, Berlingske Tidende, Information, Arbeiderbladet, Dagbladet, Arbeideren, Soerlandet.

La Prensa, Trinchera Aprista.

Index

J